Symbol	Meaning	d Item Definition bol
$\hat{A}_{ij}$	the matrix obtained by deleting the [...] j^{th} column from A	28.1
a_{ij}^{m}	the minor of a_{ij}	28.1
a_{ij}^{c}	the cofactor of a_{ij}	28.1
cof A or A^c	the matrix of cofactors of A	29.1
adj A	the adjugate (or classical adjoint) of A	29.1
A[j; **b**]	the result of substituting **b** for the jth column of A	Problem 29.1
Δ	a pseudodeterminant	30.1
${}_{\gamma}[T]_{\beta}$	the matrix that represents T with respect to β and γ	31.8
${}_{\gamma}[I]_{\beta}$	the change-of-basis matrix from β to γ	31.13
$\mathbb{F}_{n\times n}$	the set of $n \times n$ matrices over the field $\mathbb{F}$	34.1
E_{λ}	the eigenspace of λ	34.4
$\sigma(A)$	the spectrum of A	34.4
$\chi(A)$ or $\chi_A(\lambda)$	the characteristic polynomial of A	35.3
Re z	the real part of the complex number z	36.1
Im z	the imaginary part of the complex number z	36.1
$\bar{z}$	the complex conjugate of the complex number z	36.1
$\lvert z\rvert$	the modulus of the complex number z	36.3
Arg z	the principal argument of the complex number z	36.3
$\bar{A}$	the complex conjugate of the matrix A	36.11
A^H	the conjugate transpose of the matrix A	36.15
$\mathbf{u} \cdot \mathbf{v}$	the Euclidean dot product of **u** and **v**	38.4
$\lVert\mathbf{u}\rVert$	the norm of **u** (usually the Euclidean norm)	38.11
$\mathbb{E}^n$	$\mathbb{R}^n$ with the Euclidean inner product	38.18
$\mathbf{u} \times \mathbf{v}$	the cross product of **u** and **v**	Problem 38.3
$S^{\perp}$	the orthogonal complement of S	39.1
GS(S)	the output of the Gram–Schmidt algorithm applied to S	39.14
$\text{proj}_W \mathbf{u}$	the orthogonal projection of **u** onto W	39.23
$\mathbf{u} \oplus \mathbf{v}$	vector addition in an arbitrary vector space	41.2
$t \otimes \mathbf{v}$	scalar multiplication in an arbitrary vector space	41.2
V^*	the dual space of the vector space V	Problem 41.5
β^*	the basis dual to the basis β	Problem 41.5
$C(\mathbb{R})$	the space of continuous functions from $\mathbb{R}$ to $\mathbb{R}$	42.5
$\mathscr{P}_n$	the space of polynomials of degree $\leq n$	42.7
$\langle \cdot, \cdot \rangle$	an inner product	45.1
$\langle \cdot, \cdot \rangle_{\beta}$	the inner product induced by the basis β	45.14
$\lVert \cdot \rVert_{\beta}$	the norm induced by the basis β	45.14
$\mathbb{H}^n$	$\mathbb{C}^n$ with the Euclidean inner product	45.26
${}_{\beta}[p]_{\gamma}$	the matrix that represents the bilinear form p with respect to β and γ	47.8
${}_{\beta}[q]_{\beta}$	the matrix that represents the quadratic form q with respect to β	48.12
q_p	the quadratic form derived from the bilinear form p	48.14
T^*	the adjoint of the linear operator T (when it exists)	50.11
$A \oplus B$ or diag(A, B)	the direct sum of the matrices A and B	51.3
J[k; c]	the elementary $k \times k$ Jordan matrix with constant diagonal entry c	51.9
E_c^k	the nullspace of $(T - cI)^k$	51.16
J(A)	the Jordan form of the matrix A	51.29

Linear Algebra

A Comprehensive Introduction

Donald H. Pelletier

Department of Mathematics
York University
Toronto, Ontario
Canada

A Reston Book
Prentice-Hall
Englewood Cliffs, New Jersey 07632

Library of Congress Cataloging-in-Publication Data

Pelletier, Donald H.
Linear algebra.

"A Reston book."
Bibliography: p.
Includes index.
1. Algebra, Linear. I. Title.
QA184.P45 1986 512'.5 86-4934
ISBN 0-8359-4064-0

Cover art courtesy of Megatek Corp.

ISBN 0-8359-4064-0

A Reston Book
Published by Prentice-Hall
A Division of Simon & Schuster, Inc.
Englewood Cliffs, New Jersey 07632

10 9 8 7 6 5 4 3 2 1

Printed in the United States of America

Contents

CHAPTER III
Computational Aspects of Linear Algebra

CHAPTER IV
Theoretical Aspects of Linear Algebra

CHAPTER V
Determinants

CHAPTER VI
Linear Transformations and Diagonalizability

CHAPTER VII
Inner-Product Space Concepts for $\mathbb{R}^n$

CHAPTER VIII
Vector Spaces

CHAPTER IX
Miscellaneous Advanced Topics

Preface

This book's subtitle, *A Comprehensive Introduction*, is intended to convey its content, goals, and orientation.

The text is written primarily for second-year college students, though a typical class may contain some who are in their first, third, or fourth year. Some students may be majoring in mathematics, but the intended audience is much wider and includes students majoring in other disciplines where the concepts and techniques of Linear Algebra are applied.

A mastery of high-school algebra is assumed, but calculus is *not* a prerequisite. Calculus is an occasional source of examples, but these can be safely omitted without loss of continuity.

The three main goals of the book are:

(i) To present the basic subject matter of Linear Algebra in a manner that is simultaneously introductory *and* comprehensive. Most textbooks tend to be one or the other.

(ii) In a recent article discussing the CUPM '81 recommendations for the undergraduate mathematics curriculum, B. Hodgson and J. Poland write: "It should be recalled in this context that the Science Council study of mathematical sciences in Canada [the Coleman report, 1976] found 'almost all mathematics professors allege that their highest ambition in undergraduate teaching is to convey not specific content but rather a way of thinking,' a way of thinking that even our colleagues in other disciplines consider important and wish their students to undergo when taking our courses." I firmly believe this and have devoted many paragraphs in my text to explaining how mathematicians think.

(iii) "Think *before* you calculate!" One of my goals is that this motto will be adopted by every reader of this book.

Linear Algebra is a tool. My purpose is to teach students not merely how to use the tool (for this, a compendium of examples is quite sufficient) but to have them understand how and why it works. Thus every theorem that is used is accompanied by a complete proof. Naturally, the instructor has the option of omitting certain proofs; but that option does not exist if the author has already chosen to omit the proof. The fact that a proof is included does not imply that valuable time in the classroom should be spent on it. Time does not permit the classroom presentation of very many proofs; it is all the more important, therefore, that the textbook contain proofs that the student can read and understand. Linear Algebra is an ideal subject for introducing students to rigorous mathematical thinking. Omitting proofs from an introductory text leaves the mistaken impression that the proofs are difficult. Students should be able to read the proofs and decide for themselves. True, a few are difficult; but most are quite accessible, provided some care is taken in the logical arrangement of the theorems. In fact, it is this aspect of my book which I consider among its unique characteristics.

The order in which the topics are presented is unconventional. It is designed to develop the habit of thinking *before* calculating. Consequently, before teaching students *how* to calculate, we teach them *what it is* that they will be calculating. We emphasize the ability to distinguish questions that can be answered immediately on theoretical grounds without calculations from those that require calculations. The purpose of any computation is to extract information from data. The *initial* focus should be on the exact nature of the information desired and the relationship between that information and the data; only then is it relevant to ask which computation will provide the information in the most economical manner. The text does not, therefore, begin with the Gauss–Jordan algorithm and with the solution of systems of linear equations. The fact that the mechanics of the algorithm are easy is not a reason for presenting them first. The reduction of other topics in Linear Algebra to the "familiar" subject of systems of linear equations becomes a crutch that the weaker students are never able to discard.

Still, I recognize that the order in which topics are covered is very much a matter of personal taste. A textbook is not like a novel; it need not be read in a linear fashion, page $n + 1$ following page n. I have therefore deliberately written the book in a way that encourages flexibility. The individual chapters are as independent from one another as possible.

A principle that guides some authors in ordering topics is not to introduce a concept Y until it is "needed" in its own right. My own feeling is this: If concept X has just been introduced, and if the proofs of simple properties of concept Y require straightforward applications of concept X, then this is the perfect opportunity to introduce concept Y. This develops manipulative skills with concept X in a useful context; also, when concept Y is eventually reintroduced in its own properly motivated setting, attention need not then be diverted to establishing its simple properties—focus on the real meaning can be maintained.

The exercises and problem sets are essential ingredients of this textbook and have been carefully designed with the preceding principle in mind. They include both routine computational problems, intended for the development of skills, and so-called "theoretical" questions that require an understanding of the definitions and theorems. The purpose of several of the problems is to develop the student's ability to use theorems *in combination*. Often, a student understands Theorem A and can use it *by itself* to arrive at various conclusions; likewise for Theorems B and C. But when faced with a question whose answer requires even the simplest logical combination of Theorems A, B and C, the student is hopelessly lost. There are also many worked-out examples in the text and several of the exercises are really examples that students are expected to work out on their own. The habit of solving problems by blindly imitating a worked-out example is discouraged: problem sets do not consist of dozens of instances of easy problem-types; one or two of each type suffice. An overdose of easy problems leads to intellectual laziness.

For most students, the present course will be their only one in Linear Algebra. For many, this will be insufficient. Frequently, third- and fourth-year students who had Linear Algebra with me and who are now taking statistics, or differential equations, or econometrics will return to me for help with a topic in Linear Algebra that we did not get to cover in the course, positive definite matrices or the Jordan form, for example, and that they now need to understand. These students would be sufficiently mature and motivated to delve into these topics on their own if only they could find them in their old familiar Linear Algebra book. But they are not mathematically sophisticated enough to attack an unfamiliar, advanced text, in which these topics receive a high-powered treatment accessible only to mathematics majors. Students who are not mathematics majors need to understand these results, too, since they are widely used. Hopefully, the present text is one that students will want to keep for future reference.

In writing a modern textbook in Linear Algebra, one must choose between incorporating applications into the theoretical development or separating them into a distinct publication. After an initial attempt to pursue the former course, I have decided to adopt the latter. This book should be regarded as the first volume of a projected two-volume sequence. Applications deserve better and more thorough treatment than can be provided in a loosely interwoven patchwork. In the meantime, for a course that includes a component devoted to applications, this textbook could be used in conjunction with one of several paperback supplements that specifically address these aspects.

I have not included computer programs in the text. Students who have an interest in and experience with computers will also have access to better programs than the ones I might provide, while novices would be frustrated by their unavoidable machine-dependent idiosyncracies. What I have attempted to furnish in the text is an appreciation of the nature of algorithms, a programming-oriented attitude toward them, and a critical sense of computation as an information-extracting process.

Overview of the Contents

Chapter I is an introduction to matrix algebra. The historical origins of the subject are emphasised by introducing matrix multiplication first, motivated by the change-of-variables problem. Matrix addition and multiplication of a matrix by a scalar are introduced only subsequently. The algebra of matrices is presented and contrasted with the algebra of real numbers. The distinction between right and left inverses is essential for future theoretical work and for a genuine understanding of the material, and hence is stressed.

Chapter II is the heart of the subject. It begins with a somewhat vague, discursive section on the intuitive notion of "straightness"; this section should be assigned as outside reading and is not intended for classroom presentation. The fundamental concepts of subspace, of linear dependence and independence, of basis, and of dimension are then presented in the concrete context of $\mathbb{R}^n$. Distinguishing between linear subspaces and more general affine subspaces of $\mathbb{R}^n$ makes both these concepts easier to understand. The important topic of systems of linear equations is introduced at this point only to emphasize that the set of solutions of a real linear system is an affine subspace of $\mathbb{R}^n$. The material from Chapter II is related to that from Chapter I by introduction of the concepts of row-space, column-space, and rank of a matrix.

Chapter III provides the computational means to answer questions about the concepts defined in the first two chapters. The Gauss–Jordan algorithm is presented not only as a method for solving systems of linear equations but as an all-purpose tool for answering questions about linear dependence and independence, dimension, rank, bases, etc., and for computing the inverse of an invertible matrix.

Chapter IV is independent of Chapter III, with the exception of Section 20 which summarizes the two chapters. A real $m \times n$ matrix is now viewed as a function from $\mathbb{R}^n$ into $\mathbb{R}^m$, matrix multiplication is reinterpreted as composition of functions, and the fundamental theorems of Linear Algebra are proved. We emphasize (see Theorem 19.5) the ability to switch among the three points of view given in Chapters I, II, and IV.

Chapter V contains a classical treatment of determinants. There is a reason for delaying their treatment to this point. It is only *after* matrix multiplication is viewed as composition of functions that the physical interpretation of the determinant can be used to give intuitive plausibility to the important theorem that the determinant of a product is the product of the determinants. A slight variant of the Gauss–Jordan algorithm is used to compute determinants. The chapter concludes with an optional section containing a proof of the Cofactor Expansion Theorems.

Chapter VI considers linear transformations from $\mathbb{R}^n$ to $\mathbb{R}^m$ and their representations, with respect to a given pair of bases, by matrices. We view change-of-basis matrices as those that represent the identity operator, thereby providing a unified treatment of a subject that is often (mis)treated as a separate topic. A search for the "simplest" among the matrices representing a given linear operator motivates the concept of diagonalizability, the computational aspect of which is then given a thorough treatment. The need for complex numbers arises naturally in this context; they are introduced and a discussion of the Fundamental Theorem of Algebra and of the role it plays in Linear Algebra is included.

Chapter VII presents the Euclidean dot-product in dimension n, after first motivating it with a review of familiar results for dimensions two and three. The principal tool is the Cauchy–Schwarz Inequality. The Gram–Schmidt algorithm is viewed as a proof of the important theorem that every linear subspace of Euclidean n-space has an orthonormal basis. These ideas are applied to give a purely geometric (i.e., calculus-free) presentation of least-squares approximations. Finally, the study of operators that preserve the dot product leads to the concepts of orthogonality for operators and for matrices.

Chapter VIII introduces abstract vector spaces. The emphasis here is on examples; most of the proofs are trivial modifications of those given in detail in Chapter II for $\mathbb{R}^n$ and are left as exercises. The concept of isomorphism is presented. Infinite-dimensional vector spaces are also discussed. Anticipating Chapter IX, there is a treatment of direct-sum decompositions.

Chapter IX is difficult to summarize; it is a collection of advanced topics with varying prerequisites, presented from an introductory point of view. Throughout, we give separate treatments of the real and complex cases of important theorems.

It is not realistic only to encounter problems whose methods of solution are related in an obvious way to the material in the section in which they occur. For this reason, we have attached appendixes containing random lists of problems not immediately identified with individual sections of the book. To do these, the student must assimilate larger amounts of material, and is forced to develop a global outlook on the subject. The instructor has the option of overriding this feature by assigning problems from the appendixes as if they were placed in the Problem Sets to which they correspond.

Acknowledgments

I would like to take this opportunity to thank in print the following persons and institutions:

Doris Rippington, from the Secretarial Services of the Faculty of Arts at York University, for supervising the typing of the initial draft of a large portion of the manuscript;
Martin Lansche, Anita Sharifipour, and Douglas Swift, all former students of mine, for assistance with the problem sets;
Johann Wagner, H. A. Gleason Jr., and David F. Stermole, all of Gutenberg Software Limited, for producing and supporting a remarkable product that has freed me from the clutches of the Big Blue Machine;
some of my colleagues at York University, especially I. Kleiner and W. Tholen, for class-testing or critically reading portions of the manuscript;
York University itself, for the 1984–85 sabbatical year during which the textbook was finally completed;
the Université de Montréal, for providing warm hospitality and a productive working environment during the 1984–85 academic year;
Theodore Buchholz, mathematics editor at Reston Publishing Company, for understanding my reluctance to introduce changes designed to "make this book more like all the other linear algebra books," as was suggested by a few reviewers;
Linda Zuk, production editor at Reston Publishing Company, for her patience, guidance, and thoughtfulness throughout a period when a corporate shuffle gave her many more things to worry about than my book.

Notes to the Reader

This book is divided into chapters and sections. The internal numbering system makes no reference to the chapters; these merely provide broad, descriptive headings. Each section contains various types of numbered items; these may be definitions, theorems, corollaries, remarks, exercises, examples, and so on. Items in section x, regardless of their type, are consecutively numbered x.1, x.2, x.3, and so on. If the item numbered x.y involves several parts, these will be labeled x.y(i), x.y(ii), x.y(iii), and so on with lower case Roman numerals. The sections are not intended to be of uniform length or level of difficulty; they correspond rather to natural divisions of the subject matter.

There is a distinction between *exercises* and *problems*. Exercises are placed in the body of the text while problems are grouped together at the ends of sections. The reader should do the exercises *immediately* when they are encountered, before continuing to read the rest of the section. Exercises usually involve verifying easy consequences of a definition that has just been given or easy corollaries of a theorem that has just been proved. Results contained in the exercises are routinely used in the remainder of the text. The exercises are really the most important problems.

The footnotes are gathered in a separate section at the back of the book. Within Section x, footnotes are numbered consecutively as [F]x.1, [F]x.2, [F]x.3, and so on. It is safe, and perhaps preferable, to omit the footnotes on first reading since they contain, for the most part, material that would otherwise disrupt the flow of the text.

The symbol ■ is placed in the right margin to indicate the end of a proof.

The answer or a hint for the solution is given at the back of the book for all problems marked with an asterisk; this is the case for about 80 percent of the problems.

Dependence Among Chapters

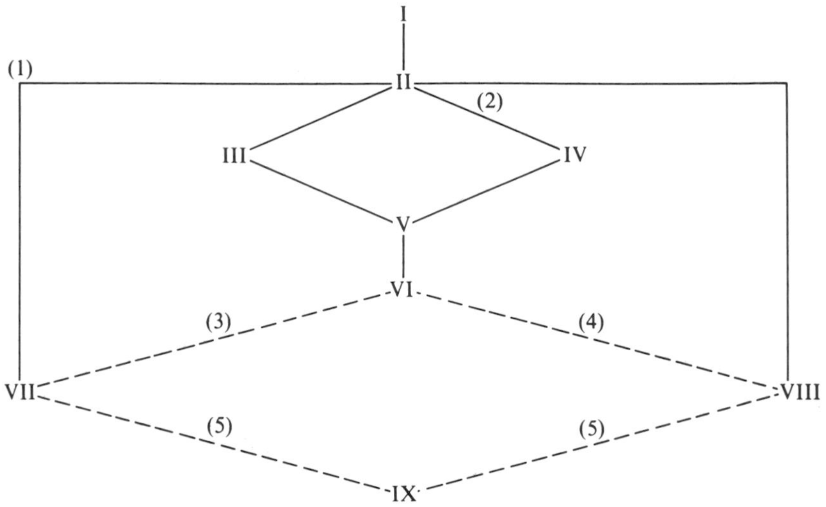

[1] For $n \times n$ matrices A and B, if $AB = I_n$, then $BA = I_n$. This fact is proved in Chapter IV. By assuming it temporarily, Chapter VII may be read independently of Chapter IV.
[2] Section 20 summarizes Chapters III and IV.
[3] Sections 38 and 39 do not depend on Chapter VI.
[4] Sections 41, 42, and 43 do not depend on Chapter VI.
[5] Different sections from Chapter IX have different prerequisites; consult the chart below for details.

Prerequisite Structure for Chapter IX

We may consider that Section 46 is divided in two parts: §46$\mathbb{R}$ for real matrices and §46$\mathbb{C}$ for complex matrices. Section 50 is divided in a similar fashion into §50$\mathbb{R}$ and §50$\mathbb{C}$.

	Prerequisite							
Section	*Chapter VI*	*Chapter VII*	*Chapter VIII*	§45	§46$\mathbb{R}$	§46$\mathbb{C}$	§47	§48
§45		×	×					
§46$\mathbb{R}$	×	×						
§46$\mathbb{C}$	×	×	×	×				
§47	×	×	×	×				
§48	×	×	×	×	×		×	
§49	×	×	×	×	×		×	×
§50$\mathbb{R}$	×	×			×			
§50$\mathbb{C}$	×	×	×	×	×	×		
§51	×		×	plus the definitions, but not the theorems from §50				

1

Matrices — Where They Come from and How to Multiply Them

A common problem in mathematics is the following:

1.1 Example

Suppose you know the x's in terms of the y's:

$$x_1 = 4y_1 - 3y_2$$
$$x_2 = 2y_1 + 5y_2$$

and that you know the y's in terms of the z's:

$$y_1 = -6z_1 + 7z_2$$
$$y_2 = 8z_1 - 9z_2$$

How do you express the x's in terms of the z's? Of course, the natural way to solve this problem is *by substitution*. Replace the y's in the first system of equations by their *values* in terms of the z's that are given in the second system of equations. Let us carry this out in full detail.

$$
\begin{aligned}
x_1 &= 4y_1 - 3y_2 = 4(-6z_1 + 7z_2) - 3(8z_1 - 9z_2) \\
&= (4)(-6)z_1 + (4)(7)z_2 + (-3)(8)z_1 + (-3)(-9)z_2 \\
&= [(4)(-6) + (-3)(8)]\,z_1 + [(4)(7) + (-3)(-9)]\,z_2 \\
&= -48z_1 + 55z_2 \\
x_2 &= 2y_1 + 5y_2 = 2(-6z_1 + 7z_2) + 5(8z_1 - 9z_2) \\
&= (2)(-6)z_1 + (2)(7)z_2 + (5)(8)z_1 + (5)(-9)z_2 \\
&= [(2)(-6) + (5)(8)]\,z_1 + [(2)(7) + (5)(-9)]\,z_2 \\
&= 28z_1 - 31z_2
\end{aligned}
$$

Thus the equations

$$
\begin{aligned}
x_1 &= -48z_1 + 55z_2 \\
x_2 &= 28z_1 - 31z_2
\end{aligned}
$$

express the x's in terms of the z's.

Let us consider another example of the same type.

1.2 Example

Suppose that the system of equations

$$
\begin{aligned}
x_1 &= 2y_1 - 3y_2 \\
x_2 &= 4y_1 + 5y_2 \\
x_3 &= -6y_1 + 7y_2
\end{aligned}
$$

expresses the x's in terms of the y's and that

$$
\begin{aligned}
y_1 &= 8z_1 - 9z_2 + 10z_3 + 11z_4 \\
y_2 &= -12z_1 + 13z_2 - 14z_3 + 15z_4
\end{aligned}
$$

expresses the y's in terms of the z's. To express the x's in terms of the z's, we substitute to obtain

$$
\begin{aligned}
x_1 &= 2y_1 - 3y_2 = 2(8z_1 - 9z_2 + 10z_3 + 11z_4) - 3(-12z_1 + 13z_2 - 14z_3 + 15z_4) \\
&= [(2)(8) + (-3)(-12)]\,z_1 + [(2)(-9) + (-3)(13)]\,z_2 \\
&\quad + [(2)(10) + (-3)(-14)]\,z_3 + [(2)(11) + (-3)(15)]\,z_4 \\
&= 52z_1 - 57z_2 + 62z_3 - 23z_4
\end{aligned}
$$

Similarly,

$$\begin{aligned} x_2 &= [(4)(8) + (5)(-12)]\, z_1 + [(4)(-9) + (5)(13)]\, z_2 \\ &\quad + [(4)(10) + (5)(-14)]\, z_3 + [(4)(11) + (5)(15)]\, z_4 \\ &= -28z_1 + 29z_2 - 30z_3 + 119z_4 \\ x_3 &= [(-6)(8) + (7) - (12)]\, z_1 + [(-6)(-9) + (7)(13)]\, z_2 \\ &\quad + [(-6)(10) + (7)(-14)]\, z_3 + [(6)(11) + (7)(15)]\, z_4 \\ &= -132z_1 + 145z_2 - 158z_3 + 39z_4 \end{aligned}$$

Thus the equations

$$\begin{aligned} x_1 &= 52z_1 - 57z_2 + 62z_3 - 23z_4 \\ x_2 &= -28z_1 + 29z_2 - 30z_3 + 119z_4 \\ x_3 &= -132z_1 + 145z_2 - 158z_3 + 39z_4 \end{aligned}$$

express the x's in terms of the z's.

For a problem of this type (i.e., given the x's in terms of the y's and the y's in terms of the z's, express the x's in terms of the z's) to make sense, there is an obvious constraint on the relative numbers of the three kinds of variables involved: if the first system of equations expresses some number of x's in terms of n-many y's, say, $y_1, y_2, \ldots, y_n$, then the second system of equations must consist of n-many equations, one for each of the y's. For example, if the first system of 17 equations expresses each of $x_1, x_2, \ldots, x_{17}$ in terms of $y_1, y_2, \ldots, y_{32}$, then the second system must consist of 32 equations expressing each of y_1 through y_{32} as functions of, say, $z_1, z_2, \ldots, z_{29}$; then when you "solve for the x's in terms of the z's" by substituting as in Examples 1.1 and 1.2, the result is a system of 17 equations expressing each of $x_1, x_2, \ldots, x_{17}$ in terms of $z_1, z_2, \ldots, z_{29}$.

In 1858, Arthur Cayley got tired of writing down all these x's, y's, and z's. He realized that in the system of equations that results from the substitution there is a pattern that determines, for example, in the equation for x_5, what the coefficient of z_9 will be.

To understand what this pattern is, let us return to Example 1.1.

$$\begin{aligned} x_1 &= 4y_1 - 3y_2 \qquad & y_1 &= -6z_1 + 7z_2 \\ x_2 &= 2y_1 + 5y_2 \qquad & y_2 &= 8z_1 - 9z_2 \end{aligned}$$

Let us save effort and paper by writing down only the coefficients, displayed in rows and columns, as they would be if we refused to write the x's, y's, and z's.

$$\begin{array}{rr} 4 & -3 \\ 2 & 5 \end{array} \qquad \begin{array}{rr} -6 & 7 \\ 8 & -9 \end{array}$$

We have seen that the answer is:

$$x_1 = -48z_1 + 55z_2 = [(4)(-6) + (-3)(8)]z_1 + [(4)(7) + (-3)(-9)]z_2$$
$$x_2 = 28z_1 - 31z_2 = [(2)(-6) + (5)(8)]z_1 + [(2)(7) + (5)(-9)]z_2$$

If we write down the coefficients of the answer, we have

$$\begin{array}{rr} -48 & 55 \\ 28 & -31 \end{array}$$

Note that the entry -48 in the first row and first column of this array equals $(4)(-6) + (-3)(8)$, which is obtained by taking the entries in the first row of the array on the left and pairing them up with the entries in the first column of the array on the right, and then taking the sum of the products of the pairs.

Similarly, the entry in the second row and first column of the resulting array, 28, is obtained by repeating this pattern using the second row of the left array and the first column of the right array [i.e., $28 = (2)(-6) + (5)(8)$].

If we leave off the x's, y's, and z's from Example 1.2, we are led to the following arrays:

$$\begin{array}{rr} 2 & -3 \\ 4 & 5 \\ -6 & 7 \end{array} \qquad \begin{array}{rrrr} 8 & -9 & 10 & 11 \\ -12 & 13 & -14 & 15 \end{array}$$

$$\begin{array}{rrrr} 52 & -57 & 62 & -23 \\ -28 & 29 & -30 & 119 \\ -132 & 145 & -158 & 39 \end{array}$$

Once again, the pattern that is followed to obtain 119, the coefficient of z_4 in the equation that expresses x_2 in terms of the z's, is to pair the entries in the second row of the left array with those from the fourth column of the right array and take the sum of the products of the pairs [i.e., $119 = (4)(11) + (5)(15)$].

The point of these examples is that problems of the form "given the x's in terms of the y's and the y's in terms of the z's, how do you express the x's in terms of the z's?" can be solved with much less effort by remembering the pattern that determines how the coefficients in the answer depend on the coefficients in the given equations. The pattern is simply this: the entry in the ith row, j^{th} column of the answer is obtained by pairing together the entries from the ith row of the left array with those from the jth column of the right array and computing the sum of the products of the pairs.

An additional point should be made here: it is important that we write the coefficients in *rectangular* arrays, displayed as rows and columns. If in the discussion of Example 1.2 we simply *listed* the coefficients, for example,

$$2 \ -3 \ 4 \ 5 \ -6 \ 7 \ 8 \ -9 \ 10 \ 11 \ -12 \ 13 \ -14 \ 15$$
$$52 \ -57 \ 62 \ -23 \ -28 \ 29 \ -30 \ 119 \ -132 \ 145 \ -158 \ 39$$

with or without separating them by commas, then the pattern which determines that $145 = (-6)(-9) + (7)(13)$ would no longer be visible. In other words, some essential information is actually retained by displaying the coefficients as a rectangular array—specifically, knowledge of which coefficient belongs in which equation with which variable. This *place-holding* aspect of the notation is its essential feature.

1.3 Exercise

Let

$$x_1 = 3y_1 - 5y_2 + \frac{1}{2}y_3 + 4y_4$$

$$x_2 = 2y_1 + 4y_2 - 6y_3 - \frac{1}{2}y_4$$

and let

$$\begin{aligned} y_1 &= 4z_1 - 2z_2 + 3z_3 \\ y_2 &= z_1 + z_2 - 2z_3 \\ y_3 &= 8z_1 + 2z_2 - 4z_3 \\ y_4 &= -2z_1 - 2z_2 + 6z_3 \end{aligned}$$

Find equations that express the x's in terms of the z's.
See [F]1.1 after completing this exercise.

With these examples to motivate us, we continue with the formal definitions of matrices and matrix multiplication. First, a remark concerning our use of the word *scalar*. For now, and until Section 41, *scalar* is merely a synonym for *real number*. What we do for the real numbers in Sections 1 through 40 will be redone in a more general context in Sections 41 to the end; this generalization is motivated by what we do in Sections 1 through 40 for the special case of the real numbers. In this more general context, the word *scalars* is used to denote those objects that behave like the real numbers do in the special case. Since one of the ultimate goals of the book is to study this generalization (called an *abstract vector space*), a useful purpose is served by using the appropriate terminology from the beginning.

Remember, until Section 41 *scalar* just means *real number*.

1.4 Definition

A *matrix* is a rectangular array of scalars.

We will use the uppercase Roman letters A, B, C, D, E, I, J, M, N, P, Q, X, Y, and Z to denote matrices.

1.5 Definition

A matrix that has m rows and n columns is said to have *size* m *by* n.

We usually write $m \times n$ rather than m by n. The number of rows is always given first, followed by the number of columns. Thus a 5×7 matrix has 5 rows and 7 columns, whereas a 7×5 matrix has 7 rows and 5 columns. An $m \times n$ matrix has m rows and n columns, whereas an $n \times p$ matrix has n rows and p columns. Although the letters A, B, C, D, E, I, J, M, N, P, and Q refer to matrices of arbitrary size, the letters X, Y, and Z will usually denote matrices of size $m \times 1$ or $1 \times n$.

It is important not to use the word *dimension* when referring to the size of matrices. If you are familiar with certain computer programming languages, you may already be in the habit of saying that the 3×5 matrix A has dimension 3×5 and of writing DIM A (m, n) to express that A has size $m \times n$. *Dimension* will acquire a specific technical meaning in Section 8, so we wish to avoid overusing it.

Another word in common usage for this concept is *order*. An $m \times n$ matrix is sometimes said to have order $m \times n$; an $n \times n$ matrix is said to have order n.

1.6 Definition

The scalar in the ith row and j^{th} column of the matrix A is called *the* $(i, j)^{th}$ *entry of A* and is denoted by a_{ij}.

It is customary that if a matrix is denoted by a given uppercase letter, then the entries of this matrix are denoted by the corresponding lowercase letter with appropriate subscripts. Thus b_{25} is the entry in the second row, fifth column of the matrix B; p_{4j} is the entry in the fourth row, j^{th} column of the matrix P; z_{71} is the entry in the seventh row, first column of the matrix Z; c_{i2} is the entry in the ith row, second column of the matrix C; and so on.

If a matrix is given by displaying all its entries, it is customary to enclose the rectangular array with square brackets.

We will not usually distinguish between a scalar b and the 1×1 matrix whose only entry is b.

1.7 Example

If

$$A = \begin{bmatrix} 3 & 5 & 9 \\ 2 & 6 & 7 \end{bmatrix}, \qquad B = \begin{bmatrix} 2 \\ 0 \\ \sqrt{3} \end{bmatrix},$$

$$C = \begin{bmatrix} 0 & \pi & -1 \\ e & \sqrt{2} & \cos(\pi/6) \\ \ln 17 & 1 & 0 \end{bmatrix}$$

then A is a 2×3 matrix, B is a 3×1 matrix, and C is a 3×3 matrix. Also, $a_{23} = 7$, $b_{31} = \sqrt{3}$, b_{13} does not exist since B does not have a third column, $c_{33} = 0$, $c_{23} = \cos(\pi/6)$, and, if $i = 1$ and $j = 3$, then $c_{ij} = -1$.

1.8 Definition (Equality for Matrices)

If A and B are matrices, A = B if and only if A and B have the same size and $a_{ij} = b_{ij}$ for all i and j.

Thus, for example,

$$\begin{bmatrix} 3 & 5 & 9 \\ 2 & 6 & 7 \end{bmatrix} \neq \begin{bmatrix} 3 & 5 & 9 \\ 2 & 6 & 7 \\ 0 & 0 & 0 \end{bmatrix}$$

since these matrices do not have the same size. Also,

$$A = \begin{bmatrix} 3 & 5 & 9 \\ 2 & 6 & 7 \end{bmatrix} \neq \begin{bmatrix} 3 & 5 & 8 \\ 2 & 6 & 7 \end{bmatrix} = D$$

since $a_{13} \neq d_{13}$.

1.9 Definition

(i) A matrix that has the same number of rows as columns is called a *square matrix*.

(ii) If A is a square matrix of size $n \times n$, then the entries $a_{11}, a_{22}, \ldots, a_{nn}$ constitute the *main diagonal* of A.

(iii) A square matrix A is called a *diagonal matrix* if the only nonzero entries in A occur on the main diagonal (i.e., if $a_{ij} = 0$ whenever $i \neq j$).

(iv) A square matrix A is called *upper triangular* if the only nonzero entries in A occur on or above the main diagonal (i.e., if $a_{ij} = 0$ whenever $i > j$).

(v) A square matrix A is called *lower triangular* if the only nonzero entries in A occur on or below the main diagonal (i.e., if $a_{ij} = 0$ whenever $i < j$).

(vi) A diagonal matrix whose entries on the main diagonal are all equal to 1 is called an *identity matrix*.

(vii) A matrix whose entries are all equal to 0 is called a *zero matrix*.

1.10 Examples

$$A = \begin{bmatrix} 1 & 2 & 3 \\ 4 & 5 & 6 \\ 7 & 8 & 9 \end{bmatrix} \qquad B = \begin{bmatrix} -1 & 0 & 0 \\ 0 & 0 & 0 \\ 0 & 0 & 4 \end{bmatrix}$$

$$C = \begin{bmatrix} 2 & 4 & 6 \\ 0 & -2 & 4 \\ 0 & 0 & 2 \end{bmatrix} \qquad D = \begin{bmatrix} 3 & 0 & 0 \\ -3 & 0 & 0 \\ 6 & 1 & 4 \end{bmatrix}$$

The main diagonal of A consists of the entries 1, 5, 9; B is a diagonal matrix; C is an upper triangular matrix; D is a lower triangular matrix.

For every positive integer n, the $n \times n$ identity matrix will be denoted by I_n. Thus

$$I_1 = 1, \qquad I_2 = \begin{bmatrix} 1 & 0 \\ 0 & 1 \end{bmatrix}, \qquad I_3 = \begin{bmatrix} 1 & 0 & 0 \\ 0 & 1 & 0 \\ 0 & 0 & 1 \end{bmatrix}$$

and so on.

The $m \times n$ matrix whose entries are all equal to 0 will be denoted by $0_{m\times n}$. The $n \times n$ diagonal matrix whose entries on the main diagonal are $d_1, d_2, \ldots, d_n$ will be denoted by $\operatorname{diag}(d_1, d_2, \ldots, d_n)$. Thus the preceding matrix B is $\operatorname{diag}(-1, 0, 4)$.

If we refer back to the discussion of 1.1 and 1.2, we see that these examples motivate *two* concepts simultaneously: not just that of matrices themselves, but also the concept of an operation or algorithm on matrices that uses certain ordered pairs of matrices as input and produces a third matrix as output. This operation produces the $(i, j)^{\text{th}}$ entry of the output by pairing the entries in the ith row of the matrix on the left with those from the jth column of the matrix on the right and taking the sum of the products of the pairs; this operation can be performed as long as the pairing of the appropriate entries can take place (i.e., as long as the number of rows of the matrix on the right equals the number of columns of the matrix on the left). This operation is called *matrix multiplication* because, as we will see below and in 3.1 and 3.2, it has many (but not all) of the properties possessed by ordinary multiplication of real numbers.

1.11 Definition (Multiplication for Matrices)

If A is an $m \times n$ matrix and B is an $n \times p$ matrix, then the *product*, AB, is an $m \times p$ matrix; the $(i, j)^{\text{th}}$ entry of AB is the sum of the products of the pairs that are obtained when the entries from the ith row of the left factor, A, are paired with those from the jth column of the right factor, B.

Note that we are using juxtaposition rather than, say, $A \cdot B$ or $A \times B$ or $A * B$, to denote multiplication of the matrices A and B. This convention is familiar since juxtaposition is also used to denote multiplication of real numbers, for example, in algebraic expressions such as $b^2 - 4ac$ and $3xy - 4xz + 5yz$.

The reason for calling I_n an identity matrix in Definition 1.9(vi) should now be clear. These matrices behave like identity elements for matrix multiplication, i.e., for any $m \times n$ matrix A, $I_m A = A$ and $AI_n = A$.

The point of the next observation is that the jth column of the product AB does not depend on any of the entries in B other than those in the jth column; similarly, the only entries in A that have any effect on the ith row of the product AB are those from the ith row of A.

When we consider the problem that motivated the definition of matrix multiplication, this point is obvious: the entries in the ith row of AB are the coefficients that express x_i in terms of the z's; clearly these depend only on how x_i itself, and not any of the other x's, was expressed in terms of the y's.

1.12 Observation

If A is an $m \times n$ matrix and B is an $n \times p$ matrix, then

(i) $$\begin{bmatrix} j^{\text{th}} \\ \text{column} \\ \text{of} \\ \text{AB} \end{bmatrix} = A \begin{bmatrix} j^{\text{th}} \\ \text{column} \\ \text{of} \\ \text{B} \end{bmatrix}$$

(ii) $[i^{\text{th}} \text{ row of AB}] = [i^{\text{th}} \text{ row of A}]B$

Proof of (i):

$$AB = \begin{bmatrix} a_{11} & a_{12} & \cdots & a_{1j} & \cdots & a_{1n} \\ a_{21} & a_{22} & \cdots & a_{2j} & \cdots & a_{2n} \\ \vdots & \vdots & & \vdots & & \vdots \\ a_{i1} & a_{i2} & \cdots & a_{ij} & \cdots & a_{in} \\ \vdots & \vdots & & \vdots & & \vdots \\ a_{m1} & a_{m2} & \cdots & a_{mj} & \cdots & a_{mn} \end{bmatrix} \begin{bmatrix} b_{11} & b_{12} & \cdots & b_{1j} & \cdots & b_{1p} \\ b_{21} & b_{22} & \cdots & b_{2j} & \cdots & b_{2p} \\ \vdots & \vdots & & \vdots & & \vdots \\ b_{i1} & b_{i2} & \cdots & b_{ij} & \cdots & b_{ip} \\ \vdots & \vdots & & \vdots & & \vdots \\ b_{n1} & b_{n2} & \cdots & b_{nj} & \cdots & b_{np} \end{bmatrix}$$

Thus the $(i, j)^{\text{th}}$ entry in AB is $a_{i1}b_{1j} + a_{i2}b_{2j} + \cdots + a_{in}b_{nj}$. So the jth column of AB is

$$\begin{bmatrix} a_{11}b_{1j} + a_{12}b_{2j} + \cdots + a_{1n}b_{nj} \\ a_{21}b_{1j} + a_{22}b_{2j} + \cdots + a_{2n}b_{nj} \\ \vdots \\ a_{i1}b_{1j} + a_{i2}b_{2j} + \cdots + a_{in}b_{nj} \\ \vdots \\ a_{m1}b_{1j} + a_{m2}b_{2j} + \cdots + a_{mn}b_{nj} \end{bmatrix} = \begin{bmatrix} a_{11} & a_{12} & \cdots & a_{1n} \\ a_{21} & a_{22} & \cdots & a_{2n} \\ \vdots & & & \vdots \\ a_{i1} & a_{i2} & \cdots & a_{in} \\ \vdots & & & \vdots \\ a_{m1} & a_{m2} & \cdots & a_{mn} \end{bmatrix} \begin{bmatrix} b_{1j} \\ b_{2j} \\ \vdots \\ b_{nj} \end{bmatrix}$$

$$= A \begin{bmatrix} j^{\text{th}} \\ \text{column} \\ \text{of} \\ B \end{bmatrix}$$

The proof of (ii) is similar and is left as an exercise. ■

1.13 Corollary

(i) The product of two matrices will have an all-zero column if the right factor has an all-zero column.
(ii) The product of two matrices will have an all-zero row if the left factor has an all-zero row.

The proof is left as an exercise. ■

As we mentioned earlier, matrix multiplication does not have all the algebraic properties of multiplication of real numbers. Before leaving this section, we should emphasize the most important differences.

First, any two real numbers s, t can be multiplied, and the answer st does not depend on the order of the factors: $st = ts$. For matrices A and B, however, it can happen that none, exactly one, or both of AB and BA are defined depending on the sizes of A and B. Moreover, even when both AB and BA are defined, they may not be equal.

If A is $m \times n$ and B is $p \times q$, then AB is defined only if $n = p$, and in this case AB is an $m \times q$ matrix. Similarly, BA is defined only if $q = m$, and in this case BA is a $p \times n$ matrix. (This is not surprising when we recall Examples 1.1 and 1.2 that motivated the definition of matrix multiplication.) Thus, let A be a 2×4 matrix; if B is a 5×3 matrix, then neither AB nor BA is defined; if B is a 4×5 matrix, then AB is defined and is a 2×5 matrix, whereas BA is undefined; if B is a 3×2 matrix, then BA is defined and is a 3×4 matrix, whereas AB is undefined; finally, if B is a 4×2 matrix, then both AB and BA are defined although they are clearly unequal, since AB is a 2×2 matrix while BA is a 4×4 matrix.

This last example shows that AB and BA may not be equal even when both are defined. Yet, in this example, they were unequal for a somewhat trivial reason: AB and BA were of different size. In Example 1.15 we will see that AB and BA may be different even when both are defined and have the same size.

1.14 Exercise

If A is an $m \times n$ matrix and B is a $p \times q$ matrix and if both AB and BA are defined and have the same size, then $m = n = p = q$ (i.e., then both A and B are square matrices of the same size).

1.15 Example

$$\mathrm{A} = \begin{bmatrix} 1 & 2 \\ 3 & 4 \end{bmatrix}, \qquad \mathrm{B} = \begin{bmatrix} 5 & 6 \\ 7 & 8 \end{bmatrix}$$

$$\mathrm{AB} = \begin{bmatrix} 19 & 22 \\ 43 & 50 \end{bmatrix}, \qquad \mathrm{BA} = \begin{bmatrix} 23 & 34 \\ 31 & 46 \end{bmatrix}$$

This example is typical. It seldom happens for $n \times n$ matrices A and B that AB = BA. Once again, when we recall the type of problem that motivated the definition of matrix multiplication, we realize that there is no reason to expect that AB = BA. AB gives us the coefficients for expressing the x's in terms of the z's when, for example,

$$\begin{aligned} x_1 &= y_1 + 2y_2 \\ x_2 &= 3y_1 + 4y_2 \end{aligned} \quad \text{and} \quad \begin{aligned} y_1 &= 5z_1 + 6z_2 \\ y_2 &= 7z_1 + 8z_2 \end{aligned}$$

with

$$\mathrm{A} = \begin{bmatrix} 1 & 2 \\ 3 & 4 \end{bmatrix} \quad \text{and} \quad \mathrm{B} = \begin{bmatrix} 5 & 6 \\ 7 & 8 \end{bmatrix}.$$

BA is the answer to a different question: it gives us the coefficients for expressing the x's in terms of the z's when

$$\begin{aligned} x_1 &= 5y_1 + 6y_2 \\ x_2 &= 7y_1 + 8y_2 \end{aligned} \quad \text{and} \quad \begin{aligned} y_1 &= z_1 + 2z_2 \\ y_2 &= 3z_1 + 4z_2 \end{aligned}$$

1.16 Definition

Two matrices A and B are said to *commute* if AB = BA.

The fact that this equality fails in general is often expressed by saying that matrix multiplication is not commutative.

Another important property of multiplication for real numbers that fails for matrix multiplication is this: for real numbers s and t, if $st = 0$, then either $s = 0$ or $t = 0$. This is not the case for matrices. The product can be a zero matrix with neither factor being a zero matrix. Indeed, it is possible to have $P^2 = 0_{n \times n}$ even though $P \neq 0_{n \times n}$.

1.17 Example

$$A = \begin{bmatrix} 0 & 1 \\ 0 & 1 \end{bmatrix}, \quad B = \begin{bmatrix} 1 & 1 \\ 0 & 0 \end{bmatrix}, \quad AB = \begin{bmatrix} 0 & 0 \\ 0 & 0 \end{bmatrix}$$

$$C = \begin{bmatrix} 8 & -6 \\ -4 & 3 \end{bmatrix}, \quad D = \begin{bmatrix} 3/4 & 9 \\ 1 & 12 \end{bmatrix}, \quad CD = \begin{bmatrix} 0 & 0 \\ 0 & 0 \end{bmatrix}$$

$$P = \begin{bmatrix} 2 & 4 \\ -1 & -2 \end{bmatrix}, \quad P^2 = \begin{bmatrix} 0 & 0 \\ 0 & 0 \end{bmatrix}$$

Still another algebraic law that fails for matrices is the Cancellation Law: if r, s, and t are real numbers with $rs = rt$ and $r \neq 0$, then $s = t$. For matrices, it can happen that AB = AC and $A \neq 0$, and yet $B \neq C$; so it would be incorrect to "cancel" the A and conclude that B = C.

1.18 Example

$$A = \begin{bmatrix} 3 & 9 \\ -1 & -3 \end{bmatrix}, \quad B = \begin{bmatrix} -5 & -10 \\ 1 & 2 \end{bmatrix}, \quad C = \begin{bmatrix} -8 & -7 \\ 2 & 1 \end{bmatrix}$$

Check that AB = AC.

We will eventually develop the theory that explains why these examples are possible. You will then be able to construct other examples (easily) for yourself. For the moment, the important point is to remember that there are these differences between multiplication for real numbers and multiplication for matrices. In fact, the only reason this needs to be stressed from the beginning is related to the terminology itself: historical processes led to the operation defined in 1.8 being *called* "multiplication" rather than, say, "amalgamation" or "coalition." There would be no temptation for beginning students to take the algebraic properties of this operation for granted had it been called "matrix coalition". See [F]1.2.

Reminder: As explained in the Preface, an asterisk preceding the number of a problem indicates that the answer or a hint for the solution is included at the end of the text.

Problem Set 1

1.1. Go back and do any of the exercises in Section 1 that you may have skipped.

*1.2. How many entries does an $m \times n$ matrix have?

*1.3. A matrix has 120 entries; list all its possible sizes.

*1.4. A matrix has 17 entries; list all its possible sizes.

*1.5. Write down the 3×5 matrix A whose $(i, j)^{\text{th}}$ entry is given by the formula $a_{ij} = i - j$.

*1.6. Write down the 5×4 matrix B whose $(i, j)^{\text{th}}$ entry is given by the formula $b_{ij} = (-1)^{i+j} 2(i + j)$.

*1.7. Write down the 3×3 matrix C whose $(i, j)^{\text{th}}$ entry is given by the formula $c_{ij} = (2i - j)^2$.

*1.8. A and B are 5×3 matrices; C, D, and E are 3×4, 4×5, and 3×5, respectively. Which of the following expressions are defined? What are the sizes of those that are defined?

(i) BC	(ii) AE	(iii) (BC)E
(iv) (AC)D	(v) DA	(vi) A(CD)
(vii) (DA)E	(viii) D(AE)	(ix) AB

*1.9.

$$\begin{aligned} x_1 &= 4y_1 + 3y_2, & y_1 &= 3z_1 - z_2 + 4z_3 \\ x_2 &= y_1 + 2y_2 & y_2 &= z_1 - 3z_2 + z_3 \\ x_3 &= -2y_1 + y_2 & & \end{aligned}$$

Use matrix multiplication to find equations that express the x's in terms of the z's.

*1.10.

$$\begin{aligned} x_1 &= y_1 + 2y_2 - 3y_3, & y_1 &= 3z_1 + 4z_2 \\ x_2 &= 2y_2 + 3y_3 & y_2 &= 2z_1 + 3z_2 - z_3 - 2z_4 \\ & & y_3 &= z_2 + z_3 \end{aligned}$$

Use matrix multiplication to find equations that express the x's in terms of the z's.

*1.11. Prove that the product of two upper triangular $n \times n$ matrices is also upper triangular.

*1.12.

$$\begin{aligned} a &= 3x + 4y - 2z & x &= u - 3v + w \\ b &= -x + 2y + 2z, & y &= 2u + 2v + w \\ c &= 4x + y - 3z & z &= 4u - v - 2w \end{aligned}$$

Use matrix multiplication to find the equations that express a, b, and c in terms of u, v, and w.

*1.13.

$$A = \begin{bmatrix} 4 & 5 & 6 \\ 7 & 8 & 9 \\ -1 & -2 & -3 \end{bmatrix}, \qquad B = \begin{bmatrix} 2 & 9 & 7 \\ 3 & 6 & 1 \\ -4 & 8 & 1 \\ 0 & 2 & 2 \end{bmatrix}$$

$$C = \begin{bmatrix} 8 & 1 \\ 0 & 8 \\ 6 & -2 \end{bmatrix}$$

(i) Compute BC.
(ii) Compute the (3, 2) entry in (BA)C.
(iii) Compute the first column of AC.
(iv) Compute the third row of AC.

*1.14. If

$$\begin{bmatrix} 2w + x - z & y \\ 3w - 2z & w - 3x - 2z \end{bmatrix} = \begin{bmatrix} 4 & 0 \\ 5 & 2 \end{bmatrix}$$

find the values of w, x, y, and z.

*1.15.

$$A = \begin{bmatrix} 1 & 1 \\ 0 & 1 \end{bmatrix}$$

Find *all* the 2×2 matrices that commute with A. In other words, find necessary and sufficient conditions on

$$B = \begin{bmatrix} a & b \\ c & d \end{bmatrix}$$

for AB = BA to hold.

1.16. Prove that if A and B are both $n \times n$ diagonal matrices then AB = BA.

1.17. Let A be an arbitrary $n \times n$ matrix and let $D = \text{diag}(d_1, d_2, \ldots, d_n)$. Does AD = DA?

1.18. It is only for systems of equations of a certain type that the process of substitution of variables can be modeled by matrix multiplication. This point will become clear as the text proceeds. For the moment, consider the following example. Let

$$x_1 = 4y_1^2 - 3y_2, \qquad y_1 = -6z_1 + 7z_2$$
$$x_2 = 2y_1 + 5y_2^2 \qquad y_2 = 8z_1^2 - 9z_2$$

It still makes sense to solve for the x's in terms of the z's by substitution.

(i) Verify by substitution that

$$x_1 = 120z_1^2 - 336z_1z_2 + 196z_2^2 + 27z_2$$
$$x_2 = -12z_1 + 14z_2 + 320z_1^4 - 720z_1^2z_2 + 405z_2^2$$

(ii) Observe that matrix multiplication cannot be used to express the coefficients of the answer in terms of the coefficients of the given systems.

2

Matrix Addition and Multiplication of a Matrix by a Scalar

In this section we define two more operations involving matrices, and in Section 3 we study their algebraic properties, as well as their interaction with the operation of matrix multiplication from Section 1.

2.1 Definition

If A and B are each $m \times n$ matrices, then the *sum*, A + B, is the $m \times n$ matrix whose $(i, j)^{\text{th}}$ entry is $a_{ij} + b_{ij}$.

Note that A + B is defined only if A and B have the same size. Note too that we are using the same symbol + to denote both addition of reals and addition of matrices; this does not create any confusion since the context in which a plus sign occurs clearly indicates whether addition of real numbers or of matrices is taking place: in an expression of the form C + D, + means matrix addition; in an expression of the form $s + t$, + means addition of real numbers.

Addition of matrices takes place entrywise; that is, A + B is formed by adding the entries in the corresponding positions in A and B.

2.2 Example

If

$$A = \begin{bmatrix} 2 & 3 & -1 \\ -4 & 0 & 6 \end{bmatrix} \quad \text{and} \quad B = \begin{bmatrix} 0 & 3 & 1 \\ -2 & 0 & -3 \end{bmatrix}$$

then

$$A + B = \begin{bmatrix} 2+0 & 3+3 & -1+1 \\ -4-2 & 0+0 & 6-3 \end{bmatrix} = \begin{bmatrix} 2 & 6 & 0 \\ -6 & 0 & 3 \end{bmatrix}$$

2.3 Definition (Multiplication of a Matrix by a Scalar)

If A is an $m \times n$ matrix and s is any scalar, then the result of multiplying A by s is the $m \times n$ matrix, sA, whose $(i, j)^{\text{th}}$ entry is sa_{ij}. This operation is commonly called *scalar multiplication*.

For example, if

$$A = \begin{bmatrix} 3 & 0 \\ -5 & 2 \\ 1 & \frac{1}{2} \end{bmatrix} \quad \text{and} \quad s = 4$$

then

$$sA = \begin{bmatrix} 12 & 0 \\ -20 & 8 \\ 4 & 2 \end{bmatrix}$$

The matrix (-1)A is denoted by $-$A; if B is the same size as A, then B + (-1)A is written more simply as B $-$ A. Thus if

$$B = \begin{bmatrix} -2 & 9 \\ 4 & 6 \\ 1 & 0 \end{bmatrix}$$

and A is the preceding matrix, then

$$B - A = \begin{bmatrix} -5 & 9 \\ 9 & 4 \\ 0 & -\frac{1}{2} \end{bmatrix}.$$

2.4 Remark

We cannot conclude this section without at least mentioning the operation that takes two matrices A, B of the same size, $m \times n$, as input and that outputs the $m \times n$ matrix whose $(i, j)^{\text{th}}$ entry is the product $a_{ij}b_{ij}$. Of course, it makes sense, just as it does for addition, to multiply matrices of the same size entrywise; but as we have seen in Section 1, this is *not* the natural thing to do. This operation is called the Hadamard product after the French mathematician Jacques Hadamard. The Hadamard product is of minor interest and we will encounter it in a few of the problem sets.

See [F]2.1.

Problem Set 2

*2.1. Let

$$P = \begin{bmatrix} 2 & 9 & -1 \\ 4 & 0 & 3 \\ -2 & 6 & -4 \end{bmatrix}, \qquad Q = \begin{bmatrix} 4 & 7 & 7 \\ 3 & -4 & 0 \\ 6 & \frac{1}{2} & -\frac{1}{2} \end{bmatrix},$$

$$R = \begin{bmatrix} -4 & 2 & 0 \\ 0 & 6 & \frac{2}{3} \\ 4 & \frac{3}{4} & -2 \end{bmatrix}$$

Compute the following:

(i) P + Q (ii) Q + R
(iii) P + Q + R (iv) P − Q
(v) 2Q + 4R (vi) −2Q − 3R
(vii) The Hadamard product of P and Q
(viii) The matrix product, PQ

*2.2. Given that

$$A = \begin{bmatrix} 3 & 9 \\ \frac{1}{3} & 0 \\ 4 & -4 \end{bmatrix}, \qquad A - 3B = \begin{bmatrix} 7 & -14 \\ 0 & \frac{8}{3} \\ 4 & -12 \end{bmatrix}$$

find B.

3

Matrix Arithmetic

We have seen that matrix multiplication is not commutative and that it violates certain other algebraic laws that are satisfied by multiplication of real numbers. We concentrate at the beginning of this section on proving for matrices those properties of the arithmetic of real numbers that do hold; we conclude the section by expanding our list of properties that fail.

In 3.1, the phrases in brackets at the right are usually used to describe in words the corresponding algebraic property at the left. The sequence in which the properties are listed is not haphazard; they have been grouped according to the operations they involve, and the fact that properties (i) through (viii) involve *only* matrix addition, multiplication of a matrix by a scalar, and the arithmetic of scalars is an important observation that will be used in Section 41. See [F]3.1.

3.1 Theorem (Basic Properties of Matrix Arithmetic)

Let A, B, C be arbitrary matrices and let s, t be arbitrary scalars. Assuming that the sizes of the matrices are such that the indicated operations can be performed, the following are valid.

Properties that concern matrix addition alone:

(i) A + B = B + A [matrix addition is commutative]

(ii) A + (B + C) = (A + B) + C [matrix addition is associative]

(iii) The $m \times n$ zero matrix, $0_{m \times n}$, has the property that for every $m \times n$ matrix A,

$$A + 0_{m \times n} = A$$ [there exists an identity element for matrix addition]

(iv) For every $m \times n$ matrix A, there exists an $m \times n$ matrix B with the property that

$$A + B = 0_{m \times n}$$ [there exists an "inverse" for matrix addition]

Properties that concern the interaction among matrix addition, multiplication of matrices by scalars, and the arithmetic of scalars:

(v) $s(A + B) = sA + sB$ [scalar multiplication distributes from the left over matrix addition]

(vi) $(s + t)A = sA + tA$ [scalar multiplication distributes from the right over addition of scalars]

(vii) $s(tA) = (st)A$

(viii) $1A = A$ [the scalar 1 is an identity element for scalar multiplication]

A property of matrix multiplication alone:

(ix) A(BC) = (AB)C [matrix multiplication is associative]

A property that concerns the interaction between matrix multiplication and scalar multiplication:

(x) $s(AB) = (sA)B = A(sB)$

Properties that concern the interaction between matrix multiplication and matrix addition:

(xi) A(B + C) = AB + AC [matrix multiplication distributes from the left over matrix addition]

(xii) (A + B)C = AC + BC [matrix multiplication distributes from the right over matrix addition]

The proofs of items (i) through (viii) and (x) are obvious and are left as exercises; they follow immediately from common arithmetic properties of real numbers. Item (ix) is not quite obvious. We will prove it after first proving (xi) as a warm-up exercise in manipulation of notation. The proof of (xii) is similar to that of (xi) and is also left as an exercise.

Proof of (xi): To prove a matrix identity (i.e., an equation in which the expressions on the left-hand side [LHS] and right-hand side [RHS] of the equality sign represent matrices), we must, according to definition, establish *two* facts: (1) that the size of the LHS equals the size of the RHS, and (2) that for arbitrary i and j, the $(i, j)^{\text{th}}$ entry of the LHS equals the $(i, j)^{\text{th}}$ entry of the RHS. To prove (1) in this instance, we make use of the assumption in 3.1 that the sizes of the matrices are such that the indicated operations can be performed. Suppose A is $m \times n$ and that B is $p \times q$; then C must also be $p \times q$ since B + C is assumed to be defined; then we must have $n = p$ since A(B + C) is assumed to be defined; thus A(B + C) has size $m \times q$, and so do AB and AC; hence AB + AC does as well. To prove (2), we invoke the following chain of equalities:

$$(i, j)^{\text{th}} \text{ entry of } \mathrm{A(B + C)} = [i^{\text{th}} \text{ row of A}] \begin{bmatrix} j^{\text{th}} \\ \text{column} \\ \text{of B + C} \end{bmatrix}$$

$$= [a_{i1} a_{i2} \ldots a_{in}] \begin{bmatrix} b_{1j} + c_{1j} \\ b_{2j} + c_{2j} \\ \vdots \\ b_{nj} + c_{nj} \end{bmatrix}$$

$$= a_{i1}(b_{1j} + c_{1j}) + a_{i2}(b_{2j} + c_{2j}) + \cdots + a_{in}(b_{nj} + c_{nj})$$

$$= a_{i1}b_{1j} + a_{i1}c_{1j} + a_{i2}b_{2j} + a_{i2}c_{2j} + \cdots + a_{in}b_{nj} + a_{in}c_{nj}$$

$$= (a_{i1}b_{1j} + a_{i2}b_{2j} + \cdots + a_{in}b_{nj}) + (a_{i1}c_{1j} + a_{i2}c_{2j} + \cdots + a_{in}c_{nj})$$

$$= [a_{i1} a_{i2} \ldots a_{in}] \begin{bmatrix} b_{1j} \\ b_{2j} \\ \vdots \\ b_{nj} \end{bmatrix} + [a_{i1} a_{i2} \ldots a_{in}] \begin{bmatrix} c_{1j} \\ c_{2j} \\ \vdots \\ c_{nj} \end{bmatrix}$$

$$= [i^{\text{th}} \text{ row of A}] \begin{bmatrix} j^{\text{th}} \\ \text{column} \\ \text{of B} \end{bmatrix} + [i^{\text{th}} \text{ row of A}] \begin{bmatrix} j^{\text{th}} \\ \text{column} \\ \text{of C} \end{bmatrix}$$

$$= (i, j)^{\text{th}} \text{ entry of AB} + (i, j)^{\text{th}} \text{ entry of AC}$$

$$= (i, j)^{\text{th}} \text{ entry of (AB + AC)}$$

Proof of (ix): If A is $m \times n$, then B must be $n \times p$ since AB is assumed to be defined; but then C must be $p \times q$ since BC is assumed to be defined. Thus both the LHS and the RHS have size $m \times q$.

$$(i,j)^{\text{th}} \text{ entry in (AB)C} = [i^{\text{th}} \text{ row of AB}] \begin{bmatrix} j^{\text{th}} \\ \text{column} \\ \text{of C} \end{bmatrix}$$

$$= \left[\left(\sum_{k=1}^{n} a_{ik}b_{k1}\right)\left(\sum_{k=1}^{n} a_{ik}b_{k2}\right)\cdots\left(\sum_{k=1}^{n} a_{ik}b_{kp}\right)\right]\begin{bmatrix} c_{1j} \\ c_{2j} \\ \vdots \\ c_{pj} \end{bmatrix}$$

$$= \left(\sum_{k=1}^{n} a_{ik}b_{k1}\right)c_{1j} + \left(\sum_{k=1}^{n} a_{ik}b_{k2}\right)c_{2j} + \cdots + \left(\sum_{k=1}^{n} a_{ik}b_{kp}\right)c_{pj}$$

$$= a_{i1}b_{11}c_{1j} + a_{i1}b_{12}c_{2j} + \cdots + a_{i1}b_{1p}c_{pj}$$ [This line contains all the terms in the preceding sums that correspond to $k = 1$]

$$+a_{i2}b_{21}c_{1j} + a_{i2}b_{22}c_{2j} + \cdots + a_{i2}b_{2p}c_{pj}$$ [Same for $k = 2$]

$$+ \cdots$$

$$+a_{in}b_{n1}c_{1j} + a_{in}b_{n2}c_{2j} + \cdots + a_{in}b_{np}c_{pj}$$ [Same for $k = n$]

$$= a_{i1}\left(\sum_{k=1}^{p} b_{1k}c_{kj}\right) + a_{i2}\left(\sum_{k=1}^{p} b_{2k}c_{kj}\right) + \cdots + a_{in}\left(\sum_{k=1}^{p} b_{nk}c_{kj}\right)$$

$$= [i^{\text{th}} \text{ row of A}]\begin{bmatrix} \sum_{k=1}^{p} b_{1k}c_{kj} \\ \sum_{k=1}^{p} b_{2k}c_{kj} \\ \vdots \\ \sum_{k=1}^{p} b_{nk}c_{kj} \end{bmatrix}$$

$$= [i^{\text{th}} \text{ row of A}]\begin{bmatrix} j^{\text{th}} \\ \text{column} \\ \text{of BC} \end{bmatrix}$$

$$= (i,j)^{\text{th}} \text{ entry in A(BC)}$$ ■

3.2 Corollary

Under the same assumptions as in the Theorem 3.1,

(i) A(B − C) = AB − AC
(ii) (A − B)C = AC − BC

(iii) $s(A - B) = sA - sB$
(iv) $(s - t)A = sA - tA$
(v) $A - A = 0_{m \times n}$, if A is an $m \times n$ matrix
(vi) $0_{m \times n} - A = -A$, if A is an $m \times n$ matrix
(vii) $0A = 0_{m \times n}$, if A is an $m \times n$ matrix
(viii) $s0_{m \times n} = 0_{m \times n}$
(ix) $A0_{n \times p} = 0_{m \times p}$, if A is an $m \times n$ matrix
(x) $0_{m \times n}A = 0_{m \times p}$, if A is an $n \times p$ matrix

The proofs are left as exercises.

3.3 Definition

Let A be an $m \times n$ matrix. An $n \times m$ matrix B such that $BA = I_n$ is called a *left inverse* of A. An $n \times m$ matrix C such that $AC = I_m$ is called a *right inverse* of A. If A is a square matrix of size $n \times n$ and D is an $n \times n$ matrix satisfying both $DA = I_n$ and $AD = I_n$, then D is called an *inverse* of A. To emphasize that both the equations must be satisfied, such a D is sometimes called a *two-sided inverse* of A. A square matrix is called *invertible* if it has a (two-sided) inverse. A square matrix that does not have a (two-sided) inverse is called *singular*.

See [F]3.2 and [F]3.3.

3.4 Examples

$$A = \begin{bmatrix} 2 & -2 & 3 \\ 0 & 1 & -1 \\ -1 & 3 & -4 \\ 5 & -2 & 3 \end{bmatrix}, \quad B_1 = \begin{bmatrix} -3 & 2 & -2 & 1 \\ -1 & 5 & -2 & 0 \\ 3 & 1 & 1 & -1 \end{bmatrix},$$

$$B_2 = \begin{bmatrix} 1 & -1 & 1 & 0 \\ 3 & 2 & 1 & -1 \\ -1 & 4 & -2 & 0 \end{bmatrix}$$

Check that $B_1A = I_3$ and $B_2A = I_3$; thus both B_1 and B_2 are left inverses of A.

$$P = \begin{bmatrix} 2 & -3 & 5 & 4 \\ -1 & 2 & -4 & -2 \\ 3 & -4 & 5 & 2 \end{bmatrix}, \quad Q_1 = \begin{bmatrix} -4 & 5 & 8 \\ -5 & 5 & 9 \\ -2 & 1 & 3 \\ 1 & 0 & -1 \end{bmatrix},$$

$$Q_2 = \begin{bmatrix} 6 & -5 & -2 \\ 7 & -7 & -3 \\ 2 & -3 & -1 \\ 0 & 1 & 0 \end{bmatrix}$$

Check that $PQ_1 = I_3$ and $PQ_2 = I_3$; thus both Q_1 and Q_2 are right inverses of P.

3.5 Example

$$A = \begin{bmatrix} 2 & 1 & -1 \\ -1 & 2 & 1 \\ 4 & 1 & 0 \end{bmatrix}, \qquad B = \begin{bmatrix} -1/11 & -1/11 & 3/11 \\ 4/11 & 4/11 & -1/11 \\ -9/11 & 2/11 & 5/11 \end{bmatrix}$$

Check that both $AB = I_3$ and $BA = I_3$; thus B is a two-sided inverse of A.

From 3.4, we see that it is possible for a matrix to have two different left inverses; we also see that a matrix can have two different right inverses. (In fact, the matrix P in Example 3.4 has infinitely many different right inverses, but that need not concern us for the moment.) The point of Theorem 3.6 is that this *cannot* happen for two-sided inverses of a square matrix.

3.6 Theorem (Uniqueness of the Inverse of an Invertible Matrix)

If B and C are each inverses of the $n \times n$ matrix A, then B = C.

Proof:

$$\begin{aligned} B &= BI_n && \text{[property of } I_n\text{]} \\ &= B(AC) && \text{[C is a right inverse of A]} \\ &= (BA)C && \text{[matrix multiplication is associative]} \\ &= I_nC && \text{[B is a left inverse of A]} \\ &= C && \text{[property of } I_n\text{]} \end{aligned}$$

■

Because of 3.6, it makes sense to talk about *the* inverse of an invertible matrix A and to introduce a special symbol to denote it. For reasons that will soon be clear, the standard symbol used is A^{-1}. Thus if A is an invertible $n \times n$ matrix, we have both that $AA^{-1} = I_n$ and $A^{-1}A = I_n$.

3.7 Corollary to the Proof of 3.6

If A is $n \times n$ and if both $BA = I_n$ and $AC = I_n$, then $B = C = A^{-1}$.

Proof: If we examine the justifications given in brackets for each step in the proof of 3.6, we see that the full hypothesis of 3.6 (that B and C were each two-sided inverses of A) did not get used. Only the facts that B is a left inverse and C is a right inverse are required. ■

One consequence of 3.7 is that if a square matrix A has *both* a right inverse and a left inverse, then A is invertible. It is very important to distinguish this fact, which, as we have seen, is a fairly straightforward consequence of the definition of invertibility, from a much stronger result that is true: if a square matrix A has *either* a right or a left inverse, then A is invertible. This stronger result is at the heart of linear algebra and we will prove it in Section 19.

3.8 Theorem

If A and B are each invertible $n \times n$ matrices, then so is the product AB; moreover, $(AB)^{-1} = B^{-1}A^{-1}$.

Proof: Because A and B are each invertible, it makes sense to write down expressions involving A^{-1} and B^{-1}. Now, $(B^{-1}A^{-1})(AB) = B^{-1}(A^{-1}A)B = B^{-1}I_nB = B^{-1}B = I_n$; thus $B^{-1}A^{-1}$ is a left inverse for AB. Similarly, $(AB)(B^{-1}A^{-1}) = A(BB^{-1})A^{-1} = AI_nA^{-1} = AA^{-1} = I_n$; thus $B^{-1}A^{-1}$ is also a right inverse for AB. ■

3.9 Corollary

(i) If $A_1, A_2, \cdots, A_k$ are all invertible $n \times n$ matrices, then so is the product $A_1A_2 \cdots A_k$; moreover, $(A_1A_2 \cdots A_k)^{-1} = A_k^{-1}A_{k-1}^{-1} \cdots A_2^{-1}A_1^{-1}$.

(ii) If A and P are $n \times n$ matrices and P is invertible but the product PA is not invertible, then A is not invertible.

Proof: Exercise. ■

In more familiar terms, 3.9(i) says that the inverse of a product of invertible matrices is the product of the inverses, taken in reverse order.

It is essential that the order of the factors be reversed. If one attempts to prove that $(AB)^{-1} = A^{-1}B^{-1}$, the preceding proof breaks down because it is not possible to "cancel" the A and the A^{-1} in the expression $(A^{-1}B^{-1})(AB) = A^{-1}(B^{-1}A)B$ without first permuting the factors B^{-1} and A to obtain $A^{-1}AB^{-1}B$. Because matrix multiplication is not commutative, this step is not allowed.

In 3.10 we introduce the concept of integer powers of a square matrix, and in 3.11 and 3.12 we observe that Laws of Exponents from the arithmetic of real numbers are also valid for matrices. There is one precaution to be observed: for any real number $s \neq 0$ and positive integer r, it makes sense to consider

$$s^{-r} = \underbrace{s^{-1} \cdot s^{-1} \cdot \cdots \cdot s^{-1}}_{r\text{-many factors}} = \frac{1}{s^r}$$

We describe this situation by saying that every nonzero real number is invertible. It is not true that every nonzero square matrix is invertible; thus, when in 3.10 we define the negative integer powers of A, we must emphasize that this only makes sense if A is invertible.

3.10 Definition

If A is an $n \times n$ matrix and r is an integer, define

$$A^r = \begin{cases} I_n, & \text{if } r = 0 \\ \underbrace{AA \cdots A}_{r\text{-many factors}}, & \text{if } r > 0 \\ \underbrace{A^{-1}A^{-1} \cdots A^{-1}}_{r\text{-many factors}}, & \text{if } r < 0 \text{ and A is invertible} \end{cases}$$

3.11 Theorem

If A is any $n \times n$ matrix and q and r are integers $\geqq 0$, then $A^qA^r = A^{q+r}$ and $(A^q)^r = A^{qr}$.

Proof: Exercise. ■

3.12 Theorem

If A is an invertible $n \times n$ matrix, then:

(i) For arbitrary integers q and r,

$$A^qA^r = A^{q+r} \quad \text{and} \quad (A^q)^r = A^{qr}$$

(ii) A^{-1} is invertible; moreover, $(A^{-1})^{-1} = A$.
(iii) A^r is invertible and $(A^r)^{-1} = (A^{-1})^r$
(iv) For any nonzero scalar s, sA is invertible; moreover, $(sA)^{-1} = (1/s)A^{-1}$. ■

Proof: Exercise.

In 3.13, we introduce an operation that *interchanges* the rows and columns of a matrix.

3.13 Definition

If A is an $m \times n$ matrix, the $n \times m$ matrix whose i^{th} row is the ith column of A is called the *transpose* of A and is denoted by A^T.

In other words, for all i and j, the $(i, j)^{\text{th}}$ entry of A^T is equal to the $(j, i)^{\text{th}}$ entry of A.

3.14 Examples

$$\text{If } A = \begin{bmatrix} 2 & -1 \\ 4 & 3 \end{bmatrix}, \text{ then } A^T = \begin{bmatrix} 2 & 4 \\ -1 & 3 \end{bmatrix}$$

$$\text{If } B = \begin{bmatrix} 3 & 0 & -4 \\ 6 & 2 & 0 \end{bmatrix}, \text{ then } B^T = \begin{bmatrix} 3 & 6 \\ 0 & 2 \\ -4 & 0 \end{bmatrix}$$

$$\text{If } C = \begin{bmatrix} 3 & -2 & 0 \\ -2 & 4 & 6 \\ 0 & 6 & 0 \end{bmatrix}, \text{ then } C^T = \begin{bmatrix} 3 & -2 & 0 \\ -2 & 4 & 6 \\ 0 & 6 & 0 \end{bmatrix}$$

3.15 Theorem

If s is any scalar and if the sizes of A and B are such that the indicated operations can be performed, then

(i) $(A^T)^T = A$
(ii) $(A + B)^T = A^T + B^T$
(iii) $(sA)^T = sA^T$
(iv) $(AB)^T = B^TA^T$
(v) if A is invertible, then so is A^T; moreover, $(A^T)^{-1} = (A^{-1})^T$.

The proofs of (i), (ii), and (iii) are left as easy exercises.

Proof of (iv): Assume A has size $m \times n$; then since AB is assumed to be defined, B has size $n \times p$. Thus AB has size $m \times p$, and so $(AB)^T$ has size $p \times m$. Now B^T has size $p \times n$ and A^T has size $n \times m$; hence B^TA^T also has size $p \times m$.

$$(i, j)^{\text{th}} \text{ entry of } B^TA^T = [i^{\text{th}} \text{ row of } B^T] \begin{bmatrix} j^{\text{th}} \\ \text{column} \\ \text{of} \\ A^T \end{bmatrix}$$

$$= [i^{\text{th}} \text{ column of } B] \begin{bmatrix} j^{\text{th}} \\ \text{row} \\ \text{of} \\ A \end{bmatrix} = [b_{1i} b_{2i} \dots b_{ni}] \begin{bmatrix} a_{j1} \\ a_{j2} \\ \vdots \\ a_{jn} \end{bmatrix}$$

$$= b_{1i}a_{j1} + b_{2i}a_{j2} + \cdots + b_{ni}a_{jn}$$

$$= a_{j1}b_{1i} + a_{j2}b_{2i} + \cdots + a_{jn}b_{ni}$$

$$= \sum_{k=1}^{n} a_{jk}b_{ki}$$

$$= (j, i)^{\text{th}} \text{ entry in } AB$$

$$= (i, j)^{\text{th}} \text{ entry in } (AB)^T$$

Proof of (v): Since A is invertible, A is a square matrix by definition; so $(A^T)^{-1}$ and $(A^{-1})^T$ have the same size, say $n \times n$.

$$\begin{aligned} A^T(A^{-1})^T &= (A^{-1}A)^T && \text{[by 3.15(iv), which was just proved]} \\ &= I_n^T && [\text{definition of } A^{-1}] \\ &= I_n && [\text{property of } I_n] \end{aligned}$$

so $(A^{-1})^T$ is a right inverse for A^T.

Similarly,

$$(A^{-1})^T A^T = (AA^{-1})^T = I_n^T = I_n$$

so $(A^{-1})^T$ is also a left inverse for A^T. ∎

There are standard mnemonic phrases to express 3.15(iv) and (v). 3.15(iv) says that the transpose of a product is the product of the transposes, taken in reverse order. 3.15(v) says that if A is invertible then so is A^T, and the inverse of the transpose is the transpose of the inverse.

3.16 Corollary

If the product $A_1A_2 \cdots A_k$ is defined, then

$$(A_1A_2 \cdots A_k)^T = A_k^T A_{k-1}^T \cdots A_1^T$$

Proof: Exercise. ∎

3.17 Exercise

Illustrate Theorem 3.15 using

$$A = \begin{bmatrix} 2 & 3 \\ 4 & 5 \end{bmatrix}, \; B = \begin{bmatrix} 6 & 7 \\ 8 & 9 \end{bmatrix}, \text{ and } s = 10.$$

A matrix such as C in Example 3.14, which is equal to its own transpose, satisfies an obvious symmetry with respect to its main diagonal. This type of matrix will be important in certain applications, and it is convenient to introduce the formal definition of this property at the present time.

3.18 Definition

(i) A matrix A satisfying $A^T = A$ is called *symmetric*.
(ii) A matrix A satisfying $A^T = -A$ is called *skew symmetric*.

Note that a symmetric or skew-symmetric matrix must be square.

Some of the simpler aspects of this definition will be treated in Problem Set 3.

We conclude this section by adding to our list, begun in Section 1, of the differences between the algebra of matrices and that of real numbers. In Section 1, we observed that matrix multiplication is not commutative, that it violates the cancellation law, and that a product of matrices can be a zero matrix without any of the factors being a zero matrix.

It is ultimately the lack of commutativity that explains why the following familiar identity fails for matrices. For any two real numbers s and t, $(s + t)^2 = s^2 + 2st + t^2$. For matrices, however, it can happen that $(A + B)^2 \neq A^2 + 2AB + B^2$.

3.19 Exercise

$$A = \begin{bmatrix} 1 & 2 \\ 3 & 4 \end{bmatrix}, \qquad B = \begin{bmatrix} 2 & 1 \\ -3 & -4 \end{bmatrix}$$

Show that $(A + B)^2 \neq A^2 + 2AB + B^2$.

The explanation is that

$$\begin{aligned} (A + B)^2 &= (A + B)(A + B) && \text{[by definition]} \\ &= A(A + B) + B(A + B) && \text{[by 3.1(xii)]} \\ &= A^2 + AB + BA + B^2 && \text{[by 3.1(xi)]} \end{aligned}$$

and since AB and BA are not equal in general, it would be incorrect to combine these two terms as 2AB.

For the same reason, although $(st)^2 = s^2t^2$ is a valid identity for real numbers s and t, $(AB)^2$ and A^2B^2 are different in general; $(AB)^2 = (AB)(AB) = ABAB$ and it would be incorrect, unless $AB = BA$, to change the order of the two factors in the middle and to continue the string of equalities by writing $\ldots = AABB = A^2B^2$.

Our final point is this: a positive real number s has exactly two square roots, $\pm\sqrt{s}$. For matrices, the situation is more complicated.

3.20 Exercise

Verify that each of the following matrices is a square root of I_2.

$$\begin{bmatrix} 1 & 0 \\ 0 & 1 \end{bmatrix}, \begin{bmatrix} -1 & 0 \\ 0 & -1 \end{bmatrix}, \begin{bmatrix} 1 & 0 \\ 0 & -1 \end{bmatrix}, \begin{bmatrix} -1 & 0 \\ 0 & 1 \end{bmatrix}, \begin{bmatrix} 1 & 2 \\ 0 & -1 \end{bmatrix}, \begin{bmatrix} -2 & 1 \\ -3 & 2 \end{bmatrix}$$

In fact, I_2 has infinitely many square roots. See Problem 3.7.

3.1. Go back and do any of the exercises in Section 3 that you may have skipped.

*3.2. Let P, Q, and R be the matrices from Problem 2.1. Compute the following:

(i) $P(Q + R)$ (ii) $QP + RP$

(iii) $(P + Q)^2$ (iv) $P^2 + 2PQ + Q^2$

(v) $(PQ)^2$ (vi) P^2Q^2

3.3. To convince yourself that even simple "theoretical" results such as the distributive laws [Theorem 3.1(xi) and (xii)] can have "practical" consequences, compare the amounts of work required to compute AB + AC and A(B + C), where

$$A = \begin{bmatrix} 1 & -1 & 6 \\ 1 & -2 & -1 \\ 3 & -3 & -4 \end{bmatrix}, \quad B = \begin{bmatrix} 5 & 3 \\ 2 & -4 \\ -3 & 1 \end{bmatrix},$$

$$C = \begin{bmatrix} 0 & -3 \\ -2 & 2 \\ 4 & 3 \end{bmatrix}$$

*3.4. Simplify the expression $(A^2C)^{-1}AB^{-1}C(A^{-1}B^{-1}C)^{-1}$.

*3.5. Simplify the expression $((A^{-1}B^{-1}C^T)^T)^{-1}$.

*3.6. Discover most of what there is to know about invertibility for 2×2 matrices by finding a formula that, when $\begin{bmatrix} a & b \\ c & d \end{bmatrix}$ is invertible, gives $\begin{bmatrix} a & b \\ c & d \end{bmatrix}^{-1}$. This formula will also allow you to conclude that $\begin{bmatrix} a & b \\ c & d \end{bmatrix}$ is invertible if and only if a certain arithmetic relationship holds involving a, b, c, and d.

Hint: How must w, x, y, and z be related to a, b, c, and d if both

$$\begin{bmatrix} a & b \\ c & d \end{bmatrix}\begin{bmatrix} w & x \\ y & z \end{bmatrix} = \begin{bmatrix} 1 & 0 \\ 0 & 1 \end{bmatrix}$$

and

$$\begin{bmatrix} w & x \\ y & z \end{bmatrix}\begin{bmatrix} a & b \\ c & d \end{bmatrix} = \begin{bmatrix} 1 & 0 \\ 0 & 1 \end{bmatrix}?$$

This leads to a system of eight equations that can be solved for w, x, y, and z in terms of a, b, c, and d.

Apply the formula you found to obtain $\begin{bmatrix} -4 & 2 \\ 3 & -1 \end{bmatrix}^{-1}$.

3.7. Show that $I_2 = \begin{bmatrix} 1 & 0 \\ 0 & 1 \end{bmatrix}$ has infinitely many square roots.

Hint: What conditions on a, b, c, and d are necessary and sufficient for

$$\begin{bmatrix} a & b \\ c & d \end{bmatrix}\begin{bmatrix} a & b \\ c & d \end{bmatrix} = \begin{bmatrix} 1 & 0 \\ 0 & 1 \end{bmatrix}?$$

Add two more to the list of square roots of I_2 begun in Exercise 3.20.

3.8. Let A and P be $n \times n$ matrices and assume that P is invertible.
 (i) Prove that if $B = P^{-1}AP$ then $A = PBP^{-1}$.
 (ii) Prove that $(P^{-1}AP)^2 = P^{-1}A^2P$.
 (iii) Prove that, for any positive integer k, $(P^{-1}AP)^k = P^{-1}A^kP$.

*3.9. Prove that if A commutes with B and B is invertible then A also commutes with B^{-1}.

3.10. From the discussion following Exercise 3.19, it follows that, if A and B commute, then $(A + B)^2 = A^2 + 2AB + B^2$.
 (i) Prove that if A and B commute then $(A + B)^3 = A^3 + 3A^2B + 3AB^2 + B^3$.
 (ii) Prove by Mathematical Induction that if A and B commute then the analogue of the Binomial Theorem is valid:

$$(A + B)^k = A^k + kA^{k-1}B + \frac{k(k-1)}{2}A^{k-2}B^2 + \cdots + kAB^{k-1} + B^k$$

3.11. (i) Prove that

$$\begin{bmatrix} \cos\theta & \sin\theta \\ -\sin\theta & \cos\theta \end{bmatrix}^2 = \begin{bmatrix} \cos 2\theta & \sin 2\theta \\ -\sin 2\theta & \cos 2\theta \end{bmatrix}$$

 (ii) Prove that

$$\begin{bmatrix} \cos\theta & \sin\theta \\ -\sin\theta & \cos\theta \end{bmatrix}^3 = \begin{bmatrix} \cos 3\theta & \sin 3\theta \\ -\sin 3\theta & \cos 3\theta \end{bmatrix}$$

 (iii) Prove by Mathematical Induction that, for any positive integer k,

$$\begin{bmatrix} \cos\theta & \sin\theta \\ -\sin\theta & \cos\theta \end{bmatrix}^k = \begin{bmatrix} \cos k\theta & \sin k\theta \\ -\sin k\theta & \cos k\theta \end{bmatrix}$$

3.12. Let A be an arbitrary $m \times n$ matrix. Prove that AA^T and A^TA are each symmetric.

3.13. Let A be an arbitrary square matrix.
 (i) Prove that $A + A^T$ is symmetric.
 (ii) Prove that $A - A^T$ is skew symmetric.
 (iii) Observe that $A = \frac{1}{2}(A + A^T) + \frac{1}{2}(A - A^T)$. We may therefore conclude that every square matrix can be written in at

least one way as the sum of a symmetric and a skew-symmetric matrix.

(iv) Prove that the expression in (iii) is the unique way to write A as the sum of a symmetric and a skew-symmetric matrix; that is, prove that if $A = B + C$, where B is symmetric and C is skew symmetric, then $B = \frac{1}{2}(A + A^T)$ and $C = \frac{1}{2}(A - A^T)$.

*3.14. *True or False*. Determine whether the following general statements about matrices are true or false. For those that are true, give a proof; for those that are false, give a counterexample that illustrates this fact. Throughout this problem, A, B, C, and D are all $n \times n$ matrices. Your answer to Problem 3.6 may be used to construct some of the counterexamples.

(i) If $A + B = 0_{n \times n}$, then $A^2 + 2AB + B^2 = 0_{n \times n}$.
(ii) If $AB = 0_{n \times n}$, then $BA = 0_{n \times n}$.
(iii) $A(B + C)D = ABD + ACD$.
(iv) If A and B are each invertible, then $A + B$ is invertible.
(v) If A and B are each symmetric, then $A + B$ is symmetric.
(vi) If A and B are each symmetric, then AB is symmetric.
(vii) If A and B are each symmetric, then ABA is symmetric.
(viii) $AA^T = A^TA$.
(ix) If A is invertible and $AB = AC$, then $B = C$.
(x) If A is symmetric, then A is invertible.
(xi) If the entries on the main diagonal of A are all 0, then A is not invertible.
(xii) If A^2 is invertible, then A is invertible.

3.15. For $n \times n$ matrices A and B, the matrix $AB - BA$ is denoted by $[A, B]$. Prove for $n \times n$ matrices A, B, and C that

$$[A,[B,C]] + [B,[C,A]] + [C,[A,B]] = 0_{n \times n}$$

3.16. Assume A is an invertible $n \times n$ matrix. For each of the following properties, is it true or false that if A has the property then A^{-1} has the same property? As in Problem 3.14, if the statement is true, give a proof; if it is false, show this by giving a counterexample.

(i) Symmetric
(ii) Skew symmetric
(iii) Diagonal

*3.17. Let A be an $n \times n$ matrix. Prove that if A has either an all-zero row or an all-zero column then A is not invertible.

3.18. The *trace* of an $n \times n$ matrix A is the sum of its entries on the main diagonal:

$$\text{trace}(A) = \sum_{i=1}^{i=n} a_{ii}$$

Prove the following for all $n \times n$ matrices A and B.

(i) For every scalar s, trace(sA) $= s$ trace(A)

(ii) trace(A + B) = trace(A) + trace(B)

(iii) trace(A^{T}) = trace(A)

(iv) trace(AB) = trace(BA)

Hint for (iv): To get an idea how the proof should go in general, first prove the case when $n = 2$.

*3.19. For $n \times n$ matrices A and B, the matrix AB − BA can be different from $0_{n \times n}$; but it can *never* be equal to the identity matrix. Use properties of the trace function from Problem 3.18 to prove that AB − BA $\neq \mathrm{I}_n$.

3.20. Let A and B be symmetric $n \times n$ matrices. Prove that AB is symmetric if and only if AB = BA.

3.21. Let A and B be skew-symmetric $n \times n$ matrices.

(i) Prove that AB is symmetric if and only if AB = BA.

(ii) Prove that AB is skew symmetric if and only if AB = −BA.

3.22. (i) Show that, if B_1 and B_2 are each left inverses for A, then so is $s\mathrm{B}_1 + t\mathrm{B}_2$ provided $s + t = 1$. (Conclude that if a matrix has two distinct left inverses then it has infinitely many.)

(ii) Use the result from part (i) to obtain three more left inverses for the matrix A in Example 3.4.

3.23. Let D = diag$(d_1, d_2, \ldots, d_n)$. Prove that D is invertible if and only if $d_i \neq 0$ for $1 \leq i \leq n$, and that, if this happens, then $\mathrm{D}^{-1} = \mathrm{diag}(1/d_1, 1/d_2, \ldots, 1/d_n)$.

3.24. Let A and P be $n \times n$ matrices.

(i) Prove that if A is symmetric then so is $\mathrm{P}^{\mathrm{T}}\mathrm{AP}$.

(ii) Prove that if A is skew symmetric then so is $\mathrm{P}^{\mathrm{T}}\mathrm{AP}$.

3.25 *Matrix Partitioning and Block Multiplication:* To partition an $m \times n$ matrix A is to view it as an array of submatrices, or blocks, of possibly varying sizes. To indicate this, we use broken lines to separate the blocks from one another. Here are two of the many possible partitionings of the matrix A:

$$\mathrm{A} = \begin{bmatrix} 2 & -2 & 3 \\ 0 & 1 & -1 \\ -1 & 3 & -4 \\ 5 & -2 & 3 \end{bmatrix}, \quad \begin{bmatrix} 2 & -2 & 3 \\ 0 & 1 & -1 \\ \hdashline -1 & 3 & -4 \\ 5 & -2 & 3 \end{bmatrix},$$

$$\left[\begin{array}{cc:c} 2 & -2 & 3 \\ 0 & 1 & -1 \\ -1 & 3 & -4 \\ \hdashline 5 & -2 & 3 \end{array}\right]$$

In the context of a given partition of a matrix M, we denote the submatrices using the same uppercase letter, M, with appropriate subscripts. For example, the first of the two partitions of A involves the two submatrices

$$A_{11} = \begin{bmatrix} 2 & -2 & 3 \\ 0 & 1 & -1 \end{bmatrix} \text{ and } A_{21} = \begin{bmatrix} -1 & 3 & -4 \\ 5 & -2 & 3 \end{bmatrix}$$

The second of the partitionings involves the four submatrices

$$A_{11} = \begin{bmatrix} 2 & -2 \\ 0 & 1 \\ -1 & 3 \end{bmatrix}, \qquad A_{12} = \begin{bmatrix} 3 \\ -1 \\ -4 \end{bmatrix},$$

$$A_{21} = [5 \quad -2], \qquad A_{22} = [3]$$

Partitioning allows us to compute the product of two large matrices by computing the products of several smaller submatrices and then piecing the results together appropriately. For example, verify that the product AB of the following two partitioned matrices can be obtained as indicated.

$$\left[\begin{array}{cc|c} 2 & -2 & 3 \\ \hline 0 & 1 & -1 \\ -1 & 3 & -4 \\ \hline 5 & -2 & 3 \end{array}\right] \left[\begin{array}{cc|cc} -3 & 2 & -2 & 1 \\ -1 & 5 & -2 & 0 \\ \hline 3 & 1 & 1 & -1 \end{array}\right]$$

$$= \begin{bmatrix} A_{11}B_{11} + A_{12}B_{21} & A_{11}B_{12} + A_{12}B_{22} \\ A_{21}B_{11} + A_{22}B_{21} & A_{21}B_{12} + A_{22}B_{22} \\ A_{31}B_{11} + A_{32}B_{21} & A_{31}B_{12} + A_{32}B_{22} \end{bmatrix} = \left[\begin{array}{cc|cc} 5 & -3 & 3 & -1 \\ \hline -4 & 4 & -3 & 1 \\ -12 & 9 & -8 & 3 \\ \hline -4 & 3 & -3 & 2 \end{array}\right]$$

Note that the formula for the $(i, j)^{\text{th}}$ block in the product is given by

$$(AB)_{ij} = \sum_{k=1}^{r(i,j)} A_{ik}B_{kj}$$

which is the usual formula for the $(i, j)^{\text{th}}$ entry in the product, modified so as to refer to submatrices rather than to individual entries. Here $r(i, j)$ varies with i and j and is equal (as it must be, if the product $A_{ik}B_{kj}$ is defined) both to the number of columns in A_{ik} and to the number of rows in B_{kj}.

To use partitioning to compute the product, AB, of an arbitrary $m \times n$ matrix A and an arbitrary $n \times p$ matrix B, the partitionings of A and B must be matched: the sizes of the submatrices must be such that all the multiplications $A_{ik}B_{kj}$ are defined. This sounds messy to check, but, in fact, it is very simple: the number of vertical broken lines in A must equal the number of horizontal broken lines in B, and, if the ith (from left to right) vertical broken line in A is between columns p and

$p + 1$, then the ith (from top to bottom) horizontal broken line in B is between rows p and $p + 1$. If this condition is satisfied, the partitionings of A and B are called *conformable for AB*.

When the partitionings are comformable for AB, the number and sizes of the blocks in the resulting partition of AB are determined by the number and placement of the horizontal broken lines in A and of the vertical broken lines in B.

More specifically, if A, viewed as a matrix of blocks, has block-size $m \times n$ (i.e., A has m-many rows of blocks and n-many columns of blocks) and if B has block-size $n \times p$, then the block-size of AB in the resulting partition is $m \times p$. Moreover, the sizes of the individual blocks in this partition depend in an obvious way on the placement of the horizontal broken lines in A and the vertical broken lines in B.

Theorem (Block Multiplication) If A and B are partitioned $m \times n$ and $n \times p$ matrices, respectively, and if the partitionings are conformable for AB, then the product, AB, is equal to the partitioned matrix whose $(i, j)^{\text{th}}$ block is given by

$$(\mathrm{AB})_{ij} = \sum_{k=1}^{r(i,j)} \mathrm{A}_{ik}\mathrm{B}_{kj}$$

Proof: This is one of those results that is easy to understand and accept but whose proof leads to notational nightmares. No one ever bothers to write it out in full.

Here, then, is problem 3.25(i):

(i) Make a serious attempt to prove the Block Multiplication Theorem.

Eventually, you will get frustrated and give up; this is a foreseeable and acceptable response. The purpose of this problem (in addition to the fact that the theorem itself is very useful) is to develop an appreciation for the difference between a proof such as this one and, say, that of the Associative Law for matrix multiplication [3.1(ix)], which can and should be written down.

(ii) Prove the special case of the Block Multiplication Theorem for $n \times n$ matrices A and B, each partitioned into only four blocks, with blocks A_{11} and B_{11} of size $k \times k$ for some fixed k, $1 \leq k \leq n - 1$. (Do not give up on this one; it is accessible!) ■

3.26 Suppose that the size of A is 8×9 and that the size of B is 9×12. Suppose that A is partitioned with broken lines between rows 2 and 3, between rows 5 and 6, between columns 3 and 4, and between columns 6 and 7. Suppose that B is partitioned with broken lines between rows 3 and 4, between rows 6 and 7, and between columns 8 and 9. Note that these two partitionings are conformable for AB.

(i) Into how many blocks does the resulting partition of AB break AB?

(ii) What is the size of the (2, 2)-block in AB?

(iii) Where $m_i \times n_j$ denotes the size of the $(i, j)^{\text{th}}$ block in AB, what is the maximum value of the product $m_i n_j$?

3.27 Matrix partitioning can be used most effectively when some of the resulting blocks have the form $0_{p\times q}$ or I_k. With this in mind, compute the products AB and BA where A and B are respectively

$$\begin{bmatrix} 0 & 0 & 6 & 2 \\ 0 & 0 & -1 & 3 \\ 2 & 3 & 1 & 0 \\ -4 & 1 & 0 & 1 \\ 1 & 1 & 0 & 0 \end{bmatrix} \text{ and } \begin{bmatrix} 0 & 0 & 1 & 0 & 2 \\ 0 & 0 & 0 & 1 & 4 \\ 1 & 0 & 1 & 0 & 3 \\ 0 & 1 & 0 & 0 & -1 \end{bmatrix}$$

4

Descriptive Geometry of $\mathbb{R}^n$

The purpose of this section is to discuss the geometrical motivation behind much of what will be done in the sections that follow. Because none of the proofs we give in later sections will ever depend on geometrical considerations, this section could be skipped insofar as the strictly logical development of the book is concerned.

However, it is extremely important to be able to "think geometrically" about what is happening in a given linear algebraic context. Very often, the answer to a theoretical question in linear algebra will be obvious when the correct geometrical interpretation of the question is made.

This section proceeds at an informal, intuitive level. We will use words such as *line*, *plane*, *space*, *subspace*, *dimension*, *parameter*, *dependent*, *independent*, and others in a familiar, informal way. Most of these words will be given precise, technical definitions in later sections. Of course, the usual purpose of a technical definition is to capture the essence of an intuitive concept. So although there is no logical dependence of the later sections on this one, there is an essential psychological dependence.

Do not be concerned if you encounter something in Section 4 that you do not understand. The purpose of this section is to generate a vague, intuitive feeling for what is going on; it does not contain specific knowledge that must be assimilated before continuing.

We assume a familiarity with Cartesian coordinate systems. We let $\mathbb{R}^n$ denote the set of all ordered n-tuples of real numbers. When $n = 1$, we omit the superscript 1 from $\mathbb{R}^1$ and simply write $\mathbb{R}$ for the set of real numbers.

For $n = 1$, the points in $\mathbb{R}$ are identified with the points on a straight line; for $n = 2$, the points in $\mathbb{R}^2$ are identified with the points on a plane; for $n = 3$, the points in $\mathbb{R}^3$ are identified with the points in three-dimensional space; and so on.

Most of our discussions, however, will not involve a fixed value for n. Our primary concern is with n-dimensional space, where n is an arbitrary positive integer. Because our sensory experience is with $\mathbb{R}^3$, $\mathbb{R}^2$, and $\mathbb{R}$, we have fairly well developed intuitions about what can happen in three or fewer dimensions. In developing our intuition about higher-dimensional spaces, we will often argue "by analogy" with the case when $n = 3$.

We begin with the problem of describing in mathematical terms the subsets of $\mathbb{R}^n$ that we are interested in. We will use standard set-theoretic notation. $\in$ is the membership relation, $\subset$ is the subset relation, and $\varnothing$ denotes the empty set. If X is a set, then the subset of X consisting of those points in X satisfying a certain property P is denoted by $\{x \in X\colon x \text{ has property } P\}$ or by $\{x \in X\colon P(x)\}$. $\cup$ denotes union and $\cap$ denotes intersection. The number of elements in a set X is called the *cardinality* of X and is denoted by $|X|$. If the reader is not familiar with these concepts, this situation can and should be remedied by spending at most a half-hour with the first dozen pages of any elementary set theory book.

We will also use "iff" as an abbreviation for "if and only if."

In our familiar three-dimensional world, lines and planes have in common a property that we will temporarily call *straightness*. The word *straight* is used here in opposition to *curved*.

If we were to ask the naive question, What pieces of $\mathbb{R}^3$ are straight?, the answer given by the person in the street would be "lines and planes." In an equally naive vein, the straight pieces of $\mathbb{R}^2$ are just the lines.

We would like to analyze this intuitive geometrical notion of straightness. Our goal is to extend the scope of this notion and to make sense of the phrase "straight piece of $\mathbb{R}^n$."

Let us focus initially on the following analogy: a piece of $\mathbb{R}^n$ will be straight if it is to $\mathbb{R}^n$ as a line or plane is to $\mathbb{R}^3$ (or as a line is to $\mathbb{R}^2$). To develop this analogy, we will begin by reviewing how lines and planes in $\mathbb{R}^2$ and $\mathbb{R}^3$ are described mathematically.

First, there is a notational matter to consider. It is customary (because it avoids cumbersome subscripting) to use the letters x and y to denote the axes when working with $\mathbb{R}^2$, and to use x, y, and z when working with $\mathbb{R}^3$. In the context of a given discussion about $\mathbb{R}^n$, n is a fixed but unspecified positive integer, and the coordinate axes will be labeled with the variables $x_1, x_2, \ldots, x_n$. Because our aim is to generalize to the n-dimensional case certain notions that are familiar in the two- and three-dimensional cases, we will insist, contrary to custom, on using x_1 and x_2 in the context of $\mathbb{R}^2$ and

x_1, x_2, and x_3 in the context of $\mathbb{R}^3$. This insistence will contribute to the development of the appropriate analogies.

4.1 Examples of Lines (or Line Segments) in $\mathbb{R}^2$

(i) $\{(x_1, x_2) \in \mathbb{R}^2: x_1 = x_2\}$ is the 45° line.
(ii) $\{(x_1, x_2) \in \mathbb{R}^2: x_1 = 0\}$ is the x_2 axis.
(iii) $\{(x_1, x_2) \in \mathbb{R}^2: x_2 = 7\}$ is the line 7 units above and parallel to the x_1 axis; it is perpendicular to the x_2 axis and intersects it at the point (0, 7).
(iv) $\{(x_1, x_2) \in \mathbb{R}^2: x_2 = 4x_1 - 3,\ -1 \le x_1 \le 2\}$ is the line segment in the plane joining the two points $(-1, -7)$ and $(2, 5)$.
(v) $\{(x_1, x_2) \in \mathbb{R}^2: x_2 = 4x_1 - 3\}$ is the line segment from (iv) extended to infinity in both directions.
(vi) A small amount of algebra shows that $\{(x_1, x_2) \in \mathbb{R}^2: x_1 = \frac{1}{4}x_2 + \frac{3}{4}\}$ is another description of the line in (v).
(vii) $\{(x_1, x_2) \in \mathbb{R}^2: x_2 = mx_1 + b\}$ is the line in the plane whose slope is m and which intersects the x_2 axis at the point $(0, b)$.
(viii) Provided a_1 and a_2 are not both 0, the set $\{(x_1, x_2) \in \mathbb{R}^2: a_1x_1 + a_2x_2 = b\}$ is a line in the plane. This lines goes through the origin, (0, 0), iff $b = 0$. [Note, in the exceptional case when a_1 and a_2 are both 0, that the set described in braces is equal either to $\varnothing$, if $b \neq 0$, or to $\mathbb{R}^2$, if $b = 0$.]

4.2 Examples of Curves in $\mathbb{R}^2$

(i) $\{(x_1, x_2) \in \mathbb{R}^2: x_2 = x_1^2\}$
(ii) $\{(x_1, x_2) \in \mathbb{R}^2: x_2 = \sin x_1\}$
(iii) $\{(x_1, x_2) \in \mathbb{R}^2: x_2 = e^{x_1}\}$
(iv) $\{(x_1, x_2) \in \mathbb{R}^2: x_2 = x_1^3 - 5\}$

These curves are sketched in Figure 4.1.

4.3 Examples of Planes in $\mathbb{R}^3$

(i) $\{(x_1, x_2, x_3) \in \mathbb{R}^3: x_1 = x_2\}$ is the plane in $\mathbb{R}^3$ that is parallel to the x_3 axis and whose intersection with the x_1x_2 plane is the 45° line in that plane.
(ii) $\{(x_1, x_2, x_3) \in \mathbb{R}^3: x_3 = 0\}$ is the plane in $\mathbb{R}^3$ determined by the x_1 and x_2 axes.
(iii) $\{(x_1, x_2, x_3) \in \mathbb{R}^3: x_3 = 7\}$ is the plane in $\mathbb{R}^3$ that is 7 units above and parallel to the plane determined by the x_1 and x_2 axes. It is perpendicular to the x_3 axis and intersects it at the point (0, 0, 7).

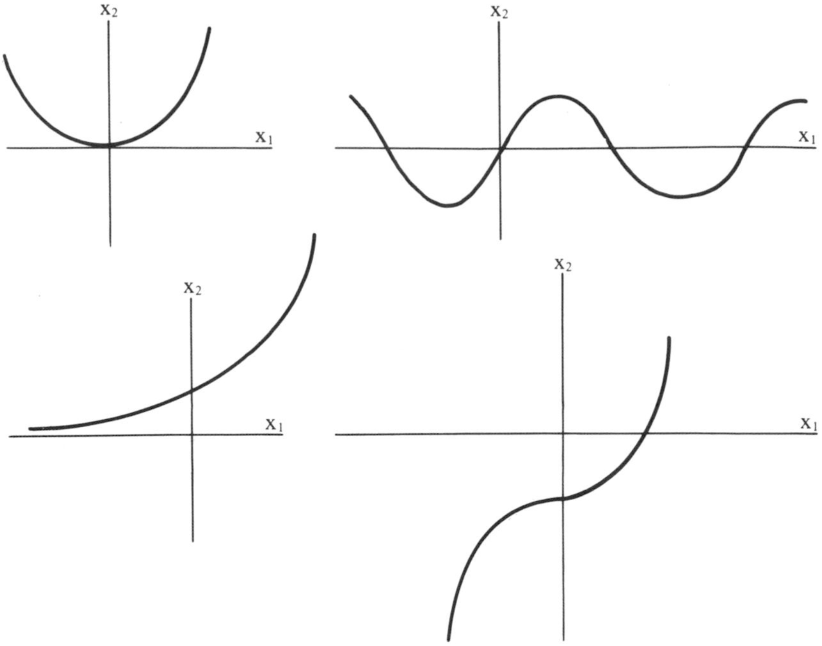

Figure 4.1

(iv) $\{(x_1, x_2, x_3) \in \mathbb{R}^3\colon x_1 + x_2 + x_3 = 1\}$ is the plane in $\mathbb{R}^3$ determined by the three points $(1, 0, 0)$, $(0, 1, 0)$, and $(0, 0, 1)$.

(v) $\{(x_1, x_2, x_3) \in \mathbb{R}^3\colon x_2 = 4x_1 - 3\}$ is the plane $\mathbb{R}^3$ that is parallel to the x_3 axis and whose intersection with the x_1x_2 plane is the line described in Example 4.1(v).

(vi) $\{(x_1, x_2, x_3) \in \mathbb{R}^3\colon x_2 = 4x_1 + 2x_3 - 3\}$ is the plane in $\mathbb{R}^3$ determined by the three points $(\frac{3}{4}, 0, 0)$, $(0, -3, 0)$, and $(0, 0, \frac{3}{2})$.

(vii) $\{(x_1, x_2, x_3) \in \mathbb{R}^3\colon x_3 = -2x_1 + \frac{1}{2}x_2 + \frac{3}{2}\}$ is the same plane as in (vi).

(viii) $\{(x_1, x_2, x_3) \in \mathbb{R}^3\colon x_1 = \frac{1}{4}x_2 - \frac{1}{2}x_3 + \frac{3}{4}\}$ is still another description of the same plane as in (vi) and (vii).

(ix) Provided a_1, a_2, and a_3 are not all zero, the set $\{(x_1, x_2, x_3) \in \mathbb{R}^3\colon a_1x_1 + a_2x_2 + a_3x_3 = b\}$ is a plane in $\mathbb{R}^3$. This plane contains the origin iff $b = 0$.

4.4 Examples of Surfaces in $\mathbb{R}^3$

(i) $\{(x_1, x_2, x_3) \in \mathbb{R}^3\colon x_3 = x_1^2 + x_2^2 - 2\}$ is an infinite bowl-shaped surface resting at the point $(0, 0, -2)$.

(ii) $\{(x_1, x_2, x_3) \in \mathbb{R}\colon x_3 = \sin x_2 \text{ and } -1 < x_1 < 1\}$ is the infinite corrugated strip of width 2 sketched in Figure 4.2.

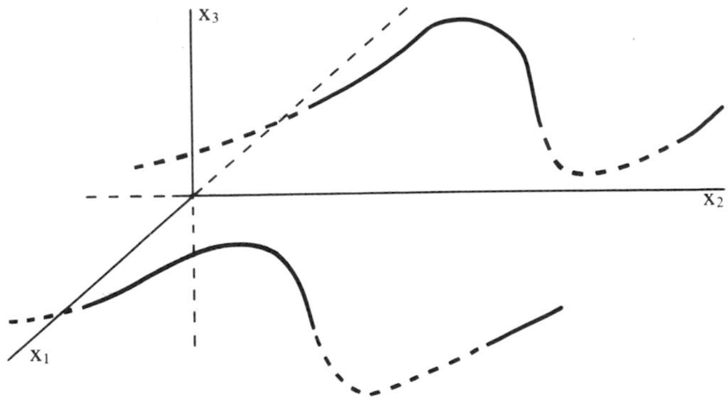

Figure 4.2

(iii) $\{(x_1, x_2, x_3) \in \mathbb{R}^3\colon\ x_3 = \sin x_2\}$ is the surface formed by extending the strip in the previous example to infinity in the positive and negative x_1 directions.

(iv) $\{(x_1, x_2, x_3) \in \mathbb{R}^3\colon\ x_3 = x_2^2 - x_1^2 + 4\}$ is a saddle surface with saddle point $(0, 0, 4)$.

(v) $\{(x_1, x_2, x_3) \in \mathbb{R}^3\colon\ x_3 = x_1 x_2\}$

4.5 Examples of Lines (or Line Segments) in $\mathbb{R}^3$

There are two standard ways to describe a line in $\mathbb{R}^3$: one is to give parametric equations for it; the other is to give symmetric equations for it. The reader is perhaps familiar with these from high-school mathematics. These ways are both well suited to problems in $\mathbb{R}^3$, but are less appropriate for developing the analogy with $\mathbb{R}^n$. Thus we will concentrate here on describing lines in $\mathbb{R}^3$ in a way that will make the notion of "a line in $\mathbb{R}^n$" easier to understand.

The eventual analogy will be clearer if we think of a line in $\mathbb{R}^3$ arising as the intersection of two planes in $\mathbb{R}^3$.

(i) $\{(x_1, x_2, x_3) \in \mathbb{R}^3\colon\ x_1 = x_2 \text{ and } x_3 = 7\}$ is the line in $\mathbb{R}^3$ formed by intersecting the two planes described in Examples 4.3(i) and (iii).

(ii) $\{(x_1, x_2, x_3) \in \mathbb{R}^3\colon\ x_1 + x_2 + x_3 = 1 \text{ and } x_2 = 4x_1 - 3\}$ is the line in $\mathbb{R}^3$ formed by intersecting the two planes described in Examples 4.3(iv) and (v).

(iii) A small amount of algebra shows that $\{(x_1, x_2, x_3) \in \mathbb{R}^3\colon\ x_2 = 4x_1 - 3 \text{ and } x_3 = -5x_1 + 4\}$ also describes the previous line.

(iv) $\{(x_1, x_2, x_3) \in \mathbb{R}^3\colon\ x_1 = \frac{1}{4}x_2 + \frac{3}{4} \text{ and } x_3 = -\frac{5}{4}x_2 + \frac{1}{4}\}$ is another description of the line in (ii) and (iii).

(v) $\{(x_1, x_2, x_3) \in \mathbb{R}^3\colon\ x_1 = -\frac{1}{5}x_3 + \frac{4}{5} \text{ and } x_2 = -\frac{6}{5}x_3 + \frac{1}{5}\}$ is still another description of the line in (ii), (iii), and (iv).

(vi) $\{(x_1, x_2, x_3) \in \mathbb{R}^3\colon\ x_2 = 4x_1 - 3$ and $x_3 = -5x_1 + 4$ and $-1 \leq x_1 \leq 2\}$ is that segment of the line from the four preceding examples joining the points $(-1, -7, 9)$ and $(2, 5, -6)$.

4.6 Exercise

Verify the claim that 4.5(ii), (iii), (iv), and (v) all describe the same line in $\mathbb{R}^3$.

4.7 Examples of Curves in $\mathbb{R}^3$

(i) $\{(x_1, x_2, x_3) \in \mathbb{R}^3\colon\ x_1 = \cos x_3$ and $x_2 = \sin x_3\}$ is the helix shown in Figure 4.3.
(ii) $\{(x_1, x_2, x_3) \in \mathbb{R}^3\colon\ x_1 = x_2$ and $x_3 = x_2^2\}$ is the curve sketched in Figure 4.4.

We will now use these examples to guide us toward the notion of "a straight piece of $\mathbb{R}^n$." Once again, we should remind ourselves that this section consists of intuitive, informal material and that precise definitions, theorems, and proofs will follow in later sections.

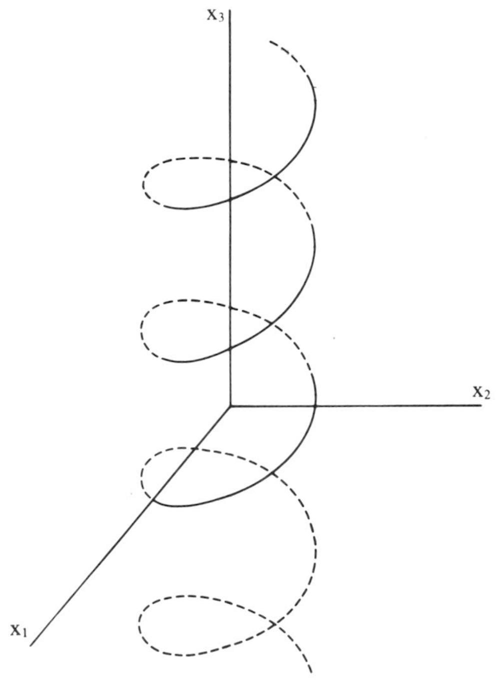

Figure 4.3

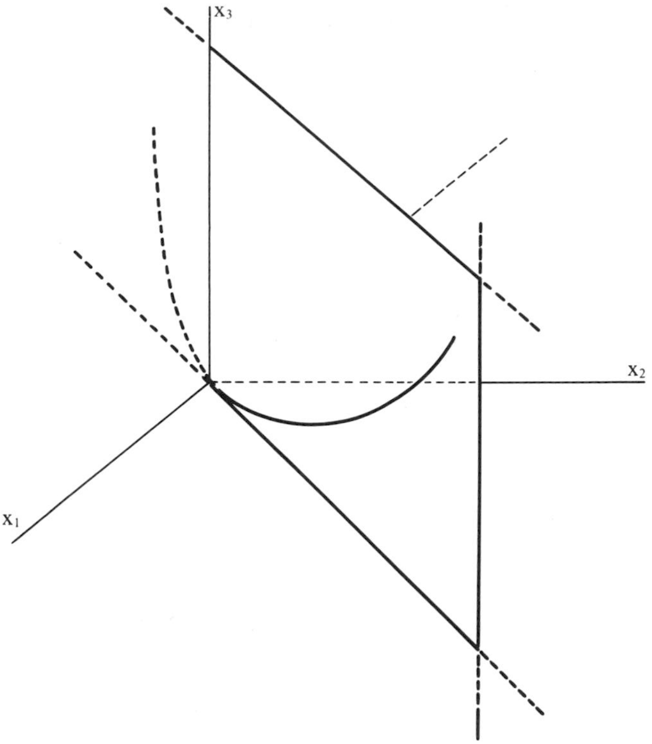

Figure 4.4

In $\mathbb{R}^2$, we spoke only about lines and curves. In $\mathbb{R}^3$, we considered planes and surfaces as well as lines and curves.

The essential difference between a line in $\mathbb{R}^3$ and a plane in $\mathbb{R}^3$ is that a plane in $\mathbb{R}^3$ cuts $\mathbb{R}^3$ in a straight way into two distinct regions. A line in $\mathbb{R}^3$ does not divide $\mathbb{R}^3$ in this sense. A line in $\mathbb{R}^2$, however, does cut $\mathbb{R}^2$ in a straight way into two distinct regions.

Thus the analogy is this: a line in $\mathbb{R}^2$ is to $\mathbb{R}^2$ as a plane in $\mathbb{R}^3$ is to $\mathbb{R}^3$.

To extend this analogy to $\mathbb{R}^n$, therefore, we should consider subsets of $\mathbb{R}^n$ that cut $\mathbb{R}^n$ in a straight way into two distinct regions.

Lines in $\mathbb{R}^2$ are sets of the form $\{(x_1, x_2) \in \mathbb{R}^2\colon a_1x_1 + a_2x_2 = b\}$, where a_1 and a_2 are not both 0.

Planes in $\mathbb{R}^3$ are sets of the form $\{(x_1, x_2, x_3) \in \mathbb{R}^3\colon a_1x_1 + a_2x_2 + a_3x_3 = b\}$, where a_1, a_2, and a_3 are not all 0.

When discussing spaces of dimension greater than 3, the prefix *hyper* is often attached to words denoting analogous lower-dimensional concepts. (Video-game addicts will be familiar with this usage.)

By analogy with the preceding, a subset of $\mathbb{R}^n$ of the form $\{(x_1, x_2, \ldots, x_n) \in \mathbb{R}^n\colon a_1x_1 + a_2x_2 + \cdots + a_nx_n = b\}$, where $a_i \neq 0$ for at least one i, will be called *a hyperplane in* $\mathbb{R}^n$.

Note that the concept being defined is not "hyperplane" but "hyperplane in $\mathbb{R}^n$." The word hyperplane by itself has no fixed meaning; it is only in the context of $\mathbb{R}^n$ for some fixed n that the phrase "hyperplane in $\mathbb{R}^n$" has a meaning. Thus, a hyperplane in $\mathbb{R}^3$ is the same thing as an ordinary plane in $\mathbb{R}^3$; a hyperplane in $\mathbb{R}^2$ is the same thing as a line in $\mathbb{R}^2$. A hyperplane in $\mathbb{R}^4$ can only be described as something that does to $\mathbb{R}^4$ what a line in $\mathbb{R}^2$ does to $\mathbb{R}^2$ or that does to $\mathbb{R}^4$ what a plane in $\mathbb{R}^3$ does to $\mathbb{R}^3$.

The relevant mental picture of a hyperplane in $\mathbb{R}^n$ is that of a subset of $\mathbb{R}^n$ that cuts $\mathbb{R}^n$ in a straight way into two distinct regions.

Note that the hyperplane in $\mathbb{R}^n$ just described contains the origin $(0, 0, \ldots, 0)$, if and only if $b = 0$.

We return now to the remaining "straight" pieces of $\mathbb{R}^3$, the lines in $\mathbb{R}^3$. Any two planes in $\mathbb{R}^3$ that are not parallel will have a line in $\mathbb{R}^3$ as their intersection. This geometrically obvious fact about $\mathbb{R}^3$ suggests that an intersection of two hyperplanes in $\mathbb{R}^4$ should be classified among the "straight" pieces of $\mathbb{R}^4$. Pushing our geometric intuition further in this direction suggests that an intersection of three hyperplanes in $\mathbb{R}^4$ should also be a "straight" piece of $\mathbb{R}^4$. Note that two distinct lines in $\mathbb{R}^2$ will intersect in the empty set, $\varnothing$, if they are parallel, and in a single point otherwise. Similarly, the intersection of three distinct planes in $\mathbb{R}^3$ can be the empty set, $\varnothing$, a single point, or a line.

Eventually, a general principle evolves. To state this principle in a simple way requires that we classify, by convention, the empty set, $\varnothing$, and sets consisting of a single point, such as $\{(u_1, u_2, \ldots, u_n)\}$, among the "straight" pieces of $\mathbb{R}^n$; this convention allows us to dispense with such qualifying phrases as "provided no two of the hyperplanes are parallel."

General Principle: An intersection of any number of "straight" pieces of $\mathbb{R}^n$ is a "straight" piece of $\mathbb{R}^n$.

Thus the phrase "straight piece of $\mathbb{R}^n$" has now been fully explained at an intuitive, geometrical level. The straight pieces of $\mathbb{R}^n$ are precisely the hyperplanes in $\mathbb{R}^n$ together with the subsets of $\mathbb{R}^n$ that can be obtained as intersections of two or more hyperplanes in $\mathbb{R}^n$.

This intuitive concept will be formalized in Section 5 and will at that time receive its official mathematical name—affine subspace of $\mathbb{R}^n$. In the meantime, we will use this name unofficially. At this point, it is possible to anticipate a good portion of the rest of the text.

Being an intersection of hyperplanes in $\mathbb{R}^n$, an affine subspace, S, of $\mathbb{R}^n$ is a subset of $\mathbb{R}^n$ of the form

$$S = \Big\{(x_1, x_2, \ldots, x_n) \in \mathbb{R}^n\colon \begin{array}{l} a_{11}x_1 + a_{12}x_2 + \cdots + a_{1n}x_n = b_1 \\ a_{21}x_1 + a_{22}x_2 + \cdots + a_{2n}x_n = b_n \\ \vdots \\ a_{m1}x_1 + a_{m2}x_2 + \cdots + a_{mn}x_n = b_m \end{array}\Big\}$$

That is, each equation of the form $a_{i1}x_1 + a_{i2}x_2 + \cdots + a_{in}x_n = b_i$ (for $i = 1, 2, \ldots, m$) determines a hyperplane in $\mathbb{R}^n$, and S consists precisely of those n-tuples $(u_1, u_2, \ldots, u_n)$ that simultaneously satisfy each of the m-many equations (i.e., S consists of the points in $\mathbb{R}^n$ that belong to the intersection of these m-many hyperplanes). (The reader may recognize that S is the set of solutions of a system of m-many simultaneous linear equations in n-many unknowns; the details of this are in Section 6.)

Some descriptions of affine subspaces are preferable to others. Consider the following, for example.

4.8 Example

$$S' = \{(x_1, x_2, x_3, x_4, x_5) \in \mathbb{R}^5 \colon \begin{aligned} 2x_1 \qquad\quad + 14x_3 + 4x_4 - 2x_5 &= 32 \\ -3x_1 + x_2 - 25x_3 \qquad\quad + 5x_5 &= -10 \\ x_1 + x_2 + 3x_3 + 8x_4 + x_5 &= 54 \\ -4x_1 + 4x_2 - 44x_3 + 5x_4 + 12x_5 &= 22\} \end{aligned}$$

Does the point $(-2, 14, 1, 5, -1)$ belong to S' or not?

The description of S' provided in Example 4.8 is not too useful since one cannot tell from it at a glance whether a given n-tuple belongs to S' or not. One would have to substitute the n-tuple into each of the equations in turn to see if it satisfies the equations. In Sections 14 through 16, we will present an algorithm for converting a description such as that in Example 4.8 into a useful description.

4.9 Example

Here is another description (obtained using the algorithm to which we just referred) of the same affine subspace S' from Example 4.8.

$$S' = \{(x_1, x_2, x_3, x_4, x_5) \in \mathbb{R}^5 \colon \begin{aligned} x_1 &= -7x_3 + x_5 + 4 \\ x_2 &= 4x_3 - 2x_5 + 2 \\ x_4 &= 6\} \end{aligned}$$

From this description, it is clear that the point $(-2, 14, 1, 5, -1)$ does not belong to S' since its fourth coordinate is not equal to 6.

Descriptions having the format of the one from Example 4.9 are the most desirable since they provide complete information about the affine subspace in question. We conclude Section 4 with a discussion of the salient features of this format.

If $S \subset \mathbb{R}^n$, then S is a set of n-tuples, and a description of S will normally involve a list of constraints or properties $P_1, P_2, \ldots, P_k$ that a given n-tuple must satisfy in order to belong to S. For example,

$$S = \{(x_1, x_2, \ldots, x_n) \in \mathbb{R}^n\text{: } P_1(x_1, x_2, \ldots, x_n) \text{ and } P_2(x_1, x_2, \ldots, x_n) \text{ and} \ldots \text{and } P_k(x_1, x_2, \ldots, x_n)\}$$

Examples 4.1 through 4.6 are all of this type.

For the moment, we wish to focus on the particular nature of the constraints. As an intuitive rough guide, by contrasting Examples 4.1, 4.3, and 4.5 with Examples 4.2, 4.4, and 4.7, we observe that if any of the constraints involve products or powers of variables, such as x_2x_3 or x_5^3, or involve trigonometric or transcendental functions of variables, such as $\cos x_4$ or e^{x^2}, then the resulting set will not be an affine subspace. This will be made precise in later sections.

For the purpose of developing our intuition further, let us temporarily call an expression of the form $a_1x_1 + a_2x_2 + \cdots + a_kx_k + b$, where $a_1, a_2, \ldots, a_k$ and b are constants, an *affine function of the variables* $x_1, x_2, \ldots, x_k$.

The distinguishing feature of the format in Example 4.9 is the splitting of the set of variables $\{x_1, x_2, x_3, x_4, x_5\}$ into two disjoint subsets $\{x_1, x_2, x_4\}$ and $\{x_3, x_5\}$ and having, for each of the variables in the first subset, a constraint that expresses that variable as an affine function of the variables from the second subset. That is, there are three constraints, one for x_1, one for x_2, and one for x_4; moreover, each of these constraints is an affine function of the variables x_3 and x_5.

In the general case, for a subset S of $\mathbb{R}^n$, the set of variables $\{x_1, x_2, \ldots, x_n\}$ is split into two disjoint collections: $\{x_{d_1}, x_{d_2}, \ldots, x_{d_r}\}$ are the dependent variables and $\{x_{p_1}, x_{p_2}, \ldots, x_{p_k}\}$ are the independent variables or parameters. Here k is an integer between 0 and n and, of course, $r + k = n$. The number of parameters, k, is called the *dimension of* S. The general format is therefore

$$S = \left\{(x_1, x_2, \ldots, x_n)\text{: } \begin{aligned} x_{d_1} &= a_{d_1p_1}x_{p_1} + a_{d_1p_2}x_{p_2} + \cdots + a_{d_1p_k}x_{p_k} + b_{d_1} \\ x_{d_2} &= a_{d_2p_1}x_{p_1} + a_{d_2p_2}x_{p_2} + \cdots + a_{d_2p_k}x_{p_k} + b_{d_2} \\ &\vdots \\ x_{d_r} &= a_{d_rp_1}x_{p_1} + a_{d_rp_2}x_{p_2} + \cdots + a_{d_rp_k}x_{p_k} + b_{d_r} \end{aligned}\right\}$$

Example 4.9 fits into this general format by taking $d_1 = 1$, $d_2 = 2$, $d_3 = 4$, $p_1 = 3$, and $p_2 = 5$.

In the two extreme cases, when $k = 0$ and when $k = n$, the affine subspaces described previously are very simple.

If $k = 0$, there are no parameters; each of $x_1, x_2, \ldots, x_n$ is a dependent variable. Thus $S = \{(x_1, x_2, \ldots, x_n) \in \mathbb{R}^n\text{: } x_1 = b_1, x_2 = b_2, \ldots, x_n = b_n\}$ is just the single point $\{(b_1, b_2, \ldots, b_n)\}$.

If $k = n$, there are no dependent variables; in other words, there are no constraints at all on membership in S. Thus $S = \mathbb{R}^n$.

If $k = 1$, the resulting set will be called a line in $\mathbb{R}^n$. We have previously encountered lines in $\mathbb{R}^2$ and $\mathbb{R}^3$. Descriptions that fit this general format are Examples 4.1(i), (ii), (iii), (v), (vi), and (vii), as well as 4.5(i), (iii), (iv), and (v).

Note the possibility that $a_{ij} = 0$ for some or all i and j; thus the parameter x_{p_j} may or may not actually appear in the equation for x_{d_i}.

4.10 Example

$$\{(x_1, x_2, x_3, x_4, x_5) \in \mathbb{R}^5: x_1 = 3x_2 + 5, x_3 = -x_2 - 1, x_4 = 0, x_5 = -7x_2\}$$

is a line in $\mathbb{R}^5$.

If $k = 2$, the resulting set will be called a plane in $\mathbb{R}^n$. Examples 4.3(i), (ii), (iii), (v), (vi), (vii) and (viii) are planes in $\mathbb{R}^3$; Example 4.9 is a plane in $\mathbb{R}^5$.

4.11 Example

$\{(x_1, x_2, x_3, x_4) \in \mathbb{R}^4$: $x_1 = 3x_2 + x_4 - 5, x_3 = 2x_2 + 17\}$ is a plane in $\mathbb{R}^4$.

If $k = n - 1$, the resulting set is a hyperplane in $\mathbb{R}^n$. To see this, recall that a hyperplane in $\mathbb{R}^n$ is a set S of the form $\{(x_1, x_2, \ldots, x_n) \in \mathbb{R}^n$: $a_1x_1 + a_2x_2 + \cdots + a_nx_n = b\}$, where $a_i \neq 0$ for at least one i. By rewriting this constraint, we can provide an alternate description of S that fits this general format.

$$S = \left\{(x_1, x_2, \ldots, x_n) \in \mathbb{R}^n: x_i = \frac{-a_1}{a_i}x_1 + \cdots + \frac{-a_{i-1}}{a_i}x_{i-1} \right.$$
$$\left. + \frac{-a_{i+1}}{a_i}x_{i+1} + \cdots + \frac{-a_n}{a_i}x_n + \frac{b}{a_i}\right\}$$

4.12 Summary of Synonyms

Technical Name	Colloquial Name
Zero-dimensional affine subspace of $\mathbb{R}^n$	Single point in $\mathbb{R}^n$
One-dimensional affine subspace of $\mathbb{R}^n$	Line in $\mathbb{R}^n$
Two-dimensional affine subspace of $\mathbb{R}^n$	Plane in $\mathbb{R}^n$
$(n - 1)$-dimensional affine subspace of $\mathbb{R}^n$	Hyperplane in $\mathbb{R}^n$

Because our intuition is based on our experience in lower dimensions, we do not have any colloquial names for k-dimensional subspaces of $\mathbb{R}^n$ when $3 \leq k \leq n - 2$.

There is one final aspect of all this that we should treat at an intuitive level. From Examples 4.1(v) and (vi), 4.3(vi), (vii), and (viii), as well as 4.5(ii), we observe that a given affine subspace can have several different descriptions, all of which fit the general format. These descriptions differ merely in the choice of roles (i.e., dependent variable or parameter) to be played by the variables. It is an important result, to be formalized and proved later in this chapter, that for any given affine subspace the number of parameters (and hence the number of dependent variables) will be the same in any description of the space. But in most cases it does not matter which variables play which roles. For instance, there is no real reason to prefer any one of 4.3(vi), (vii), or (viii) over the other two.

There are two important cases, however, in which we are not at liberty to choose the role played by x_i—that is when the subspace in question, S, is either perpendicular to or parallel to the x_i axis. We will eventually generalize the notions of "perpendicularity" and "parallelism" to affine subspaces of $\mathbb{R}^n$, but for now we will just illustrate the problems with some simple examples.

The first situation, in which the subspace is perpendicular to the x_i axis, corresponds to the fact that one of the constraining equations takes the form x_i equals a constant, b_i. In this case, x_i is perforce a dependent variable; it must also be a dependent variable in any other description of the same space because the constraint $x_i = b_i$ cannot be rewritten in any other way.

Recall Example 4.3(iii), $\{(x_1, x_2) \in \mathbb{R}^2\colon x_2 = 7\}$. This line in $\mathbb{R}^2$ is perpendicular to the x_2 axis. There is no alternate description of this line in which x_1 is the dependent variable and x_2 is the parameter.

4.13 Example

$\{(x_1, x_2, x_3) \in \mathbb{R}^3\colon x_1 = x_3 + 1, x_2 = 7\}$ is a line in $\mathbb{R}^3$. This description uses x_1 and x_2 as dependent variables and x_3 as parameter. There is an alternate description, $\{(x_1, x_2, x_3) \in \mathbb{R}^3\colon x_3 = x_1 - 1, x_2 = 7\}$, which involves x_2 and x_3 as dependent variables and x_1 as parameter; but there can be no description that has x_2 as parameter since this line lies in the plane $\{(x_1, x_2, x_3) \in \mathbb{R}^3\colon x_2 = 7\}$, which is perpendicular to the x_2 axis.

In the second situation, where S is parallel to the x_i axis, the variable x_i *must* be a parameter. When S is parallel to the x_i axis, membership of an n-tuple $(x_1, x_2, \ldots, x_i, \ldots, x_n)$ in S must be entirely independent of the value of its ith coordinate; thus x_i cannot be a dependent variable. Moreover, in the expressions for the dependent variables as affine functions of the parameters, the coefficient of the parameter x_i must be 0. We could say, in this situation, that x_i is a "hidden" parameter because when we look quickly at the equations that determine membership in S, we tend to overlook the fact that x_i is a parameter.

Recall Example 4.3(iii), $\{(x_1, x_2) \in \mathbb{R}^2\colon x_2 = 7\}$ once more. This line in $\mathbb{R}^2$ is parallel to the x_1 axis. There is no alternate description of this line in which x_1 is the dependent variable and x_2 is the parameter.

(i) $\{(x_1, x_2, x_3, x_4) \in \mathbb{R}^4: x_1 = 3x_4 - 7, x_2 = 6x_4 + 1\}$ is a plane in $\mathbb{R}^4$ that is parallel to the x_3 axis.

(ii) $\{(x_1, x_2, x_3, x_4, x_5) \in \mathbb{R}^5: x_3 = 2x_4 + 1, x_5 = 6\}$ is a three-dimensional affine subspace of $\mathbb{R}^5$ that is perpendicular to the x_5 axis and is parallel to both the x_1 axis and the x_2 axis.

Problem Set 4

4.1. Go back and do any of the exercises in Section 4 that you may have skipped.

*4.2. Identify by inspection the geometric nature of each of the following affine subspaces. You may use the colloquial names summarized in 4.12 where they are appropriate.

(i) $\{(x_1, x_2, x_3) \in \mathbb{R}^3: x_1 = 4, x_2 = 2x_3 - 6\}$

(ii) $\{(x_1, x_2, x_3, x_4) \in \mathbb{R}^4: x_1 = 2x_2 - 4x_4 + 5, x_3 = 7x_4 - 5\}$

(iii) $\{(x_1, x_2) \in \mathbb{R}^2: x_1 = -2, x_2 = 3\}$

(iv) $\{(x_1, x_2, x_3, x_4) \in \mathbb{R}^4: x_1 = 3x_2 + 4x_3 + 5x_4 + 6\}$

(v) $\{(x_1, x_2, x_3, x_4, x_5) \in \mathbb{R}^5: x_1 = 2x_4 + 1, x_3 = 2x_2 + 5x_4 - x_5\}$

(vi) $\{(x_1, x_2, x_3, x_4, x_5) \in \mathbb{R}^5: x_1 = 2x_4 + 1, x_3 = 2x_2 + 5x_4\}$

(vii) $\{(x_1, x_2, x_3) \in \mathbb{R}^3: x_1 = x_2, x_3 = x_2\}$

4.3. Write an expression of the form{_______: _________} in which the two blanks are filled in such a way that the resulting set is:

(i) A line in $\mathbb{R}^3$

(ii) A hyperplane in $\mathbb{R}^5$

(iii) A plane in $\mathbb{R}^4$

(iv) A plane in $\mathbb{R}^3$ perpendicular to the x_2 axis

(v) A plane in $\mathbb{R}^4$ that does not contain the origin

(vi) A single point in $\mathbb{R}^4$

(vii) A hyperplane in $\mathbb{R}^4$ that contains the point $(1, -1, 0, 2)$

(viii) A line in $\mathbb{R}^2$ containing the point $(-3, 4)$

(ix) A plane in $\mathbb{R}^3$ containing the two points $(2, 0, 4)$ and $(-1, -2, 6)$

4.4. For each of your answers to Problem 4.2, write down, if possible, a different expression of the form {_______: _________} that describes the same set; if an alternate description is not possible, explain why it is not. For example, 4.2(v) describes a three-dimensional affine subspace of $\mathbb{R}^5$. Is there an alternate description of this space that uses x_2 and x_4 as dependent variables and x_1, x_3, and x_5 as parameters?

5

The "Original" Vector Space(s)

In Section 4 we dealt with $\mathbb{R}^n$ in an informal, descriptive manner; in this section, we present a formal treatment. What we do in this section for $\mathbb{R}^n$ will provide the concrete examples that motivate the generalized concept of an abstract vector space that will be introduced in Section 41.

First, a remark concerning our use of the word *vector*. Just as we are now using the word *scalar* as a synonym for "real number," the word *vector* will be used, until Section 41, as a synonym for "n-tuple of real numbers" or "element of $\mathbb{R}^n$," where n is a positive integer whose value is normally unspecified, although it may be fixed in certain contexts. In Section 41, the word *vector* will be used to denote those objects that behave like the elements of $\mathbb{R}^n$ do in the special cases to be considered in the present section.

We use boldface lowercase letters $\mathbf{a}, \mathbf{b}, \mathbf{c}, \ldots, \mathbf{u}, \mathbf{v}, \mathbf{w}, \mathbf{x}, \mathbf{y}, \mathbf{z}$ to denote vectors.

5.1 Definition

If $\mathbf{u} = (u_1, u_2, \ldots, u_n)$ is a vector, the real numbers $u_1, u_2, \ldots, u_n$ are called the *coordinates of the vector* $\mathbf{u}$. u_i is the ith coordinate of $\mathbf{u}$.

(i) When $n = 1$, it is not worth the bother to distinguish between the real number u and the 1-tuple $\mathbf{u} = (u)$ whose single coordinate is u. Vector concepts for $n = 1$ are of only occasional interest and in practice we will avoid considering 1-tuples altogether. See [F]5.1.

(ii) Strictly speaking, we must distinguish between the n-tuple $(u_1, u_2, \ldots, u_n)$, the $1 \times n$ matrix $[u_1\, u_2 \ldots u_n]$, and the $n \times 1$ matrix $\begin{bmatrix} u_1 \\ u_2 \\ \vdots \\ u_n \end{bmatrix}$. But in most contexts, this distinction is not worth making explicitly; thus we will use the same symbol, $\mathbf{u}$, to denote any of the three possibilities. The reader can infer from the context to which of these three the symbol $\mathbf{u}$ is referring. On those rare occasions when we wish to emphasize the distinction, we will do so terminologically: the n-tuple $(u_1, u_2, \ldots, u_n) = \mathbf{u}$ is the *vector* $\mathbf{u}$; the $1 \times n$ matrix $[u_1\, u_2 \ldots u_n] = \mathbf{u}$ is the *row vector* $\mathbf{u}$; and $\begin{bmatrix} u_1 \\ u_2 \\ \vdots \\ u_n \end{bmatrix} = \mathbf{u}$ is the *column vector* $\mathbf{u}$. Thus the phrase *row vector* is a synonym for $1 \times n$ *matrix*; the phrase *column vector* is a synonym for $n \times 1$ *matrix*.

(iii) We must frequently discuss sets such as $S = \{\mathbf{u}_1, \mathbf{u}_2, \ldots, \mathbf{u}_m\}$ whose elements are vectors. In this case it is necessary to use double subscripts to denote the coordinates of these vectors. Thus, for example, $\mathbf{u}_1 = (u_{11}, u_{12}, \ldots, u_{1n})$, $\mathbf{u}_2 = (u_{21}, u_{22}, \ldots, u_{2n})$, and so on; u_{ij} is the jth coordinate of the vector $\mathbf{u}_i$.

(iv) The ability to distinguish vectors from scalars in typeset copy by using boldface is an extremely convenient device. It is essential to maintain this distinction in handwritten and typewritten work as well. In the manuscript copy of this text, this distinction was made by using *underlined* lowercase letters to denote vectors. With this convention, the preceding paragraph (iii) would appear as follows:

> We must frequently discuss sets such as $S = \{\underline{u}_1, \underline{u}_2, \ldots, \underline{u}_m\}$ whose elements are vectors. In this case it is necessary to use double subscripts to denote the coordinates of these vectors. Thus, for example, $\underline{u}_1 = (u_{11}, u_{12}, \ldots, u_{1n})$, $\underline{u}_2 = (u_{21}, u_{22}, \ldots, u_{2n})$, and so on; u_{ij} is the jth coordinate of the vector $\underline{u}_i$.

The importance of developing this (or some equivalent) habit from the beginning cannot be overstressed. The added effort will prevent innumerable errors.

5.3 Definition (Addition for Vectors)

If $\mathbf{u} = (u_1, u_2, \ldots, u_n)$ and $\mathbf{v} = (v_1, v_2, \ldots, v_n)$ are vectors, then their *vector sum*, $\mathbf{u} + \mathbf{v}$, is the vector $(u_1 + v_1, u_2 + v_2, \ldots, u_n + v_n)$.

Note that addition is only defined for pairs of vectors having the same number of coordinates. The phrase "vectors are added coordinatewise" is in common use to express the content of Definition 5.3.

5.4 Examples

If $\mathbf{u} = (2, 4, -1, 0)$ and $\mathbf{v} = (5, 0, 6, -2)$, then $\mathbf{u} + \mathbf{v} = (7, 4, 5, -2)$. If $\mathbf{a} = (1, -4, a_3, a_4, 6)$ and $\mathbf{b} = (-1, 2, 4, b_4, b_5)$, then $\mathbf{a} + \mathbf{b} = (0, -2, a_3 + 4, a_4 + b_4, 6 + b_5)$.

5.5 Definition (Multiplication of a Vector by a Scalar)

If $\mathbf{u} = (u_1, u_2, \ldots, u_n)$ is a vector and t is a scalar, then the result, denoted by $t\mathbf{u}$, of multiplying $\mathbf{u}$ by t is the vector $(tu_1, tu_2, \ldots, tu_n)$.

The shorter name, *scalar multiplication*, for this operation will also be used.

5.6 Examples

If $\mathbf{u} = (3, -2)$ and $t = 4$, then $t\mathbf{u} = (12, -8)$. If $\mathbf{a} = (0, -1, \pi)$ and $s = 3$, then $s\mathbf{a} = (0, -3, 3\pi)$. If $\mathbf{b} = (b_1, 6, 0)$ and $r = -1$, then $(-1)\mathbf{b} = (-b_1, -6, 0)$.

5.7 Notation

We will use $-\mathbf{v}$ to denote the vector $(-1)\mathbf{v}$ and, similarly, $\mathbf{u} - \mathbf{v}$ is an abbreviation for $\mathbf{u} + (-1)\mathbf{v}$. Also, since $1\mathbf{u} = \mathbf{u}$, we will omit writing the scalar 1 in this case.

The two operations just introduced have familiar geometric interpretations in $\mathbb{R}^2$ and $\mathbb{R}^3$. We mention them here because these interpretations can be extended to $\mathbb{R}^n$ as well, although the relevant pictures are beyond our ability to draw them in the plane. At the same time, we would like to emphasize that, although geometrical considerations may motivate the intro-

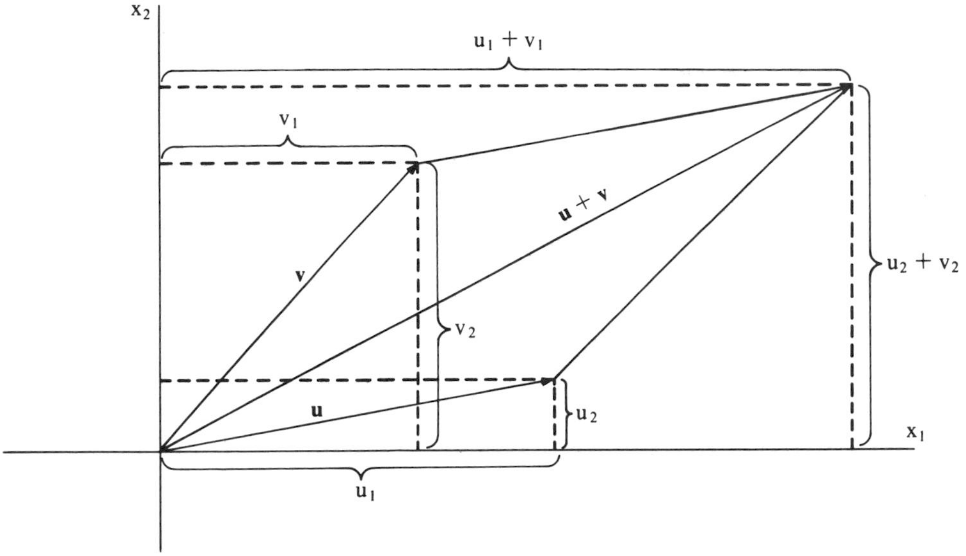

Figure 5.1

duction of certain concepts, our formal development of $\mathbb{R}^n$ will *never* depend on geometrical arguments.

Vector addition in $\mathbb{R}^2$ can be pictured by interpreting the vector $\mathbf{u} = (u_1, u_2)$ as the directed line segment in the plane whose tail is at the origin $(0, 0)$ and whose head is at the point (u_1, u_2). With this interpretation, the vector sum of two vectors **u** and **v** is the vector that is obtained by the parallelogram law; that is, $\mathbf{u} + \mathbf{v}$ is the vector whose tail is at the origin and whose head is at the opposite corner of the parallelogram determined by **u** and **v**. See Figure 5.1.

Multiplying a vector **u** by a scalar t changes the length of the vector by a factor of $|t|$, where $|t|$ is the absolute value of t; the direction is unchanged if $t > 0$ and is reversed if $t < 0$. See Figure 5.2.

5.8 Definition

A *linear combination of the vectors* $\mathbf{u}_1, \mathbf{u}_2, \ldots, \mathbf{u}_m$ is a vector of the form $t_1\mathbf{u}_1 + t_2\mathbf{u}_2 + \cdots + t_m\mathbf{u}_m$, where $t_1, t_2, \ldots, t_m$ are scalars. A linear combination is called *nontrivial* if the vectors $\mathbf{u}_1, \mathbf{u}_2, \ldots, \mathbf{u}_m$ are distinct and $t_i \neq 0$ for at least one i.

5.9 Examples

If $\mathbf{u}_1 = (9, 5)$, $\mathbf{u}_2 = (-8, 0)$, and $\mathbf{u}_3 = (1, 1)$, and if $t_1 = 3$, $t_2 = 8$, and $t_3 = 2$, then $\mathbf{v}_1 = t_1\mathbf{u}_1 + t_2\mathbf{u}_2 + t_3\mathbf{u}_3 = 3(9, 5) + 8(-8, 0) + 2(1, 1)$ is a linear combination of $\mathbf{u}_1$, $\mathbf{u}_2$, and $\mathbf{u}_3$. If $s_1 = 0$, $s_2 = 1$, and $s_3 = -1$, then

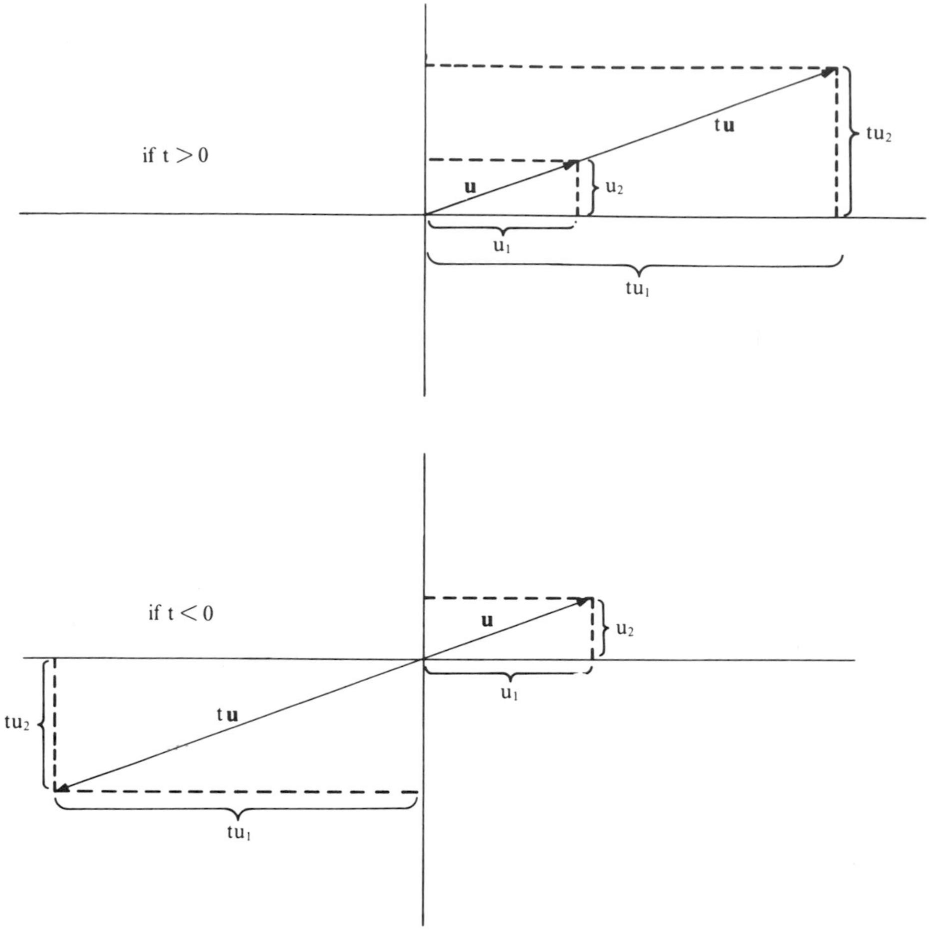

Figure 5.2

$\mathbf{v}_2 = s_1\mathbf{u}_1 + s_2\mathbf{u}_2 + s_3\mathbf{u}_3 = 0(9,5) + 1(-8,0) + (-1)(1,1)$, which we should write as $0(9,5) + (-8,0) - (1,1)$ according to 5.7, is also a linear combination of $\mathbf{u}_1$, $\mathbf{u}_2$, and $\mathbf{u}_3$. Using Definitions 5.3 and 5.5, we can simplify the expressions for $\mathbf{v}_1$ and $\mathbf{v}_2$ to find $\mathbf{v}_1 = (27 - 64 + 2, 15 + 0 + 2) = (-35, 17)$ and $\mathbf{v}_2 = (0 - 8 - 1, 0 + 0 - 1) = (-9, -1)$.

5.10 Definition

(i) For any nonempty subset S of $\mathbb{R}^n$, the set $\mathscr{L}(S) = \{\mathbf{x} \in \mathbb{R}^n$: there exist vectors $\mathbf{u}_1, \mathbf{u}_2, \ldots, \mathbf{u}_m \in S$ and scalars $t_1, t_2, \ldots, t_m$ such that $\mathbf{x} = t_1\mathbf{u}_1 + t_2\mathbf{u}_2 + \cdots + t_m\mathbf{u}_m\}$ is called the (linear) *span of S*.

(ii) $\mathscr{L}(\varnothing) = \{\mathbf{0}\}$.

In words, for $S \neq \varnothing$, $\mathscr{L}(S)$ is the set of all those vectors that can be expressed as linear combinations of vectors from S. The word *span* and the phrase *linear span* are both common names for this concept. We will only use the word *span*, but we mention the alternative as an explanation for using the symbol $\mathscr{L}(S)$ to denote the span of S. It does not really matter how we define the span of the empty set, $\varnothing$; defining $\mathscr{L}(\varnothing) = \{\mathbf{0}\}$ is a convenience that simplifies the statements of some of the results to follow.

It should be noted that in Definition 5.10 we only consider linear combinations of finitely many vectors at a time. Thus, even when S is an infinite subset of $\mathbb{R}^n$, $\mathscr{L}(S)$ consists, by definition, of those vectors that can be written as linear combinations of finitely many vectors from S. The study of infinite linear combinations $t_1\mathbf{u}_1 + t_2\mathbf{u}_2 + \cdots + t_i\mathbf{u}_i + \cdots = \Sigma_i t_i \mathbf{u}_i$ involves problems of convergence and belongs to the realm of analysis rather than algebra.

5.11 Exercise

If S_1 and S_2 are subsets of $\mathbb{R}^n$ with $S_1 \subset S_2$, then $\mathscr{L}(S_1) \subset \mathscr{L}(S_2)$.

A slight strengthening of the previous exercise leads to the standard technique for showing that two subsets S_1 and S_2 of $\mathbb{R}^n$ have the same span.

5.12 Exercise

If $S_1 \subset \mathscr{L}(S_2)$, then $\mathscr{L}(S_1) \subset \mathscr{L}(S_2)$.

Consequently, to conclude that $\mathscr{L}(S_1) = \mathscr{L}(S_2)$, it is sufficient to show that $S_1 \subset \mathscr{L}(S_2)$ and that $S_2 \subset \mathscr{L}(S_1)$.

5.13 Examples

(i) If $S = \{\mathbf{u}\}$, that is, if S consists of the single vector $\mathbf{u}$, and $\mathbf{u} \neq \mathbf{0}$, then $\mathscr{L}(S)$ is just the set of all scalar multiples of $\mathbf{u}$; geometrically, $\mathscr{L}(S)$ is the line in $\mathbb{R}^n$ determined by the vector $\mathbf{u}$. For example, if $\mathbf{u} = (3, 5, -6)$, then $\mathscr{L}(\{\mathbf{u}\})$ is the line in $\mathbb{R}^3$ connecting the origin $(0, 0, 0)$ with the point $(3, 5, -6)$ extended to infinity in both directions.

(ii) If S contains exactly two nonzero vectors, say $S = \{\mathbf{u}_1, \mathbf{u}_2\}$, then, provided $\mathbf{u}_2$ is not a scalar multiple of $\mathbf{u}_1$, $\mathscr{L}(\{\mathbf{u}_1, \mathbf{u}_2\})$ is the plane in $\mathbb{R}^n$ determined by the vectors $\mathbf{u}_1$ and $\mathbf{u}_2$. For example, if $\mathbf{u}_1 = (3, 5, -6)$ and $\mathbf{u}_2 = (1, 0, 4)$, then $\mathscr{L}(\{\mathbf{u}_1, \mathbf{u}_2\})$ is the plane in $\mathbb{R}^3$ determined by the three points $(0, 0, 0)$, $(3, 5, -6)$, and $(1, 0, 4)$. Recall that any three noncollinear points in $\mathbb{R}^3$ determine a unique plane, and that if $\mathbf{u}_2$ is not a scalar multiple of $\mathbf{u}_1$, then the three points $(0, 0, 0)$, (u_{11}, u_{12}, u_{13}), and (u_{21}, u_{22}, u_{23}) are noncollinear. If $\mathbf{v}_1 = (2, 0, -2, 4, 0)$ and $\mathbf{v}_2 = (2, 0, -2, 5, 0)$, then $\mathscr{L}(\{\mathbf{v}_1, \mathbf{v}_2\})$ is the plane in $\mathbb{R}^5$ determined by the three points $(0,0,0,0,0)$, $(2,0,-2,4,0)$, and $(2,0,-2,5,0)$.

5.14 Remark

The next definition is the one, promised in Section 4, that characterizes in a precise, mathematical way the intuitive notion of "a straight piece of $\mathbb{R}^n$." Although the defining property turns out to be extremely simple, there is a deceptively subtle aspect of this definition involving the special role played by the zero vector, $\mathbf{0}$. The route we follow is first to characterize, with properties 5.15 (i) and (ii), those "straight pieces of $\mathbb{R}^n$ that contain the origin," and second, in 5.16, to characterize "arbitrary straight pieces of $\mathbb{R}^n$" as those subsets of $\mathbb{R}^n$ that can be obtained by "shifting a straight piece that contains the origin."

5.15 Definition

A subset W of $\mathbb{R}^n$ is called a (linear) *subspace* of $\mathbb{R}^n$ if both the following conditions are satisfied:

(i) For any two vectors $\mathbf{u}, \mathbf{v} \in W$, the vector sum $\mathbf{u} + \mathbf{v} \in W$.
[*W is closed under vector addition*.]

(ii) For any vector $\mathbf{w} \in W$ and any scalar t, the vector $t\mathbf{w} \in W$.
[*W is closed under multiplication of a vector by a scalar*.]

Here the phrases in brackets at the right are used to describe the corresponding properties.

5.16 Definition

A subset H of $\mathbb{R}^n$ is called an *affine subspace* of $\mathbb{R}^n$ if $H = \varnothing$ or if there is a linear subspace W of $\mathbb{R}^n$ and a fixed vector $\mathbf{b} \in \mathbb{R}^n$ such that $H = \{\mathbf{w} + \mathbf{b} \colon \mathbf{w} \in W\}$, in which case we say that H is the *translate of W by $\mathbf{b}$*.

Affine subspaces of $\mathbb{R}^n$ are also known in the literature as *linear manifolds* or *flats*.

The letters G, H, K, and S will usually denote arbitrary subsets of $\mathbb{R}^n$, while the letters U, V, and W will be reserved for subsets of $\mathbb{R}^n$ that are closed under vector addition and scalar multiplication (i.e., for linear subspaces). We have enclosed the word *linear* in parentheses in Definition 5.15 because, once the distinction between linear and affine subspaces has been sufficiently emphasized, we will drop the word *linear* and use the word *subspace* by itself to mean *linear subspace*.

5.17 Remark

(i) An immediate consequence of 5.15(ii), obtained by letting $t = 0$, is that for any $\mathbf{w} \in W$, $0\mathbf{w} = \mathbf{0} \in W$ [i.e., a nonempty linear subspace of $\mathbb{R}^n$ must contain the origin, $\mathbf{0} = (0, 0, \ldots, 0)$]. Note that the subset $\{\mathbf{0}\}$ of $\mathbb{R}^n$ consisting of the zero vector alone is a subspace; it is called the *trivial subspace*. A nonempty affine subspace is a translate of a linear subspace; an affine subspace need not contain the origin, and, in fact, the linear subspaces are precisely those affine subspaces that do contain the origin.

(ii) Another immediate consequence of 5.15(ii) is that a nontrivial linear subspace W of $\mathbb{R}^n$ must contain infinitely many vectors: for if $\mathbf{0} \neq \mathbf{w} \in W$, then for all real numbers t, $t\mathbf{w} \in W$. Geometrically, if the linear subspace W contains the nonzero vector $\mathbf{w}$, it must contain the entire line through the origin determined by $\mathbf{w}$.

5.18 Exercise

A subset W of $\mathbb{R}^n$ is a subspace of $\mathbb{R}^n$ iff for all $\mathbf{w}_1, \mathbf{w}_2, \ldots, \mathbf{w}_m \in W$ and for all scalars $t_1, t_2, \ldots, t_m$, the linear combination $t_1\mathbf{w}_1 + t_2\mathbf{w}_2 + \cdots + t_m\mathbf{w}_m \in W$.

The content of 5.18 is often expressed as follows: a subspace of $\mathbb{R}^n$ is a subset that is closed under arbitrary linear combinations.

We wish to argue that the formal definitions of linear and affine subspaces from 5.15 and 5.16 serve their intended purpose of capturing the intuitive notion of straightness; as a first test of this claim, we should be able to show that what we know to be a straight line in $\mathbb{R}^2$ is indeed an affine subspace of $\mathbb{R}^2$, in the sense of Definition 5.16. The next example illustrates how this is done.

5.19 Example

If H is the line in $\mathbb{R}^2$ whose equation is $x_2 = 2x_1 + 5$, H is an affine subspace of $\mathbb{R}^2$. To prove this, we must exhibit a linear subspace W of $\mathbb{R}^2$ and a vector $\mathbf{b} \in \mathbb{R}^2$ such that H is the translate of W by $\mathbf{b}$. Figure 5.3 suggests one way to do this.

Step 1: Let $W = \{(x_1, x_2) \in \mathbb{R}^2\colon x_2 = 2x_1\}$. W is the line through the origin parallel to H. If $\mathbf{u} = (u_1, u_2) \in W$ and $\mathbf{v} = (v_1, v_2) \in W$, then $u_2 = 2u_1$ and $v_2 = 2v_1$; hence $\mathbf{u} + \mathbf{v} = (u_1 + v_1, u_2 + v_2) = (u_1 + v_1, 2u_1 + 2v_1) = (u_1 + v_1, 2(u_1 + v_1))$, and so $\mathbf{u} + \mathbf{v} \in W$. This proves that W is closed under vector addition. If $\mathbf{w} = (w_1, w_2) \in W$ and $t \in \mathbb{R}$, then, because

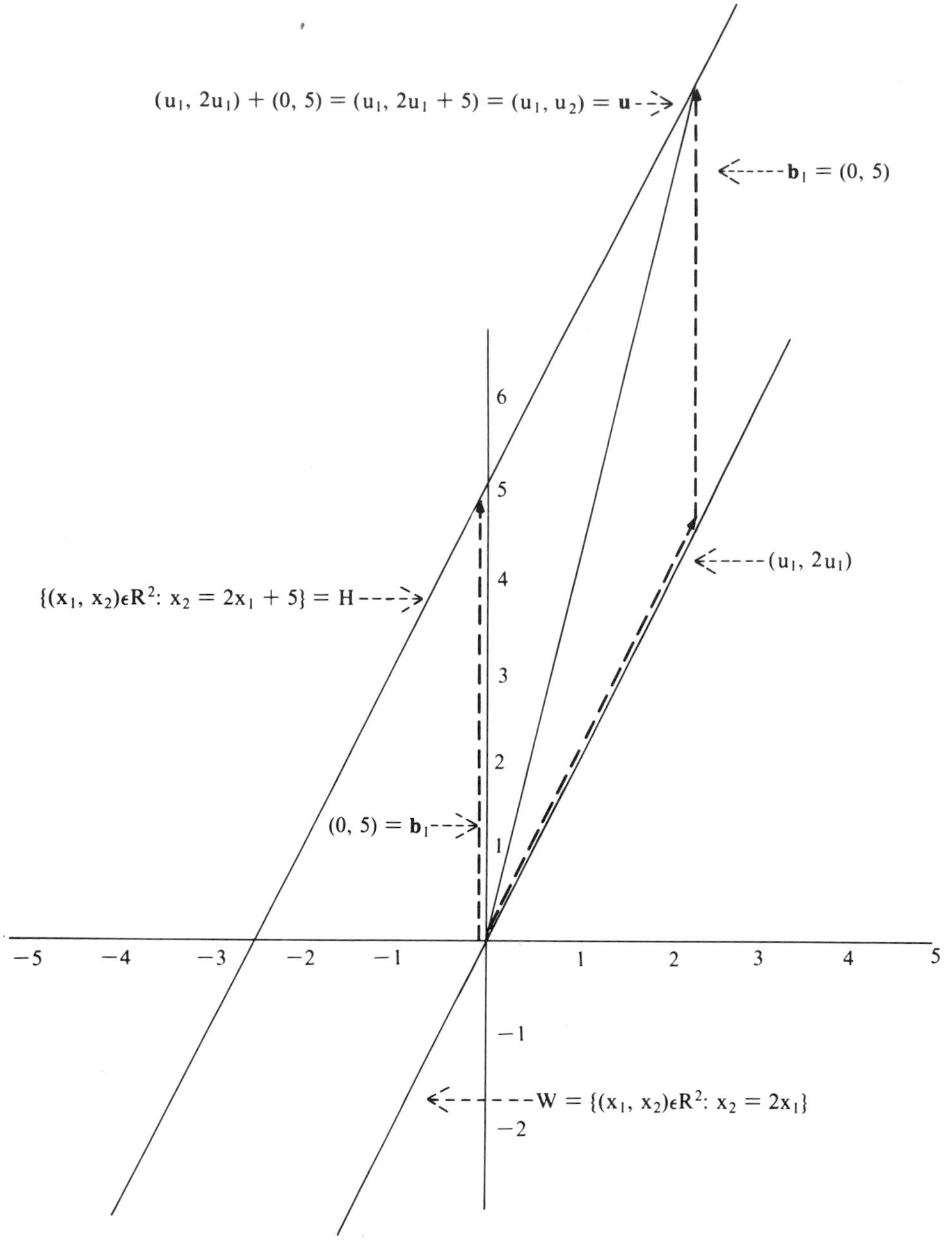

(u₁, 2u₁) + (0, 5) = (u₁, 2u₁ + 5) = (u₁, u₂) = u
b₁ = (0, 5)
6
5
4
3
2
1
(u₁, 2u₁)
{(x₁, x₂)εR²: x₂ = 2x₁ + 5} = H
(0, 5) = b₁
−5
−4
−3
−2
−1
1
2
3
4
5
−1
−2
W = {(x₁, x₂)εR²: x₂ = 2x₁}

Figure 5.3

$w_2 = 2w_1$, we have $t\mathbf{w} = (tw_1, tw_2) = (tw_1, t(2w_1)) = (tw_1, 2(tw_1)) \in W$; so W is also closed under scalar multiplication.

Step 2: Having proved that W is a linear subspace of $\mathbb{R}^2$, we show next that $H = \{(x_1, x_2) \in \mathbb{R}^2 : x_2 = 2x_1 + 5\}$ is the translate of W by the vector $\mathbf{b}_1 = (0, 5)$; that is, we show that $H = \{\mathbf{w} + (0,5) : \mathbf{w} \in W\}$.

Suppose $\mathbf{u} = (u_1, u_2) \in H$ (i.e., $u_2 = 2u_1 + 5$); then $\mathbf{u} = (u_1, u_2) = (u_1, 2u_1 + 5) = (u_1, 2u_1) + (0, 5)$. Since $(u_1, 2u_1) \in W$, this shows that $\mathbf{u} \in \{\mathbf{w} + (0,5) : \mathbf{w} \in W\}$.

Conversely, if $\mathbf{u} \in \{\mathbf{w} + (0,5) : \mathbf{w} \in W\}$, then for some v_1 and v_2 with $v_2 = 2v_1$, $\mathbf{u} = (v_1, v_2) + (0,5) = (v_1, 2v_1) + (0,5) = (v_1, 2v_1 + 5)$; hence $\mathbf{u} \in H$.

5.20 Remark Concerning the Previous Example

Figure 5.4 suggests that H is also the translate of W by the vector $\mathbf{b}_2 = (1, 7)$; that is, $H = \{\mathbf{w} + (1,7) : \mathbf{w} \in W\}$. Figure 5.5 suggests the same for $\mathbf{b}_3 = (-3, -1)$. Figure 5.6 shows Figures 5.3 and 5.4 superimposed.

To see that this is indeed the case, suppose $\mathbf{u} = (u_1, 2u_1 + 5) \in H$; then $\mathbf{u} = ((u_1 - 1), 2(u_1 - 1)) + (1, 7)$. Thus $\mathbf{u} \in \{\mathbf{w} + (1,7) : \mathbf{w} \in W\}$.

Conversely, if $\mathbf{u} \in \{\mathbf{w} + (1,7) : \mathbf{w} \in W\}$, then, for some v_1 and v_2 with $v_2 = 2v_1$, we have $\mathbf{u} = (v_1, v_2) + (1, 7)$. Thus $\mathbf{u} = (v_1, 2v_1) + (1,7) = (v_1 + 1, 2v_1 + 7) = (v_1 + 1, 2(v_1 + 1) + 5)$; hence $\mathbf{u} \in H$.

5.21 Example

We will prove that $H = \{(x_1, x_2, x_3, x_4, x_5) \in \mathbb{R}^5 : x_1 = 2x_2 + 3x_4 - x_5 + 4, x_3 = -5x_4 + 6x_5 - 7\}$ is an affine subspace of $\mathbb{R}^5$.

Step 1: Let $W = \{(x_1, x_2, x_3, x_4, x_5) \in \mathbb{R}^5 : x_1 = 2x_2 + 3x_4 - x_5, x_3 = -5x_4 + 6x_5\}$. Note that W is obtained from H by setting the constant terms equal to zero in the affine functions for the dependent variables x_1 and x_3. (In Example 5.19, W was obtained from H in the same way.) We will show that W is a linear subspace of $\mathbb{R}^5$. If $\mathbf{u}$ and $\mathbf{v}$ belong to W, then

$$\mathbf{u} = (2u_2 + 3u_4 - u_5, u_2, -5u_4 + 6u_5, u_4, u_5) \quad \text{and}$$
$$\mathbf{v} = (2v_2 + 3v_4 - v_5, v_2, -5v_4 + 6v_5, v_4, v_5);$$

so
$$\begin{aligned} \mathbf{u} + \mathbf{v} &= (2u_2 + 3u_4 - u_5 + 2v_2 + 3v_4 - v_5, u_2 + v_2, -5u_4 + 6u_5 \\ &\qquad -5v_4 + 6v_5, u_4 + v_4, u_5 + v_5) \\ &= (2(u_2 + v_2) + 3(u_4 + v_4) - (u_5 + v_5), u_2 + v_2, -5(u_4 + v_4) \\ &\qquad +6(u_5 + v_5), u_4 + v_4, u_5 + v_5), \end{aligned}$$

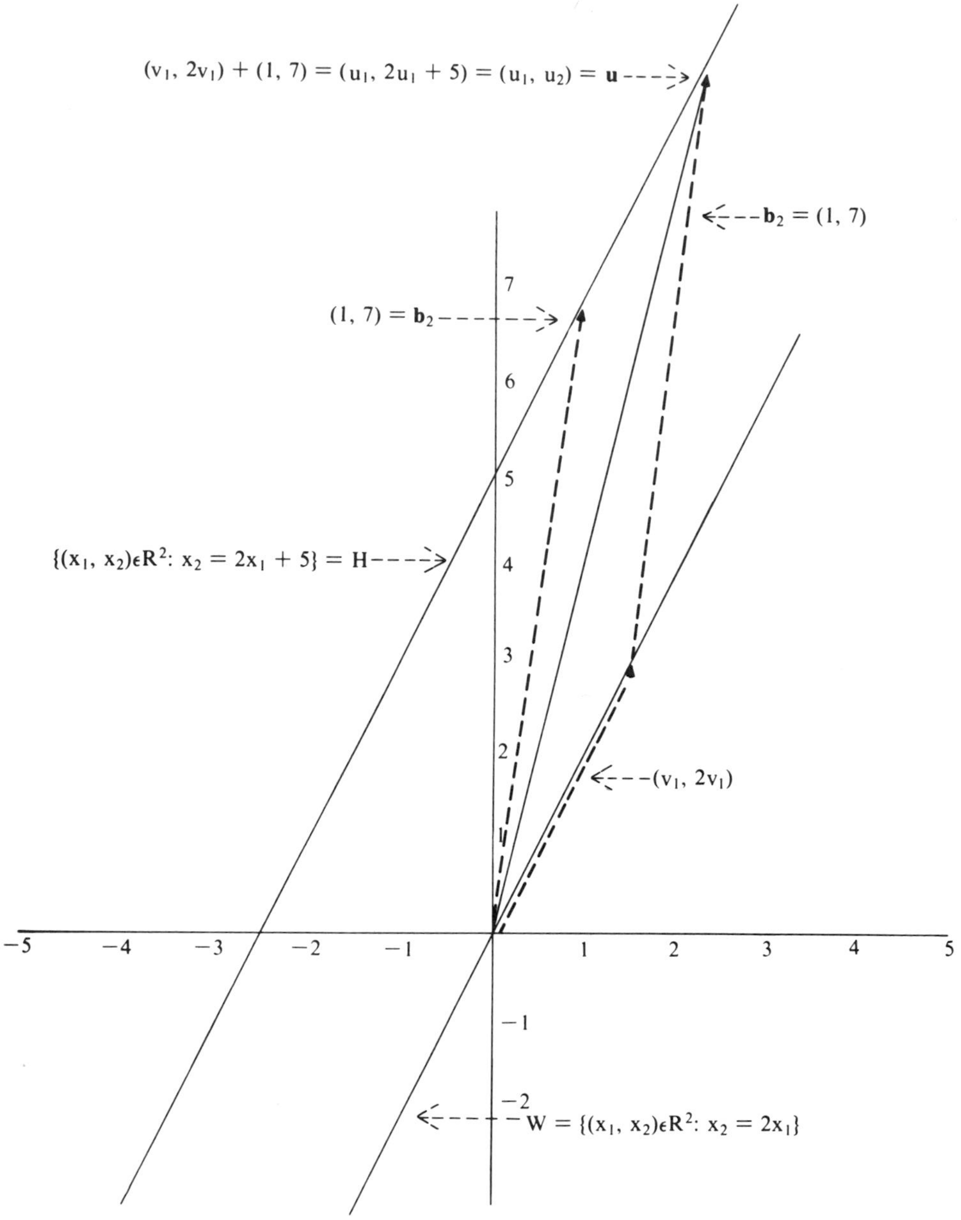
(v1, 2v1) + (1, 7) = (u1, 2u1 + 5) = (u1, u2) = u
b2 = (1, 7)
(1, 7) = b2
{(x1, x2)εR2: x2 = 2x1 + 5} = H
(v1, 2v1)
W = {(x1, x2)εR2: x2 = 2x1}
−5
−4
−3
−2
−1
1
2
3
4
5
7
6
5
4
3
2
1
−1
−2

Figure 5.4

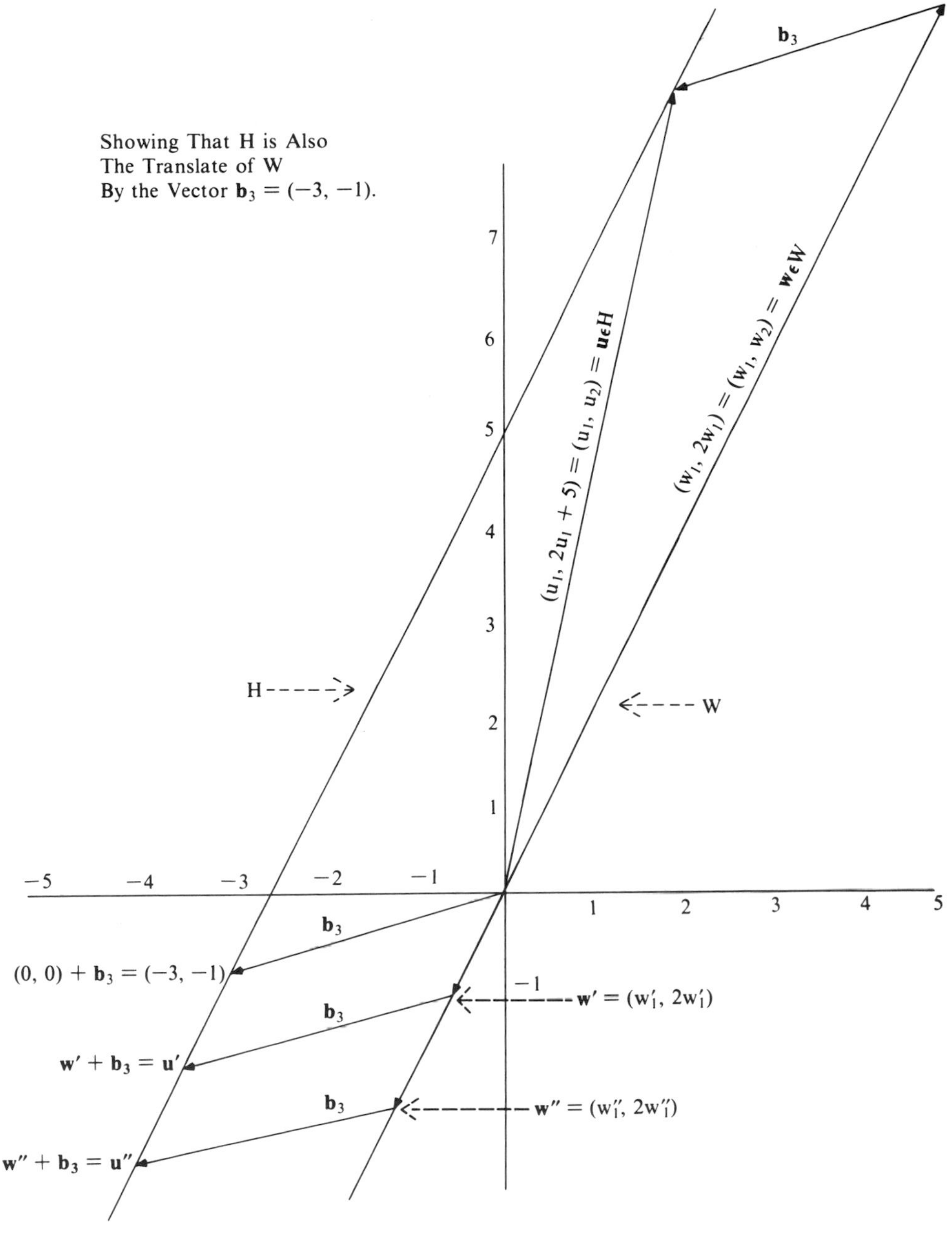

Figure 5.5 Showing that H is also the translate of W by the vector $\mathbf{b}_3 = (-3, -1)$.

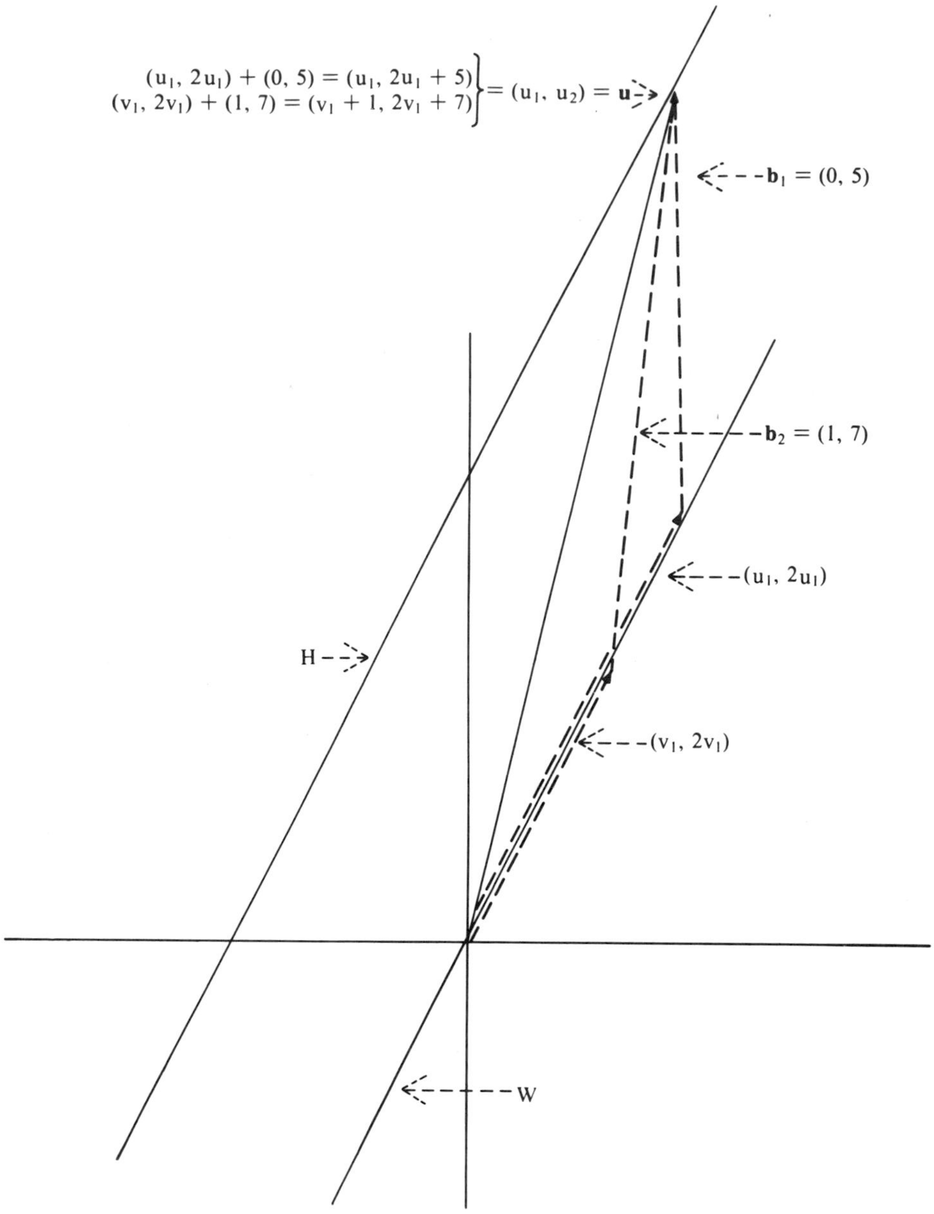

Figure 5.6 Figures 5.3 and 5.4 superimposed.

which shows that $\mathbf{u} + \mathbf{v} \in W$. Similarly, if $\mathbf{u} \in W$ and $t \in \mathbb{R}$, then, because

$$\begin{aligned}
\mathbf{u} &= (2u_2 + 3u_4 - u_5, u_2, -5u_4 + 6u_5, u_4, u_5), \\
t\mathbf{u} &= (t(2u_2 + 3u_4 - u_5), tu_2, t(-5u_4 + 6u_5), tu_4, tu_5) \\
&= (2(tu_2) + 3(tu_4) - (tu_5), tu_2, -5(tu_4) + 6(tu_5), tu_4, tu_5)
\end{aligned}$$

thus $t\mathbf{u} \in W$.

Step 2: We show next that H is the translate of W by the vector $\mathbf{b} = (4, 0, -7, 0, 0)$. Note that $\mathbf{b}$ is the vector in $\mathbb{R}^5$ obtained by taking the constant terms in the affine functions for the dependent variables x_1 and x_3 as the first and third coordinates of $\mathbf{b}$ and by setting the remaining coordinates of $\mathbf{b}$ equal to 0. (In Example 5.19, $\mathbf{b}_1$ was obtained in a similar fashion.)

To see that $H = \{\mathbf{w} + \mathbf{b} : \mathbf{w} \in W\}$, suppose $\mathbf{u} \in H$; then $u = (2u_2 + 3u_4 - u_5 + 4, u_2, -5u_4 + 6u_5 - 7, u_4, u_5)$. Thus $\mathbf{u} = (2u_2 + 3u_4 - u_5, u_2, -5u_4 + 6u_5, u_4, u_5) + (4, 0, -7, 0, 0)$, which is a vector of the form $\mathbf{w} + \mathbf{b}$, where $\mathbf{w} \in W$.

Conversely, if $\mathbf{u} \in \{\mathbf{w} + \mathbf{b} : \mathbf{w} \in W\}$, then for some v_2, v_4, and v_5, $\mathbf{u} = (2v_2 + 3v_4 - v_5, v_2, -5v_4 + 6v_5, v_4, v_5) + (4, 0, -7, 0, 0)$. Thus $\mathbf{u} = (2v_2 + 3v_4 - v_5 + 4, v_2, -5v_4 + 6v_5 - 7, v_4, v_5)$; so $\mathbf{u} \in H$.

In 5.19 and 5.20, we saw that a given affine subspace H can be the translate of a given linear subspace W by more than one vector. (In fact, such a vector is as far from unique as can be. See Problem 5.3 for the details.) Our geometric instinct tells us, however, that the linear subspace involved is unique: an arbitrary line in $\mathbb{R}^2$ is a translate of a unique line through the origin; an arbitrary plane in $\mathbb{R}^3$ is parallel to (i.e., is a translate of) a unique plane through the origin. Our first and only theorem in this section confirms for arbitrary affine subspaces of $\mathbb{R}^n$ what we instinctively know about $\mathbb{R}^2$ and $\mathbb{R}^3$. Indeed, if we were unable to prove such a theorem, we would have to withdraw our claim that the formal definitions that we have given of linear and affine subspace capture the intuitive notion of "straight piece of $\mathbb{R}^n$."

5.22 Theorem

Let H be a nonempty affine subspace of $\mathbb{R}^n$; then there is a unique linear subspace W of $\mathbb{R}^n$ such that H is a translate of W.

Proof: Suppose that H is a translate of W by $\mathbf{a}$ and of W' by $\mathbf{b}$, where W and W' are both linear subspaces of $\mathbb{R}^n$. We wish to prove that $W = W'$. We will prove the inclusion $W \subset W'$; the reverse inclusion $W' \subset W$ is proved similarly and this is left as an exercise. Our hypothesis implies that $H = \{\mathbf{w} + \mathbf{a}: \mathbf{w} \in W\} = \{\mathbf{w}' + \mathbf{b}: \mathbf{w}' \in W'\}$. Choose an arbitrary vector $\mathbf{w} \in W$; we want to conclude that $\mathbf{w} \in W'$. Since W is a nonempty linear subspace, 5.17 implies that $\mathbf{0} \in W$, and hence that $\mathbf{0} + \mathbf{a} = \mathbf{a} \in H$. Now since $\mathbf{a} \in H = \{\mathbf{w}' + \mathbf{b}: \mathbf{w}' \in W'\}$, there exists a vector $\mathbf{u} \in W'$ such that $\mathbf{a} = \mathbf{u} + \mathbf{b}$. Also, because $\mathbf{w} + \mathbf{a} \in H = \{\mathbf{w}' + \mathbf{b}: \mathbf{w}' \in W'\}$, there exists a vector $\mathbf{v} \in W'$ such that $\mathbf{w} + \mathbf{a} = \mathbf{v} + \mathbf{b}$. Because W' is closed under vector addition and scalar multiplication, $\mathbf{v} - \mathbf{u} \in W'$. But $\mathbf{v} - \mathbf{u} = (\mathbf{w} + \mathbf{a} - \mathbf{b}) - (\mathbf{a} - \mathbf{b}) = \mathbf{w}$; hence $\mathbf{w} \in W'$. ■

If you find this concept a bit abstract, and would like to anticipate its importance, you should peek ahead to Theorem 6.9.

From now on, the word *subspace* by itself will mean linear subspace.

Problem Set 5

5.1. Go back and do any of the exercises in Section 5 that you may have skipped.

*5.2. For each of the subsets S, answer the following questions.
  (a) Is S a linear subspace?
  (b) If not, is S an affine subspace?
  (c) If S is an affine subspace, find the linear subspace W and a vector $\mathbf{b}$ such that S is the translate of W by $\mathbf{b}$.
  (i) $\{(x_1, x_2, x_3) \in \mathbb{R}^3: x_1 = x_2,\ x_3 = 0\}$
  (ii) $\{(x_1, x_2, x_3) \in \mathbb{R}^3: x_1 = x_2,\ x_3 = 17\}$
  (iii) $\{(x_1, x_2) \in \mathbb{R}^2: x_2 = x_1^3\}$
  (iv) $\{(x_1, x_2, x_3) \in \mathbb{R}^3: x_1 = 1\}$
  (v) $\{(x_1, x_2, x_3) \in \mathbb{R}^3: x_1 = 0\}$
  (vi) $\{(x_1, x_2, x_3) \in \mathbb{R}^3: x_3 = x_1 x_2\}$
  (vii) $\{(x_1, x_2, x_3) \in \mathbb{R}^3: x_1 + x_2 + x_3 = 1\}$
  (viii) $\{(x_1, x_2) \in \mathbb{R}^2: x_1 \text{ and } x_2 \text{ are integers}\}$
  (ix) $\{(x_1, x_2, x_3, x_4) \in \mathbb{R}^4: x_1 + x_3 = x_2 + x_4\}$
  (x) $\{(x_1, x_2, x_3) \in \mathbb{R}^3: x_1 = \sin x_3,\ x_2 = \cos x_3\}$
  (xi) $\{(x_1, x_2) \in \mathbb{R}^2: x_1 \text{ and } x_2 \text{ are rational numbers}\}$
  (xii) $\{(x_1, x_2, x_3, x_4, x_5) \in \mathbb{R}^5: x_3 = 3x_1 + 4x_2,\ x_4 = 0,\ x_5 = 2x_2\}$
  (xiii) $\{(x_1, x_2, x_3, x_4) \in \mathbb{R}^4: x_2 = x_1 + 2,\ x_3 = x_1 + 3\}$

*5.3. We have proved that an affine subspace $H \subset \mathbb{R}^n$ is the translate of a unique linear subspace $W \subset \mathbb{R}^n$, but that there can be different vectors $\mathbf{b}_1$ and $\mathbf{b}_2$ such that H is the translate of W by $\mathbf{b}_1$ and also by $\mathbf{b}_2$. Clarify this state of affairs by proving that if $\mathbf{b}$ is *any* vector in H then H is the translate of W by $\mathbf{b}$. Conclude that, except in the trivial case when $W = \{\mathbf{0}\}$, there are infinitely many vectors $\mathbf{b}$ such that H is the translate of W by $\mathbf{b}$.

5.4. (i) Let W_1 and W_2 be linear subspaces of $\mathbb{R}^n$. Prove that $W_1 \cap W_2$ is a linear subspace of $\mathbb{R}^n$.

(ii) Let $\{W_i : i \in I\}$ be an arbitrary collection of linear subspaces of $\mathbb{R}^n$. Prove that $\bigcap_{i \in I} W_i$ is a linear subspace of $\mathbb{R}^n$.

(iii) Let $\{H_i : i \in I\}$ be an arbitrary collection of affine subspaces of $\mathbb{R}^n$. Prove that $\bigcap_{i \in I} H_i$ is an affine subspace of $\mathbb{R}^n$.

Hint for (iii): Let W_i be the unique linear subspace of which H_i is a translate. Show that $\bigcap_{i \in I} H_i$ is the translate of $\bigcap_{i \in I} W_i$ by an arbitrary vector $\mathbf{b} \in \bigcap_{i \in I} H_i$. The result in Problem 5.3 is relevant.

6

Systems of Linear Equations

The most important application of the concepts developed so far is to the problem of solving systems of linear equations. Many physical problems give rise to systems of linear equations. In the present section, however, our only purpose is to describe the fundamental mathematical problem using the language of matrices and the concepts of linear and affine subspaces of $\mathbb{R}^n$.

6.1 Definition

A *linear equation in the n variables* $x_1, x_2, \ldots, x_n$ is an expression of the form $a_1x_1 + a_2x_2 + \cdots + a_nx_n = b$, where $a_1, a_2, \ldots, a_n$ and b are scalars. Such an equation is called *nontrivial* if $a_i \neq 0$ for at least one i.

6.2 Definition

A set of m linear equations in n variables

$$\begin{aligned} a_{11}x_1 + a_{12}x_2 + \cdots + a_{1n}x_n &= b_1 \\ a_{21}x_1 + a_{22}x_2 + \cdots + a_{2n}x_n &= b_2 \\ &\vdots \\ a_{m1}x_1 + a_{m2}x_2 + \cdots + a_{mn}x_n &= b_m \end{aligned}$$

will be called a *linear system* (or an $m \times n$ *linear system* if there is some reason to emphasize the numbers of equations and unknowns involved). If $b_1 = b_2 = \cdots = b_m = 0$, the system is called *homogeneous*. The set $\{\mathbf{u} = (u_1, u_2, \ldots, u_n) \in \mathbb{R}^n : a_{i1}u_1 + a_{i2}u_2 + \cdots + a_{in}u_n = b_i \text{ for } i = 1, 2, \ldots, m\}$ consisting of those n-tuples that simultaneously satisfy each of the m equations is called the *solution set* of the linear system.

6.3 Examples

(i) $$\begin{aligned} x_1 + 2x_2 &= 5 \\ 3x_1 + 4x_2 &= 6 \end{aligned}$$

(ii) $$\begin{aligned} 3x_1 - 7x_2 + 6x_3 - 6x_4 &= -3 \\ 2x_1 - 3x_2 - x_3 + x_4 &= 7 \\ x_1 - x_2 - 2x_3 + 2x_4 &= 5 \end{aligned}$$

Example 6.3(i) is a 2×2 linear system; 6.3(ii) is a 3×4 linear system. The vector $(\frac{58}{5}, \frac{27}{5}, 0, 0)$ is not in the solution set of 6.3(ii): though it satisfies the first two of the equations, it does not satisfy the third.

Using the $n \times 1$ matrix $\mathbf{x} = \begin{bmatrix} x_1 \\ x_2 \\ \vdots \\ x_n \end{bmatrix}$ and the $m \times 1$ matrix $\mathbf{b} = \begin{bmatrix} b_1 \\ b_2 \\ \vdots \\ b_m \end{bmatrix}$ together with the notion of matrix multiplication, the preceding linear system can be written as

$$\begin{bmatrix} a_{11} & a_{12} & \cdots & a_{1n} \\ a_{21} & a_{22} & \cdots & a_{2n} \\ \vdots & \vdots & & \vdots \\ a_{m1} & a_{m2} & \cdots & a_{mn} \end{bmatrix} \begin{bmatrix} x_1 \\ x_2 \\ \vdots \\ x_n \end{bmatrix} = \begin{bmatrix} b_1 \\ b_2 \\ \vdots \\ b_m \end{bmatrix}$$

In the next definition, we continue to refer to the linear system given in 6.2.

6.4 Definition

The $m \times n$ matrix

$$A = \begin{bmatrix} a_{11} & a_{12} & \cdots & a_{1n} \\ a_{21} & a_{22} & \cdots & a_{2n} \\ \vdots & \vdots & & \vdots \\ a_{m1} & a_{m2} & \cdots & a_{mn} \end{bmatrix}$$

is called the *coefficient matrix* of the system. The $m \times (n+1)$ matrix

$$\begin{bmatrix} a_{11} & a_{12} & \cdots & a_{1n} & b_1 \\ a_{21} & a_{22} & \cdots & a_{2n} & b_2 \\ \vdots & \vdots & & \vdots & \vdots \\ a_{m1} & a_{m2} & \cdots & a_{mn} & b_m \end{bmatrix}$$

is called the *augmented matrix* of the system.

With this notation, the linear system of m equations in n unknowns given in 6.2 can be compactly abbreviated by the single matrix equation $\mathbf{Ax} = \mathbf{b}$.

6.5 Examples

(i) $\begin{bmatrix} 2 & 3 \\ 4 & 5 \end{bmatrix}\begin{bmatrix} x_1 \\ x_2 \end{bmatrix} = \begin{bmatrix} 6 \\ 7 \end{bmatrix}$ abbreviates $\begin{aligned} 2x_1 + 3x_2 &= 6 \\ 4x_1 + 5x_2 &= 7 \end{aligned}$

(ii) $\begin{bmatrix} 3 & 0 & -2 \\ -4 & 3 & 1 \end{bmatrix}\begin{bmatrix} y_1 \\ y_2 \\ y_3 \end{bmatrix} = \begin{bmatrix} -3 \\ 2 \end{bmatrix}$ abbreviates

$$\begin{aligned} 3y_1 \qquad\; - 2y_3 &= -3 \\ 4y_1 + 3y_2 + \; y_3 &= 2 \end{aligned}$$

(iii) $\begin{bmatrix} -3 & 4 & 2 \\ 1 & -1 & 6 \\ -2 & 7 & 0 \end{bmatrix}\begin{bmatrix} z_1 \\ z_2 \\ z_3 \end{bmatrix} = \begin{bmatrix} 0 \\ 0 \\ 0 \end{bmatrix}$ abbreviates

$$\begin{aligned} -3z_1 + 4z_2 + 2z_3 &= 0 \\ z_1 - \;z_2 + 6z_3 &= 0 \\ -2z_1 + 7z_2 \qquad &= 0 \end{aligned}$$

In Sections 10, 11, and 12, we will develop a systematic approach to the problem of finding the solution set of a linear system. For the moment, we wish to consider some very general facts about such systems.

6.6 Definition

If $\mathbf{Ax} = \mathbf{b}$ is an arbitrary $m \times n$ linear system, the homogeneous system $\mathbf{Ax} = \mathbf{0}_{m\times 1}$ that has the same coefficient matrix is called the *homogeneous system associated with the system* $\mathbf{Ax} = \mathbf{b}$.

6.7 Examples

The linear systems from Example 6.3 have the following as their associated homogeneous systems:

(i) $\begin{aligned} x_1 + 2x_2 &= 0 \\ 3x_1 + 4x_2 &= 0 \end{aligned}$ (ii) $\begin{aligned} 3x_1 - 7x_2 + 6x_3 - 6x_4 &= 0 \\ 2x_1 - 3x_2 - \;x_3 + \;x_4 &= 0 \\ x_1 - \;x_2 - 2x_3 + 2x_4 &= 0 \end{aligned}$

6.8 Definition

An $m \times n$ linear system $\mathbf{Ax} = \mathbf{b}$ is called *inconsistent* if there is no vector $\mathbf{v} \in \mathbb{R}^n$ satisfying $\mathbf{Ax} = \mathbf{b}$. Otherwise, the system is called *consistent*, that is, if $\{\mathbf{v} \in \mathbb{R}^n : \mathbf{Av} = \mathbf{b}\} \neq \varnothing$.

Note that every homogeneous $m \times n$ linear system is consistent since $A\mathbf{0}_{n\times 1} = \mathbf{0}_{m\times 1}$ for any $m \times n$ matrix A. The vector $\mathbf{0}_{n\times 1}$ is called the *trivial solution* of the $m \times n$ homogeneous system $A\mathbf{x} = \mathbf{0}_{m\times 1}$.

6.9 Theorem

(i) The solution set $\{\mathbf{w} \in \mathbb{R}^n\colon A\mathbf{w} = \mathbf{0}_{m\times 1}\}$ of an $m \times n$ homogeneous linear system $A\mathbf{x} = \mathbf{0}_{m\times 1}$ is a linear subspace of $\mathbb{R}^n$.

(ii) If $A\mathbf{x} = \mathbf{b}$ is an arbitrary $m \times n$ linear system, then its solution set $\{\mathbf{v} \in \mathbb{R}^n\colon A\mathbf{v} = \mathbf{b}\}$ is an affine subspace of $\mathbb{R}^n$; moreover, when it is nonempty, the unique linear subspace of which this affine subspace is a translate (see Theorem 5.22) is just the solution set of the associated homogeneous system $A\mathbf{x} = \mathbf{0}_{m\times 1}$.

Proof of (i): We must show that the set $S = \{\mathbf{w} \in \mathbb{R}^n\colon A\mathbf{w} = \mathbf{0}_{m\times 1}\}$ is closed under vector addition and scalar multiplication. But this is trivial: if $\mathbf{u}_1 \in S$ and $\mathbf{u}_2 \in S$, then $A\mathbf{u}_1 = \mathbf{0}_{m\times 1}$ and $A\mathbf{u}_2 = \mathbf{0}_{m\times 1}$; but it then follows from 3.1(xi) that $A(\mathbf{u}_1 + \mathbf{u}_2) = A\mathbf{u}_1 + A\mathbf{u}_2 = \mathbf{0}_{m\times 1} + \mathbf{0}_{m\times 1} = \mathbf{0}_{m\times 1}$, so $\mathbf{u}_1 + \mathbf{u}_2 \in S$. Similarly, if $\mathbf{u} \in S$ and t is any scalar, then by 3.1(x) we have that $A(t\mathbf{u}) = t(A\mathbf{u}) = t\mathbf{0}_{m\times 1} = \mathbf{0}_{m\times 1}$, so $t\mathbf{u} \in S$.

Proof of (ii): If $A\mathbf{x} = \mathbf{b}$ is inconsistent, its solution set is the empty set which was classified as an affine subspace by definition. So suppose that $A\mathbf{x} = \mathbf{b}$ is consistent. We know by 6.9(i), which was just proved, that $S = \{\mathbf{w} \in \mathbb{R}^n\colon A\mathbf{w} = \mathbf{0}_{m\times 1}\}$ is a linear subspace of $\mathbb{R}^n$. So it remains to prove that the set $S' = \{\mathbf{v} \in \mathbb{R}^n\colon A\mathbf{v} = \mathbf{b}\}$ is a translate of S. Because $A\mathbf{x} = \mathbf{b}$ is consistent, S' is nonempty, so choose any $\mathbf{u} \in S'$. We claim that S' is the translate of S by $\mathbf{u}$ (i.e., $S' = \{\mathbf{w} + \mathbf{u}\colon \mathbf{w} \in S\}$). To see this, note that if $\mathbf{w} \in S$, then $A(\mathbf{w} + \mathbf{u}) = A\mathbf{w} + A\mathbf{u} = \mathbf{0}_{m\times 1} + \mathbf{b} = \mathbf{b}$, so $\mathbf{w} + \mathbf{u} \in S'$; conversely, if $\mathbf{v} \in S'$, then $\mathbf{v} - \mathbf{u} \in S$ since $A(\mathbf{v} - \mathbf{u}) = A\mathbf{v} - A\mathbf{u} = \mathbf{b} - \mathbf{b} = \mathbf{0}_{m\times 1}$. So the expression $\mathbf{v} = (\mathbf{v} - \mathbf{u}) + \mathbf{u}$ shows that $\mathbf{v}$ can be written as the sum of $\mathbf{u}$ plus a vector from S. ■

Because of Theorem 6.9, the set $\{\mathbf{v} \in \mathbb{R}^n\colon A\mathbf{v} = \mathbf{b}\}$, which was called the *solution set* of the system $A\mathbf{x} = \mathbf{b}$ in Definition 6.2, will from now on be called the *solution space* of the linear system $A\mathbf{x} = \mathbf{b}$.

6.10 Corollary

An arbitrary linear system $A\mathbf{x} = \mathbf{b}$ has either no solution, exactly one solution, or infinitely many solutions.

Proof: The solution space of $A\mathbf{x} = \mathbf{b}$ may be empty; if not, the solution space of the associated homogeneous system either consists of the single point $\{\mathbf{0}\}$, or contains some nonzero vector, in which case the solution space contains infinitely many points as was observed in 5.17(ii). ■

It is worth mentioning an important special case in which a linear system is guaranteed to have exactly one solution: this is the case when the system $Ax = b$ involves the same number of equations as unknowns and the coefficient matrix, A, is invertible.

6.11 Theorem

If the $n \times n$ matrix A is invertible, then for every vector $\mathbf{b} \in \mathbb{R}^n$ the $n \times n$ linear system $A\mathbf{x} = \mathbf{b}$ has the unique solution $A^{-1}\mathbf{b}$.

Proof: $A^{-1}\mathbf{b}$ is certainly a solution of the equation $A\mathbf{x} = \mathbf{b}$ since $A(A^{-1}\mathbf{b}) = (AA^{-1})\mathbf{b} = I_n\mathbf{b} = \mathbf{b}$. Let $\mathbf{c}$ be any (other?) solution of $A\mathbf{x} = \mathbf{b}$ (i.e., assume $A\mathbf{c} = \mathbf{b}$); then $\mathbf{c} = I_n\mathbf{c} = (A^{-1}A)\mathbf{c} = A^{-1}(A\mathbf{c}) = A^{-1}\mathbf{b}$. ■

The converse of 6.11 is also true; it will be proved later as part of Theorem 19.5.

6.12 Example

The coefficient matrix of the system from Example 6.3(i) is $A = \begin{bmatrix} 1 & 2 \\ 3 & 4 \end{bmatrix}$, which, according to the answer to Problem 3.6, is invertible; its inverse is

$$A^{-1} = \frac{1}{(1)(4) - (2)(3)} \begin{bmatrix} 4 & -2 \\ -3 & 1 \end{bmatrix} = \begin{bmatrix} -2 & 1 \\ 3/2 & -1/2 \end{bmatrix}$$

The unique solution to the system is therefore

$$A^{-1}\begin{bmatrix} 5 \\ 6 \end{bmatrix} = \begin{bmatrix} -2 & 1 \\ 3/2 & -1/2 \end{bmatrix}\begin{bmatrix} 5 \\ 6 \end{bmatrix} = \begin{bmatrix} -4 \\ 9/2 \end{bmatrix}$$

There are few fundamental ideas in linear algebra. The difficulty of the subject is that each of these ideas can be expressed in several very different but equivalent ways. The same mathematical object can have many aliases. It is the *variety* of vocabulary and context in which these few fundamental ideas reappear that causes bewilderment to some beginning students. But this is precisely the source of the subject's strength and utility; mastery of the subject involves the ability to exploit this variety by switching back and forth among the many equivalent points of view toward a problem.

We will close this section with two mathematical examples of this phenomenon. The first, in Definition 6.13, does not involve much contrast, as it results in a mere renaming; but the second, in Observation 6.14, relates two concepts which superficially appear quite distinct.

When we wish to think of the solution space of a homogeneous linear system as an object that is associated with the coefficient matrix, we give it another name.

6.13 Definition

If A is an $m \times n$ matrix, the set $\{\mathbf{v} \in \mathbb{R}^n : A\mathbf{v} = \mathbf{0}_{m\times 1}\}$ is called the *nullspace of A* and is denoted by NS(A).

We assume that you have encountered 2×2 and 3×3 systems in high school and that in learning to solve these special cases you have already developed a naive point of view toward the general problem of solving linear systems. As we attempted to stress in Section 4, the most useful point of view for intuitively understanding the qualitative aspects of the subject is the geometric one. To solve a system of m linear equations in n unknowns is, geometrically, to come up with a description of the common intersection of m hyperplanes in $\mathbb{R}^n$; if the system is homogeneous, we know that each of the m hyperplanes contains the origin. It is in the spirit of presenting equivalent points of view that we make the following observation that relates the material in Section 6 with that in Section 5.

6.14 Observation

A system of m linear equations in n unknowns

$$\begin{aligned} a_{11}x_1 + a_{12}x_2 + \cdots + a_{1n}x_n &= b_1 \\ a_{21}x_1 + a_{22}x_2 + \cdots + a_{2n}x_n &= b_n \\ &\vdots \\ a_{m1}x_1 + a_{m2}x_2 + \cdots + a_{mn}x_n &= b_m \end{aligned}$$

can be written as

$$x_1 \begin{bmatrix} a_{11} \\ a_{21} \\ \vdots \\ a_{m1} \end{bmatrix} + x_2 \begin{bmatrix} a_{12} \\ a_{22} \\ \vdots \\ a_{m2} \end{bmatrix} + \cdots + x_n \begin{bmatrix} a_{1n} \\ a_{2n} \\ \vdots \\ a_{mn} \end{bmatrix} = \begin{bmatrix} b_1 \\ b_2 \\ \vdots \\ b_m \end{bmatrix}$$

This trivial but essential observation leads to an alternative way to interpret what we are doing when we solve a linear system. A solution $\mathbf{u} = (u_1, u_2, \ldots, u_n)$ of the $m \times n$ linear system $A\mathbf{x} = \mathbf{b}$ consists of coefficients $u_1, u_2, \ldots, u_n$, which allow us to write the $m \times 1$ column vector

$$\mathbf{b} = \begin{bmatrix} b_1 \\ b_2 \\ \vdots \\ b_m \end{bmatrix}$$

as a linear combination

$$u_1 \begin{bmatrix} a_{11} \\ a_{21} \\ \vdots \\ a_{m1} \end{bmatrix} + u_2 \begin{bmatrix} a_{12} \\ a_{22} \\ \vdots \\ a_{m2} \end{bmatrix} + \cdots + u_n \begin{bmatrix} a_{1n} \\ a_{2n} \\ \vdots \\ a_{mn} \end{bmatrix}$$

of the columns of A.

6.15 Example

To say that the point $(\frac{17}{11}, \frac{2}{11})$ is a solution of the linear system

$$\begin{aligned} 3x_1 + 2x_2 &= 5 \\ 4x_1 - x_2 &= 6 \end{aligned}$$

is equivalent to saying that

$$\begin{bmatrix} 5 \\ 6 \end{bmatrix} = \frac{17}{11}\begin{bmatrix} 3 \\ 4 \end{bmatrix} + \frac{2}{11}\begin{bmatrix} 2 \\ -1 \end{bmatrix}$$

or that $(5, 6) = 17/11(3, 4) + 2/11(2, -1)$.

Problem Set 6

6.1. Go back and do any of the exercises in Section 6 that you may have skipped.

6.2. Turn to Section 12 and for each of the linear systems in Problems 12.2 through 12.23:
  (i) Write down the coefficient matrix.
  (ii) Write down the augmented matrix.
  (iii) Rewrite the system as a single matrix equation.

*6.3. Let

$$A = \begin{bmatrix} 2 & 0 & 4 & -3 & 6 \\ 1 & -2 & 9 & 1 & 4 \\ 0 & 3 & -4 & 1 & 0 \\ 4 & 1 & 7 & 7 & -3 \end{bmatrix}$$

(i) Write down the 4×5 homogeneous linear system whose coefficient matrix is A.

(ii) Write down the 4×4 linear system whose augmented matrix is A.

6.4. Repeat 6.3 with the matrix

$$A = \begin{bmatrix} 1 & -3 & 8 & 6 & 5 \\ 0 & 0 & 1 & 2 & -4 \\ 0 & 0 & 0 & 0 & 1 \\ 0 & 0 & 0 & 0 & 0 \end{bmatrix}$$

*6.5. Use your answer to Problem 3.6 together with Theorem 6.11 to solve the following 2×2 linear systems:

(i) $2x_1 + 3x_2 = 4$ $\quad$ (ii) $7x_1 + 6x_2 = 2$

$-4x_1 - 2x_2 = 8$ $\quad$ $9x_1 + 8x_2 = 2$

*6.6. Use your answers to problem 6.5:

(i) To express the vector $(4, 8)$ as a linear combination of the vectors $(2, -4)$ and $(3, -2)$.

(ii) To express the vector $(2, 2)$ as a linear combination of the vectors $(7, 9)$ and $(6, 8)$.

*6.7. Use the ideas involved in Problems 6.5 and 6.6 to express the vector $(6, 7)$ as a linear combination of the vectors $(3, 3)$ and $(6, 5)$.

6.8. Prove the following extension of Theorem 6.11. If the $n \times n$ matrix A is invertible, then for every $n \times p$ matrix B the matrix equation $AX = B$ has the unique solution $A^{-1}B$.

7

Linear Dependence, Linear Independence, and Bases

7.1 Definition

If W is a subspace of $\mathbb{R}^n$ and if S is a subset of W such that $\mathscr{L}(S) = W$, we say that S *spans* W.

It is important to distinguish between the use of the word *span* as a noun (in Definition 5.10) and as a verb (in Definition 7.1). When you do the next exercise, you will find that the phrase "S spans the span of S" actually makes sense.

7.2 Exercise

For any subset of S of $\mathbb{R}^n$, $\mathscr{L}(S)$ is a subspace of $\mathbb{R}^n$ and S spans $\mathscr{L}(S)$.

7.3 Definition

A subset S of $\mathbb{R}^n$ is *linearly dependent* iff for some positive integer m there exist distinct vectors $\mathbf{v}_1, \mathbf{v}_2, \ldots, \mathbf{v}_m$ in S and there exist scalars $t_1, t_2, \ldots, t_m$ with $t_k \neq 0$ for at least one k, such that $\mathbf{0} = t_1\mathbf{v}_1 + t_2\mathbf{v}_2 + \cdots + t_m\mathbf{v}_m$.

The content of this definition is usually expressed in words as follows: a subset S is linearly dependent iff it is possible to write the zero vector as a nontrivial linear combination of vectors from S.

7.4 Exercise

If $\mathbf{0} \in S$, then S is linearly dependent.

7.5 Caution

The statement $\mathbf{0} \in S$ (which may or may not be true for a given set S) must not be confused with the statement $\mathbf{0} \in \mathscr{L}(S)$ (which is always true for any set S).

7.6 Definition

A subset S of $\mathbb{R}^n$ is called *linearly independent* iff it is not linearly dependent.

7.7 Exercise

A subset S of $\mathbb{R}^n$ is linearly independent iff for any distinct vectors $\mathbf{v}_1, \mathbf{v}_2, \ldots, \mathbf{v}_m \in S$, if $\mathbf{0} = t_1\mathbf{v}_1 + t_2\mathbf{v}_2 + \cdots + t_m\mathbf{v}_m$, then $t_1 = t_2 = \cdots = t_m = 0$.

In words, 7.7 asserts that a set S is linearly independent iff the only way to write the zero vector as a linear combination of vectors from S is the trivial way, in which each of the coefficients is 0.

7.8 Exercise

(i) If S is linearly independent and $S' \subset S$, then S' is linearly independent.
(ii) If S is linearly dependent and $S \subset S'$, then S' is linearly dependent.

The concepts of linear dependence and independence are easily understood for sets containing very few vectors. In the next two exercises we dispense with the cases where the cardinality of S is 0, 1, 2, or 3; the general situation is presented in Theorem 7.11.

7.9 Exercise

The empty set, $\varnothing$, is linearly independent.

7.10 Exercise

(i) If $S = \{\mathbf{u}\}$ (i.e., if S consists of the single vector $\mathbf{u}$), then S is linearly dependent iff $\mathbf{u} = \mathbf{0}$.

(ii) If $S = \{\mathbf{u}_1, \mathbf{u}_2\}$ (i.e., if S consists of just the two vectors $\mathbf{u}_1$ and $\mathbf{u}_2$), then S is linearly dependent iff $\mathbf{u}_1 = \mathbf{0}$, or $\mathbf{u}_2 = \mathbf{0}$, or $\mathbf{u}_1 = t\mathbf{u}_2$ for some scalar $t \neq 0$ (in which case $\mathbf{u}_2 = s\mathbf{u}_1$ where $s = 1/t$).

(iii) If $S = \{\mathbf{u}_1, \mathbf{u}_2, \mathbf{u}_3\}$ with $\mathbf{u}_1 \neq \mathbf{0}$, $\mathbf{u}_2 \neq \mathbf{0}$, and $\mathbf{u}_3 \neq \mathbf{0}$, then S is linearly dependent iff there exist scalars a_2 and a_3 such that $\mathbf{u}_1 = a_2\mathbf{u}_2 + a_3\mathbf{u}_3$, or there exist scalars b_1 and b_3 such that $\mathbf{u}_2 = b_1\mathbf{u}_1 + b_3\mathbf{u}_3$, or there exist scalars c_1 and c_2 such that $\mathbf{u}_3 = c_1\mathbf{u}_1 + c_2\mathbf{u}_2$.

Theorem 7.11 makes precise the general fact that is suggested by Exercise 7.10.

A linearly dependent set, S, is one that contains a member $\mathbf{u} \in S$ that, colloquially, "depends on its fellow members, $S - \{\mathbf{u}\}$." In other words, S contains a vector, $\mathbf{u}$, that can be written as a linear combination of other vectors in S. The effect of this is that, when such a vector $\mathbf{u}$ is removed from S, the span, $\mathscr{L}(S - \{\mathbf{u}\})$, of the vectors that remain when $\mathbf{u}$ is removed from S is still equal to the span of the original set S. This is because any linear combination of vectors from S that specifically involves $\mathbf{u}$ can be replaced by an equivalent linear combination that does not involve $\mathbf{u}$. This simple idea is all that is needed to prove the following theorem.

7.11 Theorem

A subset S of $\mathbb{R}^n$ is linearly dependent iff for some vector $\mathbf{u} \in S$, $\mathscr{L}(S - \{\mathbf{u}\}) = \mathscr{L}(S)$.

Proof: ($\rightarrow$) If S is linearly dependent, choose distinct vectors $\mathbf{v}_1, \mathbf{v}_2, \ldots, \mathbf{v}_m \in S$ and scalars $t_1, t_2, \ldots, t_m$ such that $\mathbf{0} = t_1\mathbf{v}_1 + t_2\mathbf{v}_2 + \cdots + t_m\mathbf{v}_m$ is a nontrivial way of expressing the zero vector. Suppose $t_k \neq 0$; this implies that

$$\mathbf{v}_k = \frac{-t_1}{t_k}\mathbf{v}_1 + \cdots + \frac{-t_{k-1}}{t_k}\mathbf{v}_{k-1} + \frac{-t_{k+1}}{t_k}\mathbf{v}_{k+1} + \cdots + \frac{-t_m}{t_k}\mathbf{v}_m$$

We will now show that $\mathscr{L}(S - \{\mathbf{v}_k\}) = \mathscr{L}(S)$. Because $S - \{\mathbf{v}_k\} \subset S$, the inclusion $\mathscr{L}(S - \{\mathbf{v}_k\}) \subset \mathscr{L}(S)$ is trivial, as was observed in 5.11. To show $\mathscr{L}(S) \subset \mathscr{L}(S - \{\mathbf{v}_k\})$, let $\mathbf{x} \in \mathscr{L}(S)$; then there exist distinct vectors $\mathbf{u}_1, \mathbf{u}_2, \ldots, \mathbf{u}_p \in S$ and scalars $s_1, s_2, \ldots, s_p$ such that $\mathbf{x} = s_1\mathbf{u}_1 + s_2\mathbf{u}_2 + \cdots + s_p\mathbf{u}_p$. If each $\mathbf{u}_j \in S - \{\mathbf{v}_k\}$, the proof is complete. If not, then for some unique j, $\mathbf{u}_j = \mathbf{v}_k$. But then, we can write

$$\mathbf{x} = s_1\mathbf{u}_1 + \cdots + s_{j-1}\mathbf{u}_{j-1}$$
$$+ s_j\left(\frac{-t_1}{t_k}\mathbf{v}_1 + \cdots + \frac{-t_{k-1}}{t_k}\mathbf{v}_{k-1} + \frac{-t_{k+1}}{t_k}\mathbf{v}_{k+1} + \cdots + \frac{-t_m}{t_k}\mathbf{v}_m\right)$$
$$+ s_{j+1}\mathbf{u}_{j+1} + \cdots + s_p\mathbf{u}_p$$

which shows that $\mathbf{x} \in \mathscr{L}(S - \{\mathbf{v}_k\})$.

($\leftarrow$) Let $\mathbf{u} \in S$ be a vector such that $\mathscr{L}(S - \{\mathbf{u}\}) = \mathscr{L}(S)$. If $\mathbf{u} = \mathbf{0}$, then S is linearly dependent by 7.4. If $\mathbf{u} \neq \mathbf{0}$, then $S - \{\mathbf{u}\} \neq \varnothing$ because $\mathscr{L}(\varnothing) = \{\mathbf{0}\}$, whereas $\mathbf{0} \neq \mathbf{u} \in \mathscr{L}(S) = \mathscr{L}(S - \{\mathbf{u}\})$. Thus, there exist distinct vectors $\mathbf{w}_1, \mathbf{w}_2, \ldots, \mathbf{w}_m \in S - \{\mathbf{u}\}$ and scalars $r_1, r_2, \ldots, r_m$ such that $\mathbf{u} = r_1\mathbf{w}_1 + r_2\mathbf{w}_2 + \cdots + r_m\mathbf{w}_m$. Thus S is linearly dependent since $\mathbf{0} = -\mathbf{u} + r_1\mathbf{w}_1 + r_2\mathbf{w}_2 + \cdots + r_m\mathbf{w}_m$ is a nontrivial (because the coefficient of $\mathbf{u}$ is -1) way to write $\mathbf{0}$ as a linear combination of vectors from S. ■

7.12 Corollary to the Proof of 7.11

For any linearly dependent subset S of $\mathbb{R}^n$ and any vector $\mathbf{u} \in S$ such that $\mathbf{u} \in \mathscr{L}(S - \{\mathbf{u}\})$, $\mathscr{L}(S - \{\mathbf{u}\}) = \mathscr{L}(S)$. ■

7.13 Exercise

(i) Show that $S' = \{(-4, 2, -1, -5, -2), (-3, 0, -2, 1, 4), (2, 1, 3, -6, -4), (-5, 1, -5, 5, -4)\}$ is linearly dependent by verifying that $(-5, 1, -5, 5, -4) = 2(-4, 2, -1, -5, -2) - 3(-3, 0, -2, 1, 4) - 3(2, 1, 3, -6, -4)$.

(ii) Show that $S'' = \{(2, -1, 4), (5, 8, 1), (-1, 4, -5)\}$ is linearly dependent by verifying that $(5, 8, 1) = 4(2, -1, 4) + 3(-1, 4, -5)$.

7.14 Example

If S' and S'' are the sets defined in Exercise 7.13, then $\mathscr{L}(S') = \mathscr{L}(\{(-4, 2, -1, -5, -2), (-3, 0, -2, 1, 4), (2, 1, 3, -6, -4)\})$ and $\mathscr{L}(S'') = \mathscr{L}(\{(2, -1, 4), (-1, 4, -5)\})$.

It is clear that if S is linearly independent, and if a strictly larger set $S' = S \cup \{\mathbf{u}\}$ is produced by adding to S a vector $\mathbf{u}$ that is not in S but is in $\mathscr{L}(S)$, then S' is linearly dependent because $\mathbf{u} \in \mathscr{L}(S' - \{\mathbf{u}\})$. The content of the next theorem is that adding to a linearly independent set S a vector $\mathbf{u}$ that is outside the span of S produces a strictly larger set $S \cup \{\mathbf{u}\}$ that is *still* linearly independent.

7.15 Theorem

If the subset S of $\mathbb{R}^n$ is linearly independent and if $\mathbf{u} \notin \mathscr{L}(S)$, then $S \cup \{\mathbf{u}\}$ is linearly independent.

Proof: We will prove that the only way to write $\mathbf{0}$ as a linear combination of vectors from $S \cup \{\mathbf{u}\}$ is the trivial way. So assume that $\mathbf{0} = t_1\mathbf{v}_1 + \cdots + t_p\mathbf{v}_p + s\mathbf{u}$, where $\{\mathbf{v}_1, \mathbf{v}_2, \ldots, \mathbf{v}_p\} \subset S$. Because $\mathbf{u} \notin \mathscr{L}(S)$, we must have $s = 0$; for if not, then $\mathbf{u} = (-t_1/s)\mathbf{v}_1 + \cdots + (-t_p/s)\mathbf{v}_p$, contradicting $\mathbf{u} \notin \mathscr{L}(S)$. Now that we know $s = 0$, we have $\mathbf{0} = t_1\mathbf{v}_1 + t_2\mathbf{v}_2 + \cdots + t_p\mathbf{v}_p$, and because S is linearly independent, we conclude from 7.7 that $t_1 = t_2 = \cdots = t_p = 0$ as well. ■

It is clear that the linear dependence or independence of an arbitrary subset S of $\mathbb{R}^n$ is not related to any particular ordering of the elements of S. However, we wish to turn our attention to ordered sets of distinct vectors. We will retain the customary notation, writing $S = \{\mathbf{u}_1, \mathbf{u}_2, \ldots, \mathbf{u}_m\}$ for the unordered set S, and $S = (\mathbf{u}_{j_1}, \mathbf{u}_{j_2}, \ldots, \mathbf{u}_{j_m})$ for the same set S in which $\mathbf{u}_{j_1}$ is identified as the first element of S, $\mathbf{u}_{j_2}$ as the second, and so on. Here $j_1, j_2, \ldots, j_m$ is some arbitrary permutation of the integers $1, 2, \ldots, m$. The reason for considering ordered sets of vectors is this: the ordering of the elements of S has no effect on whether S is linearly dependent; but paying attention to the ordering does, in the case when S is linearly dependent, provide us with some *additional* information. If the unordered set $S = \{\mathbf{u}_1, \mathbf{u}_2, \ldots, \mathbf{u}_m\}$ is linearly dependent, we know only that for some i, $1 \le i \le m$, $\mathbf{u}_i \in \mathscr{L}(S - \{\mathbf{u}_i\})$, but we cannot be more precise about how $\mathbf{u}_i$ depends on $S - \{\mathbf{u}_i\}$; if the ordered set $S = (\mathbf{u}_1, \mathbf{u}_2, \ldots, \mathbf{u}_m)$ is linearly dependent, then either $\mathbf{u}_1 = \mathbf{0}$ or for some i, $2 \le i \le m$, $\mathbf{u}_i$ is a linear combination of its predecessors in S. The next theorem gives a precise statement of this result.

7.16 Theorem

An ordered set $S = (\mathbf{u}_1, \mathbf{u}_2, \ldots, \mathbf{u}_m)$ is linearly dependent iff $\mathbf{u}_1 = \mathbf{0}$ or for some $i \ge 2$, $\mathbf{u}_i \in \mathscr{L}(\mathbf{u}_1, \mathbf{u}_2, \ldots, \mathbf{u}_{i-1})$.

Proof: ($\leftarrow$) This direction is a trivial consequence of 7.4 and 7.11.

($\rightarrow$) By applying 7.3 to the linearly dependent set $\{\mathbf{u}_1, \mathbf{u}_2, \ldots, \mathbf{u}_m\}$, we find scalars $t_1, t_2, \ldots, t_m$ with $t_k \ne 0$ for at least one k, such that $\mathbf{0} = t_1\mathbf{u}_1 + t_2\mathbf{u}_2 + \cdots + t_m\mathbf{u}_m$. Let i be the largest integer such that $t_i \ne 0$.

Case 1. $i = 1$. Then $t_2 = \cdots = t_m = 0$ and $\mathbf{0} = t_1\mathbf{u}_1$ with $t_1 \neq 0$; hence $\mathbf{u}_1 = \mathbf{0}$.

Case 2. $i \geq 2$. Then $t_{i+1} = \cdots = t_m = 0$ and $\mathbf{0} = t_1\mathbf{u}_1 + \cdots + t_i\mathbf{u}_i$. Hence $\mathbf{u}_i = (-t_1/t_i)\mathbf{u}_1 + \cdots + (-t_{i-1}/t_i)\mathbf{u}_{i-1}$. ■

7.17 Exercise

If the subset S of $\mathbb{R}^n$ is linearly dependent and $\mathbf{0} \notin S$, then no matter how the elements of S are permuted, there will always be some vector that is a linear combination of its predecessors.

7.18 Exercise

An ordered set $S = (\mathbf{u}_1, \mathbf{u}_2, \ldots, \mathbf{u}_m)$ with $\mathbf{0} \notin S$ is linearly independent iff no element of S is a linear combination of its predecessors.

7.19 Exercise

An arbitrary set $S = \{\mathbf{u}_1, \mathbf{u}_2, \ldots, \mathbf{u}_m\}$ with $\mathbf{0} \notin S$ is linearly independent iff S can be ordered in such a way that no element is a linear combination of its predecessors.

The next concept is one of the most important in linear algebra.

7.20 Definition

Let W be a subspace of $\mathbb{R}^n$. A subset S of W is called a *basis* for W if S spans W and S is linearly independent.

Note that we only talk about bases for (linear) subspaces of $\mathbb{R}^n$ and not for arbitrary subsets of $\mathbb{R}^n$. The reason for this is inherent in the definition: if S is a basis for the sub*set* H, then S spans H [i.e., $H = \mathscr{L}(S)$], so by 7.2 H is a sub*space*.

7.21 Example

Let $W = \mathscr{L}(S'')$, where S'' is the set defined in Exercise 7.13(ii). Recall that $S'' = \{(2,-1,4), (5,8,1), (-1,4,-5)\}$. In 7.13 we saw that $(5,8,1) = 4(2,-1,4) + 3(-1,4,-5)$. We claim that the set $S = \{(2,-1,4), (-1,4,-5)\}$ is a basis for W: S is linearly independent by 7.10(ii) and S spans $W = \mathscr{L}(S'')$ by 7.12.

It is clear that whether a subset S of a subspace W is a basis for W or not is unrelated to any ordering of the elements of S. But our experience so far suggests that certain additional features of a basis might be related to an ordering of its elements. This is indeed the case, and the custom in linear algebra is to work mostly with ordered bases rather than with arbitrary unordered bases. The distinction is usually not made explicitly but is determined by context: it will be obvious which properties of a basis are related to its ordering and which are not; when the property under discussion is one that does, it is usually assumed without explicit mention that the basis or bases are ordered.

7.22 Notation

$\mathbf{e}_j^m$ denotes the m-tuple whose j^{th} coordinate equals 1 and whose other coordinates are all equal to 0. That is,

$$\mathbf{e}_j^m = \underbrace{(0,\ldots,0,1,0,\ldots,0)}_{j\text{th coordinate}}$$

α^m denotes either the ordered set $(\mathbf{e}_1^m, \mathbf{e}_2^m, \ldots, \mathbf{e}_m^m)$ or the unordered set $\{\mathbf{e}_1^m, \mathbf{e}_2^m, \ldots, \mathbf{e}_m^m\}$. For example, $\mathbf{e}_2^5 = (0,1,0,0,0)$, $\mathbf{e}_1^3 = (1,0,0)$, and $\mathbf{e}_4^4 = (0,0,0,1)$. $\alpha^2 = ((1,0),(0,1))$, $\alpha^3 = ((1,0,0),(0,1,0),(0,0,1))$, and so on.

7.23 Theorem

α^n is a basis for $\mathbb{R}^n$.

Proof: α^n spans $\mathbb{R}^n$, for if $\mathbf{u} = (u_1, u_2, \ldots, u_n)$ is an arbitrary vector in $\mathbb{R}^n$, then

$$\begin{aligned}\mathbf{u} &= u_1(1,0,\ldots,0) + u_2(0,1,0,\ldots,0) + \cdots + u_n(0,\ldots,0,1)\\ &= u_1\mathbf{e}_1^n + u_2\mathbf{e}_2^n + \cdots + u_n\mathbf{e}_n^n.\end{aligned}$$

α^n is linearly independent, for if

$$\mathbf{0} = (0,0,\ldots,0) = t_1(1,0,\ldots,0) + t_2(0,1,0,\ldots,0) + \cdots + t_n(0,\ldots,0,1)$$

then by equating the first coordinates of each side of this equation, we conclude $t_1 = 0$; by equating second coordinates, we conclude $t_2 = 0$. We eventually conclude $t_1 = t_2 = \cdots = t_n = 0$; thus α^n is linearly independent by 7.7. ■

7.24 Definition

α^n is called the *standard basis* for $\mathbb{R}^n$.

A slight digression is appropriate at this point. The preceding proof that α^n is linearly independent was so trivial that we wish to exploit it further. We wish to argue that sets of vectors having a certain property are "obviously" linearly independent. To state this property, we need some additional notation.

7.25 Notation

If $\mathbf{u} = (u_1, u_2, \ldots, u_n)$ is an n-tuple, and $P = \{i_1, i_2, \ldots, i_k\}$ is a k-element subset of $\{1, 2, \ldots, n\}$, let $\mathbf{u} \restriction P = (u_{i_1}, u_{i_2}, \ldots, u_{i_k})$ be the k-tuple obtained from $\mathbf{u}$ by retaining the jth coordinates of $\mathbf{u}$ for all $j \in P$. The symbol $\mathbf{u} \restriction P$ should be read "$\mathbf{u}$ restricted to P." For example, if $\mathbf{u} = (3, 8, -9, 7, -12)$, $\mathbf{v} = (4, -5, 1, 16, -8, 7, -1)$, $P = \{2, 4, 5\}$, and $Q = \{1, 2, 6\}$, then $\mathbf{u} \restriction P = (8, 7, -12)$, $\mathbf{v} \restriction P = (-5, 16, -8)$, $\mathbf{u} \restriction Q$ is undefined, and $\mathbf{v} \restriction Q = (4, -5, 7)$.

If $S \subset \mathbb{R}^n$ and $P \subset \{1, 2, \ldots, n\}$, let $S \restriction P = \{\mathbf{u} \restriction P\colon \mathbf{u} \in S\}$. In other words, $S \restriction P$ is the set of all restrictions to P of vectors in S.

For example, if $S_1 = \{(3, 0, 2, -9), (4, -1, 6, 8), (1, 1, -7, 0)\}$, $S_2 = \{(0, 1, -4, 0), (-9, 0, 6, 1), (2, 1, -1, 0)\}$, $S_3 = \{(5, 1, 2, 0), (-1, 0, 3, 1)\}$, and $P = \{2, 4\}$, then $S_1 \restriction P = \{(0, -9), (-1, 8), (1, 0)\}$, $S_2 \restriction P = \{(1, 0), (0, 1)\}$, and $S_3 \restriction P = \{(1, 0), (0, 1)\}$.

Note that S_3 is linearly independent, for if $\mathbf{0} = (0, 0, 0, 0) = t_1(5, 1, 2, 0) + t_2(-1, 0, 3, 1)$, then, by equating the second coordinates of both sides, we conclude $0 = (t_1)(1) + (t_2)(0) = t_1$, and by equating the fourth coordinates, we conclude $0 = (t_1)(0) + (t_2)(1) = t_2$.

We are now in position to state the property we anticipated earlier.

7.26 Definition

A set of k-many n-tuples $S = \{\mathbf{u}_1, \mathbf{u}_2, \ldots, \mathbf{u}_k\} \subset \mathbb{R}^n$ is called *reduced* if there exists a k-element subset $\mathrm{P} = \{i_1, \ldots, i_k\}$ of $\{1, 2, \ldots, n\}$ such that $S \restriction \mathrm{P} = \alpha^k$.

For example, the set S_3 just mentioned is reduced since $S_3 \restriction \{2, 4\} = \{(1, 0), (0, 1)\} = \alpha^2$. S_1 and S_2 are not reduced since there is no three-element subset P of $\{1, 2, 3, 4\}$ such that $S_1 \restriction \mathrm{P} = \alpha^3$ or $S_2 \restriction \mathrm{P} = \alpha^3$.

7.27 Theorem

If $S = \{\mathbf{u}_1, \mathbf{u}_2, \ldots, \mathbf{u}_k\} \subset \mathbb{R}^n$ is a reduced set of n-tuples, then S is linearly independent.

Proof: Let $P = \{i_1, i_2, \ldots, i_k\}$ be a k-element subset of $\{1, 2, \ldots, n\}$ such that $S \upharpoonright P = \alpha^k = \{\mathbf{e}_1^k, \mathbf{e}_2^k, \ldots, \mathbf{e}_k^k\}$. Suppose that $\mathbf{0} = t_1\mathbf{u}_1 + t_2\mathbf{u}_2 + \cdots + t_k\mathbf{u}_k$. By equating the i_1-coordinates of both sides, we conclude $0 = (t_1)(1) + (t_2)(0) + \cdots + (t_k)(0) = t_1$; by equating the i_2-coordinates of both sides, we conclude $0 = (t_1)(0) + (t_2)(1) + (t_3)(0) + \cdots + (t_k)(0) = t_2$; we eventually conclude $t_1 = t_2 = \cdots = t_k = 0$. Thus S is linearly independent by 7.7. ∎

This theorem will nearly always be applied "by inspection" so to speak, as in the next example.

7.28 Example

The set $S = \{(1, 4, 0, 0, -6), (0, -1, 1, 0, 0), (0, 0, 0, 1, -5)\}$ is linearly independent by Theorem 7.27 since S is a set of three vectors from $\mathbb{R}^5$ such that $S \upharpoonright \{1, 3, 4\} = \alpha^3$.

The next theorem is perhaps the most important one in the book. It is the foundation for the concept of dimension (to follow in Section 8). The guts of the theory of linear algebra are in its proof. The proof is not hard (partly because the groundwork has been carefully laid in 7.12 and 7.16), and we give it in full detail.

7.29 Theorem

(i) If W is a subspace of $\mathbb{R}^n$ and if $S = \{\mathbf{u}_1, \mathbf{u}_2, \ldots, \mathbf{u}_p\}$ spans W, then any set $S' = \{\mathbf{v}_1, \mathbf{v}_2, \ldots, \mathbf{v}_q\} \subset W$ with $p < q$ is linearly dependent.

(ii) If W is a subspace of $\mathbb{R}^n$ and if $S' = \{\mathbf{v}_1, \mathbf{v}_2, \ldots, \mathbf{v}_q\} \subset W$ is linearly independent, then any set $S = \{\mathbf{u}_1, \mathbf{u}_2, \ldots, \mathbf{u}_p\}$ with $p < q$ does not span W.

Before proving 7.29, we would like to rephrase it in words. Part (i) says that, if a subspace W can be spanned by a set of p-many elements, then any subset of W containing strictly *more* than p-many elements is linearly dependent. Part (ii) says that, if a subspace W includes a linearly independent subset with q-many elements, then no subset of W with strictly *fewer* than q-many elements can span W.

It may not be obvious from these rephrasings, but (ii) is just a logically equivalent form of (i): where (i) has the form P implies Q, (ii) is simply not Q implies not P. Thus it suffices to prove (i).

Proof of (i): Assume that $S = \{\mathbf{u}_1, \mathbf{u}_2, \ldots, \mathbf{u}_p\}$ spans W and that $S' = \{\mathbf{v}_1, \mathbf{v}_2, \ldots, \mathbf{v}_q\} \subset W$ with $p < q$. Note that because $p < q$ it makes sense to talk about $\mathbf{v}_{p+1}$. We will conclude that S' is linearly dependent by obtaining a contradiction from the assumption that S' is linearly independent. Since $\mathbf{v}_1 \in W = \mathscr{L}(S)$, the ordered set $Q_1 = (\mathbf{v}_1, \mathbf{u}_1, \mathbf{u}_2, \ldots, \mathbf{u}_p)$ also spans W and is linearly dependent since $\mathbf{v}_1 \in \mathscr{L}(\mathbf{u}_1, \mathbf{u}_2, \ldots, \mathbf{u}_p) = W$. Also, $\mathbf{v}_1 \neq \mathbf{0}$ since we are assuming that S' is linearly independent. So by 7.16 there exists a k such that $\mathbf{u}_k \in \mathscr{L}(\mathbf{v}_1, \mathbf{u}_1, \ldots, \mathbf{u}_{k-1})$, and by 7.12 we have that $Q_1' = (\mathbf{v}_1, \mathbf{u}_1, \ldots, \mathbf{u}_{k-1}, \mathbf{u}_{k+1}, \ldots, \mathbf{u}_p)$ still spans W.

We now reapply this same argument to the ordered set $Q_2 = (\mathbf{v}_2, \mathbf{v}_1, \mathbf{u}_1, \mathbf{u}_2, \ldots, \mathbf{u}_{k-1}, \mathbf{u}_{k+1}, \ldots, \mathbf{u}_p)$ that is obtained from Q_1' in the same way that Q_1 was obtained from S. Since Q_1' spans W and $\mathbf{v}_2 \in W$, the ordered set Q_2 also spans W and is linearly dependent since $\mathbf{v}_2 \in W = \mathscr{L}(Q_1')$. Since $\mathbf{v}_2 \neq \mathbf{0}$, Theorem 7.16 implies that some vector $\mathbf{w}$ in Q_2 is a linear combination of its predecessors. Because S' is linearly independent, it must be that $\mathbf{w}$ is one of the remaining $\mathbf{u}$'s in Q_2. Thus there is some ordered set

$$Q_2' = (\mathbf{v}_2, \mathbf{v}_1, \mathbf{u}_1, \ldots, \mathbf{u}_{k-1}, \mathbf{u}_{k+1}, \ldots, \mathbf{u}_{j-1}, \mathbf{u}_{j+1}, \ldots, \mathbf{u}_p)$$

that still spans W.

Repeat this argument p-many times; that is, at the ith stage, we obtain Q_i from Q_{i-1}' by adding $\mathbf{v}_i$ to it as a new first element. Each time, the vector that is deleted from Q_i to produce Q_i' (because it is a linear combination of its predecessors) must be one of the remaining $\mathbf{u}$'s since S' is linearly independent.

The pth argument results in deleting the last of the $\mathbf{u}$'s, and the conclusion of the pth argument is that the set $Q_p' = (\mathbf{v}_p, \mathbf{v}_{p-1}, \ldots, \mathbf{v}_1)$ still spans W. But this implies that $\mathbf{v}_{p+1} \in W = \mathscr{L}(Q_p')$, contradicting the assumed linear independence of S'. ■

7.30 Corollary

(i) Any set S of more than n-many n-tuples is linearly dependent.
(ii) Any set S of fewer than n-many n-tuples does not span $\mathbb{R}^n$.

Proof: This is immediate from 7.23 and 7.29 since the n-element set $\alpha^n = (\mathbf{e}_1^n, \mathbf{e}_2^n, \ldots, \mathbf{e}_n^n)$ both spans $\mathbb{R}^n$ and is linearly independent. ■

7.31 Examples

$S = \{(4, 5, -2), (6, -1, 0), (-2, 0, 8), (-1, 3, 7)\}$ is linearly dependent because it is a set of four 3-tuples. $S' = \{(3, 8, 0, -4, 2),\ (4, -1, 8, 6, 1), (7, 3, -2, -3, 0), (-3, 2, 6, -1, -3)\}$ does not span $\mathbb{R}^5$ since it is a set of four 5-tuples.

The point of the last theorem in this section is that we can always find a basis for any subspace W of $\mathbb{R}^n$.

7.32 Theorem

If W is a subspace of $\mathbb{R}^n$, then W has a basis; moreover, this basis has cardinality less than or equal to n.

Proof: If W is the trivial subspace, $\{\mathbf{0}\}$, then the empty set, $\varnothing$, is a basis for W according to 5.10 (ii) and 7.9. If W is nontrivial, let $\mathbf{u}_1 \neq \mathbf{0}$ be any nonzero vector in W. Then $S_1 = \{\mathbf{u}_1\}$ is linearly independent; so if S_1 also spans W, we are done. If S_1 does not span W, choose any vector $\mathbf{u}_2 \in W$ such that $\mathbf{u}_2 \notin \mathscr{L}(S_1)$. Then by 7.15, $S_2 = \{\mathbf{u}_1, \mathbf{u}_2\}$ is linearly independent; so if S_2 spans W, we are done. If not, continue this process, producing at the kth stage a linearly independent set S_k of cardinality k. This process must come to a halt at the kth stage for some $k \leq n$, producing a basis S_k for W [otherwise, at stage $n + 1$, S_{n+1} would be a linearly independent set of more than n-many n-tuples, contradicting 7.29(i)]. ■

Problem Set 7

7.1. Go back and do any of the exercises from Section 7 that you may have skipped.

*7.2. Decide by inspection whether the following sets are linearly dependent or independent. Justify your answers.

(i) $\{(3,0),(0,-5)\}$

(ii) $\{(2,1,6),(0,0,0),(8,-1,4)\}$

(iii) $\{(2,-5),(-6,15)\}$

(iv) $\{(0,1,-3,0),(0,0,-2,1),(1,0,-1,0)\}$

(v) $\{(0,1,-3,0),(0,0,-2,1),(1,0,-1,0),(2,0,1,0),(0,-1,0,2)\}$

(vi) $\{(1,-1)\}$

*7.3. Let $\{\mathbf{u}_1, \mathbf{u}_2, \ldots, \mathbf{u}_m\} \subset \mathbb{R}^n$ be linearly independent and let A be any invertible $n \times n$ matrix. Prove that $\{\mathrm{A}\mathbf{u}_1, \mathrm{A}\mathbf{u}_2, \ldots, \mathrm{A}\mathbf{u}_m\}$ is also linearly independent.

8

Dimension

8.1 Theorem

If W is a subspace of $\mathbb{R}^n$ and if $S = \{\mathbf{u}_1, \mathbf{u}_2, \ldots, \mathbf{u}_p\}$ and $S' = \{\mathbf{v}_1, \mathbf{v}_2, \ldots, \mathbf{v}_q\}$ are any two bases for W, then $p = q$.

Proof: Because S spans W and S' is linearly independent, 7.29(i) implies that $q \leq p$. Because S' spans W and S is linearly independent, 7.29(i) implies that $p \leq q$. ■

Theorem 8.1 provides the justification for talking about *the* number of elements in a basis for a subspace W of $\mathbb{R}^n$ and for introducing the next definition.

8.2 Definition

The number of elements in a basis for a subspace W of $\mathbb{R}^n$ is called the *dimension* of W and is denoted by $\dim W$.

In the proof of 7.32 we observed that the trivial subspace, $\{\mathbf{0}\}$, of $\mathbb{R}^n$ has the empty set, $\varnothing$, as a basis; thus the trivial subspace, $\{\mathbf{0}\}$, has dimension 0. Of course, $\dim \mathbb{R}^n = n$ since, as we have seen, α^n is a basis for $\mathbb{R}^n$; and if W is any subspace of $\mathbb{R}^n$, then $0 \leq \dim W \leq n$.

Strictly speaking, the concept of dimension has only been defined for linear subspaces of $\mathbb{R}^n$. But we can usefully extend the scope of the definition to affine subspaces of $\mathbb{R}^n$. We proved in 5.22 that a nonempty affine subspace is a translate of a unique linear subspace. Thus, if H is a nonempty affine subspace of $\mathbb{R}^n$, then by $\dim H$ we mean the dimension of the unique linear subspace W of which H is a translate. For example, if H consists of a single point, say $H = \{\mathbf{u}\}$, then $\dim H = 0$.

There is one precaution to be taken when using this terminology. It does not make sense to talk about a basis for an arbitrary affine subspace of $\mathbb{R}^n$; only linear subspaces of $\mathbb{R}^n$ have bases. For if S is a basis for a subspace of $\mathbb{R}^n$, then that subspace is equal to $\mathscr{L}(S)$ and so must be a linear subspace by 7.2.

To illustrate the concept of dimension, let us return to Examples 5.19 and 5.21.

8.3 Example 5.19 Continued

We showed in 5.19 that $H = \{(x_1, x_2) \in \mathbb{R}^2\colon x_2 = 2x_1 + 5\}$ is the translate of the linear subspace $W = \{(x_1, x_2) \in \mathbb{R}^2\colon x_2 = 2x_1\}$ by the vector $(0, 5)$.

To show that $\dim H = \dim W = 1$, we exhibit a basis, S, for W consisting of exactly one vector. Let $S = \{(1,2)\}$. S spans W, for if $\mathbf{u} = (u_1, u_2) \in W$, then $u_2 = 2u_1$ and hence $\mathbf{u} = u_1(1, 2)$. S is also linearly independent by 7.10(i).

8.4 Example 5.21 Continued

We showed in 5.21 that $H = \{(x_1, x_2, x_3, x_4, x_5) \in \mathbb{R}^5\colon x_1 = 2x_2 + 3x_4 - x_5 + 4,\ x_3 = -5x_4 + 6x_5 - 7\}$ is the translate of the linear subspace $W = \{(x_1, x_2, x_3, x_4, x_5) \in \mathbb{R}^5\colon x_1 = 2x_2 + 3x_4 - x_5, x_3 = -5x_4 + 6x_5\}$ by the vector $(4, 0, -7, 0, 0)$.

To show that $\dim H = \dim W = 3$, we exhibit a basis S for W consisting of three vectors. We construct S in two stages. Noting that the general form of S must be $\{(*, *, *, *, *), (*, *, *, *, *), (*, *, *, *, *)\}$, and that the variables involved as parameters in the description of W are x_2, x_4 and x_5, the first stage is to let $S \restriction \{2, 4, 5\} = \alpha^3$. That is, $S = \{(*, 1, *, 0, 0), (*, 0, *, 1, 0), (*, 0, *, 0, 1)\}$, where the positions filled with asterisks are yet to be determined. The second stage involves replacing the asterisks with real numbers in such a way that the resulting 5-tuples are elements of W. There is only one way to do this, for if $\mathbf{u} =$

$(u_1, u_2, u_3, u_4, u_5) \in W$, then u_1 and u_3 are determined by the values of u_2, u_4, and u_5. Thus $S = \{(2,1,0,0,0),(3,0,-5,1,0),(-1,0,6,0,1)\}$. It is easy to show that S is a basis for W. S is linearly independent by Theorem 7.27 since S is, by construction, a reduced set of vectors. S spans W, for if $\mathbf{u} = (u_1, u_2, u_3, u_4, u_5) \in W$, then $u_1 = 2u_2 + 3u_4 - u_5$ and $u_3 = -5u_4 + 6u_5$, and so $\mathbf{u} = u_2(2,1,0,0,0) + u_4(3,0,-5,1,0) + u_5(-1,0,6,0,1)$.

The next exercise is subject to frequent misuse by beginning students. It is important to understand precisely what it says. Exercise 8.5 asserts that, if the subspace W of $\mathbb{R}^n$ is known to have dimension p, then to prove that a subset S of W of cardinality p is a basis for W, it is sufficient to prove either (1) that S is linearly independent, or (2) that S spans W. Normally, to prove that a subset S of W is a basis for W, we have to prove both that S is linearly independent and that S spans W. The essence of 8.5 is that in the special circumstance when S is a subset of cardinality p from a subspace W *whose dimension is known to equal* p, then S is linearly independent if and only if S spans W. Hence to prove that S is a basis, it suffices to prove that S has one or the other of these properties.

8.5 Exercise

If W is a subspace of $\mathbb{R}^n$ and $\dim W = p$ and if $S \subset W$ with $|S| = p$, then

(i) if S is linearly independent, then S also spans W (and is hence a basis for W); and
(ii) if S spans W, then S is also linearly independent (and is hence a basis for W).
Hint: For (i), use 7.15 and 7.29(i); for (ii), use 7.11 and 7.29(ii).

8.6 Exercise

If W is a subspace of $\mathbb{R}^n$ and $\dim W = n$, then $W = \mathbb{R}^n$.

When restated with the proper emphasis, 8.6 says that the only n-dimensional subspace of $\mathbb{R}^n$ is the whole of $\mathbb{R}^n$ itself.

8.7 Exercise

If W is a subspace of $\mathbb{R}^n$ and if S is any linearly independent subset of W, then there exists a set S' with $S \subset S' \subset W$ such that S' is a basis for W. *Hint:* Mimic the proof of 7.32.

The result in this exercise is frequently used. It says that any linearly independent subset S of a subspace W of $\mathbb{R}^n$ can be expanded to a basis for W.

The next exercise is in the same spirit. It says that any set S containing a nonzero vector can be shrunk to a linearly independent subset S' whose span is the same as that of the original set S.

8.8 Exercise

If S is any subset of $\mathbb{R}^n$ containing at least one nonzero vector, then there exists a linearly independent subset $S' \subset S$ such that $\mathscr{L}(S') = \mathscr{L}(S)$. *Hint:* Apply 7.12 as many times as necessary.

In Section 15, we will give an algorithm for implementing the result in this exercise when S is an arbitrary set of m-many n-tuples.

It is occasionally useful to be able to think of the dimension of a subspace W of $\mathbb{R}^n$ in the following terms:

(i) $\dim W$ is the maximum cardinality of a linearly independent subset of W.
(ii) $\dim W$ is the minimum cardinality of a subset of W that spans W.

The next exercise makes this precise.

8.9 Exercise

Let W be a subspace of $\mathbb{R}^n$ and let $S \subset W$. Prove that S is a basis for W if and only if either

(i) S is linearly independent and for any S' such that $S \subsetneq S' \subset W$, S' is linearly dependent, or
(ii) S spans W and for any $S' \subsetneq S$, S' does not span W.

Restated in words, 8.9(i) says that S is a basis for W if and only if S is a *maximal* linearly independent subset of W; the word *maximal* in this context means that any subset S' of W that is strictly larger than S cannot be linearly independent. Similarly, 8.9(ii) says that S is a basis for W if and only if S is a *minimal* spanning set for W; the word *minimal* in this context means that no subset S' that is strictly smaller than S can span W.

Until now, in our theoretical development of the concepts of basis and dimension, we have been using the letters S, S', S'', and so on. Now that the concept of basis is firmly established, we will begin using lowercase Greek letters, α, β, γ, and so on, to denote bases.

It is also appropriate at this point to formalize our use of the phrases line, plane, and hyperplane in $\mathbb{R}^n$. These were used informally in Section 4.

8.10 Definition

(i) A *line in* $\mathbb{R}^n$ is a one-dimensional affine subspace of $\mathbb{R}^n$.
(ii) A *plane in* $\mathbb{R}^n$ is a two-dimensional affine subspace of $\mathbb{R}^n$.
(iii) A *hyperplane in* $\mathbb{R}^n$ is an $(n-1)$-dimensional affine subspace of $\mathbb{R}^n$.

We conclude this section with one of the most useful characterizations of the concept of basis.

8.11 Theorem

The ordered set $\beta = (\mathbf{u}_1, \mathbf{u}_2, \ldots, \mathbf{u}_p) \subset W$ is a basis for the subspace $W \subset \mathbb{R}^m$ if and only if every vector $\mathbf{w} \in W$ can be expressed in a unique way as a linear combination, $\mathbf{w} = t_1\mathbf{u}_1 + t_2\mathbf{u}_2 + \cdots + t_p\mathbf{u}_p$, of the vectors in β.

Proof: ($\rightarrow$) Assume $\beta = (\mathbf{u}_1, \mathbf{u}_2, \ldots, \mathbf{u}_p) \subset W$ is an ordered basis for W. Because β spans W, every $\mathbf{w} \in W$ is expressible in at least one way as a linear combination of $\mathbf{u}_1, \mathbf{u}_2, \ldots, \mathbf{u}_p$. Suppose next that $\mathbf{w} = s_1\mathbf{u}_1 + s_2\mathbf{u}_2 + \cdots + s_p\mathbf{u}_p$ and $\mathbf{w} = t_1\mathbf{u}_1 + t_2\mathbf{u}_2 + \cdots + t_p\mathbf{u}_p$ are two different ways to express some $\mathbf{w} \in W$ as a linear combination of the vectors from β. Because we are viewing β as an ordered set, the fact that these expressions are different results from the existence of an i with $1 \le i \le p$ such that $s_i \ne t_i$; thus, subtracting the second equation for $\mathbf{w}$ from the first, we obtain $\mathbf{w} - \mathbf{w} = \mathbf{0} = (s_1 - t_1)\mathbf{u}_1 + \cdots + (s_i - t_i)\mathbf{u}_i + \cdots + (s_p - t_p)\mathbf{u}_p$. Because $s_i - t_i \ne 0$, this is a nontrivial way to write $\mathbf{0}$ as a linear combination of the vectors from β, contradicting the linear independence of β.

($\leftarrow$) Conversely, if β is not a basis for W, then either β does not span W, in which case there is some $\mathbf{w} \in W$ that is not expressible in even a single way as a linear combination of $\mathbf{u}_1, \mathbf{u}_2, \ldots, \mathbf{u}_p$, or else β is linearly dependent, in which case $\mathbf{0}$ can be written both as the trivial linear combination and as some nontrivial linear combination of $\mathbf{u}_1, \mathbf{u}_2, \ldots, \mathbf{u}_p$. ■

8.12 Definition

Because of 8.11, it makes sense, if $\beta = (\mathbf{u}_1, \mathbf{u}_2, \ldots, \mathbf{u}_p)$ is an ordered basis for W, to talk about *the* coordinates of the vector $\mathbf{w}$ with respect to the basis β; in this context, if $\mathbf{w} = t_1\mathbf{u}_1 + t_2\mathbf{u}_2 + \cdots + t_p\mathbf{u}_p$, we will denote the vector $(t_1, t_2, \ldots, t_p)$ by $(\mathbf{w})_\beta$ and call it the *β-coordinate vector of* $\mathbf{w}$. The associated $p \times 1$ matrix $\begin{bmatrix} t_1 \\ t_2 \\ \vdots \\ t_p \end{bmatrix}$ will be denoted by $[\mathbf{w}]_\beta$ and called the *β-coordinate matrix of* $\mathbf{w}$.

We have frequently used the trivial observation that two n-tuples $\mathbf{u} = (u_1, u_2, \ldots, u_n)$ and $\mathbf{v} = (v_1, v_2, \ldots, v_n)$ are equal if and only if they are equal in each coordinate (i.e., if and only if $u_1 = v_1$ and $u_2 = v_2$ and ... and $u_n = v_n$). This observation is really the special case for $W = \mathbb{R}^n$ and $\beta = \alpha^n$ of the following general fact, which follows easily from 8.11.

8.13 Corollary

If W is a subspace of $\mathbb{R}^n$ and if $\beta = (\mathbf{w}_1, \mathbf{w}_2, \ldots, \mathbf{w}_p)$ is any ordered basis for W, then for any two vectors $\mathbf{u}, \mathbf{v} \in W$, $\mathbf{u} = \mathbf{v}$ iff $(\mathbf{u})_\beta = (\mathbf{v})_\beta$.

Proof: Exercise. ■

Problem Set 8

8.1. Go back and do any of the exercises in Section 8 that you may have skipped.

*8.2. Return to Problem 5.2, and for each of the subsets S answer the following questions:
 (i) If S is a linear subspace, find the dimension of S and a basis for S?
 (ii) If S is an affine subspace but is not a linear subspace, let W be the linear subspace such that S is the translate of W, find the dimension of W, and exhibit a basis for W.

*8.3. This problem assumes that you know how to solve a system of three equations in two unknowns. A method for solving arbitrary linear systems will be presented in the next chapter.
 (i) Show that $(4, 7, 6) \in \mathcal{L}(\{\mathbf{u}, \mathbf{v}\})$, where $\mathbf{u} = (1, 2, 1)$ and $\mathbf{v} = (2, 3, 4)$.
 (ii) Show that $(2, 9, 5) \notin \mathcal{L}(\{\mathbf{u}, \mathbf{v}\})$.
 (iii) Find a vector $\mathbf{w}$ such that $\{\mathbf{u}, \mathbf{v}, \mathbf{w}\}$ is a basis for $\mathbb{R}^3$. Justify your answer.

*8.4. Give a justified *yes* or *no* answer to the following questions, based only on definitions and theorems; no calculations requiring pen and paper are permitted.
 (i) Does $(4, -18)$ belong to the subspace of $\mathbb{R}^2$ spanned by the set $\{(-3, 9), (2, -5)\}$?
 (ii) Is there a four-dimensional subspace of $\mathbb{R}^4$ that does not contain the vector $(2, -2, 3, 8)$?
 (iii) If $\beta = \{\mathbf{v}_1, \mathbf{v}_2, \ldots, \mathbf{v}_8\}$ is a basis for an eight-dimensional subspace of $\mathbb{R}^{10}$ and if W is a subspace of $\mathbb{R}^{10}$ that contains the vectors $\mathbf{v}_1$, $\mathbf{v}_2$ and $\mathbf{v}_5$, then the dimension of W is greater than or equal to 3.
 (iv) Is there a basis for $\mathbb{R}^3$ that contains the vectors $(1, 2, 2)$ and $(2, 4, 4)$?

*8.5. Under what conditions on a, b, c, d will the set $\{(a, c), (b, d)\}$ be a basis for $\mathbb{R}^2$?

*8.6. Let W be a linear subspace of $\mathbb{R}^{18}$. Consider the following three subsets of W:

$$A = \{\mathbf{u}_1, \mathbf{u}_2, \ldots, \mathbf{u}_5\}$$
$$B = \{\mathbf{u}_1, \mathbf{u}_2, \ldots, \mathbf{u}_5, \mathbf{u}_6, \ldots, \mathbf{u}_9\}$$
$$C = \{\mathbf{v}_1, \mathbf{v}_2, \ldots, \mathbf{v}_{14}\}$$

Note that the first five elements of B are precisely the elements of A but that there are no other known relationships among the elements of A, B, and C.

Here is a list of possible responses to the questions that follow:

(a) ... spans W.
(b) ... is a basis for W.
(c) ... is linearly independent.
(d) ... is linearly dependent.
(e) ... does not span W.
(f) ... the hypothesis provides no information about the set mentioned in the conclusion.

Fill in the blanks with a list of *all* those possible responses that can be justifiably concluded. Justify all your answers.

(i) If B is a basis for W, then C ____________ .
(ii) If B is a basis for W, then A ____________ .
(iii) If C spans W, then A ____________ .
(iv) If C is a basis for W, then B ____________ .
(v) If A is a basis for W, then B ____________ .
(vi) If B is linearly independent, then A ____________ .

8.7. Prove that the set

$$\{(x_1, \ldots, x_7) \in \mathbb{R}^7 \colon x_1 = -4,\ x_2 = 4x_3 + 5x_6 - 2x_7 - 3,\ x_4 = -2x_5 + x_7 + 3\}$$

is a four-dimensional affine subspace of $\mathbb{R}^7$.

Hint: To show it is an affine subspace, invoke Theorem 6.9; to show it is four-dimensional, follow the pattern suggested by Example 8.4.

8.8. In Section 4, we informally used the phrase "k-dimensional affine subspace of $\mathbb{R}^n$" to refer to sets of the form

$$S = \Big\{(x_1, x_2, \ldots, x_n) \in \mathbb{R}^n \colon$$
$$x_{d_1} = a_{d_1 p_1} x_{p_1} + a_{d_1 p_2} x_{p_2} + \cdots + a_{d_1 p_k} x_{p_k} + b_{d_1}$$
$$x_{d_2} = a_{d_2 p_1} x_{p_2} + a_{d_2 p_2} x_{p_2} + \cdots + a_{d_2 p_k} x_{p_k} + b_{d_2}$$
$$\vdots$$
$$x_{d_r} = a_{d_r p_1} x_{p_1} + a_{d_r p_2} x_{p_2} + \cdots + a_{d_r p_k} x_{p_k} + b_{d_r}\Big\}$$

where the variables $x_1, x_2, \ldots, x_n$ are split into two disjoint collections, the dependent variables $\{x_{d_1}, x_{d_2}, \ldots, x_{d_r}\}$ and the parameters $\{x_{p_1}, x_{p_2}, \ldots, x_{p_k}\}$ (thus $r + k = n$), and where for $i = 1, 2, \ldots, r$ and for $j = 1, 2, \ldots, k$, $a_{d_i j}$ and b_{d_i} are scalars. Prove that such a set really is an affine subspace of $\mathbb{R}^n$ in the sense of Definition 5.16 and that the number of parameters, k, really is its dimension in the sense of Definition 8.2.

Hint: A trivial rewriting of the equations shows that S is the solution space of an $r \times n$ linear system, so S is an affine subspace by Theorem 6.9. To show that S is k-dimensional, obtain a basis for the solution space, S_0, of the associated homogeneous system in the following way. For $j = 1, 2, \ldots, k$ let $\mathbf{w}_j$ be the n-tuple whose p_j^{th} coordinate is 1, whose d_i^{th} coordinate for $i = 1, 2, \ldots, r$ is $a_{d_i j}$, and whose remaining coordinates are all 0. Prove that $\{\mathbf{w}_1, \mathbf{w}_2, \ldots, \mathbf{w}_k\}$ is a basis for S_0.

*8.9. Let $\mathbf{u}_1, \mathbf{u}_2, \ldots, \mathbf{u}_m, \mathbf{v} \in \mathbb{R}^n$. Prove that the following are equivalent:

(i) $\mathbf{v} \in \mathscr{L}(\mathbf{u}_1, \mathbf{u}_2, \ldots, \mathbf{u}_m)$

(ii) $\mathscr{L}(\mathbf{u}_1, \mathbf{u}_2, \ldots, \mathbf{u}_m) = \mathscr{L}(\mathbf{u}_1, \mathbf{u}_2, \ldots, \mathbf{u}_m, \mathbf{v})$

(iii) The two spaces $\mathscr{L}(\mathbf{u}_1, \mathbf{u}_2, \ldots, \mathbf{u}_m)$ and $\mathscr{L}(\mathbf{u}_1, \mathbf{u}_2, \ldots, \mathbf{u}_m, \mathbf{v})$ have the same dimension.

8.10. There is a convenient way to restate the results of 8.1 and 8.5 combined. Let $S \subset W \subset \mathbb{R}^n$, where W is known to be a p-dimensional linear subspace of $\mathbb{R}^n$. Prove that if S has any two of the following three properties, then it also has the third:

(i) $|S| = p$; that is, S contains exactly p-many vectors.

(ii) S spans W.

(iii) S is linearly independent.

8.11. Show that if $\{\mathbf{u}_1, \mathbf{u}_2, \mathbf{u}_3\}$ is linearly dependent and $\mathbf{u}_3 \notin \mathscr{L}(\{\mathbf{u}_1, \mathbf{u}_2\})$, then for some scalar t, $\mathbf{u}_1 = t\mathbf{u}_2$.

8.12. Let W_1 and W_2 be subspaces of $\mathbb{R}^n$.

(i) Let $W_1 + W_2 = \{\mathbf{u} + \mathbf{v}\colon \mathbf{u} \in W_1, \mathbf{v} \in W_2\}$. That is, $W_1 + W_2$ consists of all those vectors in $\mathbb{R}^n$ that can be written as the sum of two vectors, one from W_1, the other from W_2. Prove that $W_1 + W_2$ is also a subspace of $\mathbb{R}^n$. It is called the *sum* of W_1 and W_2. Note that $W_1 \cap W_2$ is not empty since the zero vector, $\mathbf{0}$, belongs to both W_1 and W_2. The situation in which $W = W_1 + W_2$ and $\mathbf{0}$ is the *only* vector in $W_1 \cap W_2$ is an important one, for the following reason:

(ii) By definition, every vector $\mathbf{w} \in W = W_1 + W_2$ can be written in at least one way as a sum, $\mathbf{w} = \mathbf{w}_1 + \mathbf{w}_2$, where $\mathbf{w}_1 \in W_1$ and $\mathbf{w}_2 \in W_2$. The effect, in this context, of the additional condition $W_1 \cap W_2 = \{\mathbf{0}\}$ is to make the expression unique. If $W = W_1 + W_2$ and $W_1 \cap W_2 = \{\mathbf{0}\}$, we say that W is the *direct sum* of W_1 and W_2, and we write $W = W_1 \oplus W_2$ to denote this fact. Prove that if $W = W_1 \oplus W_2$ and if $\mathbf{w} = \mathbf{w}_1 + \mathbf{w}_2$ and $\mathbf{w} = \mathbf{w}_1' + \mathbf{w}_2'$, where $\mathbf{w}_1, \mathbf{w}_1' \in W_1$ and $\mathbf{w}_2, \mathbf{w}_2' \in W_2$, then $\mathbf{w}_1 = \mathbf{w}_1'$ and $\mathbf{w}_2 = \mathbf{w}_2'$. Prove, conversely, that if every $\mathbf{w} \in W = W_1 + W_2$ can be written *uniquely* in the form $\mathbf{w} = \mathbf{w}_1 + \mathbf{w}_2$ with $\mathbf{w}_i \in W_i$, then $W = W_1 \oplus W_2$.

(iii) $\dim(W_1 + W_2) = \dim W_1 + \dim W_2 - \dim(W_1 \cap W_2)$. *Hint:* Let $\{\mathbf{x}_1, \mathbf{x}_2, \ldots, \mathbf{x}_k\}$ be a basis for $W_1 \cap W_2$. Extend this, if necessary, to a basis $\{\mathbf{x}_1, \ldots, \mathbf{x}_k, \mathbf{y}_1, \ldots, \mathbf{y}_p\}$ for W_1 and to a basis $\{\mathbf{x}_1, \ldots, \mathbf{x}_k, \mathbf{z}_1, \ldots, \mathbf{z}_q\}$ for W_2. Prove that $\{\mathbf{x}_1, \ldots, \mathbf{x}_k, \mathbf{y}_1, \ldots, \mathbf{y}_p, \mathbf{z}_1, \ldots, \mathbf{z}_q\}$ is a basis for $W_1 + W_2$. (The results from Problem 8.12 can be generalized to any finite collection of subspaces. This will be done in Section 43.)

9

Row Space, Column Space, and Rank of a Matrix

There are two very natural subspaces related to an arbitrary $m \times n$ matrix A.

9.1 Definition

Let

$$\begin{bmatrix} a_{11} & a_{12} & \cdots & a_{1j} & \cdots & a_{1n} \\ a_{21} & a_{22} & \cdots & a_{2j} & \cdots & a_{2n} \\ \vdots & \vdots & & \vdots & & \vdots \\ a_{i1} & a_{i2} & \cdots & a_{ij} & \cdots & a_{in} \\ \vdots & \vdots & & \vdots & & \vdots \\ a_{m1} & a_{m2} & \cdots & a_{mj} & \cdots & a_{mn} \end{bmatrix} = \mathrm{A}$$

For $i = 1, 2, \ldots, m$, let $\mathbf{r}_i = (a_{i1}, a_{i2}, \ldots, a_{in})$ denote the ith row of A considered as a vector in $\mathbb{R}^n$, and for $j = 1, 2, \ldots, n$, let $\mathbf{c}_j = (a_{1j}, a_{2j}, \ldots, a_{mj})$ denote the jth column of A considered as a vector in $\mathbb{R}^m$. The subspace $\mathscr{L}(\mathbf{r}_1, \mathbf{r}_2, \ldots, \mathbf{r}_m)$ of $\mathbb{R}^n$ spanned by the rows of A is called the *row space* of A and is denoted by RS(A). The subspace $\mathscr{L}(\mathbf{c}_1, \mathbf{c}_2, \ldots, \mathbf{c}_n)$ of $\mathbb{R}^m$ spanned by the columns of A is called the *column space* of A and is denoted by CS(A).

Our use of the word *space* in the expressions *row space* and *column space* is justified by 7.2.

It is clear, if $m \neq n$, that for an $m \times n$ matrix A, RS(A) and CS(A) are different since RS(A) is a subspace of $\mathbb{R}^n$ while CS(A) is a subspace of $\mathbb{R}^m$. We would like to emphasize that even when $m = n$, RS(A) and CS(A) are, in general, different subspaces of $\mathbb{R}^n$.

9.2 Examples

Let

$$\begin{bmatrix} 2 & -2 & 8 \\ -1 & -9 & 6 \\ 4 & 13 & -1 \\ 3 & 2 & 7 \end{bmatrix} = \mathrm{A} \qquad \begin{bmatrix} 8 & 2 & 4 \\ 5 & 6 & -7 \\ 5 & 2 & 1 \end{bmatrix} = \mathrm{B} \qquad \begin{bmatrix} 2 & 0 & 4 \\ 3 & 0 & 4 \\ 5 & 0 & 4 \end{bmatrix} = \mathrm{C}$$

$$\begin{aligned}
\mathrm{RS(A)} &= \mathscr{L}(\{(2,-2,8),(-1,-9,6),(4,13,-1),(3,2,7)\}) \\
\mathrm{CS(A)} &= \mathscr{L}(\{(2,-1,4,3),(-2,-9,13,2),(8,6,-1,7)\}) \\
\mathrm{RS(B)} &= \mathscr{L}(\{(8,2,4),(5,6,-7),(5,2,1)\}) \\
\mathrm{CS(B)} &= \mathscr{L}(\{(8,5,5),(2,6,2),(4,-7,1)\}) \\
\mathrm{RS(C)} &= \mathscr{L}(\{(2,0,4),(3,0,4),(5,0,4)\}) \\
\mathrm{CS(C)} &= \mathscr{L}(\{(2,3,5),(0,0,0),(4,4,4)\})
\end{aligned}$$

Note that RS(C) $\neq$ CS(C) since, for example, $(1,1,1) \in$ CS(C) but $(1,1,1) \notin$ RS(C) since any linear combination of the rows of C must have 0 as its second coordinate.

9.3 Observation

For any $m \times n$ matrix A,

$$\mathrm{RS}(\mathrm{A}^{\mathrm{T}}) = \mathrm{CS}(\mathrm{A}) \quad \text{and} \quad \mathrm{CS}(\mathrm{A}^{\mathrm{T}}) = \mathrm{RS}(\mathrm{A}). \qquad \blacksquare$$

This observation is a trivial consequence of the definition of a transpose and needs no further comment.

9.4 Observation

If A is the coefficient matrix of an $m \times n$ linear system $\mathrm{A}\mathbf{x} = \mathbf{b}$, and if we allow the vector $\mathbf{b}$ on the right side to vary over all possible m-tuples, it follows from 6.14 that we can write

$$\mathrm{CS(A)} = \{\mathbf{b} \in \mathbb{R}^m \colon \mathrm{A}\mathbf{x} = \mathbf{b} \text{ is consistent}\}. \qquad \blacksquare$$

The result in this observation will be frequently used. It provides us with another illustration of how a single idea in linear algebra disguises itself with different vocabulary. The equivalence between the statements

1: $\mathbf{b} \in \mathrm{CS}(\mathrm{A})$, and
2: $\mathrm{A}\mathbf{x} = \mathbf{b}$ is consistent

implies that

Question 1: describe the column-space of A, and
Question 2: what conditions, if any, must the column vector $\mathbf{b}$ satisfy in order for the system of equations $\mathrm{A}\mathbf{x} = \mathbf{b}$ to have a solution?

are really the same question.

The next exercise is a trivial consequence of 7.30.

9.5 Exercise

Let A be an $m \times n$ matrix.

(i) If $m < n$, then the columns of A are linearly dependent.
(ii) If $n < m$, then the rows of A are linearly dependent.

The next theorem is one whose importance cannot really be appreciated until Section 19, where it will be used in the proof of one of the main theorems. For the moment, it can be viewed as a technical result concerning the concepts introduced in this section.

We emphasized earlier that the row space and column space of an $m \times n$ matrix A are, in general, *different* spaces, even when $m = n$. The point of the next theorem is that these two spaces have the same dimension.

9.6 Theorem

For any $m \times n$ matrix A, $\dim \mathrm{RS}(\mathrm{A}) = \dim \mathrm{CS}(\mathrm{A})$.

Proof: Assume, temporarily, that the following statement has been established:

$$(*) \text{ For any matrix B, } \dim \mathrm{CS}(\mathrm{B}) \leq \dim \mathrm{RS}(\mathrm{B})$$

Then, by applying $(*)$ to A, we conclude that $\dim \mathrm{CS}(\mathrm{A}) \leq \dim \mathrm{RS}(\mathrm{A})$; and by applying $(*)$ together with 9.3 to A^{T}, we conclude that $\dim \mathrm{RS}(\mathrm{A}) = \dim \mathrm{CS}(\mathrm{A}^{\mathrm{T}}) \leq \dim \mathrm{RS}(\mathrm{A}^{\mathrm{T}}) = \dim \mathrm{CS}(\mathrm{A})$. Thus the proof will be complete once $(*)$ has been established. So suppose that

$$\begin{bmatrix} b_{11} & b_{12} & \cdots & b_{1q} \\ b_{21} & b_{22} & \cdots & b_{2q} \\ \vdots & \vdots & & \vdots \\ b_{p1} & b_{p2} & \cdots & b_{pq} \end{bmatrix} = \mathrm{B}$$

that $\dim RS(B) = k$, and that $\beta = \{\mathbf{u}_1, \mathbf{u}_2, \ldots, \mathbf{u}_k\}$ is a basis for RS(B), where $\mathbf{u}_i = (u_{i1}, u_{i2}, \ldots, u_{iq})$. To prove $(*)$, it suffices by 7.29(i) to exhibit a set S of k-many p-tuples that spans CS(B).

Because β is a basis for RS(B), there exist scalars s_{ij} such that

$$\begin{aligned}
(b_{11}, b_{12}, \ldots, b_{1j}, \ldots, b_{1q}) &= s_{11}\mathbf{u}_1 + s_{12}\mathbf{u}_2 + \cdots + s_{1k}\mathbf{u}_k \\
(b_{21}, b_{22}, \ldots, b_{2j}, \ldots, b_{2q}) &= s_{21}\mathbf{u}_1 + s_{22}\mathbf{u}_2 + \cdots + s_{2k}\mathbf{u}_k \\
&\vdots \\
(b_{p1}, b_{p2}, \ldots, b_{pj}, \ldots, b_{pq}) &= s_{p1}\mathbf{u}_1 + s_{p2}\mathbf{u}_2 + \cdots + s_{pk}\mathbf{u}_k
\end{aligned}$$

Having written down the hypothesis in detail, we find a set of k-many p-tuples staring us in the face. Let

$$S = \{(s_{11}, s_{21}, \ldots, s_{p1}), (s_{12}, s_{22}, \ldots, s_{p2}), \ldots, (s_{1k}, s_{2k}, \ldots, s_{pk})\}.$$

To see that S spans CS(B), observe that when we equate the jth components of each side of each of the preceding equations, we obtain

$$\begin{aligned}
b_{1j} &= s_{11}u_{1j} + s_{12}u_{2j} + \cdots + s_{1k}u_{kj} \\
b_{2j} &= s_{21}u_{1j} + s_{22}u_{2j} + \cdots + s_{2k}u_{kj} \\
&\vdots \\
b_{pj} &= s_{p1}u_{1j} + s_{p2}u_{2j} + \cdots + s_{pk}u_{kj}
\end{aligned}$$

which, as in 6.14, can be rewritten as

$$u_{1j}\begin{bmatrix} s_{11} \\ s_{21} \\ \vdots \\ s_{p1} \end{bmatrix} + u_{2j}\begin{bmatrix} s_{12} \\ s_{22} \\ \vdots \\ s_{p2} \end{bmatrix} + \cdots + u_{kj}\begin{bmatrix} s_{1k} \\ s_{2k} \\ \vdots \\ s_{pk} \end{bmatrix} = \begin{bmatrix} b_{1j} \\ b_{2j} \\ \vdots \\ b_{pj} \end{bmatrix}$$

which shows that an arbitrary column of B belongs to $\mathscr{L}(S)$. ∎

The number that is the common dimension of both RS(A) and CS(A) is an important feature of the matrix A and it is given a special name.

9.7 Definition

For an $m \times n$ matrix A, the integer $r = \dim RS(A) = \dim CS(A)$ is called the *rank* of A and is denoted by r(A).

Some authors call $\dim RS(A)$ the row rank of A and call $\dim CS(A)$ the column rank of A. In this terminology, Theorem 7.6 can be restated as follows: for any $m \times n$ matrix A, row rank of A = column rank of A.

9.8 Exercise

For any $m \times n$ matrix A, the rank of A does not exceed the smaller of m and n.

9.9 Definition

An $m \times n$ matrix A for which r(A) = the smaller of m and n (i.e., for which the rank of A is as large as possible) is said to have *full rank*.

The problem of computing the rank of a matrix A and of finding bases for RS(A) and CS(A) will be treated in detail in the next chapter. For the moment, a simple observation can be made.

9.10 Observation

If the nonzero rows of an $m \times n$ matrix A form a reduced set of n-tuples (see Definition 7.26), then these nonzero rows are a basis for RS(A); thus the rank of such a matrix is just the number of nonzero rows.

This observation follows from the fact that a reduced set of vectors is linearly independent.

9.11 Example

Let

$$A = \begin{bmatrix} 1 & 2 & 3 & 0 & 0 \\ 0 & 4 & 5 & 1 & 0 \\ 0 & 6 & 7 & 0 & 1 \\ 0 & 0 & 0 & 0 & 0 \end{bmatrix}$$

Because the set $S = \{(1,2,3,0,0),(0,4,5,1,0),(0,6,7,0,1)\}$ of nonzero rows of A is a reduced set of 5-tuples (note $S \restriction \{1,4,5\} = \alpha^3$), S is a basis for RS(A) and r(A) = 3.

We close this section with an easy theorem that relates the rank of a product of two matrices with the ranks of the factors.

9.12 Theorem

For any $m \times n$ matrix A and $n \times p$ matrix B,

$$\mathrm{r(AB)} \leq \min\{\mathrm{r(A)}, \mathrm{r(B)}\}.$$

Proof: The columns of AB are linear combinations of the columns of A; to see this, note that Observation 1.12(i) can be rephrased as follows:

$$\begin{bmatrix} j^{\text{th}} \\ \text{column} \\ \text{of} \\ \text{AB} \end{bmatrix} = b_{1j} \begin{bmatrix} 1^{\text{st}} \\ \text{column} \\ \text{of} \\ \text{A} \end{bmatrix} + b_{2j} \begin{bmatrix} 2^{\text{nd}} \\ \text{column} \\ \text{of} \\ \text{A} \end{bmatrix} + \cdots + b_{nj} \begin{bmatrix} n^{\text{th}} \\ \text{column} \\ \text{of} \\ \text{A} \end{bmatrix}$$

In other words, the jth column of AB is the linear combination of the columns of A that is formed using the entries from the jth column of B as coefficients. This implies, by Exercise 5.12, that $\text{CS(AB)} \subseteq \text{CS(A)}$. Hence, $\text{r(AB)} = \dim \text{CS(AB)} \leq \dim \text{CS(A)} = \text{r(A)}$.

The fact that $\text{r(AB)} \leq \text{r(B)}$ is proved in a similar fashion using Observation 1.12(ii), which can be rephrased as follows: the rows of AB are linear combinations of the rows of B. ■

The inequality in Theorem 9.12 is the best possible result. In Problem 9.6, you are asked to provide an example for which r(AB) is strictly less than both r(A) and r(B).

9.13 Some Psychology

At this point in the text, an important decision must be made, based primarily on psychological grounds. We have individual approaches to learning. At the extremes, there is the type who, arriving home with a newly purchased, complicated gadget, spends however long it takes to thoroughly read and understand the instruction manual, perhaps before even taking the gadget out of its box; there is also the other type who immediately begins tinkering with the gadget, perhaps only after discarding the box together with the instruction manual, which may have been tucked away in one of its corners. Most of us are between these extremes.

Sections 5 through 9 are the beginning of the "instruction manual" of linear algebra. Although it was tempting to keep the manual in one piece, we have, in order to avoid the extremes, split it in two and have placed the remainder of it in Chapter IV, in Sections 16 to 20. If you found Chapter II somewhat abstract and are anxious for a change of pace, you should continue with Chapter III, which begins, in Sections 10 to 12, with an algorithm (mechanical procedure) for solving linear systems and continues, in Sections 13 to 15, with procedures for answering concrete, computational questions that concern the concepts from Chapter II. But Chapters III and IV are entirely independent of one another and can be covered in either order. If you thought that Observations 6.14 and 9.4 were useless or potentially confusing, then you should definitely *not* skip ahead to Chapter IV. On the other hand, if the variety of equivalent points of view is an aspect you find intriguing, you might read more or all of the manual before getting your hands dirty. If you are already having second thoughts about why you bought this gadget in the first place, browse through a supplement on applications (for example, references [RA] or [G]), for a brief reminder.

Problem Set 9

9.1. Go back and do any of the exercises from Section 9 that you may have skipped.

*9.2. What is the maximum possible value for the rank of A if A is
(i) a 4×6 matrix?
(ii) a 17×12 matrix?
(iii) a 9×9 matrix?

*9.3. Prove that an $m \times n$ linear system $A\mathbf{x} = \mathbf{b}$ is consistent if and only if the rank of the augmented matrix, $[A \,\vdots\, \mathbf{b}]$, is the same as the rank of the coefficient matrix, A.
Hint: Problem 8.9 is relevant.

*9.4. Prove that for any $m \times n$ matrix A, $r(A) = r(A^T)$.

*9.5. Use the definition of rank together with other relevant results, such as 7.10, 9.6, and 9.10, to find the ranks of the following matrices by inspection.

(i) $\begin{bmatrix} 2 & 0 \\ 0 & 2 \end{bmatrix}$ (ii) $\begin{bmatrix} 1 & 3 & -4 \\ 2 & 6 & -8 \end{bmatrix}$

(iii) $\begin{bmatrix} 3 & -1 \\ -6 & -5 \end{bmatrix}$ (iv) $\begin{bmatrix} 3 & 4 & 0 \\ 0 & 0 & 0 \\ 0 & 4 & 3 \end{bmatrix}$

(v) $\begin{bmatrix} 0 & 1 \\ 2 & -9 \\ 1 & 0 \end{bmatrix}$ (vi) $\begin{bmatrix} 0 & -3 & 0 & 1 & 0 \\ 0 & 2 & 1 & 0 & 0 \\ 0 & 0 & 0 & 0 & 0 \\ 1 & 7 & 0 & 0 & 0 \\ 0 & -4 & 0 & 0 & 1 \end{bmatrix}$

(vii) $\begin{bmatrix} 2 & 6 \\ 5 & -4 \\ -1 & 7 \end{bmatrix}$ (viii) $\begin{bmatrix} 2 & -1 & 6 \\ -4 & 2 & -12 \\ 6 & -3 & 18 \end{bmatrix}$

9.6. Find 2×2 matrices A and B that are counterexamples to (and thereby disprove) the following:
(i) $r(A + B) = r(A) + r(B)$
(ii) $r(AB) = r(BA)$
(iii) $r(AB) = \text{minimum of } \{r(A), r(B)\}$
(iv) If $r(A) = r(B)$, then $CS(A) = CS(B)$.

*9.7. Suppose that A, B, and C are $n \times n$ matrices and A is invertible. Prove that $r(AC) = r(C)$ and that $r(BA) = r(B)$.

10

Linear Systems Whose Solution Spaces Are Obvious

In this section, we begin to treat the quantitative aspects of "solving" systems of linear equations. The word *solving* is temporarily in quotes because the first order of business is to agree on what constitutes an answer to the question: Solve the following system of linear equations.

The qualitative aspects of this problem were discussed in Sections 4 through 8, and we will begin by reviewing briefly the main points from these earlier sections. The most important points were these:

A. That solving an $m \times n$ linear system corresponds, at the intuitive geometric level, to describing the common intersection of m-many hyperplanes in $\mathbb{R}^n$ (Sections 4 and 6 generally).

B. That matrix concepts allow us to rewrite a system of m linear equations in n unknowns as a single matrix equation. (Definition 6.4).

C. That the set of solutions of an $m \times n$ linear system is an affine subspace of $\mathbb{R}^n$ (Theorem 6.9).

D. That a given affine subspace of $\mathbb{R}^n$ might have several different but equivalent descriptions [Exercise 4.7 and Example 4.14].

E. That a subset S of $\mathbb{R}^n$ is a k-dimensional affine subspace of $\mathbb{R}^n$ if it has a description of the form

$$S = \Big\{(x_1, x_2, \ldots, x_n) \in \mathbb{R}^n:$$
$$x_{d_1} = a_{d_1 p_1} x_{p_1} + a_{d_1 p_2} x_{p_2} + \cdots + a_{d_1 p_k} x_{p_k} + b_{d_1}$$
$$x_{d_2} = a_{d_2 p_1} x_{p_2} + a_{d_2 p_2} x_{p_2} + \cdots + a_{d_2 p_k} x_{p_k} + b_{d_2}$$
$$\vdots$$
$$x_{d_r} = a_{d_r p_1} x_{p_1} + a_{d_r p_2} x_{p_2} + \cdots + a_{d_r p_k} x_{p_k} + b_{d_r}\Big\}$$

where the variables $x_1, x_2, \ldots, x_n$ are split into two disjoint collections, the dependent variables $\{x_{d_1}, x_{d_2}, \ldots, x_{d_r}\}$ and the parameters $\{x_{p_1}, x_{p_2}, \ldots, x_{p_k}\}$ (thus $r + k = n$), and where for $i = 1, 2, \ldots, r$ and for $j = 1, 2, \ldots, k$, $a_{d_i j}$ and b_{d_i} are scalars (Problem 8.8).

These main points suggest the kind of information that one might have or seek about the solution space, S, of an $m \times n$ linear system. Since S is an affine subspace of $\mathbb{R}^n$, we will consider, in order of increasing strength, possible amounts of information about a general affine subspace H of $\mathbb{R}^n$. We will simultaneously illustrate these ideas taking H to be the solution space, S_0, of the 3×5 linear system from Example 10.1. **Do not concern yourself for the moment with how the answers to these questions about S_0 were obtained. That is indeed the subject of Sections 11 and 12.**

10.1 Example

$$\begin{aligned} 2x_1 - 4x_2 + 10x_3 + 3x_4 - 8x_5 &= 4 \\ x_1 + 2x_2 + x_3 - 2x_4 &= 7 \\ 3x_1 + 5x_2 + 4x_3 - 10x_4 - x_5 &= 10 \end{aligned}$$

S_0 denotes the solution space of this 3×5 linear system.

Question 1: Is $H \neq \varnothing$ or is $H = \varnothing$?

Asking this question about S_0 corresponds to asking whether the linear system in 10.1 is consistent or inconsistent. The answer to this specific question is that $S_0 \neq \varnothing$. Of course, this gives us very little information about S_0, but in some contexts this could be all we care to know about S_0.

Question 2: What is the dimension of H?

For the linear system from 10.1, the answer is that $\dim S_0 = 2$. Again, knowledge that S_0 is a two-dimensional affine subspace of $\mathbb{R}^5$ does not yield information about which specific vectors belong to S_0, but this is more than knowing simply that $S_0 \neq \varnothing$.

Question 3: Can you find a specific vector $\mathbf{b} \in H$ and a basis for the unique linear subspace W such that H is the translation of W by $\mathbf{b}$? Of course, if $\dim H \geqq 1$, there are infinitely many correct ways to answer this question.

One possible answer for S_0 is that $S_0 = \{\mathbf{w} + \mathbf{b}\colon \mathbf{w} \in W\}$, where $\mathbf{b} = (5,3,0,2,0)$ and $\{(-1,0,1,0,1), (1,-1,1,0,2)\}$ is a basis for W. This information is often rewritten in the following form:

$$S_0 = \{s(-1,0,1,0,1) + t(1,-1,1,0,2) + (5,3,0,2,0)\colon s, t \in \mathbb{R}\}.$$

Although this is a complete description of S_0 (in the sense that we can decide from this information whether an arbitrary 5-tuple belongs to S_0 or not), we hesitate to be satisfied with this as an answer to the question of solving the system of equations from 10.1. The reason is that it does not provide us with an *easy* way to decide whether a given vector belongs to S_0 or not. For example, is $(-3,6,2,2,-1)$ a solution of the equations from 10.1? To answer this question with the information at hand, we must discover whether there exist scalars s and t such that

$$(-3,6,2,2,-1) = s(-1,0,1,0,1) + t(1,-1,1,0,2) + (5,3,0,2,0)$$

This is scarcely any easier than plugging the vector $(-3,6,2,2,-1)$ into the original equations from 10.1 to see if all of them are satisfied.

Question 4: Can you find a nice property P that determines H, in the sense that $H = \{(x_1, x_2, \ldots, x_n) \in \mathbb{R}^n\colon P(x_1, x_2, \ldots, x_n)\}$, and which is nice in the sense that it is easy to decide whether a given n-tuple satisfies property P or not?

One such property for S_0 (actually it is a list of three properties) is that

$$\begin{aligned} S_0 = \{(x_1, x_2, x_3, x_4, x_5) \in \mathbb{R}^5\colon x_1 &= 2x_2 - 5x_3 - \tfrac{3}{2}x_4 + 4x_5 + 2 \\ x_2 &= x_3 + \tfrac{7}{8}x_4 - x_5 + \tfrac{5}{4} \\ x_4 &= 2\} \end{aligned}$$

To decide from this whether $\mathbf{v} = (v_1, v_2, v_3, v_4, v_5) = (-3,6,2,2,-1)$ belongs to S_0, we observe that

$$\begin{aligned} 2 &= v_4 \\ 2 + (\tfrac{7}{8})(2) - (-1) + \tfrac{5}{4} = 6 &= v_2 \end{aligned}$$

and that

$$(2)(6) - (5)(2) - (\tfrac{3}{2})(2) + (4)(-1) + 2 = -3 = v_1$$

and so conclude that the answer is *yes*.

A minor defect of this description of S_0 is the amount of arithmetic that still needs to be done to decide whether $(-3, 6, 2, 2, -1) \in S_0$.

Question 5: Can you find a nicer condition that determines H and that provides an even easier way to decide whether or not a given n-tuple belongs to H?

One such condition for S_0 is that

$$S_0 = \{(x_1, x_2, x_3, x_4, x_5) \in \mathbb{R}^5 : \begin{aligned} x_1 &= -3x_3 + 2x_5 + 5 \\ x_2 &= x_3 - x_5 + 3 \\ x_4 &= 2\} \end{aligned}$$

To conclude from this description of S_0 that $\mathbf{v} = (-3, 6, 2, 2, -1) \in S_0$, we observe that

$$\begin{aligned} (-3)(2) + (2)(-1) + 5 &= -3 = v_1 \\ 2 - (-1) + 3 &= 6 = v_2 \end{aligned}$$

and that

$$2 = v_4$$

Although the difference between this description of S_0 and the previous one is minor, this one is preferable; the arithmetic required to check whether $\mathbf{v} \in S_0$ is reduced to a minimum because the expressions for the dependent variables x_1, x_2, and x_4 are functions of the parameters alone.

10.2 Definition

A *reduced description* of an affine subspace H of $\mathbb{R}^n$ is an expression of the form

$$H = \{(x_1, x_2, \ldots, x_n) \in \mathbb{R}^n :$$

$$x_{d_1} = a_{d_1p_1}x_{p_1} + a_{d_1p_2}x_{p_2} + \cdots + a_{d_1p_k}x_{p_k} + b_{d_1}$$

$$x_{d_2} = a_{d_2p_1}x_{p_2} + a_{d_2p_2}x_{p_2} + \cdots + a_{d_2p_k}x_{p_k} + b_{d_2}$$

$$\vdots$$

$$x_{d_r} = a_{d_rp_1}x_{p_1} + a_{d_rp_2}x_{p_2} + \cdots + a_{d_rp_k}x_{p_k} + b_{d_r}\}$$

where the variables $x_1, x_2, \ldots, x_n$ are split into two disjoint collections, the dependent variables $\{x_{d_1}, x_{d_2}, \ldots, x_{d_r}\}$ and the parameters $\{x_{p_1}, x_{p_2}, \ldots, x_{p_k}\}$ (thus $r + k = n$), and where for $i = 1, 2, \ldots, r$ and for $j = 1, 2, \ldots, k$, a_{d_ij} and b_{d_i} are scalars.

We are now in position to state what it means *to solve a system of linear equations*.

We know from Theorem 6.9 that the set S of solutions to a system of m linear equations in n unknowns is an affine subspace of $\mathbb{R}^n$. *To solve an* $m \times n$ *linear system* is to provide a reduced description of its solution space.

10.3 Definition

In Sections 11 and 12, we will present a systematic method, called the Gauss–Jordan algorithm, for solving linear systems. In the remainder of this section, we will motivate this algorithm by examining a class of linear systems whose solution spaces are obvious.

10.4 Examples

(i)

$$\begin{aligned} x_1 \qquad\qquad &= 2 \\ x_2 \qquad &= \tfrac{1}{3} \\ x_3 &= -6 \end{aligned}$$

The solution to this system is obviously the single point $\{(2, \frac{1}{3}, -6)\}$.

(ii)

$$\begin{aligned} x_1 \qquad + 3x_3 &= 29 \\ x_2 \qquad &= 14 \end{aligned}$$

The solution to this system is the line in $\mathbb{R}^3$ described by $\{(x_1, x_2, x_3) \in \mathbb{R}^3\colon x_1 = -3x_3 + 29,\ x_2 = 14\}$.

(iii) The set of points in $\mathbb{R}^5$ satisfying the 4×5 system

$$\begin{aligned} x_1 \qquad + 2x_4 \qquad &= -2 \\ x_2 \quad - 5x_4 \qquad &= 3 \\ x_3 \qquad\qquad &= 4 \\ x_5 &= -1 \end{aligned}$$

is the line $\{(x_1, x_2, x_3, x_4, x_5) \in \mathbb{R}^5\colon x_1 = -2x_4 - 2,\ x_2 = 5x_4 + 3,\ x_3 = 4,\ x_5 = -1\}$.

(iv) The set of points in $\mathbb{R}^4$ satisfying the 2×4 system

$$\begin{aligned} x_1 + 3x_2 \quad + 7x_4 &= 12 \\ x_3 - 2x_4 &= -3 \end{aligned}$$

is the plane $\{(x_1, x_2, x_3, x_4) \in \mathbb{R}^4\colon x_1 = -3x_2 - 7x_4 + 12,\ x_3 = 2x_4 - 3\}$.

When we examine these four systems together with the reduced descriptions of their solution spaces to understand what features they have in common, we are led to the following essential property (P):

(P): For each of the dependent variables, the system contains exactly one equation which involves that variable, with coefficient 1, and which does not involve any of the other dependent variables.

The systematic procedure for solving linear systems that we will develop in Sections 13 and 14 operates on the augmented matrix of the system. Let us examine the augmented matrices of the systems from Example 10.4.

10.5 Examples

(i) $$\begin{bmatrix} 1 & 0 & 0 & 2 \\ 0 & 1 & 0 & 1/3 \\ 0 & 0 & 1 & -6 \end{bmatrix}$$

(ii) $$\begin{bmatrix} 1 & 0 & 3 & 29 \\ 0 & 1 & 0 & 14 \end{bmatrix}$$

(iii) $$\begin{bmatrix} 1 & 0 & 0 & 2 & 0 & -2 \\ 0 & 1 & 0 & -5 & 0 & 3 \\ 0 & 0 & 1 & 0 & 0 & 4 \\ 0 & 0 & 0 & 0 & 1 & -1 \end{bmatrix}$$

(iv) $$\begin{bmatrix} 1 & 3 & 0 & 7 & 12 \\ 0 & 0 & 1 & -2 & -3 \end{bmatrix}$$

We have circled some of the columns in these matrices to emphasize how property (P) is reflected by the augmented matrix. The essential feature is this: if x_j is one of the dependent variables, then one of the entries in the jth column of the augmented matrix is 1 and all the other entries in that column are 0.

There is another aspect of these matrices that is not essential, although it needs to be discussed. When solving a system of equations, we are quite content to provide just one reduced description of the solution space. There is usually no reason to be concerned with other possible reduced descriptions of the same space. For example, any one of the three descriptions provided in 4.3(vi), (vii), and (viii) or in 4.5(iii), (iv), and (v) is as good as any other; they differ merely in the choice of variables used as parameters.

Where there are several ways of doing something that differ only inessentially, mathematicians adopt conventions in order to facilitate communication. A convention is a mutual agreement that a specific one among inessentially different ways is to be considered preferable.

We make the following convention concerning the choice of variables used as parameters in the description of affine subspaces of $\mathbb{R}^n$.

10.6 Convention

Assume that the variables are consecutively numbered $x_1, x_2, \ldots, x_n$. Where a choice is possible between the roles that x_i and x_j can play in the descriptions of an affine subspace of $\mathbb{R}^n$, the preferred description is the one in which the variable with the higher subscript is the parameter and the one with the lower subscript is the dependent variable.

10.7 Example

The description of the plane given in Example 10.4(iv) uses x_1 and x_3 as the dependent variables, while x_2 and x_4 are parameters. Another description of this same plane that uses x_2 and x_3 as parameters is

$$\left\{(x_1, x_2, x_3, x_4) \in \mathbb{R}^4: x_1 = \tfrac{3}{2} - 3x_2 - \tfrac{7}{2}x_3,\ x_4 = \tfrac{3}{2} + \tfrac{1}{2}x_3\right\}$$

The description from 10.4(iv) is preferred over this one since the variable with the lower subscript (x_3 rather than x_4) is being used as a dependent variable. Similarly, among the descriptions of the plane provided in Examples 4.3(vi), (vii), and (viii), (viii) is preferred since it has x_1 as the dependent variable. As well, among 4.5(iii), (iv), and (v), (v) is preferred since it uses x_1 and x_2 as the dependent variables.

10.8 Rationale for Making Convention 10.6

Any systematic procedure that, given an arbitrary linear system, provides a specific reduced description of its solution space *must*, where several such descriptions are possible, make a choice among them. The standard procedure is to be presented in Sections 11 and 12; it is called the Gauss–Jordan algorithm. The choice that is given in Convention 10.6 is the one that is made by the Gauss–Jordan algorithm.

10.9 Definition

An $m \times n$ matrix, A, is said to be in *row-echelon form* if it has the first three of the following properties:

1. For some integer r, where $0 \leq r \leq m$, each of the first r rows of A contains at least one nonzero entry, while the last $m - r$ rows are all zero.

The effect of this condition is simply to group together at the bottom of the matrix any rows whose entries are all zero. See [F]10.1.

2. Each row that contains one or more nonzero entries has 1 as its first nonzero entry.

It is convenient to call the 1's referred to in this condition the *leading 1's*. It is also convenient to have a word denoting the columns in which the leading 1's occur; these will be called *pivot columns*.

3. If the leading 1's in rows 1 through r occur in columns $j_1, j_2, \ldots, j_r$, respectively, then $j_1 \lneqq j_2 \lneqq \cdots \lneqq j_r$.

In other words, for any two nonzero rows, the leading 1 in the lower row must occur strictly to the right of the leading 1 in the upper row.

A matrix in row-echelon form satisfying the following additional condition is said to be in *reduced row-echelon form*.

4. Each column that contains a leading 1 has *all* its other entries equal to zero.

It is easy to see that if a matrix is in row-echelon form, then any column that contains a leading 1 must contain zeros *below* that leading 1. The entries *above* the leading 1's, however, are arbitrary. The general $m \times n$ matrix in row-echelon form is shown in Figure 10.1; the asterisks represent arbitrary entries.

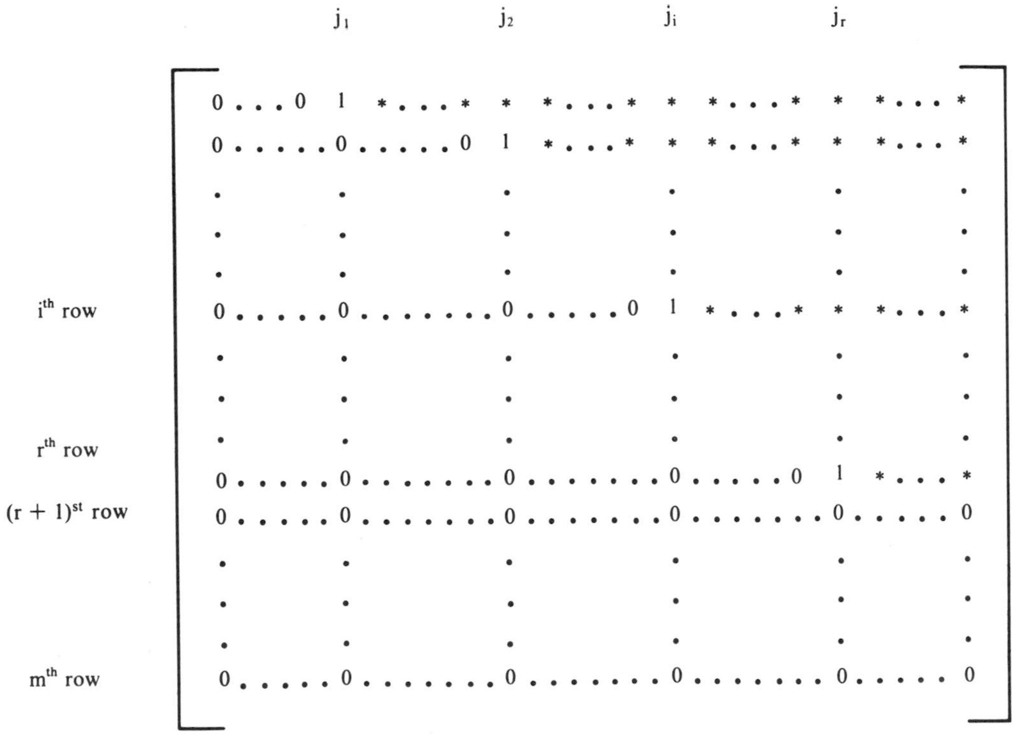

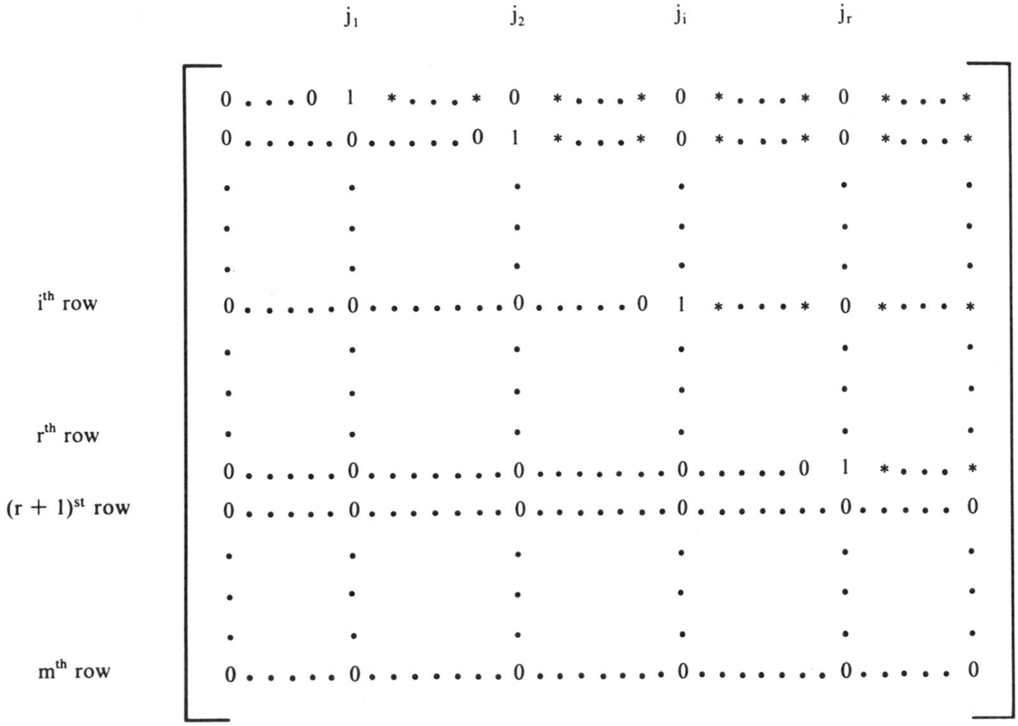

Figure 10.2

The general $m \times n$ matrix in reduced row-echelon form is shown in Figure 10.2; again, the positions containing asterisks can be filled arbitrarily.

The relation between this concept and the examples from 10.4 and 10.5 is now clear. If an $m \times n$ matrix A is in reduced row-echelon form, then we can obtain, by inspection, a reduced description of the solution spaces for both of the following:

(i) The $m \times (n-1)$ linear system whose augmented matrix is A.
(ii) The $m \times n$ homogeneous linear system whose coefficient matrix is A.

10.10 Example

Let

$$A = \begin{bmatrix} 1 & -2 & 0 & 0 & 1 & 0 \\ 0 & 0 & 1 & 0 & 5 & 4 \\ 0 & 0 & 0 & 1 & 0 & 4 \\ 0 & 0 & 0 & 0 & 0 & 0 \end{bmatrix}$$

Note that A is in reduced row-echelon form.

(i) The 4 × 5 linear system having A as its augmented matrix is

$$\begin{aligned} x_1 - 2x_2 \qquad\qquad\qquad + \;\; x_5 &= 0 \\ x_3 \qquad + 5x_5 &= 4 \\ x_4 \qquad\quad &= 4 \\ 0x_1 + 0x_2 + 0x_3 + 0x_4 + 0x_5 &= 0 \end{aligned}$$

A reduced description of its solution space, S, is

$$S = \{(x_1, x_2, x_3, x_4, x_5) \in \mathbb{R}^5: \\ x_1 = 2x_2 - x_5,\ x_3 = -5x_5 + 4,\ x_4 = 4\}$$

(ii) The 4 × 6 homogeneous linear system having A as its coefficient matrix is

$$\begin{aligned} x_1 - 2x_2 \qquad\qquad\qquad + \;\; x_5 \qquad\quad &= 0 \\ x_3 \qquad + 5x_5 + 4x_6 &= 0 \\ x_4 \qquad\qquad + 4x_6 &= 0 \\ 0x_1 + 0x_2 + 0x_3 + 0x_4 + 0x_5 + 0x_6 &= 0 \end{aligned}$$

A reduced description of its solution space, S', is

$$S' = \{(x_1, x_2, x_3, x_4, x_5, x_6) \in \mathbb{R}^6: x_1 = 2x_2 - x_5, \\ x_3 = -5x_5 - 4x_6,\ x_4 = -4x_6\}$$

10.11 Example

Let

$$B = \begin{bmatrix} 1 & -2 & 0 & 0 & 1 & 0 \\ 0 & 0 & 1 & 0 & 5 & 0 \\ 0 & 0 & 0 & 1 & 0 & 0 \\ 0 & 0 & 0 & 0 & 0 & 1 \end{bmatrix}$$

Note that B is in reduced row-echelon form.

(i) The 4 × 5 linear system having B as its augmented matrix is

$$\begin{aligned} x_1 - 2x_2 \qquad\qquad\qquad + \;\; x_5 &= 0 \\ x_3 \qquad + 5x_5 &= 0 \\ x_4 \qquad\quad &= 0 \\ 0x_1 + 0x_2 + 0x_3 + 0x_4 + 0x_5 &= 1 \end{aligned}$$

The solution space of this system is obviously the empty set, since no 5-tuple can satisfy the fourth equation.

(ii) The 4×6 homogeneous linear system having B as its coefficient matrix is

$$\begin{aligned} x_1 - 2x_2 \qquad\qquad + x_5 \qquad &= 0 \\ x_3 \qquad + 5x_5 \qquad &= 0 \\ x_4 \qquad\qquad &= 0 \\ x_6 &= 0 \end{aligned}$$

A reduced description of its solution space, S'', is

$$S'' = \{(x_1, x_2, x_3, x_4, x_5, x_6) \in \mathbb{R}^6 \colon x_1 = 2x_2 - x_5, \; x_3 = -5x_5, \; x_4 = 0, \; x_6 = 0\}$$

If the augmented matrix of a linear system is in row-echelon form but not in reduced row-echelon form, then we can still obtain a reduced description of its solution space. But we cannot, in this case, obtain it by inspection; some algebraic manipulations will be required as well. This process is often called *back substitution* and is best explained with an example.

10.12 Example

The augmented matrix of the linear system

$$\begin{aligned} x_1 - 2x_2 + 5x_3 + \tfrac{3}{2}x_4 - 4x_5 &= 2 \\ x_2 - x_3 - \tfrac{7}{8}x_4 + x_5 &= \tfrac{5}{4} \\ x_4 &= 2 \end{aligned}$$

is

$$\begin{bmatrix} 1 & -2 & 5 & \tfrac{3}{2} & -4 & 2 \\ 0 & 1 & -1 & -\tfrac{7}{8} & 1 & \tfrac{5}{4} \\ 0 & 0 & 0 & 1 & 0 & 2 \end{bmatrix}$$

This matrix is in row-echelon form but is not in reduced row-echelon form. There are some nonzero entries above the leading 1's in columns 2 and 4.

A description of the solution space, S_0, that uses x_1, x_2, and x_4 as dependent variables is easily obtained:

$$S_0 = \{(x_1, x_2, x_3, x_4, x_5) \in \mathbb{R}^5 \colon x_1 = 2x_2 - 5x_3 - \tfrac{3}{2}x_4 + 4x_5 + 2, \; x_2 = x_3 + \tfrac{7}{8}x_4 - x_5 + \tfrac{5}{4}, \; x_4 = 2\}$$

But this description is not reduced: the dependent variables are not expressed as affine functions of the parameters alone. The equation for x_1 involves x_2 and x_4, and the equation for x_2 involves x_4. What we *can* obtain, from the fact that the augmented matrix is in row-echelon form, is a description of the solution space in which the equation for the ith dependent variable, x_{d_i}, although it may involve other dependent variables, will only involve those with higher subscript. This is the effect of the second and third conditions in Definition 10.9.

This is what makes back substitution possible. The last equation will therefore express the dependent variable with the highest subscript, x_{d_r}, as an affine function of the parameters alone. This value for x_{d_r} can then be substituted into the next-to-last equation to obtain an equation for $x_{d_{r-1}}$ as an affine function of the parameters alone.

In the preceding example, substituting $x_4 = 2$ into the equation $x_2 = x_3 + \frac{7}{8}x_4 - x_5 + \frac{5}{4}$ yields $x_2 = x_3 - x_5 + 3$.

These values for $x_{d_{r-1}}$ and x_{d_r} can then be substituted into the equation for $x_{d_{r-2}}$ to obtain an equation that will express it as an affine function of the parameters alone. This process is continued until, eventually, we substitute into the first equation to obtain an equation for x_{d_1} as a function of the parameters alone.

Continuing with the same example, substituting $x_4 = 2$ and $x_2 = x_3 - x_5 + 3$ into the equation $x_1 = 2x_2 - 5x_3 - \frac{3}{2}x_4 + 4x_5 + 2$ yields $x_1 = -3x_3 + 2x_5 + 5$.

A reduced description of S_0 is given, therefore, by

$$
\begin{aligned}
S_0 = \{(x_1, x_2, x_3, x_4, x_5) \in \mathbb{R}^5\colon x_1 &= -3x_3 + 2x_5 + 5, \\
x_2 &= x_3 - x_5 + 3, \\
x_4 &= 2\}
\end{aligned}
$$

Note that the set S_0 used in this example was used earlier in Example 10.1; the difference between the nonreduced and the reduced descriptions of S_0 in Example 10.12 is analogous to the different answers to Questions 4 and 5 in Example 10.1.

The process of back substitution can also be used to get a reduced description of the solution space of a homogeneous linear system whose coefficient matrix is in row-echelon form.

Problem Set 10

10.1. Go back and do any of the exercises from Section 10 that you may have skipped.

*10.2. Let A denote any of the matrices in the following list.

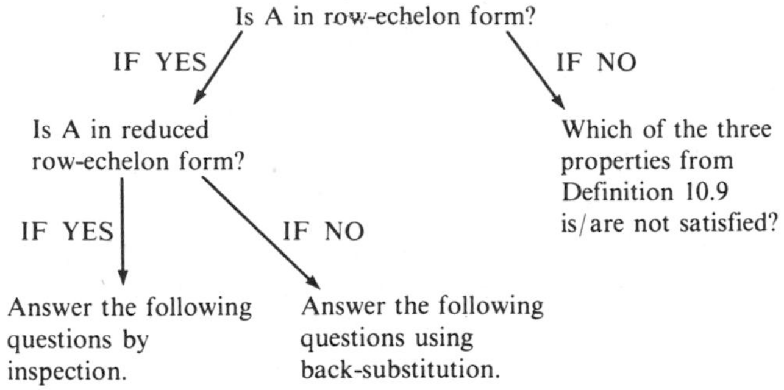

Is A in row-echelon form?
IF YES
IF NO
Is A in reduced row-echelon form?
Which of the three properties from Definition 10.9 is/are not satisfied?
IF YES
IF NO
Answer the following questions by inspection.
Answer the following questions using back-substitution.

Figure 10.3

Find reduced descriptions of the solution spaces of both:

(i) The linear system whose augmented matrix is A.

(ii) The homogeneous linear system whose coefficient matrix is A.

Also, identify the geometric nature of these solution spaces (see Problem 4.2).

(i) $\begin{bmatrix} 1 & 0 & 0 & 1 \\ 0 & 1 & 0 & 2 \\ 0 & 0 & 1 & 3 \end{bmatrix}$ (ii) $\begin{bmatrix} 1 & 0 & 0 & 1 \\ 0 & 2 & 0 & 2 \\ 0 & 0 & 3 & 3 \end{bmatrix}$

(iii) $\begin{bmatrix} 1 & 0 & 0 \\ 0 & 1 & 0 \\ 0 & 0 & 1 \end{bmatrix}$ (iv) $\begin{bmatrix} 1 & 0 & 18 \\ 0 & 1 & 17 \\ 0 & 0 & 0 \end{bmatrix}$

(v) $\begin{bmatrix} 1 & 0 & 2 & 0 & 4 \\ 0 & 1 & 3 & 0 & 6 \\ 0 & 0 & 0 & 1 & -1 \end{bmatrix}$ (vi) $\begin{bmatrix} 1 & 0 & 5 \\ 0 & 1 & -5 \end{bmatrix}$

(vii) $\begin{bmatrix} 1 & 3 & 2 & -1 & 4 \\ 0 & 1 & 3 & 2 & -1 \\ 0 & 0 & 0 & 0 & 0 \\ 0 & 0 & 0 & 1 & 6 \end{bmatrix}$ (viii) $\begin{bmatrix} 1 & 0 & 8 & 3 \\ 0 & 1 & 2 & -6 \\ 0 & 0 & 0 & 1 \end{bmatrix}$

(ix) $\begin{bmatrix} 1 & -1 & 10 & -4 & 4 \\ 0 & 1 & 0 & 0 & 17 \\ 0 & 0 & 1 & 4 & 2 \end{bmatrix}$ (x) $\begin{bmatrix} 1 & 4 & -1 & 0 & 2 & 2 \\ 0 & 0 & 0 & 1 & -2 & 1 \end{bmatrix}$

(xi) $\begin{bmatrix} 1 & 4 & 0 & 0 & 2 & 2 \\ 0 & 0 & 0 & 1 & -2 & 1 \end{bmatrix}$ (xii) $\begin{bmatrix} 1 & 4 & -1 & 0 & 2 & 0 \\ 0 & 0 & 0 & 1 & -2 & 0 \end{bmatrix}$

(xiii) $\begin{bmatrix} 0 & 1 & 4 & 0 & 2 & 2 \\ 0 & 0 & 0 & 1 & -2 & 1 \end{bmatrix}$ (xiv) $\begin{bmatrix} 1 & 0 & 0 & 0 & \frac{3}{2} \\ 0 & 1 & 7 & 0 & 0 \\ 0 & 0 & 0 & 1 & 0 \end{bmatrix}$

(xv) $\begin{bmatrix} 1 & 0 & 8 & 0 & 2 \\ 0 & 0 & 0 & 1 & 8 \\ 0 & 1 & 2 & 0 & 9 \end{bmatrix}$ (xvi) $\begin{bmatrix} 1 & 3.5 & 2.4 & -1.1 \\ 0 & 1 & 3.6 & 2.7 \\ 0 & 0 & 1 & 4.4 \end{bmatrix}$

(xvii) $\begin{bmatrix} 1 & 0 & 2 & 0 \\ 0 & -1 & 3 & 0 \\ 0 & 0 & 0 & 1 \\ 0 & 0 & 0 & 0 \end{bmatrix}$ (xviii) $\begin{bmatrix} 1 & 0 & 0 & 4 \\ 0 & 1 & 0 & 3 \\ 0 & 0 & 1 & 2 \\ 0 & 0 & 1 & 1 \end{bmatrix}$

10.3. Let A be an $m \times n$ matrix in reduced row-echelon form, and let B be the $m \times p$ matrix consisting of the first p-many columns of A. Prove that B is also in reduced row-echelon form.

*10.4. Prove that if A is a square matrix (say A is $n \times n$) in reduced row-echelon form then either A has an all-zero row or else $A = I_n$.

*10.5. A is a nonzero 2×2 matrix in reduced row-echelon form. The sum of the entries in A is zero. What is A?

11

Remarks about the Gauss – Jordan Algorithm

The Gauss–Jordan algorithm accepts as input an arbitrary $m \times n$ matrix A. The output of the algorithm will be an $m \times n$ matrix in reduced row-echelon form and will be denoted by GJ(A).

This algorithm is the principal tool of the subject. It was invented for solving systems of linear equations. As with many tools, however, it turns out to be useful for several other functions than the one for which it was originally designed. We will use the Gauss–Jordan algorithm:

1. For solving systems of linear equations.
2. For discovering whether a square matrix A is invertible, and, if it is, for finding A^{-1} (see Section 14).
3. For finding the dimensions of and bases for the various subspaces, RS(A), CS(A), NS(A), and $NS(A^T)$, associated with an arbitrary matrix A (see Section 15).
4. For calculating determinants (see Section 24).

Because of this, we wish to emphasize from the very beginning the role of the Gauss–Jordan algorithm *as a tool*. There is an unfortunate tendency among beginning students to focus primarily on the algorithm *itself*, rather than on how the algorithm can be used. In and of itself, a tool has little intrinsic worth; its real value is only manifested in the hands of a master craftsperson who knows how and when to use it.

(I myself am totally unskilled in ceramics or plumbing; having a potter's wheel or a welding torch in my basement would merely add to the clutter. The same can be said for having the Gauss–Jordan algorithm in your brain.)

Do not get the illusion that you have acquired something *substantial* once you have understood the basic operation of the algorithm. In fact, this is precisely the part of linear algebra that can be, and usually is, automated. That is, the algorithm is implemented on a computer. What remains for the human problem solver are the difficult parts:

1. To analyze the problem and decide what matrix to input to the computer.
2. To analyze and properly interpret, in relation to the problem that was to be solved, the matrix that is output from the computer.

To use this tool effectively, you need to be familiar with how it works and to be convinced that it does perform as claimed. You should also at least be aware of the most common potential sources of malfunctions and misuses, or even abuses. The remainder of Section 11 and Section 12 address these questions.

Of course, if you are interested in comparing various brands of the same tool, or if you are concerned with tool design, modification, or improvement, then you will need to know much more about the tool's inner workings; this is often the subject of a separate course in numerical analysis.

Since the Gauss–Jordan algorithm was invented for solving systems of linear equations, we will motivate the details of the algorithm (to follow in Section 12) by examining its intended use.

11.1 Definition

A system of linear equations will be called *reduced* if its augmented matrix is in reduced row-echelon form.

11.2 Exercise

Show that a homogeneous system is reduced if and only if its coefficient matrix is in reduced row-echelon form.

Using this terminology, we can summarize the essence of Section 10: a reduced system of equations can be solved by inspection.

11.3 Definition

Two systems of linear equations in n variables are *equivalent* if they have the same solution space.

Consider an arbitrary system of m linear equations in n variables:

$$\begin{aligned}
a_{11}x_1 + a_{12}x_2 + \cdots + a_{1n}x_n &= b_1 \\
a_{21}x_1 + a_{22}x_2 + \cdots + a_{2n}x_n &= b_2 \\
&\vdots \\
a_{m1}x_1 + a_{m2}x_2 + \cdots + a_{mn}x_n &= b_m
\end{aligned}$$

To save space, let $EQ_1, EQ_2, \ldots, EQ_m$ denote these equations. Thus EQ_i is the equation $a_{i1}x_1 + a_{i2}x_2 + \cdots + a_{in}x_n = b_i$.

We wish to consider three types of operations that can be performed on such a system of equations. We will prove that each type of operation preserves the solution space, in the sense that if (B) denotes the system of equations *before* one of these operations is performed and (A) represents the system of equations that results *after* this operation is performed, then (B) and (A) are equivalent systems of equations.

The three types of operations are:

Type I. Multiply one of the equations by a nonzero scalar, s.
Type II. Interchange the order of any two of the equations.
Type III. Add to any one of the equations a scalar multiple of another one of the equations.

11.4 Example

Let (B) denote the 3×2 system

$$\begin{aligned}
3x_1 + 5x_2 &= 7 \\
4x_1 - 9x_2 &= -6 \\
10x_1 + x_2 &= 8
\end{aligned}$$

After the type I operation of multiplying the third equation by $\frac{1}{2}$, the resulting system (A) is

$$\begin{aligned}
3x_1 + 5x_2 &= 7 \\
4x_1 - 9x_2 &= -6 \\
5x_1 + \frac{1}{2}x_2 &= 4
\end{aligned}$$

After the type II operation of interchanging the second and third equations, the resulting system (A) is

$$\begin{aligned}
3x_1 + 5x_2 &= 7 \\
10x_1 + x_2 &= 8 \\
4x_1 - 9x_2 &= -6
\end{aligned}$$

After the type III operation of adding to the first equation -3 times the second equation, the resulting system (A) is

$$\begin{aligned} -9x_1 + 32x_2 &= 25 \\ 4x_1 - 9x_2 &= -6 \\ 10x_1 + x_2 &= 8 \end{aligned}$$

The general effect of one of these operations can be represented schematically as follows, where s and t are arbitrary scalars and $s \neq 0$.

(B)	(A)		
	Type I	*Type II*	*Type III*
EQ_1	EQ_1	EQ_1	EQ_1
EQ_2	EQ_2	EQ_2	EQ_2
$\vdots$	$\vdots$	$\vdots$	$\vdots$
EQ_i	sEQ_i	EQ_j	$EQ_i + tEQ_j$
$\vdots$	$\vdots$	$\vdots$	$\vdots$
EQ_j	EQ_j	EQ_i	EQ_j
$\vdots$	$\vdots$	$\vdots$	$\vdots$
EQ_m	EQ_m	EQ_m	EQ_m

Using this notation, we can state the following important theorem.

11.5 Theorem

For each of the preceding three types of operations, the two systems of equations (B) and (A) are equivalent.

Proof: We must show that an arbitrary n-tuple $\mathbf{u} = (u_1, u_2, \ldots, u_n)$ is in the solution space of (B) if and only if it is in the solution space of (A). More precisely, what we must prove, according to Definition 6.4, is that an arbitrary vector $\mathbf{u}$ satisfies *all* the equations in (B) if and only if $\mathbf{u}$ satisfies *all* the equations in (A). The proof breaks into three cases, depending on the type of operation performed.

Type I: Assume that $\mathbf{u}$ satisfies all the equations in (B). Then certainly $\mathbf{u}$ satisfies all the equations in (A) except perhaps the ith equation, sEQ_i, from (A). But sEQ_i is the equation

$$sa_{i1}x_1 + sa_{i2}x_2 + \cdots + sa_{in}x_n = sb_{in}$$

This equation, too, is satisfied by $\mathbf{u}$, since by assumption

$$a_{i1}u_1 + a_{i2}u_2 + \cdots + a_{in}u_n = b_i$$

Thus

$$sa_{i1}u_1 + sa_{i2}u_2 + \cdots + sa_{in}u_n = sb_i$$

Conversely, assume that **u** satisfies all the equations in (A). In particular, we are assuming that **u** satisfies sE_i; that is,

$$sa_{i1}u_1 + sa_{i2}u_2 + \cdots + sa_{in}u_n = sb_i \qquad (*)$$

Then, as before, **u** trivially satisfies all the equations in (B) except perhaps the ith equation, EQ_i, in (B). But this too is satisfied by **u** because $s \neq 0$; thus we can multiply both sides of $(*)$ by $1/s$ to conclude

$$a_{i1}u_1 + a_{i2}u_2 + \cdots + a_{in}u_n = b_i$$

Type II: This case is obvious since belonging to the solution space of a linear system does not depend on the order in which the equations are listed.

Type III: Assume that **u** satisfies all the equations in (B). Again, trivially, **u** satisfies all the equations in (A) except perhaps the ith equation, $EQ_i + tEQ_j$, from (A). But $EQ_i + tEQ_j$ is the equation

$$(a_{i1} + ta_{j1})x_1 + (a_{i2} + ta_{j2})x_2 + \cdots + (a_{in} + ta_{jn})x_n = b_i + tb_j$$

By assumption, $a_{i1}u_1 + a_{i2}u_2 + \cdots + a_{in}u_n = b_i$ and $a_{j1}u_1 + a_{j2}u_2 + \cdots + a_{jn}u_n = b_j$. Thus, if we multiply the latter equation by t and add the result to the former equation, we conclude that **u** satisfies $EQ_i + tEQ_j$.

Conversely, assume that **u** satisfies all the equations in (A). The only equation in (B) that is not already in (A) is EQ_i. To prove that **u** satisfies EQ_i, we use our assumption that **u** satisfies both $EQ_i + tEQ_j$ and EQ_j. Thus, by assumption, $(a_{i1} + ta_{j1})u_1 + (a_{i2} + ta_{j2})u_2 + \cdots + (a_{in} + ta_{jn})u_n = b_i + tb_j$ and $a_{j1}u_1 + a_{j2}u_2 + \cdots + a_{jn}u_n = b_j$. If we multiply the latter equation by $-t$ and add the result to the former equation, we conclude that **u** satisfies EQ_i. ■

11.6 Remark

It is important to realize why $s \neq 0$ is required in Theorem 11.5. The type I operation in which $s = 0$ does not preserve solution spaces. For example, if (B) were the single equation $x_1 - x_2 = 0$, (A) would be the equation $0x_1 - 0x_2 = 0$. The solution space of (B) is the 45° line in $\mathbb{R}^2$, while the solution space of (A) is the whole of $\mathbb{R}^2$.

11.7 Corollary

If (B) is a linear system and if (B′) is a system obtained from (B) by any finite sequence of operations of types I, II, or III, then the solution space of (B′) is the same as that of (B). ■

It is this Corollary that justifies the method we will now present for solving linear systems. The idea is to take an arbitrary system (B) and apply to it systematically a finite sequence of operations of types I, II, or III until the resulting system, (B′), is a reduced system of equations; from this, a reduced description of the solution space of (B′), which is also the solution space of (B), can be obtained by inspection, as explained in Section 10.

Matrices enter into this by obviating the need, while performing type I, II or III operations, to keep writing down the variables $x_1, x_2, \ldots, x_n$ each time. It is matrix notation that keeps track for us of which coefficient belongs in which equation with which variable.

In view of Theorem 11.5, we define three types of operations that can be performed on an arbitrary $m \times n$ matrix A.

11.8 Definition

An *elementary row-operation* is one of the following three types:

Type I: Multiply a single row by a nonzero scalar, s. The operation of multiplying the ith row of a matrix by s is denoted by sR_i.

Type II: Interchange two rows. The operation of interchanging the ith and j^{th} rows of a matrix is denoted by $R_i \leftrightarrow R_j$.

Type III: Add to one row a scalar multiple of another row. The operation of adding t times the jth row to the ith row is denoted by $R_i + tR_j$.

The letter $\mathcal{O}$ stands for an arbitrary elementary row-operation and $\mathcal{O}(\text{A})$ denotes the matrix that results when the operation $\mathcal{O}$ is performed on the matrix A.

11.9 Examples

Let

$$\text{A} = \begin{bmatrix} 3 & 5 & 7 \\ 4 & -9 & -6 \\ 10 & 1 & 8 \end{bmatrix}$$

(i) If $\mathcal{O}$ is the type I operation $\frac{1}{2}R_3$, then

$$\mathcal{O}(\text{A}) = \begin{bmatrix} 3 & 5 & 7 \\ 4 & -9 & -6 \\ 5 & \frac{1}{2} & 4 \end{bmatrix}$$

(ii) If $\mathcal{O}$ is the type II operation $R_2 \leftrightarrow R_3$, then

$$\mathcal{O}(\mathrm{A}) = \begin{bmatrix} 3 & 5 & 7 \\ 10 & 1 & 8 \\ 4 & -9 & -6 \end{bmatrix}$$

(iii) If $\mathcal{O}$ is the type III operation $R_1 + (-3)R_2$ (which will normally be expressed as $R_1 - 3R_2$), then

$$\mathcal{O}(\mathrm{A}) = \begin{bmatrix} -9 & 32 & 25 \\ 4 & -9 & -6 \\ 10 & 1 & 8 \end{bmatrix}$$

This example should be compared with Example 11.4.

We mentioned at the beginning of this section that the Gauss–Jordan algorithm accepts an arbitrary $m \times n$ matrix A as input, and produces, as output, an $m \times n$ matrix in reduced row-echelon form, denoted by GJ(A). It accomplishes this by performing, on A, a predetermined finite sequence of elementary row-operations. Precisely which elementary row-operations are performed will depend on the input matrix A, but this finite sequence of elementary row-operations *is* predetermined in the sense that the algorithm itself consists of a fixed set of instructions, given once and for all (here, it will be given in Section 12), for performing elementary row-operations on whatever matrix is input.

It is the fact that the algorithm uses only elementary row-operations that guarantees, in conjunction with Theorem 11.5, that the Gauss–Jordan algorithm performs as intended for solving linear systems. That is, *if* the $m \times n$ matrix A that is input to the algorithm happens to be the augmented matrix of an $m \times (n - 1)$ linear system, then the output GJ(A) will be the augmented matrix of a reduced $m \times (n - 1)$ linear system whose solution space is the same as that of the original system and can be obtained by inspection.

11.10 Example

The 3×5 system of equations from Example 10.1 is

$$\begin{aligned} 2x_1 - 4x_2 + 10x_3 + 3x_4 - 8x_5 &= 4 \\ x_1 + 2x_2 + x_3 - 2x_4 \qquad &= 7 \\ 3x_1 + 5x_2 + 4x_3 - 10x_4 - x_5 &= 10 \end{aligned}$$

Its augmented matrix is the 3×6 matrix

$$\mathrm{A} = \begin{bmatrix} 2 & -4 & 10 & 3 & -8 & 4 \\ 1 & 2 & 1 & -2 & 0 & 7 \\ 3 & 5 & 4 & -10 & -1 & 10 \end{bmatrix}$$

For the moment, we will take it on faith that

$$\text{GJ(A)} = \begin{bmatrix} 1 & 0 & 3 & 0 & -2 & 5 \\ 0 & 1 & -1 & 0 & 1 & 3 \\ 0 & 0 & 0 & 1 & 0 & 2 \end{bmatrix}$$

We conclude that the reduced system of equations

$$\begin{aligned} x_1 \quad\quad + 3x_3 \quad\quad - 2x_5 &= 5 \\ x_2 - x_3 \quad\quad + x_5 &= 3 \\ x_4 \quad\quad &= 2 \end{aligned}$$

is equivalent to the original system.

A reduced description of the solution space, S_0, of the system is therefore

$$S_0 = \{(x_1, x_2, x_3, x_4, x_5) \in \mathbb{R}^5 \colon x_1 = -3x_3 + 2x_5 + 5,\; x_2 = x_3 - x_5 + 3,\; x_4 = 2\}$$

Similarly, if the input A to the algorithm is the coefficient matrix of an $m \times n$ homogeneous linear system, then a reduced description of its solution space can be obtained by inspection from GJ(A).

11.11 Example

The 3×3 homogeneous system

$$\begin{aligned} 2x_1 - 3x_2 + x_3 &= 0 \\ -7x_1 + 6x_2 + 4x_3 &= 0 \\ -2x_1 + 12x_2 - 16x_3 &= 0 \end{aligned}$$

has

$$\text{A} = \begin{bmatrix} 2 & -3 & 1 \\ -7 & 6 & 4 \\ -2 & 12 & -16 \end{bmatrix}$$

as its coefficient matrix. Taking it on faith for the moment that

$$\text{GJ(A)} = \begin{bmatrix} 1 & 0 & -2 \\ 0 & 1 & -\frac{5}{3} \\ 0 & 0 & 0 \end{bmatrix}$$

we conclude that the reduced system of equations

$$\begin{aligned} x_1 \qquad - 2x_3 &= 0 \\ x_2 - \frac{5}{3}x_3 &= 0 \end{aligned}$$

is equivalent to the original one, and that its solution space is therefore

$$S = \left\{(x_1, x_2, x_3) \in \mathbb{R}^3 \colon x_1 = 2x_3,\ x_2 = \frac{5}{3}x_3\right\}$$

11.12 Remark

In the previous two examples, we wrote down the reduced system of equations whose augmented matrix is GJ(A). Actually, direct passage from GJ(A) to the solution space is easy to do mentally (with a bit of practice) and we will never again explicitly write down this intermediate step.

11.13 Caution

When using the Gauss–Jordan algorithm to solve linear systems, it is the user's responsibility to remember whether the $m \times n$ input matrix A is the augmented matrix of an $m \times (n-1)$ linear system or is the coefficient matrix of a homogeneous $m \times n$ linear system and to interpret the output, GJ(A), accordingly.

For example, in 11.11 we saw that if

$$\mathrm{A} = \begin{bmatrix} 2 & -3 & 1 \\ -7 & 6 & 4 \\ -2 & 12 & -16 \end{bmatrix} \quad \text{then} \quad \mathrm{GJ(A)} = \begin{bmatrix} 1 & 0 & -2 \\ 0 & 1 & -\frac{5}{3} \\ 0 & 0 & 0 \end{bmatrix}$$

and drew a certain conclusion from these data.

Another conclusion that can be drawn from these same data is that the solution space of the 3×2 system

$$\begin{aligned} 2x_1 - \ 3x_2 &= 1 \\ -7x_1 + \ 6x_2 &= 4 \\ -2x_1 + 12x_2 &= -16 \end{aligned}$$

is the single point $\{(-2, -\frac{5}{3})\}$.

It is especially important to recall this distinction when interpreting what is sometimes called the *mark of inconsistency*, that is, a row of the form $[0, 0, \ldots, 0, t]$, where $t \neq 0$. If the *augmented* matrix of an *inhomogeneous* linear system (B) contains a row that is all zero except for a nonzero last entry, then (B) is inconsistent. This is obvious since one of the equations in (B) is $0x_1 + 0x_2 + \cdots + 0x_{n-1} = t \neq 0$. When applying the Gauss–Jordan algorithm *in this context*, if the matrix obtained at some stage of the algorithm contains such a row, you may immediately conclude that the solution space of the system is the empty set, $\varnothing$.

Here is an example of this phenomenon. Given the fact that if

$$\mathrm{A} = \begin{bmatrix} 7 & 1 & 17 \\ 4 & -2 & 2 \\ 7 & 8 & 7 \end{bmatrix} \quad \text{then} \quad \mathrm{GJ(A)} = \begin{bmatrix} 1 & 0 & 0 \\ 0 & 1 & 0 \\ 0 & 0 & 1 \end{bmatrix}$$

we may conclude the following:

(i) The 3×2 system

$$\begin{aligned} 7x_1 + x_2 &= 17 \\ 4x_1 - 2x_2 &= 2 \\ 7x_1 + 8x_2 &= 7 \end{aligned}$$

is inconsistent.

(ii) The 3×3 homogeneous system

$$\begin{aligned} 7x_1 + x_2 + 17x_3 &= 0 \\ 4x_1 - 2x_2 + 2x_3 &= 0 \\ 7x_1 + 8x_2 + 7x_3 &= 0 \end{aligned}$$

has only the trivial solution $(0, 0, 0)$.

We would like to conclude this section with some observations about elementary row-operations.

11.14 Observation

Each elementary row-operation $\mathcal{O}$ is reversible in the sense that there is another elementary row-operation $\mathcal{O}'$, of the same type as $\mathcal{O}$, that undoes what $\mathcal{O}$ does, and vice versa; specifically, $\mathcal{O}'(\mathcal{O}(\mathrm{A})) = \mathrm{A}$ and $\mathcal{O}(\mathcal{O}'(\mathrm{A})) = \mathrm{A}$ for any matrix A.

Proof: If $\mathcal{O}$ is sR_i, where $s \neq 0$, then $\mathcal{O}'$ is $(1/s)R_i$. If $\mathcal{O}$ is $R_i \leftrightarrow R_j$, then $\mathcal{O}'$ is also $R_i \leftrightarrow R_j$. If $\mathcal{O}$ is $R_i + tR_j$, then $\mathcal{O}'$ is $R_i - tR_j$. ■

Let us write $\mathrm{A} \underset{r}{\sim} \mathrm{B}$ to denote that the matrix B can be obtained from the matrix A by a sequence of elementary row-operations. It is clear that $\mathrm{A} \underset{r}{\sim} \mathrm{A}$, and that if $\mathrm{A} \underset{r}{\sim} \mathrm{B}$ and $\mathrm{B} \underset{r}{\sim} \mathrm{C}$, then $\mathrm{A} \underset{r}{\sim} \mathrm{C}$. It follows from 11.14 that if $\mathrm{A} \underset{r}{\sim} \mathrm{B}$, then $\mathrm{B} \underset{r}{\sim} \mathrm{A}$; in other words, if B can be obtained from A by a sequence of elementary row-operations, then by reversing these operations A can be obtained from B by a sequence of elementary row-operations. In fancy language, if

$$B = \mathcal{O}_k(\cdots \mathcal{O}_2(\mathcal{O}_1(A)) \cdots)$$

then

$$A = \mathcal{O}_1'(\cdots \mathcal{O}_{k-1}'(\mathcal{O}_k'(B)) \cdots)$$

Thus it is reasonable to use the following name for the relation $\underset{r}{\sim}$.

11.15 Definition

The relation $\underset{r}{\sim}$ will be called *row-equivalence*. If $A \underset{r}{\sim} B$, we say that A and B are row-equivalent.

Using this terminology, the Gauss–Jordan algorithm, to be presented in Section 12, should be viewed as a proof of the following fundamental theorem.

11.16 Theorem

Every matrix is row-equivalent to a matrix in reduced row-echelon form.

See [F]11.1.

Problem Set 11

11.1. Go back and do any of the exercises in Section 11 that you may have skipped.

*11.2. Let

$$A = \begin{bmatrix} 4 & 0 & -\frac{1}{2} \\ 1 & -1 & 6 \\ -3 & \frac{1}{3} & 6 \end{bmatrix}, \qquad B = \begin{bmatrix} 2 & 1 & 0 & 1 \\ 3 & 1 & 0 & 3 \end{bmatrix},$$

$$C = \begin{bmatrix} -2 & 8 & 8 \\ 6 & 5 & 1 \\ 4 & 8 & -1 \\ 0 & 4 & 4 \end{bmatrix}$$

Let $\mathcal{O}_1$ be $R_2 - \frac{3}{2}R_1$, let $\mathcal{O}_2$ be $R_3 + 2R_1$, let $\mathcal{O}_3$ be $\frac{1}{4}R_1$, and let $\mathcal{O}_4$ be $R_2 \leftrightarrow R_3$.

(i) Except for $\mathcal{O}_2(B)$ and $\mathcal{O}_4(B)$, which are not defined, compute $\mathcal{O}_i(P)$ for $i = 1, 2, 3, 4$ and $P = A, B, C$.

(ii) Find $\mathcal{O}_2(\mathcal{O}_1(A))$ and $\mathcal{O}_1(\mathcal{O}_2(A))$.
(iii) Find $\mathcal{O}_3(\mathcal{O}_1(A))$ and $\mathcal{O}_1(\mathcal{O}_3(A))$.
(iv) Let $\mathcal{O}_i'$ denote the "reverse" of the operation $\mathcal{O}_i$, as defined in Observation 11.14. Find $\mathcal{O}_3'(A)$, $\mathcal{O}_2'(C)$, $\mathcal{O}_1'(B)$, $\mathcal{O}_2(\mathcal{O}_2'(C))$, and $\mathcal{O}_2'(\mathcal{O}_2(C))$.

*11.3. Let

$$B = \begin{bmatrix} 2 & 1 & 0 & 1 \\ 3 & 1 & 0 & 3 \end{bmatrix}$$

(i) Let $\mathcal{O}_1$ be $\frac{1}{2}R_1$, $\mathcal{O}_2$ be $R_2 - 3R_1$, $\mathcal{O}_3$ be $-2R_2$, and $\mathcal{O}_4$ be $R_1 - \frac{1}{2}R$. Find $\mathcal{O}_4(\mathcal{O}_3(\mathcal{O}_2(\mathcal{O}_1(B))))$.
(ii) Let $\mathcal{O}_1$ be $R_2 - R_1$, let $\mathcal{O}_2$ be $R_1 - 2R_2$, and let $\mathcal{O}_3$ be $R_1 \leftrightarrow R_2$. Find $\mathcal{O}_3(\mathcal{O}_2(\mathcal{O}_1(B)))$.

11.4. Prove that an all-zero column can be neither created nor destroyed by any sequence of elementary row-operations. More specifically, prove that if $A \underset{r}{\sim} B$, then, for arbitrary j, the jth column of A is all zero if and only if the jth column of B is all zero.

*11.5. Let

$$A = \begin{bmatrix} 1 & 1 & 1 & 1 \\ 4 & -1 & 0 & 3 \end{bmatrix}$$

(i) Let $\mathcal{O}_1$ be $R_2 + R_1$ and let $\mathcal{O}_2$ be $R_1 - R_2$; find $\mathcal{O}_2(\mathcal{O}_1(A))$.
(ii) Let $\mathcal{O}_1$ be $R_2 - 4R_1$, $\mathcal{O}_2$ be $R_1 + \frac{1}{4}R_2$, and $\mathcal{O}_3$ be $-\frac{1}{4}R_2$; find $\mathcal{O}_3(\mathcal{O}_2(\mathcal{O}_1(A)))$.
(iii) Let $\mathcal{O}_1$ be $R_2 - 4R_1$, $\mathcal{O}_2$ be $-\frac{1}{5}R_2$, and $\mathcal{O}_3$ be $R_1 - R_2$; find $\mathcal{O}_3(\mathcal{O}_2(\mathcal{O}_1(A)))$.
(iv) What is the relationship between this problem and Examples 4.5(ii), (iii), (iv), and (v)?

11.6. Suppose $A = [B \,\vdots\, C]$, where A is $m \times n$, B is $m \times p$, and C is $m \times (n - p)$, and that $\mathcal{O}$ is any elementary row-operation. Verify that $\mathcal{O}(A) = [\mathcal{O}(B) \,\vdots\, \mathcal{O}(C)]$.

12

Details of the Gauss – Jordan Algorithm

This section should be viewed as a proof of Theorem 11.16. The goal is to provide a set of instructions for performing, on an arbitrary matrix, a sequence of elementary row-operations in such a way that the resulting matrix is in reduced row-echelon form. The reasons for wanting to do this were covered in Sections 10 and 11; so in this section we concentrate exclusively on the details of implementation.

As might be expected, there is not just one way to accomplish this. Different textbooks will contain many variants of "the" Gauss–Jordan algorithm. The point of Theorem 11.16 is rather that a certain thing *can* be done, and we proceed to describe, in this section, *one* of the ways to do it.

We begin, not at the beginning, but with the principal subroutines, that is, with the separate building blocks from which the algorithm will eventually be constructed.

As motivation, recall that the main features of matrices in reduced row-echelon form are columns having a 1 as their single nonzero entry.

12.1 Example

Let

$$A = \begin{bmatrix} 6 & 2 & 4 \\ 4 & 2 & 1 \\ 0 & 5 & 7 \\ 7 & -4 & -6 \end{bmatrix}$$

To introduce a 1 as, say, the (2,2)-entry, with zeros *below* it, we could take $\mathcal{O}_0$ to be $\frac{1}{2}R_2$, $\mathcal{O}_1$ to be $R_3 - 5R_2$, and $\mathcal{O}_2$ to be $R_4 + 4R_2$. Thus

$$\mathcal{O}_0(\mathrm{A}) = \begin{bmatrix} 6 & 2 & 4 \\ 2 & 1 & \frac{1}{2} \\ 0 & 5 & 7 \\ 7 & -4 & -6 \end{bmatrix}, \qquad \mathcal{O}_1(\mathcal{O}_0(\mathrm{A})) = \begin{bmatrix} 6 & 2 & 4 \\ 2 & 1 & \frac{1}{2} \\ -10 & 0 & \frac{9}{2} \\ 7 & -4 & -6 \end{bmatrix},$$

and

$$\mathcal{O}_2(\mathcal{O}_1(\mathcal{O}_0(\mathrm{A}))) = \begin{bmatrix} 6 & 2 & 4 \\ 2 & 1 & \frac{1}{2} \\ -10 & 0 & \frac{9}{2} \\ 15 & 0 & -4 \end{bmatrix}.$$

12.2 Example

Again, let

$$\mathrm{A} = \begin{bmatrix} 6 & 2 & 4 \\ 4 & 2 & 1 \\ 0 & 5 & 7 \\ 7 & -4 & -6 \end{bmatrix}$$

To introduce a 1 as, say, the (1,3)-entry, with zeros *below* it, we could take $\mathcal{O}_0$ to be $\frac{1}{4}R_1$, $\mathcal{O}_1$ to be $R_2 - R_1$, $\mathcal{O}_2$ to be $R_3 - 7R_1$, and $\mathcal{O}_3$ to be $R_4 + 6R_1$. Thus

$$\mathcal{O}_0(\mathrm{A}) = \begin{bmatrix} \frac{3}{2} & \frac{1}{2} & 1 \\ 4 & 2 & 1 \\ 0 & 5 & 7 \\ 7 & -4 & -6 \end{bmatrix},$$

$$\mathcal{O}_1(\mathcal{O}_0(\mathrm{A})) = \begin{bmatrix} \frac{3}{2} & \frac{1}{2} & 1 \\ \frac{5}{2} & \frac{3}{2} & 0 \\ 0 & 5 & 7 \\ 7 & -4 & -6 \end{bmatrix},$$

$$\mathcal{O}_0(\mathcal{O}_1(\mathcal{O}_0(\mathrm{A}))) = \begin{bmatrix} \frac{3}{2} & \frac{1}{2} & 1 \\ \frac{5}{2} & \frac{3}{2} & 0 \\ -\frac{21}{2} & \frac{3}{2} & 0 \\ 7 & -4 & -6 \end{bmatrix},$$

$$\mathcal{O}_3(\mathcal{O}_2(\mathcal{O}_1(\mathcal{O}_0(\mathrm{A})))) = \begin{bmatrix} \frac{3}{2} & \frac{1}{2} & 1 \\ \frac{5}{2} & \frac{3}{2} & 0 \\ -\frac{21}{2} & \frac{3}{2} & 0 \\ 16 & -1 & 0 \end{bmatrix}.$$

12.3 Definition

To pivot downward on the $(i, j)^{th}$ entry, a_{ij}, of an $m \times n$ matrix A is to introduce a 1 as the new $(i, j)^{\text{th}}$ entry by multiplying the ith row of A by $1/a_{ij}$ (this can only be done, of course, if $a_{ij} \neq 0$) and to introduce zeros below the 1 by adding in turn, to each of the rows below, a suitable multiple of the new i^{th} row. Specifically, one forms the matrix $\mathcal{O}_{m-i}(\cdots \mathcal{O}_1(\mathcal{O}_0(\mathrm{A}))\cdots)$, where $\mathcal{O}_0$ is $(1/a_{ij})R_i$ and for $k = 1, 2, \ldots, m - i$, $\mathcal{O}_k$ is $R_{i+k} - a_{i+k,j}R_i$.

In this context a_{ij} is called the *pivot entry*. Examples 12.1 and 12.2 illustrate pivoting downward on a_{22} and on a_{13}, respectively.

We will want to introduce zeros *above* the leading 1's as well, so it is natural to define an analogue of 12.3.

The exact analogue would again involve $(1/a_{ij})R_i$ as the initial operation, as in the following example.

12.4 Example

Once more let

$$\mathrm{A} = \begin{bmatrix} 6 & 2 & 4 \\ 4 & 2 & 1 \\ 0 & 5 & 7 \\ 7 & -4 & -6 \end{bmatrix}$$

To introduce a 1 as the (3, 3)-entry and to introduce zeros above it, we perform the following sequence of elementary row-operations: $\mathcal{O}_0$ is $\frac{1}{7}R_3$, $\mathcal{O}_1$ is $R_2 - R_3$, and $\mathcal{O}_2$ is $R_1 - 4R_3$.

$$\mathcal{O}_0(\mathrm{A}) = \begin{bmatrix} 6 & 2 & 4 \\ 4 & 2 & 1 \\ 0 & \frac{5}{7} & 1 \\ 7 & -4 & -6 \end{bmatrix}, \qquad \mathcal{O}_1(\mathcal{O}_0(\mathrm{A})) = \begin{bmatrix} 6 & 2 & 4 \\ 4 & \frac{9}{7} & 0 \\ 0 & \frac{5}{7} & 1 \\ 7 & -4 & -6 \end{bmatrix}$$

and

$$\mathcal{O}_2(\mathcal{O}_1(\mathcal{O}_0(\mathrm{A}))) = \begin{bmatrix} 6 & -\frac{6}{7} & 0 \\ 4 & \frac{9}{7} & 0 \\ 0 & \frac{5}{7} & 1 \\ 7 & -4 & -6 \end{bmatrix}.$$

Since the Gauss–Jordan algorithm will only have occasion to pivot upward on entries that are already equal to 1, we will dispense with $\mathcal{O}_0$ in our definition and narrow its scope.

12.5 Definition

Assuming $a_{ij} = 1$, *to pivot upward on the* $(i, j)^{th}$ *entry*, 1, *of the* $m \times n$ *matrix* A is to form the matrix $\mathcal{O}_{i-1}(\cdots \mathcal{O}_2(\mathcal{O}_1(\mathrm{A}))\cdots)$, where for $k = 1, 2, \ldots, i-1$, $\mathcal{O}_k$ is $R_{i-k} - a_{i-k,j}R_i$.

It is clear that if the jth column of the $m \times n$ matrix A is not all zero, it can be transformed into a column having 1 as its single nonzero entry by selecting a fixed nonzero entry, say the ith, a_{ij}, from that column and pivoting first downward then upward on a_{ij}.

12.6 Exercise

Let A be the matrix from 12.1 and transform A into a matrix whose second column will be $\begin{bmatrix} 0 \\ 0 \\ 1 \\ 0 \end{bmatrix}$ by pivoting downward, then upward, on a_{32}.

The Gauss–Jordan algorithm can now be presented; its building blocks are pivoting (which involves only elementary row-operations of types I and III) and interchanging rows (type II operations). We have seen how pivoting can be used to produce one of the essential features of matrices in reduced row-echelon form. The problem is that this must be done *systematically* so that the effects of earlier operations are not undone by later operations. Notice, in Examples 12.1 and 12.2, how pivoting on a_{22} and a_{13} affects a_{31}.

It will be useful, before presenting the algorithm, to observe it at work on an example.

12.7 Notation

It is convenient to write $\mathrm{A} \overset{\mathcal{O}}{\underset{r}{\sim}} \mathrm{B}$ instead of $\mathrm{B} = \mathcal{O}(\mathrm{A})$ to express the fact that B is the matrix obtained by performing the elementary operation $\mathcal{O}$ on the matrix A. Thus, if $\mathrm{A} \overset{\mathcal{O}_1}{\underset{r}{\sim}} \mathrm{B} \overset{\mathcal{O}_2}{\underset{r}{\sim}} \mathrm{C} \overset{\mathcal{O}_3}{\underset{r}{\sim}} \mathrm{D}$, then $\mathrm{D} = \mathcal{O}_3(\mathcal{O}_2(\mathcal{O}_1(\mathrm{A})))$. When a sequence of operations is involved, we will also write

$$A \overset{\substack{\mathcal{O}_k \\ \vdots \\ \mathcal{O}_2 \\ \mathcal{O}_1}}{\underset{r}{\sim}} B$$

to stand for the fact that $\mathrm{B} = \mathcal{O}_k(\ldots \mathcal{O}_2(\mathcal{O}_1(\mathrm{A}))\ldots)$.

The next example is a concrete illustration of the Gauss–Jordan algorithm at work; the effect of each individual row-operation is shown. When reading this example, you may wonder why certain operations are performed and why they are performed in this order. You may observe that there are shorter, more efficient passages from A_0 to $GJ(A_0)$ *for this particular matrix* A_0. But that is precisely the point: for a *particular* matrix A_0, there will often be a shorter route from A_0 to $GJ(A_0)$ than the one taken by the Gauss–Jordan algorithm (see Problem 11.3, for example). We are *not*, however, concerned with particular cases; the algorithm is a set of instructions that is fixed in advance, once and for all, and that will operate on an *arbitrary* input matrix. The fact that its operation will not be optimal in particular cases is to be expected.

Example 12.8 should be skimmed once now and reread again carefully following 12.10.

12.8 Example

$$A_0 = \begin{bmatrix} 0 & 3 & 1 & 1 \\ 4 & 5 & 3 & 1 \\ -4 & 1 & -1 & 1 \\ 2 & 1 & 1 & 2 \end{bmatrix} \underset{r}{\overset{R_1 \leftrightarrow R_2}{\sim}} \begin{bmatrix} 4 & 5 & 3 & 1 \\ 0 & 3 & 1 & 1 \\ -4 & 1 & -1 & 1 \\ 2 & 1 & 1 & 2 \end{bmatrix}$$

$$\underset{r}{\overset{\frac{1}{4}R_1}{\sim}} \begin{bmatrix} 1 & \frac{5}{4} & \frac{3}{4} & \frac{1}{4} \\ 0 & 3 & 1 & 1 \\ -4 & 1 & -1 & 1 \\ 2 & 1 & 1 & 2 \end{bmatrix} \underset{r}{\overset{R_2 - 0R_1}{\sim}} \begin{bmatrix} 1 & \frac{5}{4} & \frac{3}{4} & \frac{1}{4} \\ 0 & 3 & 1 & 1 \\ -4 & 1 & -1 & 1 \\ 2 & 1 & 1 & 2 \end{bmatrix}$$

$$\underset{r}{\overset{R_3 + 4R_1}{\sim}} \begin{bmatrix} 1 & \frac{5}{4} & \frac{3}{4} & \frac{1}{4} \\ 0 & 3 & 1 & 1 \\ 0 & 6 & 2 & 2 \\ 2 & 1 & 1 & 2 \end{bmatrix} \underset{r}{\overset{R_4 - 2R_1}{\sim}} \begin{bmatrix} 1 & \frac{5}{4} & \frac{3}{4} & \frac{1}{4} \\ 0 & 3 & 1 & 1 \\ 0 & 6 & 2 & 2 \\ 0 & -\frac{3}{2} & -\frac{1}{2} & \frac{3}{2} \end{bmatrix}$$

$$\underset{r}{\overset{\frac{1}{3}R_2}{\sim}} \begin{bmatrix} 1 & \frac{5}{4} & \frac{3}{4} & \frac{1}{4} \\ 0 & 1 & \frac{1}{3} & \frac{1}{3} \\ 0 & 6 & 2 & 2 \\ 0 & -\frac{3}{2} & -\frac{1}{2} & \frac{3}{2} \end{bmatrix} \underset{r}{\overset{R_3 - 6R_2}{\sim}} \begin{bmatrix} 1 & \frac{5}{4} & \frac{3}{4} & \frac{1}{4} \\ 0 & 1 & \frac{1}{3} & \frac{1}{3} \\ 0 & 0 & 0 & 0 \\ 0 & -\frac{3}{2} & -\frac{1}{2} & \frac{3}{2} \end{bmatrix}$$

$$\underset{r}{\overset{R_4 + \frac{3}{2}R_2}{\sim}} \begin{bmatrix} 1 & \frac{5}{4} & \frac{3}{4} & \frac{1}{4} \\ 0 & 1 & \frac{1}{3} & \frac{1}{3} \\ 0 & 0 & 0 & 0 \\ 0 & 0 & 0 & 2 \end{bmatrix} \underset{r}{\overset{R_3 \leftrightarrow R_4}{\sim}} \begin{bmatrix} 1 & \frac{5}{4} & \frac{3}{4} & \frac{1}{4} \\ 0 & 1 & \frac{1}{3} & \frac{1}{3} \\ 0 & 0 & 0 & 2 \\ 0 & 0 & 0 & 0 \end{bmatrix}$$

$$\underset{r}{\overset{\frac{1}{2}R_3}{\sim}} \begin{bmatrix} 1 & \frac{5}{4} & \frac{3}{4} & \frac{1}{4} \\ 0 & 1 & \frac{1}{3} & \frac{1}{3} \\ 0 & 0 & 0 & 1 \\ 0 & 0 & 0 & 0 \end{bmatrix} \underset{r}{\overset{R_4 - 0R_3}{\sim}} \begin{bmatrix} 1 & \frac{5}{4} & \frac{3}{4} & \frac{1}{4} \\ 0 & 1 & \frac{1}{3} & \frac{1}{3} \\ 0 & 0 & 0 & 1 \\ 0 & 0 & 0 & 0 \end{bmatrix}$$

$$\underset{r}{\overset{R_2-\frac{1}{3}R_3}{\sim}} \begin{bmatrix} 1 & \frac{5}{4} & \frac{3}{4} & \frac{1}{4} \\ 0 & 1 & \frac{1}{3} & 0 \\ 0 & 0 & 0 & 1 \\ 0 & 0 & 0 & 0 \end{bmatrix} \quad \underset{r}{\overset{R_1-\frac{1}{4}R_3}{\sim}} \begin{bmatrix} 1 & \frac{5}{4} & \frac{3}{4} & 0 \\ 0 & 1 & \frac{1}{3} & 0 \\ 0 & 0 & 0 & 1 \\ 0 & 0 & 0 & 0 \end{bmatrix}$$

$$\underset{r}{\overset{R_1-\frac{5}{4}R_2}{\sim}} \begin{bmatrix} 1 & 0 & \frac{1}{3} & 0 \\ 0 & 1 & \frac{1}{3} & 0 \\ 0 & 0 & 0 & 1 \\ 0 & 0 & 0 & 0 \end{bmatrix} = \mathrm{GJ}(\mathrm{A}_0)$$

12.9 Exercise

(i) What is the solution space of the 4×3 linear system whose augmented matrix is A_0?
(ii) What is the solution space of the 4×4 homogeneous linear system whose coefficient matrix is A_0?

In 12.11, we will present the Gauss-Jordan algorithm in a very formal way, as a sort of "computer program in words." This degree of formalism is not really necessary for the student who wishes to learn only to use the algorithm; for this purpose, the following overview is provided. When read in conjunction with Examples 12.8 and 12.13, it should provide a sufficient understanding of the algorithm. It would then be possible to skip 12.11 and 12.12.

12.10 Overview of the Gauss – Jordan Algorithm

When the size of input matrix A is $m \times n$, the algorithm begins by repeating the following process up to m-many times:

(a) search for the leftmost nonzero column;
(b) by exchanging rows if necessary, bring the first nonzero entry in this column to the top;
(c) pivot downward on this nonzero entry.

This process is repeated up to m-many times. First, apply it to the matrix A itself. Second, take the matrix obtained thus far and, keeping its first row fixed, reapply the process to the submatrix consisting of the last $(m - 1)$-many rows. Third, take the matrix obtained thus far and, keeping its first two rows fixed, reapply the process to the submatrix consisting of the last $(m - 2)$-many rows; and so on.

Repeat this process m-many times or simply stop at some earlier stage if the submatrix to which the process would be applied is a zero matrix.

The result thus far is a matrix in row-echelon form called the *Gaussian form of A* and denoted by G(A). To obtain a matrix that is in reduced row-echelon form, pivot upward on each of the leading 1s in G(A), beginning with the rightmost leading 1, and proceeding left. The result is a matrix in reduced row-echelon form called the *Gauss-Jordan form of A* and denoted by GJ(A).

At this point, you should reread Example 12.8 carefully.

12.11 The Gauss-Jordan Algorithm

Figure 12.1 contains a flow chart.

At each step in the algorithm, there is an underlying $m \times n$ matrix that is row-equivalent to A; we will call this "the matrix obtained thus far."

Steps $1.k$, $2.k$, and $3.k$ constitute a loop that is successively traversed with $k = 1, 2, \ldots$ up to a possible maximum value of $k = m$. During the kth iteration of this loop, certain elementary row-operations are performed on the matrix obtained thus far. To explain precisely which ones, however, it is easier notationally to view these operations as being performed on a certain

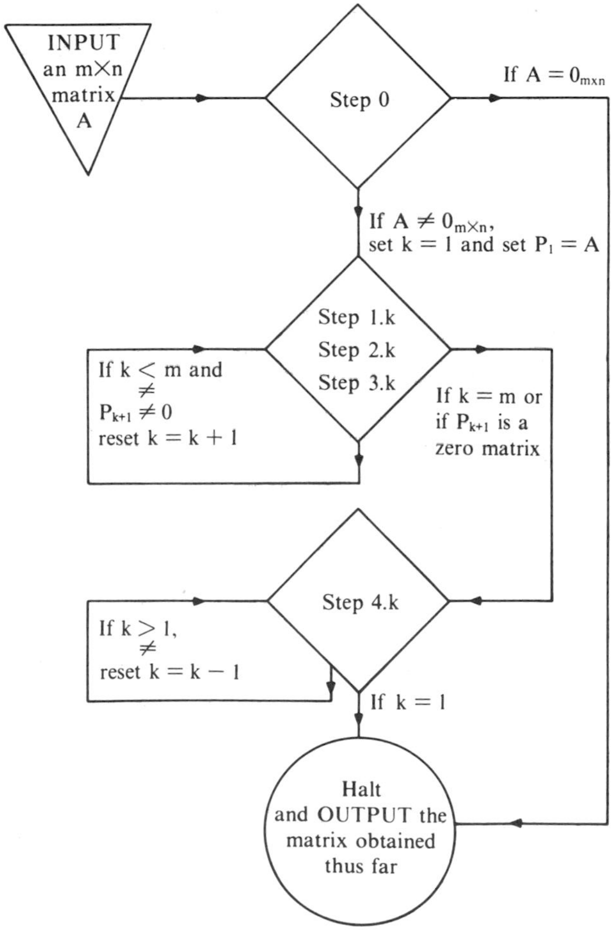

Figure 12.1

$(m - k + 1) \times n$ submatrix, P_k, of the matrix that has been obtained thus far.

We exit this loop upon performing step $3.m$ or sooner if, at some earlier stage, P_{k+1} is a zero matrix. The matrix that is obtained upon exiting from this first loop (i.e., when $k = m$ or when $P_{k+1} = 0$) will be denoted by G(A) and called the *Gaussian form of A*. G(A) is in row-echelon form, but not, generally, in reduced row-echelon form.

Step 4 is also a loop that is traversed beginning with step $4.k, 4.k - 1, 4.k - 2, \ldots, 4.3, 4.2$, and ending with step 4.1, which terminates the algorithm.

The $m \times n$ matrix obtained at this point will be denoted by GJ(A) and called the *Gauss–Jordan form of A*.

The input is an arbitrary $m \times n$ matrix, A.

Here are the details of the algorithm.

Step 0. If $A = 0_{m \times n}$, go to step 4.1. If $A \neq 0_{m \times n}$, set $k = 1$, set $P_1 = A$, and go to step 1.1.

Step $1.k$. Find, and denote by j_k, the least j such that the jth column of P_k is not all zero and go to step $2.k$.

Step $2.k$. Find the least i such that $p_{ij_k} \neq 0$. If $i = 1$, let $Q_k = P_k$ and go to step $3.k$; if $i \gneq 1$, let $\mathcal{O}$ be the operation $R_1 \leftrightarrow R_i$, let $Q_k = \mathcal{O}(P_k)$, and go to step $3.k$.

Step $3.k$. Pivot downward on the $(1, j_k)$-entry of Q_k. From the $m \times n$ matrix obtained thus far, delete the first k-many rows and let P_{k+1} denote the resulting $(m - k) \times n$ matrix. If $k = m$ or if $P_{k+1} = 0$, go to step $4.k$; if $k < m$ and $P_{k+1} \neq 0$, go to step $1.k + 1$.

Step $4.k$. If $k = 1$, halt and output the $m \times n$ matrix obtained thus far. If $k \gneq 1$, pivot upward on the (k, j_k)-entry of the $m \times n$ matrix obtained thus far; go to step $4.k - 1$.

12.12 Exercise

The following questions all refer to Example 12.8, which illustrated the passage from A_0 to $GJ(A_0)$ via the Gauss–Jordan algorithm. In that example:

(i) What are P_1, P_2, P_3, and P_4?

(ii) What elementary row-operation(s) is/are performed during step 2.3 and during step 3.2?

(iii) What is $G(A_0)$?

To save effort and paper when row-reducing by hand, we do not write down the matrix that is obtained after *each* elementary row-operation is performed. When pivoting upward or downward on the $(i, j)^{\text{th}}$ entry, the operations involved are performed in succession and only the end result of these operations is written down. We will illustrate this by rewriting Example 12.8 in this condensed fashion.

12.13 Example 12.8 Revisited

$$
A_0 = \begin{bmatrix} 0 & 3 & 1 & 1 \\ 4 & 5 & 3 & 1 \\ -4 & 1 & -1 & 1 \\ 2 & 1 & 1 & 2 \end{bmatrix}
\overset{R_1 \leftrightarrow R_2}{\underset{r}{\sim}}
\begin{bmatrix} 4 & 5 & 3 & 1 \\ 0 & 3 & 1 & 1 \\ -4 & 1 & -1 & 1 \\ 2 & 1 & 1 & 2 \end{bmatrix}
$$

$$
\overset{\frac{1}{4}R_1}{\underset{r}{\sim}}
\begin{bmatrix} 1 & \frac{5}{4} & \frac{3}{4} & \frac{1}{4} \\ 0 & 3 & 1 & 1 \\ -4 & 1 & -1 & 1 \\ 2 & 1 & 1 & 2 \end{bmatrix}
\overset{\substack{R_4 - 2R_1 \\ R_3 + 4R_1 \\ R_2 - 0R_1}}{\underset{r}{\sim}}
\begin{bmatrix} 1 & \frac{5}{4} & \frac{3}{4} & \frac{1}{4} \\ 0 & 3 & 1 & 1 \\ 0 & 6 & 2 & 2 \\ 0 & -\frac{3}{2} & -\frac{1}{2} & \frac{3}{2} \end{bmatrix}
$$

$$
\overset{\frac{1}{3}R_2}{\underset{r}{\sim}}
\begin{bmatrix} 1 & \frac{5}{4} & \frac{3}{4} & \frac{1}{4} \\ 0 & 1 & \frac{1}{3} & \frac{1}{3} \\ 0 & 6 & 2 & 2 \\ 0 & -\frac{3}{2} & -\frac{1}{2} & \frac{3}{2} \end{bmatrix}
\overset{\substack{R_4 + \frac{3}{2}R_2 \\ R_3 - 6R_2}}{\underset{r}{\sim}}
\begin{bmatrix} 1 & \frac{5}{4} & \frac{3}{4} & \frac{1}{4} \\ 0 & 1 & \frac{1}{3} & \frac{1}{3} \\ 0 & 0 & 0 & 0 \\ 0 & 0 & 0 & 2 \end{bmatrix}
$$

$$
\overset{R_3 \leftrightarrow R_4}{\underset{r}{\sim}}
\begin{bmatrix} 1 & \frac{5}{4} & \frac{3}{4} & \frac{1}{4} \\ 0 & 1 & \frac{1}{3} & \frac{1}{3} \\ 0 & 0 & 0 & 2 \\ 0 & 0 & 0 & 0 \end{bmatrix}
\overset{\frac{1}{2}R_3}{\underset{r}{\sim}}
\begin{bmatrix} 1 & \frac{5}{4} & \frac{3}{4} & \frac{1}{4} \\ 0 & 1 & \frac{1}{3} & \frac{1}{3} \\ 0 & 0 & 0 & 1 \\ 0 & 0 & 0 & 0 \end{bmatrix}
$$

$$
\overset{\substack{R_1 - \frac{1}{4}R_3 \\ R_2 - \frac{1}{3}R_3}}{\underset{r}{\sim}}
\begin{bmatrix} 1 & \frac{5}{4} & \frac{3}{4} & 0 \\ 0 & 1 & \frac{1}{3} & 0 \\ 0 & 0 & 0 & 1 \\ 0 & 0 & 0 & 0 \end{bmatrix}
\overset{R_1 - \frac{5}{4}R_2}{\underset{r}{\sim}}
\begin{bmatrix} 1 & 0 & \frac{1}{3} & 0 \\ 0 & 1 & \frac{1}{3} & 0 \\ 0 & 0 & 0 & 1 \\ 0 & 0 & 0 & 0 \end{bmatrix} = GJ(A_0)
$$

Using the result from Problem 11.6 and the definition of the Gauss-Jordan algorithm, it is easy to see that if the algorithm is applied to a partitioned matrix $[A \mid B]$, then $GJ([A \mid B])$ has the form $[GJ(A) \mid C]$. This fact will be used in 12.14 below and again in Sections 14 and 15.

12.14 Computational Procedure

Let A be an $m \times n$ matrix. To find a reduced description of the solution space, S, of the $m \times n$ linear system $A\mathbf{x} = \mathbf{b}$, compute

$$
GJ([A \mid \mathbf{b}]) = \left[\begin{array}{c|c} & c_1 \\ & c_2 \\ & \vdots \\ GJ(A) & c_r \\ & c_{r+1} \\ & \vdots \\ & c_m \end{array} \right]
$$

where $r =$ the number of nonzero rows in GJ(A). If $r \lneq m$ and c_{r+1} is nonzero, conclude that $S = \varnothing$. If either $r = m$ or if $r \lneq m$ and $c_{r+1} = 0$, in which case $c_{r+2} = \cdots = c_m = 0$ as well, conclude that

$$S = \Big\{(x_1, x_2, \ldots, x_n) \in \mathbb{R}^n\colon \begin{aligned} x_{d_1} &= s_{d_1p_1}x_{p_1} + s_{d_1p_2}x_{p_2} + \cdots + x_{d_1p_k}x_{p_k} + c_1 \\ x_{d_2} &= s_{d_2p_1}x_{p_1} + s_{d_2p_2}x_{p_2} + \cdots + s_{d_2p_k}x_{p_k} + c_2 \\ &\vdots \\ x_{d_r} &= s_{d_rp_1}x_{p_1} + s_{d_rp_2}x_{p_2} + \cdots + s_{d_rp_k}x_{p_k} + c_r \end{aligned}\Big\}$$

where the leading 1's of GJ(A) occur in columns $d_1, d_2, \ldots, d_r$, where columns $p_1, p_2, \ldots, p_k$ are the remaining columns, and where $s_{d_ip_j} = -(d_i, p_j)^{\text{th}}$ entry in GJ(A).

To solve certain types of problems, we must apply the algorithm to matrices that contain one or more unknown entries. This in no way affects the operation of the algorithm. The solution of the problem breaks up into cases; we must consider all possible ways in which the algorithm might proceed, based on assumptions about the unknown entry or entries. The next examples illustrate this.

12.15 Example

Solve the linear system

$$x_1 + 2x_2 + (a + 2)x_3 = a$$

$$ax_1 + (a - 1)x_2 - x_3 = 1$$

Since the constants and the coefficients are given as functions of the unknown, a, it is to be expected that the description of the solution space will also, in general, involve functions of a. This is the *only* difference, however; the technique for solving such a system remains the same.

$$\begin{bmatrix} 1 & 2 & a+2 & a \\ a & a-1 & -1 & 1 \end{bmatrix}$$

$$\underset{r}{\overset{R_2 - aR_1}{\sim}} \begin{bmatrix} 1 & 2 & a+2 & a \\ 0 & -a-1 & -a^2-2a-1 & 1-a^2 \end{bmatrix}$$

The next operation to be performed by the algorithm is $1/(-a - 1)R_2$. This can only be done assuming $a \neq -1$. Thus the algorithm branches, and we will have to return to treat the case $a = -1$ separately; we continue the row reduction assuming $a \neq -1$.

$$\underset{\text{if } a \neq -1}{\overset{\frac{1}{-a-1}R_2}{\underset{r}{\sim}}} \begin{bmatrix} 1 & 2 & a+2 & a \\ 0 & 1 & a+1 & a-1 \end{bmatrix}$$

$$\underset{\text{if } a \neq -1}{\overset{R_1 - 2R_2}{\underset{r}{\sim}}} \begin{bmatrix} 1 & 0 & -a & -a+2 \\ 0 & 1 & a+1 & a-1 \end{bmatrix}$$

Thus we conclude that the solution space, for the case $a \neq -1$, is the line in $\mathbb{R}^3$ described by $\{(x_1, x_2, x_3) \in \mathbb{R}^3\colon\ x_1 = ax_3 - a + 2, x_2 = -(a+1)x_3 + a - 1\}$.

To find the solution for the case $a = -1$, we return to the point where the algorithm branched, we substitute -1 for a in the matrix

$$\begin{bmatrix} 1 & 2 & a+2 & a \\ 0 & -a-1 & -a^2 - 2a - 1 & 1 - a^2 \end{bmatrix}$$

and continue row-reducing if necessary; the result is

$$\begin{bmatrix} 1 & 2 & 1 & -1 \\ 0 & 0 & 0 & 0 \end{bmatrix}$$

We conclude that when $a = -1$ the solution space is the plane in $\mathbb{R}^3$ described by $\{(x_1, x_2, x_3) \in \mathbb{R}^3\colon\ x_1 = -2x_2 - x_3 - 1\}$.

12.16 Example

Solve the linear system

$$2x_1 + 4x_2 = 1$$

$$x_1 - ax_2 = b$$

More precisely, for what value(s) of a and b does this system have (i) no solution, (ii) exactly one solution, and (iii) infinitely many solutions? In cases (ii) and (iii), describe the solution spaces.

To vary the presentation slightly, we will solve this problem by writing down the effect of the algorithm in the form of a tree, showing the branches and the assumptions that prevail along each branch.

$$\begin{bmatrix} 2 & 4 & 1 \\ 1 & -a & b \end{bmatrix} \underset{r}{\overset{\frac{1}{2}R_1}{\sim}} \begin{bmatrix} 1 & 2 & \frac{1}{2} \\ 1 & -a & b \end{bmatrix} \underset{r}{\overset{R_2 - R_1}{\sim}} \begin{bmatrix} 1 & 2 & \frac{1}{2} \\ 0 & -a-2 & b-\frac{1}{2} \end{bmatrix}$$

If $a \neq -2$:

$$\underset{r}{\overset{\frac{1}{-a-2}R_2}{\sim}} \begin{bmatrix} 1 & 2 & \frac{1}{2} \\ 0 & 1 & \dfrac{b-\frac{1}{2}}{-a-2} \end{bmatrix} \underset{r}{\overset{R_1 - 2R_2}{\sim}} \begin{bmatrix} 1 & 0 & \dfrac{4b+a}{2a+4} \\ 0 & 1 & \dfrac{b-\frac{1}{2}}{-a-2} \end{bmatrix} \quad \text{if } a \neq -2$$

If $a = -2$:

$$\underset{r}{\sim} \begin{bmatrix} 1 & 2 & \frac{1}{2} \\ 0 & 0 & b-\frac{1}{2} \end{bmatrix} \quad \text{if } a = -2$$

$$\underset{r}{\overset{\frac{1}{b-\frac{1}{2}}R_2}{\sim}} \begin{bmatrix} 1 & 2 & \frac{1}{2} \\ 0 & 0 & 1 \end{bmatrix} \quad \text{if } b \neq \tfrac{1}{2}$$

$$\underset{r}{\sim} \begin{bmatrix} 1 & 2 & \frac{1}{2} \\ 0 & 0 & 0 \end{bmatrix} \quad \text{if } b = \tfrac{1}{2}$$

We conclude from this that:

(i) If $a \neq -2$, then, for any b, the system has the single point solution $\left(\dfrac{4b+a}{2a+4}, \dfrac{b-\frac{1}{2}}{-a-2}\right)$.

(ii) If $a = -2$ and $b \neq \frac{1}{2}$, the system is inconsistent.

(iii) If $a = -2$ and $b = \frac{1}{2}$, the solution space is the line $\{(x_1, x_2) \in \mathbb{R}^2 : x_1 = -2x_2 + \frac{1}{2}\}$.

[As a parenthetical remark, let us interpret this problem geometrically. The system involves two lines in $\mathbb{R}^2$: the first is fixed, $2x_1 + 4x_2 = 1$; the second, $x_1 - ax_2 = b$, is variable, its position depending on values for a and b. To solve this system is to find the common intersection, if any, of these two lines. The answer will of course depend on a and b. These two lines have the same slope (i.e., are parallel) precisely when $a = -2$. Thus, when $a \neq -2$, the lines are skew and intersect at the point $\left(\dfrac{4b+a}{2a+4}, \dfrac{b-\frac{1}{2}}{-a-2}\right)$. When $a = -2$, the lines are parallel; they coincide if $b = \frac{1}{2}$, and they are distinct if $b \neq \frac{1}{2}$.]

The next item in this section is my own personal advice, based on years of exposure to students' errors using the Gauss–Jordan algorithm. These hints need not be followed; they are offered as "an ounce of prevention."

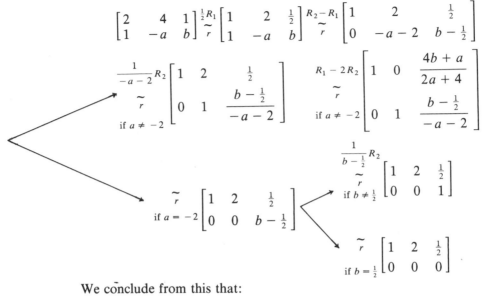

12.17 Helpful Hints Concerning Row-Reduction by Hand

(i) Try to avoid spreading a row-reduction across two pages; errors have a tendency to creep in as you copy the matrix you are working on over to the next page. Begin the problem at the top of a fresh page if necessary.

(ii) If the operation $\mathcal{O}$ to be performed on A is one that affects only the ith row (i.e., sR_i or $R_i + tR_j$), then it is suggested that you *first* write down all but the ith row of $\mathcal{O}(\text{A})$ (these will be the same as the corresponding rows of A), and *then* write down the ith row of $\mathcal{O}(\text{A})$ (which is the row that changes). For example, to perform $\frac{1}{3}R_2$ on

$$\begin{bmatrix} 1 & \frac{5}{4} & \frac{3}{4} & \frac{1}{4} \\ 0 & 3 & 1 & 1 \\ 0 & 6 & 2 & 2 \\ 0 & -\frac{3}{2} & -\frac{1}{2} & \frac{3}{2} \end{bmatrix}$$

begin by recopying the first, third, and fourth rows of the matrix

$$\begin{bmatrix} 1 & \frac{5}{4} & \frac{3}{4} & \frac{1}{4} \\ & & & \\ 0 & 6 & 2 & 2 \\ 0 & -\frac{3}{2} & -\frac{1}{2} & \frac{3}{2} \end{bmatrix}$$

Then fill in the second row to obtain

$$\begin{bmatrix} 1 & \frac{5}{4} & \frac{3}{4} & \frac{1}{4} \\ 0 & 1 & \frac{1}{3} & \frac{1}{3} \\ 0 & 6 & 2 & 2 \\ 0 & -\frac{3}{2} & -\frac{1}{2} & \frac{3}{2} \end{bmatrix}$$

To perform $R_1 - \frac{5}{4}R_2$ on

$$\begin{bmatrix} 1 & \frac{5}{4} & \frac{3}{4} & 0 \\ 0 & 1 & \frac{1}{3} & 0 \\ 0 & 0 & 0 & 1 \\ 0 & 0 & 0 & 0 \end{bmatrix}$$

begin by recopying the last three rows:

$$\begin{bmatrix} & & & \\ 0 & 1 & \frac{1}{3} & 0 \\ 0 & 0 & 0 & 1 \\ 0 & 0 & 0 & 0 \end{bmatrix}$$

Then fill in the first row to obtain

$$\begin{bmatrix} 1 & 0 & \frac{1}{3} & 0 \\ 0 & 1 & \frac{1}{3} & 0 \\ 0 & 0 & 0 & 1 \\ 0 & 0 & 0 & 0 \end{bmatrix}$$

(iii) Except for the shortcut explained in 12.13, avoid the temptation to "do two operations at once." Although such shortcuts will cause no trouble when row-reduction is being used to solve a linear system, these shortcuts are the source of serious errors when row-reduction is being used to compute determinants (see Chapter V) since the different types of operations affect the determinant of the underlying matrix in different ways. It is a good habit, therefore, to avoid such shortcuts from the beginning. For example, avoid writing

$$\begin{bmatrix} 2 & 5 & -3 \\ 0 & 0 & 3 \\ 0 & 4 & 8 \end{bmatrix} \underset{r}{\overset{\frac{1}{4}R_2}{\underset{R_3 \leftrightarrow R_2}{\sim}}} \begin{bmatrix} 2 & 5 & -3 \\ 0 & 1 & 2 \\ 0 & 0 & 3 \end{bmatrix}$$

Write this out as two separate operations. It is particularly important to avoid writing such things as

$$\begin{bmatrix} 4 & 2 & 9 \\ 2 & 1 & 3 \\ 5 & 0 & -2 \end{bmatrix} \underset{r}{\overset{2R_2-R_1}{\sim}} \begin{bmatrix} 4 & 2 & 9 \\ 0 & 0 & -3 \\ 5 & 0 & -2 \end{bmatrix}$$

First, $2R_2 - R_1$ is *not* an elementary row-operation. By writing $2R_2 - R_1$, this is perceived as a single type III operation, concealing the fact that two elementary operations, first $2R_2$, followed by $R_2 - R_1$, are being performed.

(iv) Reread item 11.13.

12.18 Remark

The Gauss–Jordan algorithm is the fundamental method for solving linear systems. For small-scale problems that are done "by hand," the algorithm in its crude form, as presented here, is sufficient. In an introductory course intended for a wide audience, these illustrative problems are adequate.

In practice, however, the algorithm is implemented on a digital computer. This is necessary, of course, because realistic problems give rise to linear systems with a large number of equations and variables; moreover, accuracy to several significant digits is desired.

When this is done, there is an assortment of complications. Because a given computer operates with a limited number of significant digits, an approximation to real arithmetic, known as *floating-point arithmetic*, is used. The effect of error introduced by round-off must be considered, and variants of the algorithm have been developed to minimize this effect. Another concern is with stability (i.e., the extent to which the output of the algorithm is sensitive to small changes in the input). An additional consideration is the relative efficiency of possible variants of the algorithm.

These are serious and difficult questions, which are handled by modifying the crude algorithm in various ways. This is a vast subject, which merits a separate, advanced course in numerical analysis or scientific computing.

A typical serious user of linear algebra will have access to extensive libraries of commercially available, prepackaged programs that incorporate the latest refinements. These are transparent to the user, whose first concern should be the selection of the input and the proper interpretation of the output.

An analogy can be made here: a typical new driver of an automobile needs to acquire basic driving skills and to learn the rules of the road; an understanding of turbochargers and electronic fuel injection can come later.

In addition to the vastness of the subject, another reason for postponing numerical linear algebra to a subsequent course is that many of the results of the past two decades are being rendered obsolete by revolutionary developments in computer architecture and in the design of algorithms. Now that modern computers can routinely handle 64 significant digits (until recently, the norm was 7 or 8), round-off error and stability are no longer the concern they once were for most medium-scale, practical problems.

Another development is parallel processing, in which a single computer has several processing units acting simultaneously on different parts of a problem; this requires a rethinking of the criteria by which traditional (sequential) algorithms are judged. The following quote is from James M. Ortega and Robert G. Voigt, Institute for Computer Applications in Science and Engineering (ICASE), ICASE Report No. 85-1, Langley Research Center, January 1985.

> Traditionally, one of the most important tools of the numerical analyst for evaluating algorithms has been computational complexity analysis, i.e. operation counts. ... This arithmetic complexity remains important for vector and parallel computers, but several other factors become equally significant. ... Many of the best sequential algorithms turn out to be unsatisfactory [for vector or parallel machines] and need to be modified or even discarded. On the other hand, many older algorithms which had been found to be less than optimal on sequential machines have had a rejuvenation because of their parallel properties.

Another recent factor is economic and sociological. The decreasing costs of raw computing power relative to the increasing costs of computer programmers' time are now at the point where trade-offs must be made: do you pay a programmer to optimize an algorithm so that a (slow) computer can solve your problem faster, or do you buy a faster computer on which to run your nonoptimal program?

To get a sense of what is being omitted here, you should consult one of the following classical references. Any one of these should be accessible to you when you have finished the present text.

Faddeev, D. K., and V. N. Faddeeva, *Computational Method of Linear Algebra*, W. H. Freeman, San Francisco, 1967.

Forsythe, G. E., and C. B. Moler, *Computer Solution of Linear Algebraic Systems*, Prentice-Hall, Englewood Cliffs, N.J., 1967.

Varga, R., *Matrix Iterative Analysis*, Prentice-Hall, Englewood Cliffs, N.J., 1962.

Young, D. M., and R. T. Gregory, *A Survey of Numerical Mathematics*, Vol. II, Addison-Wesley, Reading, Mass., 1973.

A sample of recent developments can be obtained from the following:

Heller, Don, "A Survey of Parallel Algorithms in Numerical Linear Algebra," *SIAM Review*, Vol. 20, 1978, 740–777.

Pan, Victor, "How Can We Speed Up Matrix Multiplication," *SIAM Review*, Vol. 26, 1984, 393–415.

Problem Set 12

12.1. Go back and do any of the exercises in Section 12 that you may have skipped.

Describe the solution space of each of the systems of linear equations given in Problems 12.2 through 12.23.

*12.2.
$$\begin{aligned} 3x_1 + 2x_2 - 2x_3 &= 0 \\ 3x_1 - 4x_2 + 5x_3 &= 3 \\ 3x_2 - x_3 &= -3 \\ 9x_1 - 3x_2 + 2x_3 &= 6 \end{aligned}$$

*12.3.
$$\begin{aligned} x_1 - x_2 - 2x_3 + 2x_4 &= 5 \\ 2x_1 - 3x_2 - x_3 + x_4 &= 7 \\ 3x_1 - 7x_2 + 6x_3 - 6x_4 &= -3 \end{aligned}$$

*12.4.
$$\begin{aligned} \tfrac{3}{2}x_1 - \tfrac{3}{2}x_2 - \tfrac{3}{2}x_3 &= \tfrac{2}{3} \\ \tfrac{2}{3}x_1 - \tfrac{2}{3}x_2 - \tfrac{2}{3}x_3 &= \tfrac{3}{2} \end{aligned}$$

*12.5.
$$\begin{aligned} 3x_1 \qquad - 2x_3 &= 5 \\ 2x_1 - 4x_2 \qquad &= 5 \end{aligned}$$

*12.6.
$$\begin{aligned} 2x_1 + 5x_2 - x_3 &= 8 \\ x_1 + 4x_2 - 3x_3 &= 8 \\ 2x_1 + x_2 + 4x_3 &= -2 \end{aligned}$$

*12.7.
$$\begin{aligned} 2x_1 - 2x_2 + 3x_3 &= 0 \\ 2x_1 + x_2 - 2x_3 &= 0 \\ -x_1 + 3x_2 + x_3 &= 0 \end{aligned}$$

*12.8.
$$\begin{aligned} \tfrac{1}{2}x_1 + x_2 + \tfrac{3}{2}x_3 &= 2 \\ \tfrac{1}{3}x_1 + \tfrac{2}{3}x_2 + x_3 &= \tfrac{4}{3} \\ \tfrac{3}{5}x_1 + \tfrac{6}{5}x_2 + \tfrac{9}{5}x_3 &= \tfrac{12}{5} \end{aligned}$$

*12.9.
$$\begin{aligned} x_1 + 2x_2 - x_3 + 3x_4 + 3x_5 &= 5 \\ x_1 - x_2 + 2x_3 - 3x_4 + x_5 &= 4 \\ 2x_1 + x_2 + 4x_3 \qquad + 2x_5 &= 5 \\ 3x_1 + 2x_2 \qquad + x_4 + 3x_5 &= 7 \end{aligned}$$

*12.10.
$$\begin{aligned} \tfrac{2}{3}x_1 + \tfrac{1}{3}x_2 &= 4 \\ \tfrac{1}{2}x_1 + 4x_2 &= -2 \end{aligned}$$

*12.11.
$$\begin{aligned} 2x_1 \qquad - 3x_3 &= 4 \\ 2x_1 - 3x_2 + 8x_3 &= 8 \\ 4x_1 - 3x_2 - 6x_3 &= 1 \\ 5x_1 - 3x_2 - x_3 &= 6 \end{aligned}$$

*12.12.
$$\begin{aligned} x_1 + 2x_2 + x_3 + x_4 &= -2 \\ 2x_1 + 6x_2 - 2x_3 + 3x_4 &= 7 \\ 2x_1 + 5x_2 \qquad + 3x_4 &= 4 \end{aligned}$$

*12.13.
$$\begin{aligned} x_1 - x_2 + 3x_3 &= -2 \\ 3x_1 - 2x_2 + 3x_3 &= -4 \\ 2x_1 - x_2 \qquad &= -2 \\ x_1 + 2x_2 - 15x_3 &= 4 \end{aligned}$$

*12.14.
$$\begin{aligned} x_1 - 4x_2 + 9x_3 - x_4 &= 9 \\ 2x_1 - 3x_2 + 8x_3 + 3x_4 &= 7 \\ 3x_1 - 2x_2 + 7x_3 + 7x_4 &= 7 \end{aligned}$$

*12.15.
$$\begin{aligned} x_1 + \tfrac{5}{2}x_2 + \tfrac{1}{3}x_3 + \tfrac{3}{4}x_4 &= 2 \\ \tfrac{6}{5}x_1 + \tfrac{3}{5}x_2 + \tfrac{2}{10}x_3 + \tfrac{9}{5}x_4 &= \tfrac{12}{5} \\ 4x_1 + 10x_2 + \tfrac{4}{3}x_3 + 3x_4 &= 8 \end{aligned}$$

*12.16.
$$\begin{aligned} 2x_1 - x_2 + x_3 &= 3 \\ 3x_1 + 2x_2 + 5x_3 &= -2 \\ 2x_1 - 8x_2 - 6x_3 &= -1 \end{aligned}$$

*12.17.
$$\begin{aligned}\tfrac{1}{8}x_1 + \tfrac{3}{4}x_2 + \tfrac{3}{2}x_3 &= \tfrac{3}{5}\\ \tfrac{1}{3}x_1 + 2x_2 + 4x_3 &= \tfrac{8}{5}\\ \tfrac{5}{6}x_1 + 5x_2 + 10x_3 &= 4\\ \tfrac{1}{2}x_1 + 3x_2 + 6x_3 &= \tfrac{12}{5}\end{aligned}$$

*12.18.
$$\begin{aligned}2x_1 - 2x_2 - 3x_3 &= 1\\ 2x_1 + x_2 - 2x_3 &= 1\\ -x_1 - 3x_2 + x_3 &= 1\end{aligned}$$

*12.19.
$$\begin{aligned}x_1 + 2x_2 + x_3 &= 1\\ 3x_1 + 4x_2 - 3x_3 &= 7\\ 4x_1 + 7x_2 + x_3 &= 6\end{aligned}$$

*12.20.
$$\begin{aligned}2x_1 + x_2 - 2x_3 &= 1\\ \sqrt{2}\,x_1 + 2x_2 - 4x_3 &= 2\end{aligned}$$

*12.21.
$$\begin{aligned}2x_1 + x_2 + x_3 &= -3\\ 4x_1 + 3x_2 + 2x_3 &= -11\\ 2x_1 \qquad + x_3 &= 2\end{aligned}$$

*12.22.
$$\begin{aligned}\sqrt{2}\,x_1 + 2\sqrt{2}\,x_2 + \frac{\sqrt{2}}{2}x_3 &= 2\\ \sqrt{3}\,x_1 + 2\sqrt{3}\,x_2 + \frac{\sqrt{3}}{2}x_3 &= \sqrt{6}\end{aligned}$$

*12.23.
$$\begin{aligned}2x_1 - x_2 + x_3 \qquad + 6x_5 &= 5\\ 3x_1 - 2x_2 + 2x_3 - x_4 + 12x_5 &= 7\\ 3x_1 + 5x_2 + 2x_3 - 8x_4 - 2x_5 &= 14\end{aligned}$$

*12.24. (i) Find a system of linear equations in three unknowns whose solution space is the line in $\mathbb{R}^3$ that goes through the points $(0, 0, 0)$ and $(1, 1, 1)$.
(ii) Justify your answer to (i) by solving the system of equations you gave as an answer to (i) and checking that its solution has the stated properties.

*12.25. For what value(s) of t does the following system of equations have
(i) no solution?
(ii) a unique solution?
(iii) infinitely many solutions?

In cases (ii) and (iii), describe the solution space.

$$\begin{aligned} x_1 - 2x_2 + \quad 3x_3 &= 1 \\ 2x_1 + tx_2 + \quad 6x_3 &= 6 \\ -x_1 + 3x_2 + (t-3)x_3 &= 0 \end{aligned}$$

12.26. Repeat Problem 12.25 for the system

$$\begin{aligned} x_1 + 2x_2 + \qquad x_3 &= 1 \\ 3x_1 + 4x_2 + (1-t)x_3 &= 7 \\ 4x_1 + 7x_2 + \qquad x_3 &= t + 2 \end{aligned}$$

12.27. Repeat Problem 12.25 for the system

$$\begin{aligned} x_1 + \quad x_2 + tx_3 &= 2 \\ 3x_1 + 4x_2 + 2x_3 &= t \\ 2x_1 + 3x_2 - \quad x_3 &= 1 \end{aligned}$$

12.28. Repeat Problem 12.25 for the system

$$\begin{aligned} x_1 + 2x_2 + \qquad x_3 &= 0 \\ x_1 - 4x_2 - \qquad x_3 &= 3 - t \\ 5x_1 + 4x_2 + (t^2-1)x_3 &= 1 \end{aligned}$$

12.29. A is the augmented matrix of a 4×5 linear system. You are given that the leading 1's of GJ(A) are in columns 1, 2, and 5 and that the points $(-1, 4, 1, 0, 0)$, $(-2, 1, 0, -1, 0)$, and $(-4, 5, 1, 1, 0)$ are known to be solutions of the system. Find the general solution.

12.30. For what values of a, if any, is $(a, 3, -a, -1)$ a solution of the system

$$\begin{aligned} x_1 - 3x_2 + 4x_3 - 5x_4 &= 0 \\ 2x_1 + 2x_2 - \quad x_3 + 2x_4 &= 0 \\ 3x_1 + \quad x_2 + 3x_3 - 2x_4 &= 0 \end{aligned}$$

Hint: Begin by finding the solution space; then check to see if it contains any points of the form $(a, 3, -a, -1)$.

12.31. For students who know a computer programming language, write a computer program to implement the Gauss–Jordan algorithm. Test it using the matrix from Example 12.8. Use it to do Problems 12.2 through 12.23.

12.32. Here is an amusing assignment based on an article by Michael B. Nicolai in *The Mathematics Teacher*, Volume 67, Number 5, 403–404.

(i) Solve the following linear system.

$$\begin{aligned} 2x_1 + 3x_2 &= 4 \\ 5x_1 + 6x_2 &= 7 \end{aligned}$$

(ii) Solve the following linear system.

$$\begin{aligned} 10x_1 + 11x_2 &= 12 \\ 13x_1 + 14x_2 &= 15 \end{aligned}$$

(iii) Solve the following linear system.

$$\begin{aligned} 8x_1 + 9x_2 &= 10 \\ 11x_1 + 12x_2 &= 13 \end{aligned}$$

(iv) Formulate and prove a result that explains the phenomenon observed in (i), (ii), and (iii).

(v) Solve the following linear system.

$$\begin{aligned} 9x_1 + 8x_2 &= 7 \\ 6x_1 + 5x_2 &= 4 \end{aligned}$$

(vi) Did the answer you gave to (iv) account for the result obtained in (v)? If not, generalize your answer.

(vii) Solve the following linear system.

$$\begin{aligned} 2x_1 + 4x_2 &= 6 \\ 8x_1 + 10x_2 &= 12 \end{aligned}$$

(viii) Solve the following linear system.

$$\begin{aligned} 4x_1 + 4.5x_2 &= 5 \\ 5.5x_1 + 6x_2 &= 6.5 \end{aligned}$$

(ix) Solve the following linear system.

$$\begin{aligned} 15x_1 + 12x_2 &= 9 \\ 6x_1 + 3x_2 &= 0 \end{aligned}$$

(x) Did the answer you gave to (vi) account for the results obtained in (vii), (viii), and (ix)? If not, generalize your answer even further.

(xi) Solve the following linear system.

$$\begin{aligned} x_1 + 2x_1 + 3x_3 &= 4 \\ 5x_1 + 6x_2 + 7x_3 &= 8 \\ 9x_1 + 10x_2 + 11x_3 &= 12 \end{aligned}$$

(xii) How much would you be willing to gamble, without doing any calculations, that the solution space of the linear system

$$\begin{aligned} 44x_1 + 40x_2 + 36x_3 &= 32 \\ 28x_1 + 24x_2 + 20x_3 &= 16 \\ 12x_1 + 8x_2 + 4x_3 &= 0 \end{aligned}$$

is the line $\{(x_1, x_2, x_3) \in \mathbb{R}^3\colon x_1 = x_3 - 2, x_2 = -2x_3 + 3\}$?

13

Elementary Matrices and the Reinterpretation of Elementary Row-Operations

(Note that Section 15 does not depend on Sections 13 and 14; it could be covered first, if the reader prefers.)

The purpose of this section is to examine elementary row-operations under a different light. The main benefit derived from this will be the ability to use the Gauss–Jordan algorithm to decide whether an $n \times n$ matrix A is invertible or not and, if it is, to find A^{-1}.

13.1 Definition

An $n \times n$ matrix is called *elementary* if it is of the form $\mathcal{O}(I_n)$ for some elementary row-operation $\mathcal{O}$.

That is, an elementary $n \times n$ matrix is one that can be obtained by performing a *single* elementary row-operation on the $n \times n$ identity matrix.

The letter E will be reserved to denote a matrix that is elementary.

13.2 Exercise

For each of the following matrices, decide if it is elementary or not. If it is, which elementary row-operation was performed to obtain it?

(i) $\begin{bmatrix} 3 & 0 \\ 0 & 1 \end{bmatrix}$ (ii) $\begin{bmatrix} 1 & 0 \\ -1 & 1 \end{bmatrix}$

(iii) $\begin{bmatrix} 1 & 0 & 0 \\ 0 & 1 & -5 \\ 0 & 0 & 1 \end{bmatrix}$ (iv) $\begin{bmatrix} 1 & 0 & 0 \\ 0 & 0 & 1 \\ 0 & 1 & 0 \end{bmatrix}$

(v) $\begin{bmatrix} 1 & 0 & 0 & 0 \\ 0 & 0 & 0 & 1 \\ 0 & 0 & 1 & 0 \\ 0 & 2 & 0 & 0 \end{bmatrix}$ (vi) $\begin{bmatrix} 1 & 0 & 0 & 0 \\ 0 & 1 & -2 & 0 \\ 0 & 0 & 1 & 0 \\ 0 & 0 & 0 & 1 \end{bmatrix}$

(vii) $\begin{bmatrix} 1 & 2 \\ 2 & 1 \end{bmatrix}$ (viii) $\begin{bmatrix} 1 & 0 & 0 \\ 0 & 1 & 0 \\ 0 & 0 & 1 \end{bmatrix}$

(ix) $\begin{bmatrix} 1 & 0 & 0 & 0 \\ 0 & 1 & 0 & 0 \\ 0 & 0 & 0 & 0 \\ 0 & 0 & 0 & 1 \end{bmatrix}$ (x) $\begin{bmatrix} 1 & 0 & 0 & 0 \\ 0 & 1 & 1 & 1 \\ 0 & 0 & 1 & 0 \\ 0 & 0 & 0 & 1 \end{bmatrix}$

The main technical result concerning this concept is not difficult. Let A be any $m \times n$ matrix and let $\mathcal{O}$ be any elementary row-operation. $\mathcal{O}(A)$ is the matrix that results when the operation $\mathcal{O}$ is performed on A. $\mathcal{O}(I_m)$ is the matrix that results when this same operation is performed on I_m. $\mathcal{O}(I_m)A$ is the product of the two matrices $\mathcal{O}(I_m)$ and A, in the order indicated. What the next theorem asserts is that the result of performing the elementary row-operation $\mathcal{O}$ on the $m \times n$ matrix A is the same as the result of multiplying A *on the left* by $\mathcal{O}(I_m)$.

13.3 Theorem

For any $m \times n$ matrix A and any elementary row-operation $\mathcal{O}$, $\mathcal{O}(A) = \mathcal{O}(I_m)A$.

Before giving the proof, we illustrate the theorem with a few examples.

13.4 Examples

$$A = \begin{bmatrix} 3 & 5 & 7 & 2 \\ 4 & -9 & -6 & 0 \\ 10 & 1 & 8 & -1 \end{bmatrix}$$

(i) If $\mathcal{O}$ is $\frac{1}{2}R_3$, then

$$\mathcal{O}(I_3)A = \begin{bmatrix} 1 & 0 & 0 \\ 0 & 1 & 0 \\ 0 & 0 & \frac{1}{2} \end{bmatrix} \begin{bmatrix} 3 & 5 & 7 & 2 \\ 4 & -9 & -6 & 0 \\ 10 & 1 & 8 & -1 \end{bmatrix}$$

$$= \begin{bmatrix} 3 & 5 & 7 & 2 \\ 4 & -9 & -6 & 0 \\ 5 & \frac{1}{2} & 4 & -\frac{1}{2} \end{bmatrix} = \mathcal{O}(A)$$

(ii) If $\mathcal{O}$ is $R_2 \leftrightarrow R_3$, then

$$\mathcal{O}(\mathrm{I}_3)\mathrm{A} = \begin{bmatrix} 1 & 0 & 0 \\ 0 & 0 & 1 \\ 0 & 1 & 0 \end{bmatrix} \begin{bmatrix} 3 & 5 & 7 & 2 \\ 4 & -9 & -6 & 0 \\ 10 & 1 & 8 & -1 \end{bmatrix}$$

$$= \begin{bmatrix} 3 & 5 & 7 & 2 \\ 10 & 1 & 8 & -1 \\ 4 & -9 & -6 & 0 \end{bmatrix} = \mathcal{O}(\mathrm{A})$$

(iii) If $\mathcal{O}$ is $R_1 - 3R_2$, then

$$\mathcal{O}(\mathrm{I}_3)\mathrm{A} = \begin{bmatrix} 1 & -3 & 0 \\ 0 & 1 & 0 \\ 0 & 0 & 1 \end{bmatrix} \begin{bmatrix} 3 & 5 & 7 & 2 \\ 4 & -9 & -6 & 0 \\ 10 & 1 & 8 & -1 \end{bmatrix}$$

$$= \begin{bmatrix} -9 & 32 & 25 & 2 \\ 4 & -9 & -6 & 0 \\ 10 & 1 & 8 & -1 \end{bmatrix} = \mathcal{O}(A)$$

13.5 Proof of Theorem 13.3

The proof naturally breaks into three separate cases, according to the type of the elementary row-operation $\mathcal{O}$. In each case, what must be proved is that, for arbitrary i and j, the $(i, j)^{\text{th}}$ entry of $\mathcal{O}(\mathrm{A})$ is equal to the $(i, j)^{\text{th}}$ entry of $\mathcal{O}(\mathrm{I}_m)\mathrm{A}$. We will do this in each case by computing separately the $(i, j)^{\text{th}}$ entries of $\mathcal{O}(\mathrm{A})$ and of $\mathcal{O}(\mathrm{I}_m)\mathrm{A}$, and then observing that the results are the same.

Case I: $\mathcal{O}$ is sR_p, where $s \neq 0$.

$$(i, j)^{\text{th}} \text{ entry of } \mathcal{O}(\mathrm{A}) = \begin{cases} a_{ij}, & \text{if } i \neq p \\ sa_{pj}, & \text{if } i = p \end{cases}$$

$(i, j)^{\text{th}}$ entry of $\mathcal{O}(\mathrm{I}_m)\mathrm{A}$

$$= \left[i^{\text{th}} \text{ row of } \mathcal{O}(\mathrm{I}_m)\right] \begin{bmatrix} j^{\text{th}} \\ \text{column} \\ \text{of A} \end{bmatrix}$$

$$= \begin{cases} [0 \ \cdots \ \underbrace{0 \ 1 \ 0}_{i^{\text{th}} \text{ coordinate}} \ \cdots \ 0] \begin{bmatrix} a_{1j} \\ \vdots \\ a_{mj} \end{bmatrix} = a_{ij}, & \text{if } i \neq p \\ [0 \ \cdots \ \underbrace{0 \ s \ 0}_{p^{\text{th}} \text{ coordinate}} \ \cdots \ 0] \begin{bmatrix} a_{1j} \\ \vdots \\ a_{mj} \end{bmatrix} = sa_{pj}, & \text{if } i = p \end{cases}$$

Case II: $\mathcal{O}$ is $R_p \leftrightarrow R_q$.

$$(i, j)^{\text{th}} \text{ entry of } \mathcal{O}(\text{A}) = \begin{cases} a_{ij}, & \text{if } i \neq p \text{ and } i \neq q \\ a_{qj}, & \text{if } i = p \\ a_{pj}, & \text{if } i = q \end{cases}$$

$(i, j)^{\text{th}}$ entry of $\mathcal{O}(\text{I}_m)\text{A}$

$$= \left[i^{\text{th}} \text{ row of } \mathcal{O}(\text{I}_m)\right]\begin{bmatrix} j^{\text{th}} \\ \text{column} \\ \text{of A} \end{bmatrix}$$

$$= \begin{cases} [0 \quad \cdots \quad \underbrace{0 \quad 1 \quad 0}_{i^{\text{th}} \text{ coordinate}} \quad \cdots \quad 0]\begin{bmatrix} a_{1j} \\ \vdots \\ a_{mj} \end{bmatrix} = a_{ij}, & \text{if } i \neq p \text{ and } i \neq q \\ [0 \quad \cdots \quad \underbrace{0 \quad 1 \quad 0}_{q^{\text{th}} \text{ coordinate}} \quad \cdots \quad 0]\begin{bmatrix} a_{1j} \\ \vdots \\ a_{mj} \end{bmatrix} = a_{qj}, & \text{if } i = p \\ [0 \quad \cdots \quad \underbrace{0 \quad 1 \quad 0}_{p^{\text{th}} \text{ coordinate}} \quad \cdots \quad 0]\begin{bmatrix} a_{1j} \\ \vdots \\ a_{mj} \end{bmatrix} = a_{pj}, & \text{if } i = q \end{cases}$$

Case III: $\mathcal{O}$ is $R_p + tR_q$.

$$(i, j)^{\text{th}} \text{ entry of } \mathcal{O}(\text{A}) = \begin{cases} a_{ij}, & \text{if } i \neq p \\ a_{pj} + ta_{qj}, & \text{if } i = p \end{cases}$$

$(i, j)^{\text{th}}$ entry of $\mathcal{O}(\text{I}_m)\text{A}$

$$= \left[i^{\text{th}} \text{ row of } \mathcal{O}(\text{I}_m)\right]\begin{bmatrix} j^{\text{th}} \\ \text{column} \\ \text{of A} \end{bmatrix}$$

$$= \begin{cases} [0 \quad \cdots \quad \underbrace{0 \quad 1 \quad 0}_{i^{\text{th}} \text{ coordinate}} \quad \cdots \quad 0]\begin{bmatrix} a_{1j} \\ \vdots \\ a_{mj} \end{bmatrix} = a_{ij}, & \text{if } i \neq p \\ [0 \quad \cdots \quad \underbrace{0 \quad 1 \quad 0}_{p^{\text{th}} \text{ coordinate}} \quad \cdots \quad \underbrace{0 \quad t \quad 0}_{q^{\text{th}} \text{ coordinate}} \quad \cdots \quad 0]\begin{bmatrix} a_{1j} \\ \vdots \\ a_{mj} \end{bmatrix} \\ \qquad = a_{pj} + ta_{qj}, & \text{if } i = p \end{cases}$$

■

Theorem 13.3 is of no use to us when we are *actually* performing elementary row-operations. It is much easier to form $\mathcal{O}(A)$ directly than to first form the elementary matrix $\mathcal{O}(I_m)$ and then to multiply A on the left by this elementary matrix. Theorem 13.3 is used when we wish to prove something *about* the row-reduction process.

It is in the following form that we will most often use it.

13.6 Corollary to Theorem 13.3

Suppose that A and B are $m \times n$ matrices and that B can be obtained from A by a sequence of elementary row-operations, say, $\mathcal{O}_k(\ldots \mathcal{O}_2(\mathcal{O}_1(A))\ldots) = B$. Then there is a sequence of elementary matrices $E_1, E_2, \ldots, E_k$ such that $E_k \cdots E_2E_1A = B$.

Proof: Apply 13.3 k-many times. ■

Corollary 13.6 will be important for Section 14, but already in the proof of the next theorem, 13.3 will be used.

13.7 Theorem

Elementary matrices are invertible and their inverses are themselves elementary matrices. Specifically, if $E = \mathcal{O}(I)$, then $E^{-1} = \mathcal{O}'(I)$, where $\mathcal{O}'$ is the elementary row-operation that *reverses* $\mathcal{O}$, as explained in 11.14.

Proof: Assume $E = \mathcal{O}(I)$ and let $E' = \mathcal{O}'(I)$. Then

$$\begin{aligned} EE' &= \mathcal{O}(I)\mathcal{O}'(I) \\ &= \mathcal{O}(\mathcal{O}'(I)) && \text{[by 13.3]} \\ &= I && \text{[by 11.14]} \end{aligned}$$

Similarly,

$$\begin{aligned} E'E &= \mathcal{O}'(I)\mathcal{O}(I) \\ &= \mathcal{O}'(\mathcal{O}(I)) && \text{[by 13.3]} \\ &= I && \text{[by 11.14]} \end{aligned}$$

Thus $E' = \mathcal{O}'(I) = E^{-1}$ by 3.7. ■

One consequence of this is that we will never do any written work to invert an elementary matrix: this can easily be done by inspection. It is important to develop the ability to recognize elementary matrices and to be able to invert them "in your head."

13.8 Examples

(i) If $E = \begin{bmatrix} 1 & 0 & 0 \\ 0 & 1 & 0 \\ 0 & 0 & \frac{1}{2} \end{bmatrix}$, then $E^{-1} = \begin{bmatrix} 1 & 0 & 0 \\ 0 & 1 & 0 \\ 0 & 0 & 2 \end{bmatrix}$.

(ii) If $E = \begin{bmatrix} 1 & 0 & 0 \\ 0 & 0 & 1 \\ 0 & 1 & 0 \end{bmatrix}$, then $E^{-1} = \begin{bmatrix} 1 & 0 & 0 \\ 0 & 0 & 1 \\ 0 & 1 & 0 \end{bmatrix}$.

(iii) If $E = \begin{bmatrix} 1 & -3 & 0 \\ 0 & 1 & 0 \\ 0 & 0 & 1 \end{bmatrix}$, then $E^{-1} = \begin{bmatrix} 1 & 3 & 0 \\ 0 & 1 & 0 \\ 0 & 0 & 1 \end{bmatrix}$.

All the above follow immediately from Theorem 13.7.

13.9 Exercise

Find the inverses of those matrices from Exercise 13.2 that are elementary.

Problem Set 13

13.1. Go back and do any of the exercises from Section 13 that you may have skipped.

13.2. *Elementary Column Operations*:

(i) By analogy with 11.8, define the concept of elementary column-operation.

(ii) *Definition:* An $n \times n$ matrix is called *elementary* if it is of the form $\mathcal{O}(I_n)$ for some elementary column-operation, $\mathcal{O}$.

Prove that this definition of "elementary" coincides with the one given in 13.1. In other words, prove that an $n \times n$ elementary matrix can be viewed as having been obtained in a single step from the $n \times n$ identity matrix by *either* an elementary row-operation *or* an elementary column-operation.

(iii) *Theorem:* For any $m \times n$ matrix A and any elementary column-operation $\mathcal{O}$, $\mathcal{O}(A) = A\mathcal{O}(I_n)$.

This is the analogue of Theorem 13.3. It asserts that the result of performing the elementary column-operation $\mathcal{O}$ on the $m \times n$ matrix A is the same as the result of multiplying A *on the right* by $\mathcal{O}(I_n)$. Prove this theorem by imitating the proof of Theorem 13.3.

(iv) *Remark:* The preceding theorem is of some importance. It would be possible to pursue the analogy further by introducing (1) a notion of "reduced column-echelon form," and (2) an analogue of the Gauss–Jordan algorithm that would use elementary column operations, etc. The reason that elementary row-operations play a prominent role relative to elementary column-operations is that the former, when applied to the augmented matrix of a linear system, preserve the solution space of the underlying linear system. Elementary column operations do not.

13.3. Illustrate the theorem from Problem 13.2 with the matrix

$$\begin{bmatrix} 3 & 5 & -2 & 9 \\ 8 & 1 & 4 & -3 \\ -1 & 0 & 6 & 7 \end{bmatrix}$$

and the following three elementary column-operations: (i) multiply column 3 by 7, (ii) interchange columns 1 and 4, and (iii) to column 2, add 3 times column 4.

13.4. Solve the 3 × 3 linear system whose augmented matrix is the one problem in 13.3. Convince yourself that elementary column-operations do *not* preserve the solution space of the underlying linear system by solving each of the three linear systems whose augmented matrices are those that arise as the answers to Problems 13.3(i), (ii), and (iii).

*13.5. Evaluate the product

$$\begin{bmatrix} 2 & 1 & 7 \\ -3 & 9 & 10 \\ -2 & -4 & 6 \\ 8 & 8 & -1 \end{bmatrix}\begin{bmatrix} 1 & 0 & 0 \\ 0 & 3 & 0 \\ 0 & 0 & 1 \end{bmatrix}\begin{bmatrix} 1 & 0 & -2 \\ 0 & 1 & 0 \\ 0 & 0 & 1 \end{bmatrix}$$

by inspection, using the result from Problem 13.2.

13.6. (i) Show that the transpose, E^T, of an elementary matrix E is another elementary matrix of the same type as E.

(ii) We know EA is the result of performing a certain elementary row-operation on A. Use the Theorem from Problem 13.2 to show that AE^T is the result of performing the corresponding elementary column-operation on A.

14

An Algorithm for Computing A^{-1}

In this section, we develop a method, using the Gauss–Jordan algorithm, for deciding whether a square matrix A is invertible or not and, if it is, to find A^{-1}. The method itself will be extracted as a by-product of the proof of the following important theorem.

14.1 Theorem

Let A be an $n \times n$ matrix. Then the following are equivalent:

(i) A is invertible.
(ii) $GJ(A) = I_n$.
(iii) $A \underset{r}{\sim} I_n$.
(iv) A is expressible as a product of elementary matrices.

Proof: (i) → (ii). We wish to conclude, assuming the invertibility of A, that $GJ(A) = I_n$. We know that GJ(A) is a square matrix in reduced row-echelon form and thus, from Problem 10.4, that either GJ(A) has an all-zero row or else $GJ(A) = I_n$. We will conclude that $GJ(A) = I_n$ by showing that the alternative, GJ(A) having an all-zero row, is impossible. From 13.6, we know there exists a sequence of elementary matrices $E_1, E_2, \ldots, E_p$ such that $E_p \cdots E_2E_1A = GJ(A)$. Each E_i is invertible by

13.7 and A is invertible by assumption; thus the product, GJ(A), is invertible by 3.9. In particular, GJ(A) cannot have an all-zero row. (Recall Problem 3.17.)

(ii) → (iii) is obvious.

(iii) → (iv). Again using 13.6, if $A \underset{r}{\sim} I_n$, then there is some sequence $E_1, E_2, \ldots, E_k$ of elementary matrices such that $E_k \cdots E_2E_1A = I_n$. The product $P = E_k \cdots E_2E_1$ is invertible by 13.7 and 3.9, so we may multiply both sides of $PA = I_n$ on the left by P^{-1} to conclude that $A = P^{-1}I_n = P^{-1} = E_1^{-1}E_2^{-1} \cdots E_k^{-1}$. Since each E_i^{-1} is also elementary, this shows that A is expressible as a product of elementary matrices.

(iv) → (i) follows from the facts that elementary matrices are invertible and that any product of invertible matrices is invertible. ■

14.2 Corollary to the Proof of 14.1

Let A be an $n \times n$ matrix. If $\mathcal{O}_1, \mathcal{O}_2, \ldots, \mathcal{O}_k$ is any sequence of elementary row-operations that transforms A into I_n [i.e., if $\mathcal{O}_k(\ldots\mathcal{O}_2(\mathcal{O}_1(A))\ldots) = I_n$], then A is invertible, and this same sequence of elementary row-operations will transform I_n into A^{-1} [i.e., $\mathcal{O}_k(\ldots\mathcal{O}_2(\mathcal{O}_1(I_n))\ldots) = A^{-1}$].

Proof: Let E_i be the elementary matrix corresponding to $\mathcal{O}_i$. Assuming $\mathcal{O}_k(\ldots(\mathcal{O}_2(\mathcal{O}_1(A))\ldots) = E_k \cdots E_2E_1A = I_n$, then as in the proof of 14.1, we conclude that $A = E_1^{-1}E_2^{-1} \cdots E_k^{-1}$, so A is invertible and, by 3.9, $A^{-1} = E_k \cdots E_2E_1 = E_k \cdots E_2E_1I_n = \mathcal{O}_k(\ldots\mathcal{O}_2(\mathcal{O}_1(I_n))\ldots)$. ■

Because $\mathcal{O}([B \vdots C]) = [\mathcal{O}(B) \vdots \mathcal{O}(C)]$, we can perform the same sequence of elementary row-operations on both A and I_n by performing this sequence on the partitioned matrix, $[A \vdots I_n]$.

14.3 Computational Procedure

To discover whether an $n \times n$ matrix A is invertible or not and, if it is, to find A^{-1}, compute

$$GJ([A \vdots I_n]) = [GJ(A) \vdots B]$$

Then either $GJ(A) = I_n$, in which case A is invertible and $B = A^{-1}$ (by 14.2), or else GJ(A) contains a row of zeros, in which case A is not invertible (by 14.1).

14.4 Example

Let

$$A = \begin{bmatrix} 2 & 3 \\ -1 & 5 \end{bmatrix}$$

Find A^{-1}, if it exists.

To do this, apply the Gauss–Jordan algorithm to the matrix

$$\left[\begin{array}{rr|rr} 2 & 3 & 1 & 0 \\ -1 & 5 & 0 & 1 \end{array}\right]$$

The result is

$$\left[\begin{array}{rr|rr} 1 & 0 & \frac{5}{13} & -\frac{3}{13} \\ 0 & 1 & \frac{1}{13} & \frac{2}{13} \end{array}\right]$$

From this you conclude that A is invertible and that

$$A^{-1} = \begin{bmatrix} \frac{5}{13} & -\frac{3}{13} \\ \frac{1}{13} & \frac{2}{13} \end{bmatrix}$$

We already learned in Problem 3.6 how to answer this question for 2×2 matrices, so we did not really need to use 14.3 to do Example 14.4. It is less work to invert 2×2 matrices by inspection using the result of Problem 3.6. For larger matrices, 14.3 is the correct procedure.

14.5 Example

Let

$$B = \begin{bmatrix} 2 & -2 & 3 \\ 2 & 1 & -2 \\ -1 & 3 & 1 \end{bmatrix}$$

To find B^{-1}, if it exists, compute $GJ([B \,\vdots\, I_3])$. You should verify that this yields

$$\left[\begin{array}{rrr|rrr} 1 & 0 & 0 & \frac{1}{5} & \frac{11}{35} & \frac{1}{35} \\ 0 & 1 & 0 & 0 & \frac{1}{7} & \frac{2}{7} \\ 0 & 0 & 1 & \frac{1}{5} & -\frac{4}{35} & \frac{6}{35} \end{array}\right]$$

The conclusion is that B is invertible and

$$B^{-1} = \begin{bmatrix} \frac{1}{5} & \frac{11}{35} & \frac{1}{35} \\ 0 & \frac{1}{7} & \frac{2}{7} \\ \frac{1}{5} & -\frac{4}{35} & \frac{6}{35} \end{bmatrix}$$

14.6 Example

Let

$$C = \begin{bmatrix} 1 & 0 & -1 & 3 \\ 2 & 3 & 4 & 12 \\ 3 & -1 & -5 & 7 \\ 0 & -3 & -2 & 2 \end{bmatrix}$$

Find C^{-1}, if it exists.

Applying the Gauss–Jordan algorithm to the matrix $[C \mid I_4]$ yields the matrix

$$\left[\begin{array}{cccc|cccc} 1 & 0 & 0 & 5 & 0 & \frac{13}{44} & \frac{3}{22} & \frac{1}{4} \\ 0 & 1 & 0 & -2 & 0 & -\frac{3}{22} & \frac{1}{11} & -\frac{1}{2} \\ 0 & 0 & 1 & 2 & 0 & \frac{9}{44} & -\frac{3}{22} & \frac{1}{4} \\ 0 & 0 & 0 & 0 & 1 & -\frac{1}{11} & -\frac{3}{11} & 0 \end{array}\right]$$

We conclude from this that C^{-1} does not exist.

14.7 Remark

If we are row-reducing the partitioned $n \times 2n$ matrix $[A \mid I_n]$ by hand, it may not be necessary to carry out the Gauss–Jordan algorithm to the very end. If a stage is reached at which the left half of the $n \times 2n$ matrix obtained thus far contains an all-zero row, it is correct to halt and conclude that A is not invertible.

14.8 Exercise

Justify Remark 14.7.

14.9 Exercise

In Example 14.6, $GJ([C \mid I_4])$ was obtained (to save time) with the aid of a computer. Do 14.6 by hand, with Remark 14.7 in mind.

There are circumstances in which we want merely a *yes* or *no* answer to the question, Does A^{-1} exist? We do not care, in case the answer is *yes*, to actually compute A^{-1}. Needless to say, to get less information, less work is required.

14.10 Computational Procedure

To discover whether an $n \times n$ matrix A is invertible or not, compute G(A). Then either G(A) is an upper triangular matrix with each entry on the main diagonal equal to 1, in which case A is invertible, or else G(A) contains a row of zeros, in which case A is not invertible.

14.11 Exercise

Justify 14.10.

There is another important corollary to the proof of 14.1. The proof that an invertible matrix is expressible as a product of elementary matrices is a constructive proof; that is, the proof not only demonstrates that something *can* be done but gives a method for actually doing it.

When you study Section 18, you will appreciate why it is important to be able to express an invertible matrix A as a product of elementary matrices. (This allows you to analyze the linear transformation T_A, whose behavior is possibly complicated, as the composition of several linear transformations whose behaviors are easily understood.)

The justification for the next computational procedure is immediate from the proof of 14.1.

14.12 Computational Procedure

To express an invertible $n \times n$ matrix A as a product of elementary matrices, compute GJ(A), keeping track of which elementary row-operations are performed.

If $\mathcal{O}_k(\ldots(\mathcal{O}_2(\mathcal{O}_1(A)))\ldots) = \mathrm{GJ}(A) = I_n$, then $A = E_1^{-1}E_2^{-1} \cdots E_k^{-1}$, where E_i is the elementary matrix corresponding to the operation $\mathcal{O}_i$.

14.13 Example

Express

$$A = \begin{bmatrix} 2 & 3 \\ -1 & 5 \end{bmatrix}$$

as a product of elementary matrices, if possible.

To do this, compute GJ(A), keeping a record of which elementary row-operations are performed.

$$A = \begin{bmatrix} 2 & 3 \\ -1 & 5 \end{bmatrix} \underset{r}{\overset{\frac{1}{2}R_1}{\sim}} \begin{bmatrix} 1 & \frac{3}{2} \\ -1 & 5 \end{bmatrix} \underset{r}{\overset{R_2+R_1}{\sim}} \begin{bmatrix} 1 & \frac{3}{2} \\ 0 & \frac{13}{2} \end{bmatrix}$$

$$\underset{r}{\overset{\frac{2}{13}R_2}{\sim}} \begin{bmatrix} 1 & \frac{3}{2} \\ 0 & 1 \end{bmatrix} \underset{r}{\overset{R_1-\frac{3}{2}R_2}{\sim}} \begin{bmatrix} 1 & 0 \\ 0 & 1 \end{bmatrix} = GJ(A)$$

Because $GJ(A) = I_2$, A is expressible as a product of elementary matrices.

$$A = \begin{bmatrix} 2 & 0 \\ 0 & 1 \end{bmatrix} \begin{bmatrix} 1 & 0 \\ -1 & 1 \end{bmatrix} \begin{bmatrix} 1 & 0 \\ 0 & \frac{13}{2} \end{bmatrix} \begin{bmatrix} 1 & \frac{3}{2} \\ 0 & 1 \end{bmatrix}$$

We are making use here of the fact, explained in 13.7, that elementary matrices can be inverted by inspection.

Problem Set 14

14.1. Go back and do any of the exercises from Section 14 that you may have skipped.

*14.2. Let

$$A = \begin{bmatrix} 1 & 2 & 0 & 2 \\ 2 & 3 & -1 & 4 \\ 6 & -1 & 4 & 1 \\ 0 & 2 & 1 & 7 \end{bmatrix}, \quad B = \begin{bmatrix} 1 & 2 & 0 & 2 \\ 2 & 3 & -1 & 4 \\ 6 & -1 & 4 & 1 \\ 0 & 4 & 2 & 14 \end{bmatrix},$$

$$C = \begin{bmatrix} 1 & 2 & 0 & 2 \\ 0 & -1 & -1 & 0 \\ 6 & -1 & 4 & 1 \\ 0 & 4 & 2 & 14 \end{bmatrix}$$

(i) Find elementary matrices E_1, E_2, and E_3 such that $E_1A = B$, $E_2B = C$, and $E_3C = B$.

(ii) What are E_1^{-1}, E_2^{-1}, and E_3^{-1}?

(iii) Find a matrix D such that $DA = C$.

*14.3. For each matrix given, decide if it is invertible or not and, if it is, find its inverse.

(i) $\begin{bmatrix} 1 & 6 & 1 \\ 0 & 5 & 1 \\ 3 & 5 & 2 \end{bmatrix}$ (ii) $\begin{bmatrix} 1 & 2 & -3 \\ 2 & 1 & -1 \\ 2 & -1 & 2 \end{bmatrix}$

(iii) $\begin{bmatrix} 1 & -2 & 3 \\ 3 & 2 & 9 \\ 0 & -5 & 0 \end{bmatrix}$ (iv) $\begin{bmatrix} 2 & 2 & 0 \\ 1 & 0 & 4 \\ -3 & 2 & -4 \end{bmatrix}$

(v) $\begin{bmatrix} 3 & -5 \\ -2 & \frac{10}{3} \end{bmatrix}$ (vi) $\begin{bmatrix} 4 & 2 \\ 5 & 3 \end{bmatrix}$

$$\text{(vii)} \begin{bmatrix} 1 & 0 & 0 & 3 \\ 0 & 2 & 0 & 0 \\ 0 & 1 & 2 & 1 \\ 2 & -1 & 0 & -2 \end{bmatrix} \qquad \text{(viii)} \begin{bmatrix} 1 & 0 & 1 & 1 \\ 1 & 1 & 1 & 0 \\ 1 & 0 & 1 & 1 \\ 0 & 1 & 1 & 0 \end{bmatrix}$$

*14.4. Repeat 14.3 for the coefficient matrices of the linear systems from Problems 12.6, 12.7, 12.8, 12.10, 12.16, 12.18, 12.19, and 12.21.

*14.5. Express the matrix from Problem 14.3(vi) as a product of elementary matrices.

*14.6. Let

$$\mathrm{A} = \begin{bmatrix} 1 & 0 & 2 \\ -2 & 1 & -3 \\ -1 & 2 & -1 \end{bmatrix}, \qquad \begin{aligned} x_1 \qquad\quad + 2x_3 &= y_1 \\ -2x_1 + \quad x_2 + 3x_3 &= y_2 \\ -x_1 + 2x_2 - \quad x_3 &= y_3 \end{aligned}$$

and

$$\begin{aligned} x_1 + x_2 - x_3 &= z_1 \\ -x_1 - x_2 + x_3 &= z_2. \end{aligned}$$

(i) Find A^{-1}.
(ii) Express z_1 and z_2 in terms of y_1, y_2, and y_3.
(iii) If A^{-1} did not exist, would you still have been able to answer part (ii)?

*14.7. Let A be the matrix from Problem 14.3(iv).

(i) Solve the linear systems $\mathbf{Ax} = \begin{bmatrix} 0 \\ 1 \\ 2 \end{bmatrix}$, $\mathbf{Ax} = \begin{bmatrix} 3 \\ 6 \\ 2 \end{bmatrix}$, and $\mathbf{Ax} = \begin{bmatrix} 9 \\ 6 \\ 2 \end{bmatrix}$.

(ii) Solve the linear system $\mathbf{Ax} = \begin{bmatrix} b_1 \\ b_2 \\ b_3 \end{bmatrix}$.

(The solution will of course be expressed as a function of b_1, b_2, and b_3.)

*14.8. (i) For what value(s) of t, if any, is the matrix $\mathrm{A} = \begin{bmatrix} 2 & t \\ -6 & -18 \end{bmatrix}$ invertible?
(ii) For those values of t for which A is invertible, find A^{-1}. (Note that the entries in A^{-1} will, in general, be functions of t.)
Hint: Because A is only 2×2, these questions can be answered on the basis of Problem 3.6; begin by doing 14.8 this way. Another way to answer (i) and (ii) simultaneously is to apply the Gauss–Jordan algorithm to $[\mathrm{A} \,\vdots\, \mathrm{I}_2]$ and to suitably interpret the result. You may wish to review 12.14

and 12.15 for examples of the Gauss–Jordan algorithm applied to matrices with one or more unknown entries. Redo 14.8 this other way as a warm-up exercise for 14.9 and 14.10.

*14.9. For what value(s) of t, if any, is the matrix $B = \begin{bmatrix} 2 & t & 3 \\ -3 & 2 & -5 \\ 4 & t & 5 \end{bmatrix}$ invertible, and what is B^{-1} for these value(s) of t.

*14.10. Repeat Problem 14.9 for the matrix $C = \begin{bmatrix} 1 & -2 & t \\ 3 & 2 & t^2 \\ 0 & -5 & 0 \end{bmatrix}$.

14.11. Use 14.3 to show that the inverse of an invertible upper triangular matrix is upper triangular.

14.12. If it is possible to do so, express each of the following matrices as products of elementary matrices.

(i) $\begin{bmatrix} 1 & -1 \\ 3 & -2 \end{bmatrix}$ (ii) $\begin{bmatrix} 3 & 0 \\ 2 & -4 \end{bmatrix}$ (iii) $\begin{bmatrix} 1 & 2 & 1 \\ 2 & 6 & -2 \\ 2 & 5 & 0 \end{bmatrix}$

15

More Computational Procedures

Four computational procedures involving the Gauss–Jordan algorithm have been presented thus far: in 12.14, for solving linear systems; in 14.10, for obtaining a yes or no answer to the question, Is the $n \times n$ matrix A invertible?; in 14.3, for obtaining not merely a yes or no answer to this question but for also finding A^{-1} in case the answer is yes; and 14.11, for deciding whether it is possible and, if it is, to actually write A as a product of elementary matrices.

In Chapter II we introduced many concepts related to an $m \times n$ matrix A: the nullspace of A, the row-space of A, the column-space of A, the rank of A, and so on. The theoretical interrelationships among these concepts will be thoroughly treated in Chapter IV. In the present section, we explain how the Gauss–Jordan algorithm is used as a tool to answer questions about these concepts when a specific matrix A is given.

15.1 Observation

The nonzero rows of an $m \times n$ matrix in reduced row-echelon form constitute a linearly independent set of n-tuples.

Proof:

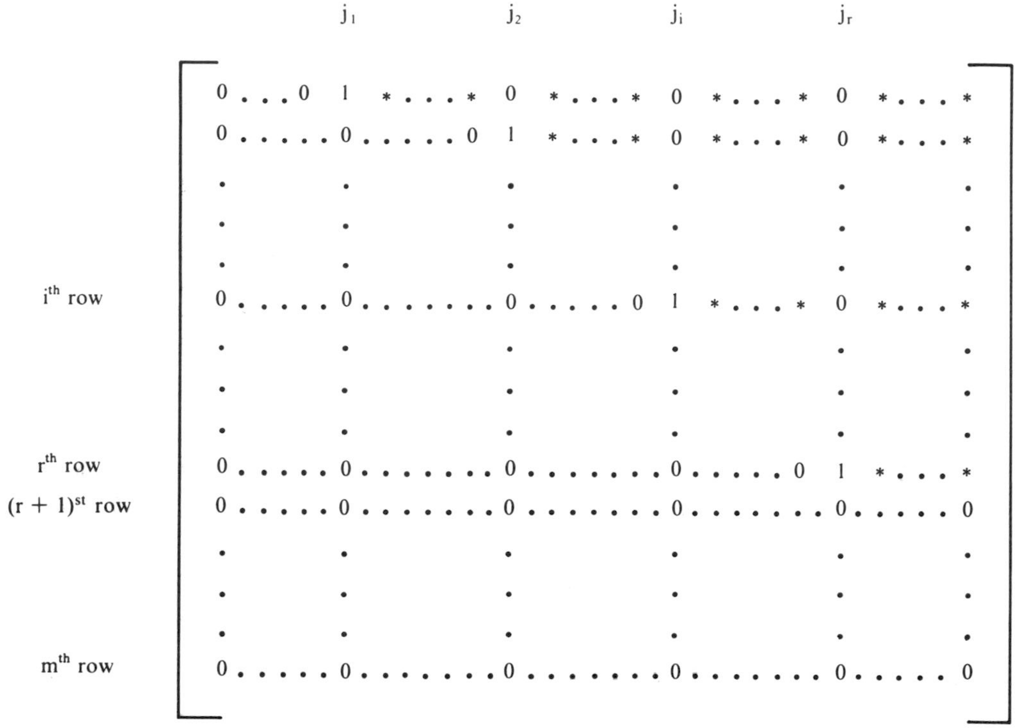

Figure 15.1

The matrix has the general form shown in Figure 15.1. It is clear that the set S of nonzero rows of A is a reduced set of n-tuples (see Definition 7.26); that is, $S \upharpoonright \{j_1, j_2, \ldots, j_r\} = \alpha^r$, so S is linearly independent by Theorem 7.27. ■

A slight improvement of this observation will be useful. We can obtain the same conclusion from the weaker assumption that the matrix is merely in row-echelon form.

15.2 Theorem

The nonzero rows of an $m \times n$ matrix in row-echelon form constitute a linearly independent set of n-tuples.

Proof: The matrix has the general form shown in Figure 15.2. Let $\mathbf{u}_i$ denote the ith row. By Exercise 7.19, $S = \{\mathbf{u}_1, \mathbf{u}_2, \ldots, \mathbf{u}_r\}$ is linearly independent if the vectors in S can be ordered in such a way that no vector is a linear combination of its predecessors. In the ordering $(\mathbf{u}_r, \mathbf{u}_{r-1}, \ldots, \mathbf{u}_i, \ldots, \mathbf{u}_1)$, this property is obvious: $\mathbf{u}_i$ cannot possibly be a linear combination of its predecessors $\mathbf{u}_r, \mathbf{u}_{r-1}, \ldots, \mathbf{u}_{i+1}$ since each of these has 0 as its j_i^{th} coordinate, while the j_i^{th} coordinate of $\mathbf{u}_i$ is 1. ■

$$
\begin{array}{r}
\\
\\
\\
i^{th}\text{ row} \\
\\
r^{th}\text{ row} \\
(r+1)^{st}\text{ row} \\
\\
m^{th}\text{ row}
\end{array}
\begin{array}{c}
\begin{array}{cccccccccc}
 & j_1 & & j_2 & & j_i & & j_r & & \\
\end{array} \\
\left[\begin{array}{cccccccccc}
0 \cdots 0 & 1 & * \cdots * & * & * \cdots * & * & * \cdots * & * & * \cdots * \\
0 \cdots & 0 & \cdots 0 & 1 & * \cdots * & * & * \cdots * & * & * \cdots * \\
\vdots & \vdots & & \vdots & & \vdots & & \vdots & \vdots \\
0 \cdots & 0 & \cdots & 0 & \cdots 0 & 1 & * \cdots * & * & * \cdots * \\
\vdots & \vdots & & \vdots & & \vdots & & \vdots & \vdots \\
0 \cdots & 0 & \cdots & 0 & \cdots & 0 & \cdots 0 & 1 & * \cdots * \\
0 \cdots & 0 & \cdots & 0 & \cdots & 0 & \cdots & 0 & \cdots 0 \\
\vdots & \vdots & & \vdots & & \vdots & & \vdots & \vdots \\
0 \cdots & 0 & \cdots & 0 & \cdots & 0 & \cdots & 0 & \cdots 0
\end{array}\right]
\end{array}
$$

Figure 15.2

■

The next observation is straightforward and has as a consequence the fact that row-equivalent matrices have the same row-space.

15.3 Observation

For any $m \times n$ matrix A and any elementary row-operation $\mathcal{O}$, RS(A) = RS($\mathcal{O}(A)$).

15.4 Exercise

Justify Observation 15.3 using Exercise 5.12 together with ideas from the proof of Theorem 11.5.

15.5 Corollary

For any pair of $m \times n$ matrices A and B, if $A \underset{r}{\sim} B$, then RS(A) = RS(B).

15.6 Theorem

If A is an $m \times n$ matrix in row-echelon form, then the nonzero rows of A constitute a basis for RS(A).

Proof: The set of nonzero rows spans RS(A) by definition; it is linearly independent by 15.2. ■

15.7 Computational Procedure

To find the rank, $r(A)$, of an $m \times n$ matrix A, compute G(A); conclude that $r(A)$ is equal to the number of nonzero rows in G(A).

Justification: $r(A)$ is (among other things) the dimension of RS(A). By 15.5, RS(A) = RS(G(A)), and by 15.6, the nonzero rows of G(A) are a basis for RS(G(A)). ■

15.8 Example

Let

$$A = \begin{bmatrix} 1 & 2 & -1 & -3 \\ -1 & 3 & -2 & 4 \\ 2 & -1 & 1 & -7 \end{bmatrix}$$

To find $r(A)$, calculate

$$G(A) = \begin{bmatrix} 1 & 2 & -1 & -3 \\ 0 & 1 & -\frac{3}{5} & \frac{1}{5} \\ 0 & 0 & 0 & 0 \end{bmatrix}$$

Conclude that $r(A) = 2$.

15.9 Computational Procedure

To find a basis, β, for the row-space, RS(A), of an $m \times n$ matrix A, let β consist of the nonzero rows of G(A).

Justification: Corollary 15.5 and Theorem 15.6. ■

15.10 Example

The set $\beta = \{(1, 2, -1, -3), (0, 1, -\frac{3}{5}, \frac{1}{5})\}$ is a basis for the row-space of the matrix A from Example 15.8.

15.11 Computational Procedure

To find a reduced basis β (i.e., a basis consisting of a reduced set of vectors in the sense of definition 7.26) for the row-space of an $m \times n$ matrix A, let β consist of the nonzero rows of GJ(A).

Justification: Corollary 15.5 and Theorem 15.6 together with the fact that the set of nonzero rows of an $m \times n$ matrix in reduced row-echelon form is a reduced set of n-tuples. ■

15.12 Example

Once again, let A be the matrix from Example 10.8. Verify that

$$\mathrm{GJ(A)} = \begin{bmatrix} 1 & 0 & \frac{1}{5} & -\frac{17}{5} \\ 0 & 1 & -\frac{3}{5} & \frac{1}{5} \\ 0 & 0 & 0 & 0 \end{bmatrix}$$

Conclude that $\beta = \{(1,0,\frac{1}{5},-\frac{17}{5}),(0,1,-\frac{3}{5},\frac{1}{5})\}$ is a reduced basis for RS(A).

The procedures developed thus far can be used to answer questions that one might ask about a set $S = \{\mathbf{u}_1, \mathbf{u}_2, \ldots, \mathbf{u}_m\} \subset \mathbb{R}^n$ of m-many n-tuples.

15.13 Example

Let $S_0 = \{(1,1,0,1),(2,5,3,-4),(1,0,-1,3)\}$. Recall that $\mathscr{L}(S_0)$ denotes the subspace spanned by S_0. Consider the following questions:

(i) What is the dimension of $\mathscr{L}(S_0)$?
(ii) Find a basis for $\mathscr{L}(S_0)$.
(iii) Find a reduced basis for $\mathscr{L}(S_0)$.
(iv) Is S_0 linearly independent?
(v) Does S_0 span $\mathbb{R}^4$?
(vi) Does the vector $(-1,2,-3,-1)$ belong to $\mathscr{L}(S_0)$?
(vii) Does the vector $(-1,2,3,-7)$ belong to $\mathscr{L}(S_0)$?
(viii) If the answer to (vi) or (vii) is yes, find coefficients that can be used to express the given vector as a linear combination of the vectors from S_0.
(ix) Are the coefficients in the answer to question (viii) unique? If not, what are all the possible answers to question (viii)?
(x) Find a reduced description of $\mathscr{L}(S_0)$.

All these questions about S_0 can be answered using the Gauss–Jordan algorithm as a tool. The algorithm will be applied either to the matrix

$$M_0 = \begin{bmatrix} 1 & 1 & 0 & 1 \\ 2 & 5 & 3 & -4 \\ 1 & 0 & -1 & 3 \end{bmatrix}$$

having the vectors from S_0 as its rows, or to the matrix

$$N_0 = \begin{bmatrix} 1 & 2 & 1 \\ 1 & 5 & 0 \\ 0 & 3 & -1 \\ 1 & -4 & 3 \end{bmatrix}$$

having the vectors from S_0 as its columns, or to some third matrix that has either M_0 or N_0 as a submatrix.

Since we wish to justify, rather than merely illustrate, the procedures involved in answering these questions, we will deal with the most general case, where $S = \{\mathbf{u}_1, \mathbf{u}_2, \ldots, \mathbf{u}_m\} \subset \mathbb{R}^n$ is an arbitrary set of m-many n-tuples. Throughout the following, M is the $m \times n$ matrix having the vectors from S as its rows and N is the $n \times m$ matrix having the vectors from S as its columns.

15.14 Computational Procedure

To determine the dimension of $\mathscr{L}(S)$, compute G(M); conclude that dim $\mathscr{L}(S)$ is equal to the number of nonzero rows in G(M).

15.15 Computational Procedure

To find a basis for $\mathscr{L}(S)$, compute G(M); the nonzero rows of G(M) constitute a basis for $\mathscr{L}(S)$.

15.16 Computational Procedure

To find a reduced basis for $\mathscr{L}(S)$, compute GJ(M); the nonzero rows of GJ(M) constitute a reduced basis for $\mathscr{L}(S)$.

These three computational procedures are justified by 15.7, 15.9, and 15.11, respectively, since $\mathscr{L}(S) = \text{RS(M)}$. ■

15.13 (i), (ii), (iii) Revisited:

$$M_0 = \begin{bmatrix} 1 & 1 & 0 & 1 \\ 2 & 5 & 3 & -4 \\ 1 & 0 & -1 & 3 \end{bmatrix}$$

Verify that

$$G(M_0) = \begin{bmatrix} 1 & 1 & 0 & 1 \\ 0 & 1 & 1 & -2 \\ 0 & 0 & 0 & 0 \end{bmatrix}$$

and that

$$GJ(M_0) = \begin{bmatrix} 1 & 0 & -1 & 3 \\ 0 & 1 & 1 & -2 \\ 0 & 0 & 0 & 0 \end{bmatrix}$$

Conclude that $\dim \mathscr{L}(S_0) = 2$, that $\{(1,1,0,1),(0,1,1,-2)\}$ is a basis for $\mathscr{L}(S_0)$, and that $\{(1,0,-1,3),(0,1,1,-2)\}$ is a reduced basis for $\mathscr{L}(S_0)$.

15.17 Computational Procedure

To determine whether $S = \{\mathbf{u}_1, \mathbf{u}_2, \ldots, \mathbf{u}_m\} \subset \mathbb{R}^n$ is linearly independent: if $m \gneq n$, conclude immediately that S is linearly dependent; if $m \leqq n$, compute G(M); conclude that S is linearly dependent if and only if G(M) has an all-zero row [i.e., if and only if $r(\mathrm{M}) \lneq m$].

Justification: The conclusion in case $m \gneq n$ is precisely Corollary 7.30. In the remaining cases, we first use 15.14 to conclude that $r(\mathrm{M}) = \dim \mathscr{L}(S)$; thus if $r(\mathrm{M}) = m$, S is linearly independent by 8.5(ii), since S is a set of m-many vectors spanning a space, $\mathscr{L}(S)$, known to have dimension m; if $r(\mathrm{M}) \lneq m$, then S is linearly dependent by 7.29(i), since S is a set of m-many vectors from a space, $\mathscr{L}(S)$, that can be spanned by fewer than m-many vectors. ■

15.18 Computational Procedure

To determine whether $S = \{\mathbf{u}_1, \mathbf{u}_2, \ldots, \mathbf{u}_m\} \subset \mathbb{R}^n$ spans $\mathbb{R}^n$: if $m \lneq n$, conclude immediately that S does not span $\mathbb{R}^n$; if $m \geq n$, compute G(M); conclude that S spans $\mathbb{R}^n$ if and only if the number of nonzero rows of G(M) is equal to n [i.e., if and only if $r(\mathrm{M}) = n$].

Justification: The conclusion in case $m \lneq n$ is precisely 7.30(ii). In the remaining cases, we again first use 15.14 to conclude that $r(\mathrm{M}) = \dim \mathscr{L}(S)$; thus if $r(\mathrm{M}) = n$, it follows from 8.6 that $\mathscr{L}(S) = \mathbb{R}^n$; if $r(\mathrm{M}) \lneq n$, it follows that $\mathscr{L}(S) \neq \mathbb{R}^n$ since spaces of different dimensions cannot be equal. ■

15.13 (iv) and (v) Revisited: Applying 15.17 and 15.18 to the set S_0 from Example 15.13 leads to the following conclusions:

(i) S_0 is linearly dependent since $G(M_0)$, which we computed earlier, has an all-zero row.

(ii) S_0 does not span $\mathbb{R}^4$ since no set of fewer than four 4-tuples can span $\mathbb{R}^4$.

Returning once again to the list of questions in Example 15.13, we note that the general form of questions (vi) and (vii) is this: does the vector $\mathbf{b} = (b_1, b_2, \ldots, b_n)$ belong to the space, $\mathscr{L}(S)$, spanned by the set $S = \{\mathbf{u}_1, \mathbf{u}_2, \ldots, \mathbf{u}_m\} \subset \mathbb{R}^n$?

Since $\mathscr{L}(S) = \text{CS(N)}$, where N is the matrix having the vectors from S as its columns, it is clear, using Observations 6.14 and 9.4, that the question of whether $\mathbf{b} \in \mathscr{L}(S)$ is equivalent to the question of whether the $n \times m$ linear system $\mathrm{N}\mathbf{x} = \mathbf{b}$ is consistent.

This settles the matter since we already have a procedure, 12.14, for answering this question about arbitrary linear systems.

15.19 Computational Procedure

To obtain a simple yes or no answer to the question of whether a given vector $\mathbf{b} = (b_1, b_2, \ldots, b_n)$ belongs to $\mathscr{L}(S)$, where $S = \{\mathbf{u}_1, \mathbf{u}_2, \ldots, \mathbf{u}_m\} \subset \mathbb{R}^n$, compute

$$\mathrm{G}([\mathrm{N} \,\vdots\, \mathbf{b}]) = \left[\begin{array}{c|c} & c_1 \\ & c_2 \\ & \vdots \\ \mathrm{G(N)} & c_r \\ & c_{r+1} \\ & \vdots \\ & c_n \end{array}\right]$$

where $r = \text{rank N} =$ the number of nonzero rows in G(N). Conclude that $\mathbf{b} \in \mathscr{L}(S)$ if and only if $c_{r+1} = c_{r+2} = \cdots = c_n = 0$.

Justification: This follows from 12.14. Note that computing $\mathrm{G}([\mathrm{N} \,\vdots\, \mathbf{b}])$, rather than $\mathrm{GJ}([\mathrm{N} \,\vdots\, \mathbf{b}])$, provides only a yes or no answer to the question. ■

15.13 (vi) and (vii) Revisited: Recall question (vi): does $(-1, 2, -3, -1)$ belong to the subspace of $\mathbb{R}^4$ spanned by the set $S_0 = \{(1,1,0,1), (2,5,3,-4), (1,0,-1,3)\}$? To answer this, form the matrix

$$[\mathrm{N}_0 \,\vdots\, \mathbf{b}] = \left[\begin{array}{rrr|r} 1 & 2 & 1 & -1 \\ 1 & 5 & 0 & 2 \\ 0 & 3 & -1 & -3 \\ 1 & -4 & 3 & -1 \end{array}\right]$$

Verify that

$$G([N_0 \mid \mathbf{b}]) = \left[\begin{array}{ccc|c} 1 & 2 & 1 & 1 \\ 0 & 1 & -\frac{1}{3} & 1 \\ 0 & 0 & 0 & 1 \\ 0 & 0 & 0 & 0 \end{array}\right] = \left[\begin{array}{c|c} & -1 \\ G(N_0) & 1 \\ & 1 \\ & 0 \end{array}\right]$$

Conclude that $(-1, 2, -3, -1) \notin \mathscr{L}(S_0)$.

Recall question (vii): does $(-1, 2, 3, -7) \in \mathscr{L}(S_0)$? To answer this, compute

$$G\left(\left[\begin{array}{ccc|c} 1 & 2 & 1 & -1 \\ 1 & 5 & 0 & 2 \\ 0 & 3 & -1 & 3 \\ 1 & -4 & 3 & -7 \end{array}\right]\right) = \left[\begin{array}{ccc|c} 1 & 2 & 1 & -1 \\ 0 & 1 & -\frac{1}{3} & 1 \\ 0 & 0 & 0 & 0 \\ 0 & 0 & 0 & 0 \end{array}\right]$$

Conclude that $(-1, 2, 3, -7) \in \mathscr{L}(S_0)$.

Questions (viii) and (ix) from 15.13 differ from (vi) and (vii) by requiring, when $\mathbf{b}$ does belong to $\mathscr{L}(S)$, additional information about the way or ways in which $\mathbf{b}$ can be written as a linear combination of the vectors from S. Computing $G([N \mid \mathbf{b}])$ provides the yes or no answer and, in case the answer is yes, the desired additional information can be obtained by continuing the row-reduction process to yield $GJ([N \mid \mathbf{b}])$. This is because, by observation 6.14, $\mathbf{b} = t_1\mathbf{u}_1 + t_2\mathbf{u}_2 + \cdots + t_m\mathbf{u}_m$ if and only if $\mathbf{t} = (t_1, t_2, \ldots, t_m)$ belongs to the solution space of the $n \times m$ linear system $N\mathbf{x} = \mathbf{b}$. This justifies the next computational procedure.

15.20 Computational Procedure

To determine whether a given vector $\mathbf{b} = (b_1, b_2, \ldots, b_n)$ belongs to $\mathscr{L}(S)$, where $S = \{\mathbf{u}_1, \mathbf{u}_2, \ldots, \mathbf{u}_m\} \subset \mathbb{R}^n$ and, if it does, to determine all possible coefficients $t_1, t_2, \ldots, t_m$ satisfying $\mathbf{b} = t_1\mathbf{u}_1 + t_2\mathbf{u}_2 + \cdots + t_m\mathbf{u}_m$, solve the $n \times m$ linear system $N\mathbf{x} = \mathbf{b}$ using 12.14. If this system is inconsistent, conclude that $\mathbf{b} \notin \mathscr{L}(S)$; if this system is consistent, conclude that $\mathbf{b} \in \mathscr{L}(S)$ and that $\mathbf{b} = t_1\mathbf{u}_1 + t_2\mathbf{u}_2 + \cdots + t_m\mathbf{u}_m$ if and only if $\mathbf{t}$ is a solution of the system $N\mathbf{x} = \mathbf{b}$. ■

15.13 (viii) and (ix) Revisited: We determined previously that $(-1, 2, 3, -7) \in \mathscr{L}(S_0)$. Compute

$$GJ\left(\begin{bmatrix} 1 & 2 & 1 & -1 \\ 1 & 5 & 0 & 2 \\ 0 & 3 & -1 & 3 \\ 1 & -4 & 3 & -7 \end{bmatrix}\right) = \begin{bmatrix} 1 & 0 & \frac{5}{3} & -3 \\ 0 & 1 & -\frac{1}{3} & 1 \\ 0 & 0 & 0 & 0 \\ 0 & 0 & 0 & 0 \end{bmatrix}$$

Conclude from this that the solution space of the linear system

$$\begin{bmatrix} 1 & 2 & 1 \\ 1 & 5 & 0 \\ 0 & 3 & -1 \\ 1 & -4 & 3 \end{bmatrix} \begin{bmatrix} x_1 \\ x_2 \\ x_3 \end{bmatrix} = \begin{bmatrix} -1 \\ 2 \\ 3 \\ -7 \end{bmatrix}$$

is the set $\{(x_1, x_2, x_3) \in \mathbb{R}^3\colon x_1 = -\frac{5}{3}x_3 - 3,\ x_2 = \frac{1}{3}x_3 + 1\}$. Thus there are infinitely many ways to express $(-1, 2, 3, -7)$ as a linear combination of the vectors $\{(1,1,0,1), (2,5,3,-4), (1,0,-1,3)\}$. Any triple (t_1, t_2, t_3) from this set can serve as coefficients.

For example, $(-1, 2, 3, -7) = -8(1, 1, 0, 1) + 2(2, 5, 3, -4) + 3(1, 0, -1, 3)$ and $(-1, 2, 3, -7) = -3(1, 1, 0, 1) + 1(2, 5, 3, -4) + 0(1, 0, -1, 3)$.

We have saved until now the question whose answer requires the most work. But the principles that justify this procedure are the same ones that we have been using.

The general form of question 15.13(x) is this: given a set $S = \{\mathbf{u}_1, \mathbf{u}_2, \ldots, \mathbf{u}_m\} \subset \mathbb{R}^n$ of m-many n-tuples, provide a reduced description of the linear subspace, $\mathscr{L}(S)$, of $\mathbb{R}^n$ spanned by S. To transform this into a question about matrices, we first note the obvious fact that $\mathscr{L}(S) = \text{CS(N)}$. It is then Observation 9.4 that permits a rephrasing of the question in a form that allows it to be settled using the Gauss–Jordan algorithm as a tool. The equivalent question is this: for which vectors $\mathbf{w} = (w_1, w_2, \ldots, w_n) \in \mathbb{R}^n$ is the $n \times m$ linear system $\mathbf{Nx} = \mathbf{w}$ consistent?

A procedure for deciding the consistency of $\mathbf{Nx} = \mathbf{b}$ for a fixed vector $\mathbf{b} = (b_1, b_2, \ldots, b_n)$ was presented in 15.19. To determine more generally all possible values of $\mathbf{w} = (w_1, w_2, \ldots, w_n)$ for which the $n \times m$ linear system $\mathbf{Nx} = \mathbf{w}$ is consistent, we apply this procedure to the $n \times (m + 1)$ matrix

$$\left[\begin{array}{c|c} & w_1 \\ & w_2 \\ N & \vdots \\ & w_n \end{array}\right]$$

having the variable entries $w_1, w_2, \ldots, w_n$ in the $(m + 1)^{\text{st}}$ column.

[If the level of abstraction in the description of the following computational procedure seems too complicated, feel free to skip ahead a few pages to Example 15.13(x) revisited, and then return to 15.21.]

To find a reduced description of $\mathscr{L}(S)$, compute $G([N \mid \mathbf{w}])$. The entries in the $(m+1)^{\text{st}}$ column of the resulting matrix will be linear functions of the variables $w_1, w_2, \ldots, w_n$.

$$G\left(\left[\begin{array}{c|c} N & \begin{matrix} w_1 \\ w_2 \\ \vdots \\ w_n \end{matrix} \end{array}\right]\right) = \left[\begin{array}{c|l} & a_{11}w_1 + a_{12}w_2 + \cdots + a_{1n}w_n \\ & a_{21}w_1 + a_{22}w_2 + \cdots + a_{2n}w_n \\ & \vdots \\ G(N) & a_{r1}w_1 + a_{r2}w_2 + \cdots + a_{rn}w_n \\ & a_{r+1,1}w_1 + a_{r+1,2}w_2 + \cdots + a_{r+1,n}w_n \\ & \vdots \\ & a_{n1}w_1 + a_{n2}w_2 + \cdots + a_{nn}w_n \end{array}\right]$$

where $r = \text{rank } N$.

By 15.19, $N\mathbf{x} = \mathbf{w}$ is consistent if and only if

$$\begin{aligned} a_{r+1,1}w_1 + a_{r+1,2}w_2 + \cdots + a_{r+1,n}w_n &= 0 \\ a_{r+2,1}w_1 + a_{r+2,2}w_2 + \cdots + a_{r+2,n}w_n &= 0 \\ &\vdots \\ a_{n1}w_1 + a_{n2}w_2 + \cdots + a_{nn}w_n &= 0 \end{aligned} \qquad (*)$$

Conclude that $\mathscr{L}(S)$ is the solution space of the $(n - r) \times n$ homogeneous system (*). A reduced description of $\mathscr{L}(S)$ can then be found by solving (*), using 12.14 if necessary.

The justification for this procedure has essentially been presented as we went along. It is based on 15.19 together with the fact that $\mathscr{L}(S) = CS(N) = \{\mathbf{w} \in \mathbb{R}^n: N\mathbf{x} = \mathbf{w} \text{ is consistent}\}$. ■

15.13 (x) Revisited: $S_0 = \{(1, 1, 0, 1), (2, 5, 3, -4), (1, 0, -1, 3)\}$. To obtain a reduced description of $\mathscr{L}(S_0)$, we calculate the Gaussian form of the matrix

$$\left[\begin{array}{ccc|c} 1 & 2 & 1 & w_1 \\ 1 & 5 & 0 & w_2 \\ 0 & 3 & -1 & w_3 \\ 1 & -4 & 3 & w_4 \end{array}\right]$$

The result is

$$\left[\begin{array}{ccc|l} 1 & 2 & 1 & w_1 \\ 0 & 1 & -\frac{1}{3} & \frac{1}{3}w_2 - \frac{1}{3}w_1 \\ 0 & 0 & 0 & w_3 - w_2 + w_1 \\ 0 & 0 & 0 & w_4 + 2w_2 - 3w_1 \end{array}\right]$$

$\mathscr{L}(S_0)$ is the solution space of the 2×4 homogeneous linear system

$$\begin{aligned} w_1 - w_2 + w_3 &= 0 \\ -3w_1 + 2w_2 + w_4 &= 0 \end{aligned} \qquad (\dagger)$$

In this particular example, we can solve (†) by inspection:

$$\mathscr{L}(S_0) = \{(w_1, w_2, w_3, w_4) \in \mathbb{R}^4 \colon w_3 = -w_1 + w_2,\ w_4 = 3w_1 - 2w_2\}$$

[This description of $\mathscr{L}(S_0)$ involves w_1 and w_2 as parameters. If we were to solve (†) using the Gauss–Jordan algorithm, we would obtain an alternate description of $\mathscr{L}(S_0)$ that would involve w_1 and w_2 as dependent variables and w_3 and w_4 as parameters.]

15.13 (vi) and (vii) Revisited Again: Note that the reduced description of $\mathscr{L}(S_0)$ given in answer to question (x) yields immediate answers to both question (vi), does $(-1, 2, -3, -1) \in \mathscr{L}(S_0)$, and question (vii), does $(-1, 2, 3, -7) \in \mathscr{L}(S_0)$? It is clear that $(-1, 2, -3, -1) \notin \mathscr{L}(S_0)$ since $-3 \neq -(-1) + 2$. It is also clear that $(-1, 2, 3, -7) \in \mathscr{L}(S_0)$ since both $3 = -(-1) + 2$ and $-7 = 3(-1) - 2(2)$.

Typically, when $S = \{\mathbf{u}_1, \mathbf{u}_2, \ldots, \mathbf{u}_m\} \subset \mathbb{R}^n$ is given and there is a list of questions of the form is $\mathbf{b}_1 \in \mathscr{L}(S)$?, is $\mathbf{b}_2 \in \mathscr{L}(S)$?, is $\mathbf{b}_3 \in \mathscr{L}(S)$?, ..., is $\mathbf{b}_k \in \mathscr{L}(S)$?, rather than applying 15.19 k-many times, it is less work to apply 15.21 once to obtain a reduced description of $\mathscr{L}(S)$, which can then be used to answer easily each of the questions in the list.

15.22 Example

Describe the subspace, $\mathscr{L}(S)$, of $\mathbb{R}^3$ spanned by the set $S = \{(3, 2, -1), (-4, 0, 6)\}$? To answer this using 15.21, verify that

$$G\left(\left[\begin{array}{rr|l} 3 & -4 & w_1 \\ 2 & 0 & w_2 \\ -1 & 6 & w_3 \end{array}\right]\right) = \left[\begin{array}{rr|l} 1 & -\frac{4}{3} & \frac{1}{3}w_1 \\ 0 & 1 & \frac{3}{8}w_2 - \frac{1}{4}w_1 \\ 0 & 0 & w_3 - \frac{7}{4}w_2 + \frac{3}{2}w_1 \end{array}\right]$$

Conclude that $\mathbf{w} = (w_1, w_2, w_3) \in \mathscr{L}(S)$ if and only if $w_3 - \frac{7}{4}w_2 + \frac{3}{2}w_1 = 0$. In other words, $\mathscr{L}(S) = \{(w_1, w_2, w_3) \in \mathbb{R}^3 \colon w_3 = \frac{7}{4}w_2 - \frac{3}{2}w_1\}$.

Just after Exercise 8.8, we promised to provide a computational procedure for extracting, from a given set $S \subseteq \mathbb{R}^n$ that contains at least one nonzero vector, a subset $S' \subseteq S$ that is a basis for $\mathscr{L}(S)$. Note that we are not seeking an *arbitrary* basis for $\mathscr{L}(S)$ (we already have a procedure for this in 15.15) but one that consists of vectors *from the original set S*.

15.23 Computational Procedure

Let $S = (\mathbf{u}_1, \mathbf{u}_2, \ldots, \mathbf{u}_m) \subset \mathbb{R}^n$ contain at least one nonzero vector. Apply the Gaussian algorithm to the $n \times m$ matrix N having the vectors from S as its columns. If the leading 1's in G(N) occur in columns $i_1, i_2, \ldots, i_r$, then the set $S' = \{\mathbf{u}_{i_1}, \mathbf{u}_{i_2}, \ldots, \mathbf{u}_{i_r}\}$ is a basis for $\mathscr{L}(S)$.

Justification: From the fact that G(N) has r-many leading 1's, we may conclude that the dim $\mathscr{L}(S) = \dim \mathrm{CS(N)} = \mathrm{rank(N)} = r$. Since S' is a set of cardinality r from a space whose dimension is known to equal r, it suffices by 8.5 to show that S' spans $\mathscr{L}(S)$. We claim, for arbitrary k, that column k of G(N) contains a leading 1 iff $\mathbf{u}_k \notin \mathscr{L}(\mathbf{u}_1, \mathbf{u}_2, \ldots, \mathbf{u}_{k-1})$: one way to see this is to let A_k denote the $n \times k$ matrix having $\mathbf{u}_1, \mathbf{u}_2, \ldots, \mathbf{u}_k$ as its columns, to note that A_k is equal to the augmented matrix $[\mathrm{A}_{k-1} \,\vdots\, \mathbf{u}_k]$, and to recall from Problem 9.3 that $\mathbf{u}_k \in \mathscr{L}(\mathbf{u}_1, \mathbf{u}_2, \ldots, \mathbf{u}_{k-1})$ iff $\mathrm{rank}(\mathrm{A}_k) = \mathrm{rank}(\mathrm{A}_{k-1})$.

Here is a picture of this situation assuming without loss of generality that $\mathbf{u}_1 \neq \mathbf{0}$. Suppose G(N) =

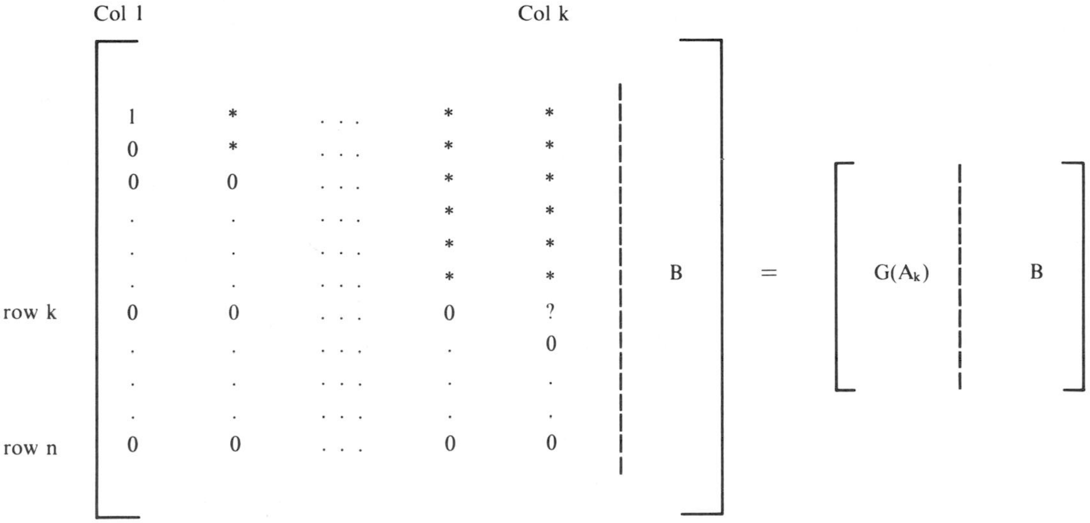

The (k, k) entry in G(N) is equal either to 0 or to 1 according as $\mathbf{u}_k$ belongs or does not belong to $\mathscr{L}(\mathbf{u}_1, \mathbf{u}_2, \ldots, \mathbf{u}_{k-1})$. It follows from this claim that $\mathscr{L}(S') = \mathscr{L}(S)$ because the vectors from $S - S'$ can be discarded without affecting the span. ■

15.24 Example

Find a subset of $S = \{(2, 1, -2, -4), (3, 4, 1, 0), (-1, 2, 5, 8), (2, 1, 0, -2), (4, 7, -8, -8)\}$ that is a basis for $\mathscr{L}(S)$.

Solution: Let

$$N = \begin{bmatrix} 2 & 3 & -1 & 2 & 4 \\ 1 & 4 & 2 & 1 & 7 \\ -2 & 1 & 5 & 0 & -8 \\ -4 & 0 & 8 & -2 & -8 \end{bmatrix}$$

You should verify that

$$G(N) = \begin{bmatrix} 1 & \frac{3}{2} & -\frac{1}{2} & 1 & 2 \\ 0 & 1 & 1 & 0 & 2 \\ 0 & 0 & 0 & 1 & -6 \\ 0 & 0 & 0 & 0 & 0 \end{bmatrix}$$

Since the leading 1's occur in the first, second, and fourth columns, you may conclude that the set $S' = \{(2, 1, -2, -4), (3, 4, 1, 0), (2, 1, 0, -2)\}$ is a basis for $\mathscr{L}(S)$.

Problem Set 15

15.1. Go back and do any of the exercises from Section 15 that you may have skipped.

*15.2. Let A be the augmented matrix of the system from Problem 12.3. Find the rank of A and a reduced basis for RS(A).

15.3. Let B be the augmented matrix of the system from Problem 12.9. Find the rank of B and a subset of the columns of B that is a basis for CS(B).

*15.4. Is the set $S = \{(-1, -3, 4), (5, 4, 1), (2, 1, 2)\}$ a basis for $\mathbb{R}^3$?

*15.5. Let $S = \{(1, 2, 2), (2, 6, 5), (1, -2, 0)\}$. Does $(1, 3, 3) \in \mathscr{L}(S)$? If it does, find all possible ways of expressing it as a linear combination of the vectors in S.

*15.6. Does $(-2, 1, -3) \in \mathscr{L}\{(-3, -2, 1), (1, 1, 1), (2, 2, -1)\}$? If it does, find all possible ways of expressing it as a linear combination of the three vectors.

*15.7. Find a reduced description of $\mathscr{L}\{(-2, -4, -2, 4), (3, 3, 0, -15), (-1, -2, -1, 2), (1, 3, 2, 1)\}$. What is its dimension?

*15.8. Does $\{(2, 4, 7), (1, -3, 1), (1, 7, 6), (1, 3, 4)\}$ span $\mathbb{R}^3$?

*15.9. Find a reduced description of $\mathscr{L}\{(2, -8, -6), (3, 2, 5), (2, -1, 1)\}$. What is its dimension?

15.10. What conditions, if any, must b_1, b_2, and b_3 satisfy for the following system of equations to be consistent?

$$\begin{aligned} x_1 + x_2 + x_3 &= b_1 \\ 2x_1 + 5x_2 - 2x_3 &= b_2 \\ -x_1 - 7x_2 + 7x_3 &= b_3 \end{aligned}$$

15.11. For what value(s) of t, if any, is the following system consistent?

$$\begin{aligned} 2x_1 - x_2 - 2x_3 &= 0 \\ 4x_1 + x_2 + 2x_3 &= t \\ -x_1 + \tfrac{1}{2}x_2 + x_3 &= t + 1 \\ -2x_1 + 2x_2 + 4x_3 &= t + 2 \end{aligned}$$

15.12. For what value(s) of t, if any, is the following system consistent?

$$\begin{aligned} x_1 + 2x_2 + 4x_3 + 5x_4 &= t \\ -x_1 + 3x_2 + x_3 &= 0 \\ 2x_1 + 4x_2 + 2x_3 + 4x_4 &= t \\ -2x_1 + x_2 - x_3 - 3x_4 &= t \\ 3x_1 + x_2 + x_3 &= t + 1 \end{aligned}$$

15.13. Let $S = \{(2, 6, -2, -1), (3, -3, 1, 3), (-1, -4, 3, 2), (4, 12, 5, \tfrac{1}{2})\}$.

(i) Which of the following are in $\mathscr{L}(S)$?

$$\mathbf{0} = (0,0,0,0), \qquad \mathbf{u} = (2,1,-1,0), \qquad \mathbf{v} = (1,1,0,0)$$

(ii) If it is possible to do so, write $\mathbf{0}$ as a nontrivial linear combination of the vectors from S.

15.14. Let $S = \{(1, -2, 2), (-4, -1, 1), (3, 5, -5)\}$.

(i) Decide whether S is linearly dependent or independent.

(ii) Let A be the matrix whose columns are the vectors from S. Find a reduced description of $\{\mathbf{b} \in \mathbb{R}^3 \colon \mathrm{A}\mathbf{x} = \mathbf{b} \text{ is consistent}\}$.

(iii) Find a basis for the column space of A.

15.15. Let

$$\mathrm{A} = \begin{bmatrix} 1 & 2 & 4 & 5 & -1 \\ -1 & 3 & -4 & 0 & 0 \\ 2 & 10 & 0 & 2 & 1 \\ 3 & 0 & 0 & 0 & 3 \end{bmatrix}$$

(i) Find a basis for the column space of A.

(ii) Find the rank of A.

(iii) Decide whether the system $\mathrm{A}\mathbf{x} = \mathbf{b}$ is consistent, where $\mathbf{b} = (1, 2, 3, 4)$.

15.16. You are *given* that when the Gauss-Jordan algorithm is applied to the 4 × 8 matrix

$$A = \begin{bmatrix} 1 & 3 & 2 & 5 & 1 & 0 & 0 & 0 \\ 1 & 2 & 2 & 6 & 0 & 1 & 0 & 0 \\ 1 & 2 & 1 & 3 & 0 & 0 & 1 & 0 \\ 1 & 2 & 3 & 9 & 0 & 0 & 0 & 1 \end{bmatrix}$$

the result is the 4 × 8 matrix

$$GJ(A) = \begin{bmatrix} 1 & 0 & 0 & 2 & -2 & 0 & \frac{5}{2} & \frac{1}{2} \\ 0 & 1 & 0 & -1 & 1 & 0 & -\frac{1}{2} & -\frac{1}{2} \\ 0 & 0 & 1 & 3 & 0 & 0 & -\frac{1}{2} & \frac{1}{2} \\ 0 & 0 & 0 & 0 & 0 & 1 & -\frac{1}{2} & -\frac{1}{2} \end{bmatrix}$$

The information just given may or may not be relevant to the following questions. Justify all your answers.

(i) What is the dimension of $\mathscr{L}\{(1, 1, 1, 1), (3, 2, 2, 2), (2, 2, 1, 3)\}$?

(ii) Does the set $\{(1, 3, 2, 5), (1, 2, 2, 6), (1, 2, 1, 3)\}$ span $\mathbb{R}^4$?

(iii) Is the set $\{(1, 1, 1, 1), (3, 2, 2, 2), (2, 2, 1, 3), (5, 6, 3, 9)\}$ linearly dependent?

(iv) Let C be the matrix

$$\begin{bmatrix} 1 & 3 & 2 & 5 \\ 1 & 2 & 2 & 6 \\ 1 & 2 & 1 & 3 \\ 1 & 2 & 3 & 9 \end{bmatrix}$$

Is C invertible?

(v) Find a reduced description of the solution space of the homogeneous system $C\mathbf{x} = \mathbf{0}$ where C is the matrix from the previous question.

(vi) Is the set $\{(1, 3, 2, 5), (1, 2, 2, 6), (1, 2, 1, 3), (1, 2, 3, 9)\}$ linearly dependent?

(vii) Does the set $\{(1, 1, 1, 1), (3, 2, 2, 2), (2, 2, 1, 3), (5, 6, 3, 9), (1, 0, 0, 0), (0, 1, 0, 0)\}$ span $\mathbb{R}^4$?

(viii) Find a reduced description of $\mathscr{L}\{(1, 1, 1, 1), (3, 2, 2, 2), (2, 2, 1, 3), (5, 6, 3, 9)\}$.

(ix) Does the vector $(5, 6, 3, 9)$ belong to $\mathscr{L}\{(1, 1, 1, 1), (3, 2, 2, 2), (2, 2, 1, 3)\}$?

(x) What is the dimension of the row-space of A?

(xi) What is the dimension of the nullspace of A?

(xii) Does the set $\{(1, 1, 1, 1), (3, 2, 2, 2), (2, 2, 1, 3), (5, 6, 3, 9), (1, 0, 0, 0)\}$ span $\mathbb{R}^4$?

(xiii) Is the matrix

$$\begin{bmatrix} -2 & 0 & \frac{5}{2} & \frac{1}{2} \\ 1 & 0 & -\frac{1}{2} & -\frac{1}{2} \\ 0 & 0 & -\frac{1}{2} & \frac{1}{2} \\ 0 & 1 & -\frac{1}{2} & -\frac{1}{2} \end{bmatrix}$$

invertible?

16

Functions

The purpose of this section is to introduce the basic facts about functions. We have included only the small amount of material that is absolutely necessary for the remainder of the book. Thus, even students for whom this material is new should find the present treatment quite accessible. Students who are familiar with these concepts are advised to skim this section briefly rather than to omit it altogether, since the notation, terminology, and outlook of this section are essential in later sections.

16.1 Definition

A *function f from the set X to the set Y* associates with each element $x \in X$ a unique element $f(x) \in Y$. This situation is described symbolically by writing either $f: X \to Y$ or $X \xrightarrow{f} Y$. The point $f(x)$ is called the *image of x under f* or, synonymously, the *value of f at x*. If $f: X \to Y$, the set X is called the *domain of f* and the set Y is called the *codomain of f*. We abbreviate these names by $\operatorname{dom} f$ and $\operatorname{codom} f$, respectively. Although a function f from X to Y associates some element $f(x) \in Y$ with every element x of its domain, it need not happen that every element y of the codomain Y is the image under f of one or more elements $x \in X$. When this does happen, the function $f: X \to Y$ is said to be *onto*. The subset of Y consisting of those elements of Y

that are the image under f of at least one $x \in X$ is called the *range of* f. This name is abbreviated by ran f. Thus ran $f = \{y \in Y$: there is at least one $x \in X$ such that $y = f(x)\}$ and $f: X \to Y$ is onto if and only if ran $f = Y$. We say that $f: X \to Y$ is *one-to-one* if f never assumes the same value twice. In other words, $f: X \to Y$ is one-to-one if, whenever x_1 and x_2 are distinct elements of X, then $f(x_1)$ and $f(x_2)$ are distinct elements of Y. The function $i_X: X \to X$ defined by $i_X(x) = x$ for all $x \in X$ is called the *identity function on* X. If $f: X \to Y$ and $S \subset X$, then $f \upharpoonright S$ denotes the function with domain S and codomain Y defined by $(f \upharpoonright S)(x) = f(x)$ for all $x \in S$; $f \upharpoonright S$ is called the *restriction of* f *to* S.

16.2 Examples

Let $W = \{0, 1, 2, 3, 4, 5\}$, $X = \{0, 2, 4, 6, 8, 10, 12\}$, $Y = \{0, 1, 3, 5, 7\}$, and $Z = \{98, 99, 100, 101, 102\}$.

$f: W \to X$ is defined by $f(0) = 2$, $f(1) = 4$, $f(2) = 12$, $f(3) = 8$, $f(4) = 10$, $f(5) = 6$.

$g: X \to Y$ is defined by $g(0) = 0$, $g(2) = 1$, $g(4) = 1$, $g(6) = 7$, $g(8) = 3$, $g(10) = 3$, $g(12) = 5$.

$h: Y \to Z$ is defined by $h(0) = 100$, $h(1) = 98$, $h(3) = 102$, $h(5) = 101$, $h(7) = 99$.

$r: Y \to Z$ is defined by $r(0) = 100$, $r(1) = 100$, $r(3) = 100$, $r(5) = 98$, $r(7) = 102$.

f is one-to-one since it never assumes the same value twice; f is not onto since $0 \in X =$ codom f, but $0 \notin$ ran f. g is not one-to-one since, for example, it assumes the value 3 twice; g is onto since ran $g = Y$. h is both one-to-one and onto. r is neither one-to-one nor onto.

As we see from these examples, a function can have none, exactly one, or both of the properties of one-to-oneness and ontoness. There is in general no connection between the two properties.

16.3 Examples

$f_1: \mathbb{R} \to \mathbb{R}$ is the function defined by the formula $f_1(x) = x^2$; $g_1: \mathbb{R} \to \mathbb{R}$ is the function defined by the formula $g_1(x) = x^3$; $h_1: \mathbb{R} \to \mathbb{R}$ is the function defined by the formula $h_1(x) = 4x + 5$.

f_1 is not one-to-one since, for example, $f_1(-2) = f_1(2) = 4$; f_1 is not onto since $-1 \notin$ ran f. g_1 and h_1 are both one-to-one and onto.

16.4 Definition

If $f: X \to Y$ and $g: Y \to Z$ are two functions such that dom $g =$ codom f, then it is possible to define a function, which we call f *composed with* g and denote by $g \circ f$, whose domain is X and whose codomain is Z. $g \circ f: X \to Z$

is defined by $(g \circ f)(x) = g(f(x))$. We should think of $\circ$ as an operation on pairs of functions; this operation is called *composition*.

Note that $g \circ f$ is defined only if dom g = codom f; thus the order of the two functions is essential: f composed with g is not the same as g composed with f. It can happen that none, exactly one, or both of $g \circ f$ and $f \circ g$ are defined depending on the domains and codomains of f and g. For both to be defined, it must be the case that if $f: X \to Y$, then $g: Y \to X$. When both are defined, they are, in general, different functions; this is clear if $X \neq Y$, since $g \circ f: X \to X$, whereas $f \circ g: Y \to Y$; but even when $X = Y$ it is rare that $g \circ f$ and $f \circ g$ will be equal.

16.5 Examples

If f, g, and h are the functions defined in 16.2, then $g \circ f$, $h \circ g$, $h \circ (g \circ f)$, and $(h \circ g) \circ f$ are all defined, whereas $f \circ g$, $g \circ h$, $h \circ f$, and $f \circ h$ are *not* defined.

If f_1, g_1, and h_1 are the functions defined in 16.3, then $g_1 \circ f_1$, $f_1 \circ g_1$, $g_1 \circ h_1$, $h_1 \circ g_1$, $h_1 \circ f_1$ and $f_1 \circ h_1$ are all defined since all three functions have the same domain and codomain.

$$(g_1 \circ h_1)(x) = g_1(h_1(x)) = g_1(4x + 5) = (4x + 5)^3$$
$$(h_1 \circ g_1)(x) = h_1(g_1(x)) = h_1(x^3) = 4x^3 + 5$$
$$(g_1 \circ f_1)(x) = g_1(f_1(x)) = g_1(x^2) = (x^2)^3 = x^6$$
$$(f_1 \circ g_1)(x) = f_1(g_1(x)) = f_1(x^3) = (x^3)^2 = x^6$$
$$(f_1 \circ h_1)(x) = f_1(h_1(x)) = f_1(4x + 5) = (4x + 5)^2$$
$$(h_1 \circ f_1)(x) = h_1(f_1(x)) = h_1(x^2) = 4x^2 + 5$$

If $f: W \to X$, $g: X \to Y$, and $h: Y \to Z$, then it makes sense to consider both $h \circ (g \circ f): W \to Z$ and $(h \circ g) \circ f: W \to Z$. The next theorem asserts that these two functions are equal.

16.6 Theorem (Composition of functions is associative)

If $f: W \to X$, $g: X \to Y$, and $h: Y \to Z$, then $h \circ (g \circ f) = (h \circ g) \circ f$.

Proof: We leave this as an easy exercise for the reader.

16.7 Exercise

Verify that if f, g, and h are the functions defined in 16.2, then $h \circ (g \circ f) = (h \circ g) \circ f$.

16.8 Definition

Let $f: X \to Y$.

(i) A function $g: Y \to X$ is a *right inverse of f* if $f \circ g: Y \to Y$ equals i_Y [i.e., if $(f \circ g)(y) = y$ for all $y \in Y$].
(ii) A function $h: Y \to X$ is a *left inverse of f* if $h \circ f: X \to X$ equals i_X [i.e., if $(h \circ f)(x) = x$ for all $x \in X$].

16.9 Examples

Let $g: X \to Y$ be the function defined in 16.2. Define $h_2: Y \to X$ by $h_2(0) = 0$, $h_2(1) = 2$, $h_2(3) = 8$, $h_2(5) = 12$, $h_2(7) = 6$; define $h_3: Y \to X$ by $h_3(0) = 0$, $h_3(1) = 4$, $h_3(3) = 10$, $h_3(5) = 12$, $h_3(7) = 6$. Then

$$(g \circ h_2)(0) = 0 = (g \circ h_3)(0)$$
$$(g \circ h_2)(1) = 1 = (g \circ h_3)(1)$$
$$(g \circ h_2)(3) = 3 = (g \circ h_3)(3)$$
$$(g \circ h_2)(5) = 5 = (g \circ h_3)(5)$$
$$(g \circ h_2)(7) = 7 = (g \circ h_3)(7)$$

Thus h_2 and h_3 are both right inverses of g.

Let $f: W \to X$ be the function defined in 16.2. Define $g_2: X \to W$ by $g_2(0) = 0$, $g_2(2) = 0$, $g_2(4) = 1$, $g_2(6) = 5$, $g_2(8) = 3$; $g_2(10) = 4$, $g_2(12) = 2$; define $g_3: X \to W$ by $g_3(0) = 5$, $g_3(2) = 0$, $g_3(4) = 1$, $g_3(6) = 5$, $g_3(8) = 3$, $g_3(10) = 4$, $g_3(12) = 2$. Then

$$(g_2 \circ f)(0) = 0 = (g_3 \circ f)(0)$$
$$(g_2 \circ f)(1) = 1 = (g_3 \circ f)(1)$$
$$(g_2 \circ f)(2) = 2 = (g_3 \circ f)(2)$$
$$(g_2 \circ f)(3) = 3 = (g_3 \circ f)(3)$$
$$(g_2 \circ f)(4) = 4 = (g_3 \circ f)(4)$$
$$(g_2 \circ f)(5) = 5 = (g_3 \circ f)(5)$$

Thus g_2 and g_3 are both left inverses of f.

From these examples, we see that a function can have more than one right inverse or left inverse. In the next theorem, we characterize those functions that can have right inverses or left inverses, and in the remainder of the section we will discuss the question of uniqueness.

Assume throughout the following that $X \neq \varnothing$.

16.10 Theorem

(i) A function $f: X \to Y$ has a right inverse if and only if f is onto.
(ii) A function $f: X \to Y$ has a left inverse if and only if f is one-to-one.

Proof of (i): Suppose $f: X \to Y$ has a right inverse $g: Y \to X$. To show that f is onto, we must show that an arbitrary $y \in Y$ is the image under f of at least one $x \in X$. But this does in fact happen, since $y = i_Y(y) = (f \circ g)(y) = f(g(y))$, and so y is the image under f of $g(y)$.

Conversely, suppose $f: X \to Y$ is onto. Then we can define a function $g: Y \to X$ that will be a right inverse of f as follows: since f is onto, every $y \in Y$ is the image under f of at least one $x \in X$; define $g(y)$ to be any such x. Then for all $y \in Y$ we have $(f \circ g)(y) = f(g(y)) = y$, where the second equality holds because $g(y)$ is by definition a point in X whose image under f is y.

Proof of (ii): Suppose $f: X \to Y$ has a left inverse $h: Y \to X$. To show that f never assumes the same value twice, suppose $f(x_1) = f(x_2)$; then $h(f(x_1)) = h(f(x_2))$. But since $h \circ f = i_X$, we conclude $x_1 = x_2$.

Conversely, suppose $f: X \to Y$ is one-to-one. Then we can construct a left inverse $h: Y \to X$ as follows: an arbitrary $y \in Y$ either belongs to ran f or it does not; if $y \in \text{ran } f$, then there is at least one $x \in X$ such that $y = f(x)$, and because f never assumes the same value twice, there is a unique such x, so we can define $h(y)$ to be the unique $x \in X$ such that $f(x) = y$. If $y \notin \text{ran } f$, it does not matter how $h(y)$ is defined, so define $h(y)$ to be any element of X. To see that h is a left inverse, note that for all $x \in X$ we have $(h \circ f)(x) = h(f(x)) = x$, where the second equality holds by the definition of h, because $f(x) \in \text{ran } f$ and x is the unique element of X whose image under f is $f(x)$. ∎

16.11 Definition

A function $g: Y \to X$ that is both a right and left inverse for the function $f: X \to Y$ is simply called an *inverse* of f. To emphasize that both properties must be satisfied, such a g is sometimes called a *two-sided inverse* of f. The function f is called *invertible* if it has a (two-sided) inverse.

16.12 Example

Let $h: Y \to Z$ be the function defined in 16.2. Define $k: Z \to Y$ by $k(98) = 1$, $k(99) = 7$, $k(100) = 0$, $k(101) = 5$, $k(102) = 3$. Then

$$\begin{aligned} (h \circ k)(98) &= 98 \\ (h \circ k)(99) &= 99 \\ (h \circ k)(100) &= 100 \\ (h \circ k)(101) &= 101 \\ (h \circ k)(102) &= 102 \end{aligned} \qquad \text{and} \qquad \begin{aligned} (k \circ h)(0) &= 0 \\ (k \circ h)(1) &= 1 \\ (k \circ h)(3) &= 3 \\ (k \circ h)(5) &= 5 \\ (k \circ h)(7) &= 7 \end{aligned}$$

Thus $h \circ k = i_Z$ and $k \circ h = i_Y$, so k is a two-sided inverse of h.

16.13 Theorem (Uniqueness of the inverse of an invertible function)

If $g: Y \to X$ and $h: Y \to X$ are each inverses of $f: X \to Y$, then $g = h$.

Proof: Let $y \in Y$. Then

$$\begin{aligned} h(y) &= h(i_Y(y)) && \text{[definition of } i_Y\text{]} \\ &= h((f \circ g)(y)) && [g \text{ is a right inverse of } f] \\ &= (h \circ (f \circ g))(y) && \text{[definition of composition]} \\ &= ((h \circ f) \circ g)(y) && \text{[composition is associative]} \\ &= (i_X \circ g)(y) && [h \text{ is a left inverse of } f] \\ &= i_X(g(y)) && \text{[definition of composition]} \\ &= g(y) && \text{[definition of } i_X\text{]} \end{aligned}$$

■

Because of 16.13, it makes sense to talk about *the* inverse of an invertible function $f: X \to Y$ and to introduce a special symbol to denote it. The standard symbol used is f^{-1}. (See [F]16.1.) Thus if $f: X \to Y$ is invertible, then $f^{-1}: Y \to X$ and we have both that $f^{-1}(f(x)) =$ for all $x \in X$ and $f(f^{-1}(y)) = y$ for all $y \in Y$.

16.14 Corollary to the Proof of 16.13

If $h: Y \to X$ is a left inverse and $g: Y \to X$ is a right inverse of $f: X \to Y$, then $g = h = f^{-1}$.

Proof: If we examine the justifications given in brackets for each step in the proof of 16.13, we see that the full assumptions of 16.13 (i.e., that g and h were each two-sided inverses of f) did not get used. Only the facts that h is a left inverse and g is a right inverse were required. ■

16.15 Theorem

A function $f: X \to Y$ is invertible if and only if f is both one-to-one and onto.

Proof: If $f: X \to Y$ has a two-sided inverse, it follows from 16.10 that f is both one-to-one and onto. Conversely, if f is both one-to-one and onto, it follows from 16.10 that f has both a left inverse and a right inverse. It then follows from 16.14 that these one-sided inverses are in fact equal, and so f is invertible. ■

16.16 Exercise

If $f: X \to Y$ and $g: Y \to Z$ are both invertible, then so is $g \circ f$; moreover $(g \circ f)^{-1} = f^{-1} \circ g^{-1}$.

If $f: X \to X$ (i.e., if codom f = dom f), then it makes sense to compose f with itself; $(f \circ f): X \to X$ is defined by $(f \circ f)(x) = f(f(x))$ for all $x \in X$. It also makes sense to compose f with $f \circ f$, and so on. Because composition is associative, there is no need to distinguish between $(f \circ f) \circ f$ and $f \circ (f \circ f)$. The functions $f \circ f$, $f \circ f \circ f, \ldots, f \circ \cdots \circ f, \ldots$, which should properly be called the *iterates of* f, are usually called the *powers of* f and are denoted by $f^2, f^3, \ldots, f^k$, and so on.

For functions whose values can be multiplied (e.g., if $f: \mathbb{R} \to \mathbb{R}$), this name is an occasional source of confusion. It is important to distinguish f^2 from the function $g: \mathbb{R} \to \mathbb{R}$ defined by $g(x) = (f(x))^2$.

16.17 Exercise

With the convention that $f^0 = i_X$, if $f: X \to X$ and if q and r are integers ≥ 0, then

(i) $f^q \circ f^r = f^{q+r}$.
(ii) $(f^q)^r = f^{qr}$.

16.18 Exercise

If f is invertible, then the iterates of f^{-1} are also defined. Show that the results from the previous exercise hold for arbitrary integers if f is invertible.

Problem Set 16

16.1. Go back and do any of the exercises from Section 16 that you may have skipped.

*16.2. Let $X = \{0, 2, 4, 6, 8, 10\}$, $Y = \{1, 2, 3, 4\}$, and let $f: X \to Y$ be defined by $f(0) = 3$, $f(2) = 4$, $f(4) = 4$, $f(6) = 3$, $f(8) = 2$, and $f(10) = 1$.
 (i) Is f one-to-one?
 (ii) Is f onto?
 (iii) Find two different right inverses for f.

*16.3. Let $X = \{1, 2, 3, 4\}$, $Y = \{0, 2, 4, 6, 8, 10\}$, and let $g: X \to Y$ be defined by $g(x) = 2x$.
 (i) Is g one-to-one?
 (ii) Is g onto?
 (iii) Find two different left inverses for g.

*16.4. Let $f: X \to Y$ and $g: Y \to Z$.

(i) Prove that $g \circ f$ is one-to-one if and only if f and $g \upharpoonright \operatorname{ran}(f)$ are both one-to-one.

(ii) Prove that $g \circ f$ is onto if and only if $g \upharpoonright \operatorname{ran}(f)$ is onto.

(iii) Give an example in which f is one-to-one, $g \circ f$ is one-to-one, but g is not one-to-one.

*16.5. Let $\mathbb{N}$ denote the natural numbers, i.e., $\mathbb{N} = \{0, 1, 2, 3, 4, \ldots\}$; define $f: \mathbb{N} \to \mathbb{N}$ by $f(n) = 2n$, and define $g: \mathbb{N} \to \mathbb{N}$ by $g(n) = n + 5$.

(i) Find $\operatorname{ran}(f)$.

(ii) Find $\operatorname{ran}(g)$.

(iii) Find $\operatorname{ran}(g \circ f)$.

(iv) Find $\operatorname{ran}(f \circ g)$.

*16.6. Repeat Problem 16.5 with $f: \mathbb{R} \to \mathbb{R}$ and $g: \mathbb{R} \to \mathbb{R}$ defined by $f(x) = |x|$ (the absolute value of x) and $g(x) = \cos x$.

16.7. The developments of Sections 3 and 16 are closely parallel. Reread Section 3 and compare the following paired items:

3.1 (ix)	—	16.6
3.3	—	16.8 and 16.11
3.6	—	16.13
3.7	—	16.14
3.8	—	16.16
3.11	—	16.17
3.12	—	16.18

17

Linear Transformations from $\mathbb{R}^n$ to $\mathbb{R}^m$

The purpose of studying functions in general in Section 16 was to set the background for studying the class of functions that is of particular interest in linear algebra.

The simplest imaginable functions from $\mathbb{R}$ to $\mathbb{R}$ are the constant functions; after that, the next simplest are perhaps those of the form $f(x) = mx$, that is, the functions whose graphs are just straight lines in $\mathbb{R}^2$ through the origin. Note that if $f(x) = mx$, then $f(x_1 + x_2) = m(x_1 + x_2) = mx_1 + mx_2 = f(x_1) + f(x_2)$; also $f(tx) = m(tx) = t(mx) = tf(x)$. The functions from $\mathbb{R}^n$ to $\mathbb{R}^m$ that interest us are just the higher-dimensional analogues of these (i.e., those whose graphs are linear subspaces of $\mathbb{R}^{n+m}$). After all, the only two operations on vectors that we have dealt with are vector addition and multiplication of a vector by a scalar. When it comes to functions whose domains and ranges are sets of vectors, the ones that interest us in particular will be those that "preserve" these two operations on vectors.

17.1 Definition

A function $T: \mathbb{R}^n \to \mathbb{R}^m$ is called a *linear transformation* if

(i) for all $\mathbf{u}, \mathbf{v} \in \mathbb{R}^n$, $T(\mathbf{u} + \mathbf{v}) = T(\mathbf{u}) + T(\mathbf{v})$ [T preserves vector addition] and

(ii) for all $\mathbf{u} \in \mathbb{R}^n$ and $t \in \mathbb{R}$, $T(t\mathbf{u}) = tT(\mathbf{u})$ [T preserves scalar multiplication].

Another name is often used for the special case of linear transformations whose domain and codomain are the same. A linear transformation T: $\mathbb{R}^n \to \mathbb{R}^n$ is also called a *linear operator on* $\mathbb{R}^n$.

Before considering examples of linear transformations, let us first note two trivial consequences of the definition. The first is that a linear transformation from $\mathbb{R}^n$ to $\mathbb{R}^m$ must send the zero vector of $\mathbb{R}^n$ to the zero vector of $\mathbb{R}^m$; the second is that linear transformations will preserve *arbitrary* linear combinations.

17.2 Exercise

If T: $\mathbb{R}^n \to \mathbb{R}^m$ is a linear transformation, then

(i) $T(\mathbf{0}) = \mathbf{0}$ and
(ii) if $\mathbf{v} = t_1\mathbf{u}_1 + t_2\mathbf{u}_2 + \cdots + t_p\mathbf{u}_p$, then

$$T(\mathbf{v}) = t_1T(\mathbf{u}_1) + t_2T(\mathbf{u}_2) + \cdots + t_pT(\mathbf{u}_p).$$

In particular, if $\beta = \{\mathbf{u}_1, \mathbf{u}_2, \ldots, \mathbf{u}_n\}$ is a basis for $\mathbb{R}^n$, then the value of $T(\mathbf{v})$ for arbitrary $\mathbf{v} \in \mathbb{R}^n$ is completely determined by the values $T(\mathbf{u}_1), T(\mathbf{u}_2), \ldots, T(\mathbf{u}_n)$ of T on the basis vectors.

The most important examples of linear transformtions from $\mathbb{R}^n$ to $\mathbb{R}^m$ will be those provided by Theorem 17.6. For the moment though, let us illustrate the definition with a few simple examples.

17.3 Examples

T_1: $\mathbb{R}^2 \to \mathbb{R}^3$ is defined by $T_1(x_1, x_2) = (x_1, x_1 + x_2, 3x_1 + 4x_2)$

T_2: $\mathbb{R}^2 \to \mathbb{R}^3$ is defined by $T_2(x_1, x_2) = (x_1 + 1, x_1 + x_2, 3x_1 + 4x_2)$

T_3: $\mathbb{R}^2 \to \mathbb{R}^3$ is defined by $T_3(x_1, x_2) = (0, x_1, x_1 + 2x_2)$

T_4: $\mathbb{R}^2 \to \mathbb{R}^3$ is defined by $T_4(x_1, x_2) = (0, x_1x_2, x_1 + 2x_2)$

T_1 and T_3 are both linear transformations, while T_2 and T_4 are not.
Here are the details for T_3: let $\mathbf{u} = (u_1, u_2)$ and $\mathbf{v} = (v_1, v_2)$; then

$$\begin{aligned} T_3(\mathbf{u} + \mathbf{v}) &= T_3(u_1 + v_1, u_2 + v_2) \\ &= (0, u_1 + v_1, u_1 + v_1 + 2(u_2 + v_2)) \\ &= (0, u_1, u_1 + 2u_2) + (0, v_1, v_1 + 2v_2) \\ &= T_3(\mathbf{u}) + T_3(\mathbf{v}) \end{aligned}$$

so T_3 preserves vector addition; similarly, if $t \in \mathbb{R}$, then $T_3(t\mathbf{u}) = T_3(tu_1, tu_2) = (0, tu_1, tu_1 + 2tu_2) = t(0, u_1, u_1 + 2u_2) = tT_3(\mathbf{u})$, so T_3 preserves scalar multiplication.

The details for T_1 are similar.

Because $T_2(0,0) = (1,0,0) \neq (0,0,0)$, we conclude from 17.2(i) that T_2 is not linear.

The result in 17.2(i) can often be used this way to conclude "by inspection" that a function is not linear. In fact, one should develop the habit always to check the zero vector first: if the function does not send the zero vector to the zero vector, then the function is *not* linear.

If the function does send the zero vector to the zero vector, then no conclusion about linearity or nonlinearity can be made from this alone, and further analysis will be required. This is the case with T_4 since, clearly, $T_4(0,0) = (0,0,0)$.

Since, for example, $T_4(1,1) = (0,1,3)$, but $T_4(2(1,1)) = T_4(2,2) = (0,4,6) \neq 2(0,1,3) = 2T_4(1,1)$, T_4 does not preserve scalar multiplication.

We could also verify (out of mere curiosity now since we already know that T_4 is nonlinear) that T_4 does not preserve vector addition either.

It is common in mathematics that the set of points where a function assumes the value zero is one of some importance, for example, the roots of a polynomial or the critical points of a function (the points at which its derivative assumes the value zero). In the present context, we introduce a special name for the set of points where a linear transformation from $\mathbb{R}^n$ to $\mathbb{R}^m$ assumes the value $\mathbf{0}$.

17.4 Definition

If $T\colon \mathbb{R}^n \to \mathbb{R}^m$ is a linear transformation, the set $\{\mathbf{u} \in \mathbb{R}^n\colon T(\mathbf{u}) = \mathbf{0}\}$ is called the *kernel* of T and is denoted by ker T.

Since linear transformations are functions, the notion of range of a linear transformation has already been defined in 16.1. The *range* of the linear transformation $T\colon \mathbb{R}^n \to \mathbb{R}^m$ is the set $\{\mathbf{v} \in \mathbb{R}^m\colon \mathbf{v} = T(\mathbf{u}) \text{ for some } \mathbf{u} \in \mathbb{R}^n\}$, and it will be denoted by ran T.

The next exercise relates these concepts to those from Section 5.

17.5 Exercise

For any linear transformation $T\colon \mathbb{R}^n \to \mathbb{R}^m$,

(i) ker T is a subspace of $\mathbb{R}^n$
(ii) ran T is a subspace of $\mathbb{R}^m$
(iii) The graph of T = Gr T =

$$\{(u_1, u_2, \ldots, u_n, v_1, v_2, \ldots, v_m) \in \mathbb{R}^{n+m}\colon T(u_1, u_2, \ldots, u_n) = (v_1, v_2, \ldots, v_m)\}$$

is a subspace of $\mathbb{R}^{n+m}$.

The most important source of linear transformations from $\mathbb{R}^n$ to $\mathbb{R}^m$ is provided by the next theorem.

17.6 Theorem

For any $m \times n$ matrix A, the function $T_A\colon \mathbb{R}^n \to \mathbb{R}^m$ defined by $T_A(\mathbf{x}) = A\mathbf{x}$ is a linear transformation.

Before giving the proof of 17.6, we should remind you of the convention that we do not distinguish notationally among n-tuples, $n \times 1$ matrices, and $1 \times n$ matrices; it is the context that is the determining factor. Thus in the equation $T_A(\mathbf{x}) = A\mathbf{x}$, the $\mathbf{x}$ on the left side is an n-tuple, while the $\mathbf{x}$ on the right side is the corresponding $n \times 1$ matrix; the matrix product, $A\mathbf{x}$, is, strictly speaking, an $m \times 1$ matrix, but should really be interpreted as the corresponding m-tuple in order for T_A to map $\mathbb{R}^n$ into $\mathbb{R}^m$.

Proof of 17.6: T_A preserves vector addition since the fact that $A(\mathbf{u} + \mathbf{v}) = A\mathbf{u} + A\mathbf{v}$ follows from 3.1(xi). T_A preserves scalar multiplication since the fact that $A(t\mathbf{u}) = tA\mathbf{u}$ follows from 3.1(x). ■

17.7 Examples

Let

$$A = \begin{bmatrix} 0 & 3 & 9 \\ 4 & 6 & -2 \end{bmatrix}$$

Then

$$A\begin{bmatrix} 1 \\ 2 \\ 3 \end{bmatrix} = \begin{bmatrix} 0 & 3 & 9 \\ 4 & 6 & -2 \end{bmatrix}\begin{bmatrix} 1 \\ 2 \\ 3 \end{bmatrix} = \begin{bmatrix} 33 \\ 10 \end{bmatrix}$$

$$A\begin{bmatrix} x_1 \\ x_2 \\ x_3 \end{bmatrix} = \begin{bmatrix} 0 & 3 & 9 \\ 4 & 6 & -2 \end{bmatrix}\begin{bmatrix} x_1 \\ x_2 \\ x_3 \end{bmatrix} = \begin{bmatrix} 3x_2 + 9x_3 \\ 4x_1 + 6x_2 - 2x_3 \end{bmatrix}$$

The linear transformation T_A maps $\mathbb{R}^3$ into $\mathbb{R}^2$. $T_A(1,2,3) = (33,10)$, and the general behavior of the function T_A on the arbitrary point (x_1, x_2, x_3) is $T_A(x_1, x_2, x_3) = (3x_2 + 9x_3, 4x_1 + 6x_2 - 2x_3)$.

It is convenient to introduce a temporary name for those linear transformations from $\mathbb{R}^n$ to $\mathbb{R}^m$ that are obtained from an $m \times n$ matrix as in 17.6.

We will temporarily call these *matrix transformations*. This label is temporary because, as we will see in Sections 31 and 32, matrix transformations are for all intents and purposes the only ones there are (and so we do not really need a special name for the ones of this form). The fact is that every linear transformation $T: \mathbb{R}^n \to \mathbb{R}^m$ can be "represented" as a matrix transformation T_A for some $m \times n$ matrix A. The problem is to explain what is meant by the word "represent" in this context. It turns out that a given linear transformation $T: \mathbb{R}^n \to \mathbb{R}^m$ can be "represented" by infinitely many "different" matrix transformations. Although the explanation of these matters is not difficult, it is premature to discuss them further at this point.

For the moment, all we wish to stress is the additional compounding of points of view provided by the ability to think of an $m \times n$ matrix as a function from $\mathbb{R}^n$ to $\mathbb{R}^m$. We pursue this point in the next three observations.

17.8 Observation

If

$$A = \begin{bmatrix} a_{11} & \cdots & a_{1j} & \cdots & a_{1n} \\ a_{21} & \cdots & a_{2j} & \cdots & a_{2n} \\ \vdots & & \vdots & & \vdots \\ a_{m1} & \cdots & a_{mj} & \cdots & a_{mn} \end{bmatrix}$$

and $\mathbf{e}_j^n = (0, \ldots, 0, 1, 0, \ldots, 0)$ then

$$T_A(\mathbf{e}_j^n) = \begin{bmatrix} a_{11} & \cdots & a_{1j} & \cdots & a_{1n} \\ a_{21} & \cdots & a_{2j} & \cdots & a_{2n} \\ \vdots & & \vdots & & \vdots \\ a_{m1} & \cdots & a_{mj} & \cdots & a_{mn} \end{bmatrix} \begin{bmatrix} 0 \\ \vdots \\ 1 \\ \vdots \\ 0 \end{bmatrix} = \begin{bmatrix} a_{1j} \\ a_{2j} \\ \vdots \\ a_{mj} \end{bmatrix}$$

In other words, the jth column of A is just the image of $\mathbf{e}_j^n$ under the transformation T_A. It will often be useful to think of the columns of an $m \times n$ matrix A in this way, that is, as the ordered set $(T_A(\mathbf{e}_1^n), T_A(\mathbf{e}_2^n), \ldots, T_A(\mathbf{e}_n^n))$ of images under T_A of the vectors in the standard ordered basis $\alpha^n = (\mathbf{e}_1^n, \mathbf{e}_2^n, \ldots, \mathbf{e}_n^n)$ for $\mathbb{R}^n$.

17.9 Observation

If A is an $m \times n$ matrix, then $T_A: \mathbb{R}^n \to \mathbb{R}^m$, and it follows from 6.14 and 9.4 that for any $\mathbf{b} \in \mathbb{R}^m$ the following are equivalent:

(i) $\mathbf{b} \in \operatorname{ran} T_A$.
(ii) $\mathbf{b} = T_A(\mathbf{u})$ for some $\mathbf{u} \in \mathbb{R}^n$.
(iii) The matrix equation $A\mathbf{x} = \mathbf{b}$ is consistent.
(iv) $\mathbf{b} \in CS(A)$.

Thus, for any $m \times n$ matrix A, the column-space of A is the same as the range of T_A.

17.10 Observation

If A is an $m \times n$ matrix, then for any $\mathbf{u} \in \mathbb{R}^n$ the following are equivalent:

(i) $\mathbf{u} \in \ker T_A$.
(ii) $\mathbf{u}$ is a solution of the homogeneous system $A\mathbf{x} = \mathbf{0}_{m \times 1}$.

Thus for any $m \times n$ matrix A, the nullspace of A, NS(A), is the same as the kernel of T_A.

The dimension of NS(A) is called the *nullity* of A, and is denoted by $n(A)$.

Because of 17.5, 17.9, and 17.10, we will extend our use of the words rank and nullity to the context of linear transformations.

If $T: \mathbb{R}^n \to \mathbb{R}^m$ is a linear transformation, then the dimension of the range of T will be called the *rank of T*; the dimension of the kernel of T will be called the *nullity of T*.

This use of the words is appropriate since, by 17.9 and 17.10, both the rank and nullity of a matrix transformation T_A are the same as the rank and the nullity, respectively, of the matrix A.

17.11 Example

(We assume in this example that Section 15 has been covered.) To illustrate these ideas, let us compute the kernel, range, rank, and nullity of $T_A:\mathbb{R}^3 \to \mathbb{R}^2$, where $T_A(x_1, x_2, x_3) = (3x_2 + 9x_3, 4x_1 + 6x_2 - 2x_3)$. Note that A is the 2×3 matrix from Example 17.7. By 17.10, $\ker T_A$ is the solution space of $A\mathbf{x} = \mathbf{0}$. You should verify that when the Gauss–Jordan algorithm is applied to A, the result is

$$GJ(A) = \begin{bmatrix} 1 & 0 & -5 \\ 0 & 1 & 3 \end{bmatrix}$$

Thus $\ker T_A = \{(x_1, x_2, x_3) \in \mathbb{R}^3: x_1 = 5x_3, x_2 = -3x_3\}$. By 17.9, $\operatorname{ran} T_A = CS(A)$; this is equal, by inspection, to $\mathbb{R}^2$. From the preceding, we conclude that the rank of T_A is 2 and that its nullity is 1.

There is an important theorem involving these concepts, which is sometimes called the *rank plus nullity theorem* or the *dimension theorem*.

17.12 Theorem

For any linear transformation $T:\mathbb{R}^n \to \mathbb{R}^m$, rank T + nullity T = dimension of dom T.

Proof: Assume nullity $T = \dim(\ker T) = p$, where $0 \leqq p \leqq n$, and let $\beta = \{\mathbf{u}_1, \mathbf{u}_2, \ldots, \mathbf{u}_p\}$ be a basis for ker T. If $p = 0$, take $\beta = \varnothing$. Since β is a linearly independent subset of $\mathbb{R}^n$, β can be expanded, by 8.7, to a basis $\gamma = \{\mathbf{u}_1, \mathbf{u}_2, \ldots, \mathbf{u}_p, \mathbf{u}_{p+1}, \ldots, \mathbf{u}_n\}$ for $\mathbb{R}^n$. We will prove that rank T = dim(ran T) $= n - p$ by showing that $\delta = \{T(\mathbf{u}_{p+1}), T(\mathbf{u}_{p+2}), \ldots, T(\mathbf{u}_n)\}$ is a basis for ran T.

Suppose $\mathbf{v} \in$ ran T and let $\mathbf{x} \in \mathbb{R}^n$ be such that $T(\mathbf{x}) = \mathbf{v}$. Then $\mathbf{x} = s_1\mathbf{u}_1 + \cdots + s_p\mathbf{u}_p + s_{p+1}\mathbf{u}_{p+1} + \cdots + s_n\mathbf{u}_n$, where $(s_1, s_2, \ldots, s_n) = (\mathbf{x})_\beta$. Thus, using the linearity of T and the fact that for $i = 1, 2, \ldots, p$, $T(\mathbf{u}_i) = \mathbf{0}$, we have $\mathbf{v} = T(\mathbf{x}) = s_1T(\mathbf{u}_1) + \cdots + s_pT(\mathbf{u}_p) + s_{p+1}T(\mathbf{u}_{p+1}) + \cdots + s_nT(\mathbf{u}_n) = s_{p+1}T(\mathbf{u}_{p+1}) + \cdots + s_nT(\mathbf{u}_n)$. This shows that δ spans ran T.

To see that δ is linearly independent, suppose $\mathbf{0} = t_{p+1}T(\mathbf{u}_{p+1}) + \cdots + t_nT(\mathbf{u}_n)$; then again using the linearity of T, we have $T(t_{p+1}\mathbf{u}_{p+1} + \cdots + t_n\mathbf{u}_n) = \mathbf{0}$, which shows that $t_{p+1}\mathbf{u}_{p+1} + \cdots + t_n\mathbf{u}_n \in \ker T$. Since β is a basis for ker T, there are scalars $t_1, t_2, \ldots, t_p$ such that $t_{p+1}\mathbf{u}_{p+1} + \cdots + t_n\mathbf{u}_n = t_1\mathbf{u}_1 + t_2\mathbf{u}_2 + \cdots + t_p\mathbf{u}_p$. Thus $-t_1\mathbf{u}_1 - t_2\mathbf{u}_2 - \cdots - t_p\mathbf{u}_p + t_{p+1}\mathbf{u}_{p+1} + t_{p+1}\mathbf{u}_{p+1} + \cdots + t_n\mathbf{u}_n = \mathbf{0}$, and since γ is linearly independent, we conclude $t_1 = t_2 = \cdots = t_p = t_{p+1} = \cdots = t_n = 0$. In particular, $t_{p+1} = \cdots = t_n = 0$, which proves that δ is linearly independent. ■

17.13 Corollary

For any $m \times n$ matrix A, rank(A) + nullity(A) $= n$.

Proof: Apply 17.12 to T_A, using the fact that the rank and nullity of the matrix transformation T_A are the same as the rank and nullity, respectively, of the matrix A. ■

17.14 Corollary

A homogeneous linear system with fewer equations than unknowns must have infinitely many solutions.

Proof: Express the system as $A\mathbf{x} = \mathbf{0}_{m\times 1}$, where A is an $m \times n$ matrix and $m \lneqq n$. Because rank (A) does not exceed the smaller of m and n (see 9.8), we know that rank (A) $\lneqq n$. So by 17.13, $n -$ rank (A) = nullity (A)

$= \dim NS(A) \geqq 1$. Thus NS(A), which is the solution space of the homogeneous system, must contain infinitely many vectors. ■

17.15 Corollary

If $\mathbf{Ax} = \mathbf{b}$ is an arbitrary linear system with fewer equations than unknowns, then its solution space, S, is either empty or else contains infinitely many vectors.

Proof: If $\mathbf{Ax} = \mathbf{b}$ is inconsistent, then $S = \varnothing$. If $\mathbf{Ax} = \mathbf{b}$ is consistent, then S is a translate (see 6.9) of the solution space of the associated homogeneous system $\mathbf{Ax} = \mathbf{0}_{m\times 1}$, which is infinite by 17.14. ■

17.16 Caution

There is an unfortunate tendency among beginning students, when stating 17.15, to omit the first alternative, $S = \varnothing$, from the conclusion.

In 17.5(iii), we saw that the graph of a linear transformation from $\mathbb{R}^n$ to $\mathbb{R}^m$ is a linear subspace of $\mathbb{R}^{n+m}$; indeed, at the beginning of this section, we motivated linear transformations from $\mathbb{R}^n$ to $\mathbb{R}^m$ by remarking that they are the higher-dimensional analogues of the functions from $\mathbb{R}$ to $\mathbb{R}$ of the form $f(x) = mx$, whose graphs are just straight lines in $\mathbb{R}^2$ through the origin.

But why should we restrict our attention to lines through the origin? It does not involve a great leap of the imagination to consider the sightly more general and more natural class of functions from $\mathbb{R}$ to $\mathbb{R}$ of the form $f(x) = mx + b$, whose graphs are straight lines in $\mathbb{R}^2$ that may (if $b = 0$) or may not (if $b \neq 0$) go through the origin. Here the higher-dimensional analogue consists of those functions from $\mathbb{R}^n$ to $\mathbb{R}^m$ whose graphs are affine subspaces of $\mathbb{R}^{n+m}$. With this in mind, we make the following definition.

17.17 Definition

A function S: $\mathbb{R}^n \to \mathbb{R}^m$ is called an *affine transformation* if there exist a linear transformation T: $\mathbb{R}^n \to \mathbb{R}^m$ and a vector $\mathbf{b} \in \mathbb{R}^m$ such that for all $\mathbf{u} \in \mathbb{R}^n$, $S(\mathbf{u}) = T(\mathbf{u}) + \mathbf{b}$.

Our first theorem about this concept asserts that there is a *unique* way in which an affine transformation S can be written as the sum of a linear transformation T and a fixed vector $\mathbf{b}$. For functions from $\mathbb{R}$ to $\mathbb{R}$, this is a familiar fact of high-school mathematics: a function f whose graph is a line in the plane has a unique *slope-intercept form*; that is, there is a unique way to write $f(x)$ as $f(x) = mx + b$, where m is the slope of the line and b is its y-intercept.

17.18 Theorem

Let $S:\mathbb{R}^n \to \mathbb{R}^m$ be an affine transformation and suppose that $S(\mathbf{u}) = T_1(\mathbf{u}) + \mathbf{b}_1$, and $S(\mathbf{u}) = T_2(\mathbf{u}) + \mathbf{b}_2$ for all $\mathbf{u} \in \mathbb{R}^n$, where T_1 and T_2 are linear transformations from $\mathbb{R}^n$ to $\mathbb{R}^m$ and $\mathbf{b}_1$ and $\mathbf{b}_2$ are vectors in $\mathbb{R}^m$; then $T_1 = T_2$ and $\mathbf{b}_1 = \mathbf{b}_2$.

Proof: $T_1(\mathbf{0}) + \mathbf{b}_1 = S(\mathbf{0}) = T_2(\mathbf{0}) + \mathbf{b}_2$ and, by 17.2(i), $T_1(\mathbf{0}) = \mathbf{0} = T_2(\mathbf{0})$; hence, $\mathbf{b}_1 = \mathbf{b}_2$. Now, using this, we obtain, for any $\mathbf{u} \in \mathbb{R}^n$, that $T_1(\mathbf{u}) + \mathbf{b}_1 = S(\mathbf{u}) = T_2(\mathbf{u}) + \mathbf{b}_2 = T_2(\mathbf{u}) + \mathbf{b}_1$; subtracting $\mathbf{b}_1$ from both sides proves that $T_1(\mathbf{u}) = T_2(\mathbf{u})$. ■

It is Theorem 17.18 that guarantees that the following definition makes sense.

17.19 Definition

If $S: \mathbb{R}^n \to \mathbb{R}^m$ is an affine transformation and if $\mathbf{b} \in \mathbb{R}^m$ and $T: \mathbb{R}^n \to \mathbb{R}^m$ is a linear transformation satisfying $S(\mathbf{u}) = T(\mathbf{u}) + \mathbf{b}$ for all $\mathbf{u} \in \mathbb{R}^n$, then T is called the *linear transformation associated with* S and $\mathbf{b}$ is called the *translation associated with* S.

Problem Set 17

17.1. Go back and do any of the exercises from Section 17 that you may have skipped.

17.2. Let S and T be linear transformations from $\mathbb{R}^n$ to $\mathbb{R}^m$ and let t be any scalar. Define $tS: \mathbb{R}^n \to \mathbb{R}^m$ by $(tS)(\mathbf{u}) = tS(\mathbf{u})$, and define $S + T: \mathbb{R}^n \to \mathbb{R}^m$ by $(S + T)(\mathbf{u}) = S(\mathbf{u}) + T(\mathbf{u})$. Prove that tS and $S + T$ are also linear transformations.

*17.3. Suppose that $S: \mathbb{R}^p \to \mathbb{R}^n$ and $T: \mathbb{R}^n \to \mathbb{R}^m$ are linear transformations. Prove that the composition $T \circ S: \mathbb{R}^p \to \mathbb{R}^m$ is also a linear transformation.

*17.4. If A is a 4×5 matrix and $\dim(\operatorname{ran} T_A) = 3$, what are $\dim(CS(A))$, $\dim(RS(A))$, and $\dim(NS(A))$?

*17.5. If A is a 3×5 matrix and $\{\mathbf{x} \in \mathbb{R}^5\colon A\mathbf{x} = \mathbf{0}_{3\times 1}\} = \{(x_1, x_2, x_3, x_4, x_5) \in \mathbb{R}^5\colon x_2 = 3x_1 + x_4, x_3 = -x_4, x_5 = 0\}$, what are $\dim(CS(A))$, $\dim(RS(A))$, and $\dim(NS(A))$?

*17.6. Let $T: \mathbb{R}^n \to \mathbb{R}^m$ be a linear transformation. Prove that T is one-to-one if and only if $\ker T = \{\mathbf{0}\}$.

*17.7. Is the function $f: \mathbb{R} \to \mathbb{R}$ defined by $f(x) = x + 2$ a linear transformation?

*17.8. Is the function $g: \mathbb{R}^2 \to \mathbb{R}$ defined by $g(x_1, x_2) = x_1x_2$ a linear transformation?

17.9. Let $S: \mathbb{R}^n \to \mathbb{R}^m$ be an affine transformation. Prove that the graph of $S = \operatorname{Gr} S = \{(u_1, u_2, \ldots, u_n, v_1, v_2, \ldots, v_m) \in \mathbb{R}^{n+m}\colon$

$S(u_1, u_2, \ldots, u_n) = (v_1, v_2, \ldots, v_m)\}$ is an affine subspace of $\mathbb{R}^{n+m}$.

Hint: We know there is a unique linear transformation $T\colon \mathbb{R}^n \to \mathbb{R}^m$ and a unique vector $\mathbf{b} = (b_1, b_2, \ldots, b_m) \in \mathbb{R}^m$ such that $S(\mathbf{u}) = T(\mathbf{u}) + \mathbf{b}$ for all $\mathbf{u} \in \mathbb{R}^n$. We know from 17.5(iii) that the graph of T, $\mathrm{Gr}\, T$, is a linear subspace of $\mathbb{R}^{n+m}$. Prove that $\mathrm{Gr}\, S$ is the translation of $\mathrm{Gr}\, T$ by the vector $(\underbrace{0, 0, \ldots, 0}_{n\text{-many zeros}}, b_1, b_2, \ldots, b_m)$.

17.10 Here is a list of possible answers to the questions below:

(a) The solution space is empty.
(b) The solution space is a single point.
(c) The system has infinitely many solutions.
(d) The information given is contradictory.
(e) The information given is insufficient to determine the number of solutions.

Which of (a)–(e) is the correct conclusion to be drawn from the following information about a linear system of the form $\mathbf{Ax} = \mathbf{b}$:

(i) The system consists of 4 equations in three unknowns, the rank of the coefficient matrix is 3, and the rank of the augmented matrix is 3.
(ii) The system consists of 4 equations in three unknowns, the rank of the coefficient matrix is 3, and the rank of the augmented matrix is 2.
(iii) The system consists of 5 equations in 6 unknowns, is homogeneous, and the rank of both the coefficient matrix and the augmented matrix is 5.
(iv) The system involves 4 unknowns, and the rank of both the coefficient matrix and the augmented matrix is 3.
(v) The system consists of 5 equations in 6 unknowns, the rank of the coefficient matrix is 4, and the rank of the augmented matrix is 5.
(vi) The system consists of 4 equations in 5 unknowns, is homogeneous, the rank of the coefficient matrix is 3, and the rank of the augmented matrix is 4.
(vii) The system consists of 7 equations in 7 unknowns and the rank of the coefficient matrix is 7.
(viii) The system consists of 6 equations in 6 unknowns and the rank of the augmented matrix is 6.
(ix) The system consists of 6 equations in 5 unknowns and the rank of the augmented matrix is 6.

18

Relation between Composition of Matrix Transformations and Matrix Multiplication

In Section 1, we defined an operation on certain pairs of matrices and gave the nineteenth-century motivation for making this definition. The operation was called matrix multiplication, although this name is not the most appropriate. We suggested in [F]1.3 that this operation should have been called "matrix composition." It is the purpose of this section to review once more the motivation for this definition, making use of the twentieth-century language developed in Sections 16 and 17.

18.1 Example

Let

$$A = \begin{bmatrix} 1 & 0 \\ 0 & 2 \\ 3 & 1 \end{bmatrix}$$

The matrix transformation T_A: $\mathbb{R}^2 \to \mathbb{R}^3$ is then defined by $T_A(x, y) = (x, 2y, 3x + y)$. Let

$$B = \begin{bmatrix} 1 & 1 & 0 \\ 0 & 2 & 1 \end{bmatrix}$$

The matrix transformation T_B: $\mathbb{R}^3 \to \mathbb{R}^2$ is then defined by $T_B(r, s, t) = (r + s, 2s + t)$. Since dom T_B = codom T_A, it makes sense to consider T_A composed with T_B. $(T_B \circ T_A)$: $\mathbb{R}^2 \to \mathbb{R}^2$ is easily computed. $(T_B \circ T_A)(x, y) = T_B(T_A(x, y)) = T_B(x, 2y, 3x + y) = (x + 2y, 4y + 3x + y) = (x + 2y, 3x + 5y)$. Note that the product

$$\mathrm{BA} = \begin{bmatrix} 1 & 1 & 0 \\ 0 & 2 & 1 \end{bmatrix} \begin{bmatrix} 1 & 0 \\ 0 & 2 \\ 3 & 1 \end{bmatrix} = \begin{bmatrix} 1 & 2 \\ 3 & 5 \end{bmatrix}$$

is such that the matrix transformation T_{BA}: $\mathbb{R}^2 \to \mathbb{R}^2$ equals $T_B \circ T_A$.

In this example, it is also possible to compose T_B with T_A to obtain $T_A \circ T_B$: $\mathbb{R}^3 \to \mathbb{R}^3$. $(T_A \circ T_B)(r, s, t) = T_A(T_B(r, s, t)) = T_A(r + s, 2s + t) = (r + s, 2(2s + t), 3(r + s) + (2s + t)) = (r + s, 4s + 2t, 3r + 5s + t)$. Note that the product

$$\mathrm{AB} = \begin{bmatrix} 1 & 0 \\ 0 & 2 \\ 3 & 1 \end{bmatrix} \begin{bmatrix} 1 & 1 & 0 \\ 0 & 2 & 1 \end{bmatrix} = \begin{bmatrix} 1 & 1 & 0 \\ 0 & 4 & 2 \\ 3 & 5 & 1 \end{bmatrix}$$

is such that matrix transformation T_{AB}: $\mathbb{R}^3 \to \mathbb{R}^3$ is equal to $T_A \circ T_B$.

If matrices and matrix multiplication had not already been discovered in the nineteenth century, they surely would have been discovered in the twentieth century by mathematicians studying linear transformations from $\mathbb{R}^n$ to $\mathbb{R}^m$ and ways of composing them, as in the following experiment.

18.2 Experiment

Let S:$\mathbb{R}^3 \to \mathbb{R}^2$ be defined by $S(x, y, z) = (x + y + z, 2x + 3z)$. Let T: $\mathbb{R}^2 \to \mathbb{R}^4$ be defined by $T(s,t) = (4s + 5t, 0, 6t, 7s)$. Then $T \circ S$: $\mathbb{R}^3 \to \mathbb{R}^4$ is given by

$$\begin{aligned}(T \circ S)(x, y, z) &= T(S(x, y, z)) = T(x + y + z, 2x + 3z)\\ &= (4(x + y + z) + 5(2x + 3z), 0, 6(2x + 3z), 7(x + y + z))\\ &= (14x + 4y + 19z, 0, 12x + 18z, 7x + 7y + 7z).\end{aligned}$$

Imagine that you know nothing about matrices or about matrix multiplication and that you are trying to find, as in Section 1, a "pattern" that describes how the numbers involved in the composition $T \circ S$ are related to those in the given transformations S and T.

18.3 Theorem

If A is an $m \times n$ matrix and B is an $n \times p$ matrix, then the composition $T_A \circ T_B$: $\mathbb{R}^p \to \mathbb{R}^m$ of the two matrix transformations T_B: $\mathbb{R}^p \to \mathbb{R}^n$ and

$T_A\colon \mathbb{R}^n \to \mathbb{R}^m$ is simply the matrix transformation T_{AB} determined by the product AB of the two matrices A and B.

$$\mathbb{R}^p \xrightarrow{T_B} \mathbb{R}^n \xrightarrow{T_A} \mathbb{R}^m$$

$$T_A \circ T_B = T_{AB}$$

Proof: For an arbitrary vector $\mathbf{x} \in \mathbb{R}^p$, we have $T_{AB}(\mathbf{x}) = (AB)\mathbf{x} = A(B\mathbf{x}) = A(T_B(\mathbf{x})) = T_A(T_B(\mathbf{x})) = (T_A \circ T_B)(\mathbf{x})$, where the second equality in this chain follows from the associative law for matrix multiplication. ■

18.4 Corollary

If the sizes of the matrices $A_1, A_2, \ldots, A_k$ are such that the product $A_1 A_2 \cdots A_k$ is defined, then the composition $T_{A_1} \circ T_{A_2} \circ \cdots \circ T_{A_k}$ of the matrix transformations $T_{A_1}, T_{A_2}, \ldots, T_{A_k}$ is defined, and

$$T_{A_1} \circ T_{A_2} \circ \cdots \circ T_{A_k} = T_{A_1 A_2 \cdots A_k}$$ ■

Despite its simple proof, the importance of Theorem 18.3 cannot be overstressed. It is one of the essential results in linear algebra. The link between Sections 3 and 17 will be made in Section 19 using Theorem 18.3.

We state the next theorem at this point merely to provide a sense of direction for Section 19.

18.5 Theorem

The following are equivalent for any $n \times n$ matrix, A:

(i) A is an invertible matrix, in the sense of Definition 3.3.
(ii) $T_A\colon \mathbb{R}^n \to \mathbb{R}^n$ is an invertible function, in the sense of Definition 16.11.

The proof of this theorem is left to the end of Section 19 as Exercise 19.7.

Problem Set 18

18.1. Go back and do any of the exercises from Section 18 that you may have skipped.

*18.2. Define $T_1\colon \mathbb{R}^3 \to \mathbb{R}^4$ by

$$T_1(x, y, z) = (3x + y, -2x - 2y + z, 4y + z, x - 4y - 3z)$$

and define $T_2\colon \mathbb{R}^4 \to \mathbb{R}^2$ by

$$T_2(p, q, r, s) = (2p + q - r, 2p - q + r - 2s)$$

(i) Verify that T_1 and T_2 are linear transformations.

(ii) Find the formula for $T_2 \circ T_1$: $\mathbb{R}^4 \to \mathbb{R}^2$ directly from the definition of composition [i.e., $(T_2 \circ T_1)(x, y, z) = T_2(T_1(x, y, z))$].

(iii) Verify that $T_1 = T_C$, where $C = \begin{bmatrix} 3 & 1 & 0 \\ -2 & -2 & 1 \\ 0 & 4 & 1 \\ 1 & -4 & -3 \end{bmatrix}$.

(iv) Verify that $T_2 = T_D$, where $D = \begin{bmatrix} 2 & 1 & -1 & 0 \\ 2 & -1 & 1 & -2 \end{bmatrix}$.

(v) Compute the product DC and verify that

$$T_{\mathrm{DC}} = T_2 \circ T_1 = T_{\mathrm{D}} \circ T_{\mathrm{C}}$$

18.3. Find linear transformations $S{:}\mathbb{R}^2 \to \mathbb{R}^2$ and $T{:}\mathbb{R}^2 \to \mathbb{R}^2$ such that $T \circ S \neq S \circ T$.

19

The Fundamental Theorems of Linear Algebra

In 3.3 we introduced the concepts of left inverse, right inverse, and two-sided inverse for arbitrary matrices. In 16.8 and 16.11, we gave these same names to three concepts that relate to functions in general, and hence to matrix transformations in particular. In view of the terminology, we could have anticipated that some essential relationships among these concepts must exist. This is indeed the case, and these relationships constitute the heart of the theory of linear algebra (see [F]19.1).

For an arbitrary $m \times n$ matrix A, we will first treat right inverses in 19.1 and then treat left inverses in 19.3. The work is then complete, and the most important result for two-sided inverses is obtained simply by combining 19.1 and 19.3.

Throughout Section 19, it is essential to keep the notational conventions explained in 5.2 firmly in mind.

19.1 Theorem

For any $m \times n$ matrix A, the following are equivalent:

(i) A has a right inverse.
(ii) The linear transformation $T_A: \mathbb{R}^n \to \mathbb{R}^m$ is onto [i.e., the function T_A has a right inverse; see Theorem 16.10(i)].

(iii) The columns of A span $\mathbb{R}^m$ [i.e., CS(A) = $\mathbb{R}^m$].
(iv) For every $\mathbf{b} \in \mathbb{R}^m$, the matrix equation $A\mathbf{x} = \mathbf{b}$ is consistent.

Proof: (i) → (ii). Let C be any right inverse of A; thus C is an $n \times m$ matrix satisfying $AC = I_m$. To show that T_A is onto, we must find, for an arbitrary vector $\mathbf{v} \in \mathbb{R}^m$, some vector $\mathbf{u} \in \mathbb{R}^n$ such that $T_A(\mathbf{u}) = \mathbf{v}$. But $C\mathbf{v}$ is such a vector since $C\mathbf{v} \in \mathbb{R}^n$ and

$$T_A(C\mathbf{v}) = A(C\mathbf{v}) = (AC)\mathbf{v} = I_m\mathbf{v} = \mathbf{v}.$$

(ii) ↔ (iv) is obvious. In view of the definition of T_A, (iv) is simply a rewording of (ii).

(iii) ↔ (iv) is a simple consequence of Observation 9.4.

(iv) → (i). We construct an $n \times m$ matrix C one column at a time as follows: as the first column of C, choose any solution to the equation $A\mathbf{x} = \mathbf{e}_1^m$; as the second column of C, choose any solution to the equation $A\mathbf{x} = \mathbf{e}_2^m$; and so on. C is clearly an $n \times m$ matrix, and the fact that $AC = I_m$ follows from 1.12(i). ■

The equivalence of (i) through (iv) means that, for an arbitrary $m \times n$ matrix A, the four properties listed are either all true or all false. When we combine this with 7.30(ii), which says that $\mathbb{R}^m$ cannot be spanned by any set of fewer than m-many vectors, we conclude that if $n \lneqq m$, then 19.1(iii) is false, and hence so are (i), (ii), and (iv). Thus, for matrices having more rows than columns, we have the following corollary.

19.2 Corollary

If A is an $m \times n$ matrix and $n \lneqq m$, then:

(i) A does not have a right inverse.
(ii) The linear transformation T_A: $\mathbb{R}^n \to \mathbb{R}^m$ is not onto.
(iii) The columns of A do not span $\mathbb{R}^m$.
(iv) For some $\mathbf{b} \in \mathbb{R}^m$, the matrix equation $A\mathbf{x} = \mathbf{b}$ is inconsistent. ■

We next turn our attention to the question of left inverses. The reason for numbering the items using primes will be clear when we arrive at 19.5.

19.3 Theorem

For any $m \times n$ matrix A, the following are equivalent:

(i)′ A has a left inverse.
(ii)′ The linear transformation T_A: $\mathbb{R}^n \to \mathbb{R}^m$ is one-to-one [i.e., the function T_A has a left inverse; see Theorem 16.10(ii)].

(iii)′ The columns of A form a linearly independent subset of $\mathbb{R}^m$.
(iv)′ The matrix equation $A\mathbf{x} = \mathbf{0}_{m\times 1}$ has only the trivial solution $\mathbf{x} = \mathbf{0}_{n\times 1}$.

Proof: (i)′ → (ii)′. Assume that B is an $n \times m$ matrix satisfying $BA = I_n$. We have to show that if $T_A(\mathbf{u}) = T_A(\mathbf{v})$ then $\mathbf{u} = \mathbf{v}$. So suppose $T_A(\mathbf{u}) = T_A(\mathbf{v})$; then $A\mathbf{u} = A\mathbf{v}$. But this implies that $\mathbf{u} = I_n\mathbf{u} = (BA)\mathbf{u} = B(A\mathbf{u}) = B(A\mathbf{v}) = (BA)\mathbf{v} = I_n\mathbf{v} = \mathbf{v}$, as desired.

(ii)′ ↔ (iv)′ is obvious: if $\mathbf{u} \neq \mathbf{0}_{n\times 1}$ were a nontrivial solution to $A\mathbf{x} = \mathbf{0}_{m\times 1}$, then T_A would send the distinct points $\mathbf{u}$ and $\mathbf{0}_{n\times 1}$ into the same point, $\mathbf{0}_{m\times 1}$; conversely, if $\mathbf{u}$ and $\mathbf{v}$ are distinct points in $\mathbb{R}^n$ such that $T_A(\mathbf{u}) = T_A(\mathbf{v})$, then $\mathbf{u} - \mathbf{v} \neq \mathbf{0}_{n\times 1}$ and, using the linearity of T_A, we would obtain $T_A(\mathbf{u} - \mathbf{v}) = T_A(\mathbf{u}) - T_A(\mathbf{v}) = \mathbf{0}_{m\times 1}$.

(iii)′ ↔ (iv)′ is a simple consequence of Observation 6.14.

(iii)′ → (i)′ is a consequence of the following chain of six implications; the justifications for each step in the chain are provided in the corresponding paragraphs that follow.

The columns of A are a linearly independent set of n-many m-tuples

$$\xrightarrow{1} \dim CS(A) \geqq n$$

$$\xrightarrow{2} \dim RS(A) = n$$

$$\xrightarrow{3} \dim CS(A^T) = n$$

$$\xrightarrow{4} \text{The columns of } A^T \text{ span } \mathbb{R}^n$$

$$\xrightarrow{5} A^T \text{ has a right inverse}$$

$$\xrightarrow{6} A \text{ has a left inverse.}$$

1. Because CS(A) includes a linearly independent subset of size n, its dimension must, by 7.29(ii), be at least n.
2. Since RS(A) is subspace of $\mathbb{R}^n$, we know that $\dim RS(A) \leqq n$. But $\dim CS(A) \geqq n$, and since $\dim RS(A) = \dim CS(A)$ by 9.6, we have that $\dim RS(A) \geqq n$ as well.
3. The trivial observation that $RS(A) = CS(A^T)$ was made in 9.3, so of course these spaces have the same dimension.
4. $CS(A^T)$, being an n-dimensional subspace of $\mathbb{R}^n$, must, by 8.6, equal $\mathbb{R}^n$.
5. Apply Theorem 19.1 to the $n \times m$ matrix A^T.
6. Assume that the $m \times n$ matrix C satisfies $A^TC = I_n$; then, by taking the transpose of both sides and applying 3.15, we obtain $C^TA = I_n$. Thus, if C is a right inverse for A^T, then C^T is a left inverse for A. ■

Because conditions (i)′ to (iv)′ are either all true or all false for an arbitrary $m \times n$ matrix, and because any set of more than m-many m-tuples is linearly dependent by 7.30(i), we obtain the following corollary for matrices having fewer rows than columns.

19.4 Corollary

If A is an $m \times n$ matrix and $m \lneq n$, then

(i) A does not have a left inverse.
(ii) The linear transformation $T_A: \mathbb{R}^n \to \mathbb{R}^m$ is not one-to-one.
(iii) The columns of A are linearly dependent.
(iv) There are nontrivial solutions to the matrix equation $A\mathbf{x} = \mathbf{0}_{m\times 1}$. ■

It is now time to consolidate all the significant ideas encountered in the text so far into a single result.

19.5 Theorem

For any *square* matrix A of size $n \times n$, the following are equivalent:

(i) A has a right inverse.
(i)′ A has a left inverse.
(ii) $T_A: \mathbb{R}^n \to \mathbb{R}^n$ is onto.
(ii)′ $T_A: \mathbb{R}^n \to \mathbb{R}^n$ is one-to-one.
(iii) The columns of A span $\mathbb{R}^n$.
(iii)′ The columns of A are linearly independent.
(iv) The matrix equation $A\mathbf{x} = \mathbf{b}$ is consistent for every $\mathbf{b} \in \mathbb{R}^n$.
(iv)′ The matrix equation $A\mathbf{x} = \mathbf{0}_{n\times 1}$ has only the trivial solution $\mathbf{x} = \mathbf{0}_{n\times 1}$.

Proof: (i), (ii), (iii), and (iv) are equivalent by 19.1; (i)′, (ii)′, (iii)′, and (iv)′ are equivalent by 19.3; since A is square, the columns of A form a set of n-many n-tuples, and we know from 8.5 that such a set spans $\mathbb{R}^n$ if and only if it is linearly independent; hence (iii) and (iii)′ are equivalent. ■

There are many important consequences of Theorem 19.5. The equivalence of (i) and (i)′ for square matrices implies the stronger result that we promised just after 3.7—that is, if a square matrix has *either* a right inverse or a left inverse, then it is invertible. Here are the details.

19.6 Corollary (Uniqueness of Right and Left Inverses for Square Matrices)

(i) If C is a right inverse for the $n \times n$ matrix A, then A is invertible and $C = A^{-1}$.

(ii) If B is a left inverse for the $n \times n$ matrix A, then A is invertible and $B = A^{-1}$.

Proof of (i): Assume $AC = I_n$; then, by 19.5, A must also have a left inverse. For any such left inverse, B, it follows from 3.7 that $B = C = A^{-1}$. The proof of (ii) is analogous. ■

Corollary 19.6 provides the justification for the standard computational shortcut: to conclude that a given pair of $n \times n$ matrices, P and Q, are inverses of one another, it suffices to check *either* that $PQ = I_n$ or that $QP = I_n$; 19.6 then guarantees that the remaining equality holds. Thus the following property is equivalent to the eight properties already listed in 19.5:

(v) A is invertible.

A more familiar way to state the equivalence of (v) and (iii) is the following:

an $n \times n$ matrix A is invertible if and only if the rank of A is equal to n.

The equivalence of (ii) and (ii)′ is also a surprising and important result. We saw in Section 16 that for functions in general there is no relationship whatever between the properties of being one-to-one and/or onto. It is a very rare occurrence in mathematics that, for a given class of functions, these two properties are related. Here we have that for matrix transformations from $\mathbb{R}^n$ into $\mathbb{R}^n$ being one-to-one is equivalent to being onto. We will explain later in the text how this phenomenon is the essence of linearity combined with finite dimensionality.

In Chapter V, which concerns determinants, we will add to Theorem 19.5 by presenting another property of $n \times n$ matrices equivalent to invertibility.

We conclude this section with an exercise that outlines a proof of Theorem 18.5.

19.7 Exercise (Proof of Theorem 18.5)

(i) Assume that A is an invertible $n \times n$ matrix in the sense of Definition 3.3. Prove that the matrix transformation $T_A: \mathbb{R}^n \to \mathbb{R}^n$ is an invertible function in the sense of Definition 16.11 by verifying that $T_{A^{-1}} = (T_A)^{-1}$.

(ii) Let A be any $n \times n$ matrix. Assume that the matrix transformation $T_A: \mathbb{R}^n \to \mathbb{R}^n$ is an invertible function in the sense of Definition 16.11 and let $S: \mathbb{R}^n \to \mathbb{R}^n$ be the (unique by Theorem 16.13) function such that $S \circ T_A = i_{\mathbb{R}^n}$ and $T_A \circ S = i_{\mathbb{R}^n}$. Note that we *do not assume* S is a matrix transformation. Prove that A is an invertible matrix in the sense of Definition 3.3 and that $S = T_{A^{-1}}$.

Problem Set 19

19.1. Go back and do any of the exercises from Section 19 that you may have skipped.

*19.2. Fredholm Alternative Theorem: Show that for any $n \times n$ matrix A exactly one of the following holds: either
(i) $\mathbf{Ax} = \mathbf{b}$ has a unique solution for every $\mathbf{b} \in \mathbb{R}^n$, or
(ii) $\mathbf{Ax} = \mathbf{0}_{n \times 1}$ has nontrivial solutions.

*19.3. (True or False) Decide whether the following statements are true or false and justify your answers.
(i) If A and B are $n \times n$ matrices and $AB = I_n$, then $BA = I_n$.
(ii) An $m \times n$ matrix A has a right inverse if and only if $r(A) = m$.
(iii) A nonzero 2×3 matrix always has at least one right inverse.
(iv) It is possible for a square matrix to have two distinct left inverses.
(v) If A is a square matrix and the linear system $\mathbf{Ax} = \mathbf{0}_{n \times 1}$ has nontrivial solutions, then A is invertible.
(vi) If A, B, and C are $n \times n$ matrices and $ABC = I_n$, then A, B, and C are all invertible.
(vii) There is a one-to-one linear transformation from $\mathbb{R}^2 \to \mathbb{R}^2$ whose range does not contain the vector $(3, -5)$.
(viii) If A is an $m \times n$ matrix and B is any left inverse for A, then for every $\mathbf{b} \in \mathbb{R}^m$, the linear system $\mathbf{Ax} = \mathbf{b}$ has the unique solution $\mathbf{x} = \mathrm{B}\mathbf{b}$.

*19.4. Let A be an $m \times n$ matrix.
(i) Prove that A has a right inverse if and only if $r(A) = m$.
(ii) Prove that A has a left inverse if and only if $r(A) = n$.
[These results provide a useful form in which to remember the essential ingredients in Theorems 19.1 and 19.3. They also provide an explanation for footnote 3.2: it follows from Problem 19.4 that if an $m \times n$ matrix A has both a right inverse (i.e., $AC = I_m$ for some $n \times m$ matrix C) and a left inverse (i.e., $BA = I_n$ for some $n \times m$ matrix B) then in fact $m = n$ and A must be a square matrix.]

*19.5. Let A and B be $n \times n$ matrices. Prove that if the product AB is invertible then both A and B are invertible.

19.6. *Project*: It follows from Corollary 19.2 that no matrix transformation $T_A: \mathbb{R} \to \mathbb{R}^2$ (or $T_B: \mathbb{R}^2 \to \mathbb{R}^3$, etc.) can be onto. To appreciate the significance of this, you should be aware of the fact that there do exist functions from $\mathbb{R} \to \mathbb{R}^2$ that are onto.

Consider the following intuitively appealing statement: if Y is a space of higher dimension than X, then Y should have more elements than X. This statement relates two intuitive concepts, "dimension" and "number of elements in a set."

In the 1870s, Georg Cantor gave mathematical precision to the intuitive concept of "number of elements in a set." He argued that it is the existence of a function $f: X \to Y$ that is both one-to-one and onto that captures the essence of the intuitive idea that X and Y "have the same number of elements." After many years of trying to show that $\mathbb{R}^2$ has "more" elements than $\mathbb{R}$, Cantor shocked both himself and the mathematical world with his discovery of a function $f:\mathbb{R} \to \mathbb{R}^2$ that is both one-to-one and onto. This showed, among other things, that the intuitive concept of dimension was much more complicated than expected and that it could not be explained in terms of cardinality alone.

Later there was another shock when Giuseppe Peano discovered in 1890 a function $g: \mathbb{R} \to \mathbb{R}^2$ that is both continuous and onto. Such functions are more familiarly known as space-filling curves. Eventually, the intuitive concept of dimension was given the precise mathematical formulation that we have seen in 8.2.

If $h:\mathbb{R} \to \mathbb{R}^2$ is onto, what additional properties can h have? We mentioned previously that h could be either one-to-one or continuous. It is a deep result of early twentieth-century mathematics that such an h cannot be both one-to-one and continuous. It follows from 19.2 that such an h cannot be a matrix transformation, and thus from Sections 31 to 32 that h cannot be a linear transformation.

Learn more about this and related topics by reading the following:

Duda, R., "The Origins of the Concept of Dimension," *Colloquium Mathematicum*, v. XLII (1979), 95–110.

Hurewicz, W., and H. Wallman, *Dimension Theory*, Chapter I, Princeton University Press, Princeton, N.J., 1948.

Kaplansky, I., *Set Theory and Metric Spaces*, Allyn and Bacon, Boston, 1972.

Sanderson, D. E., "Advanced Plane Topology from an Elementary Standpoint," *Mathematics Magazine*, v. 53, no. 2 (1980) 81–89.

20

List of Equivalences of Invertibility

This short section is intended as a convenient summary of material from the first four chapters. It integrates the computational aspects treated in Chapter III with the theoretical aspects from Chapter IV.

20.1 Theorem

The following sixteen properties are equivalent for any $n \times n$ matrix A:

(i) A has a right inverse.
(i)′ A has a left inverse.
(ii) $T_A: \mathbb{R}^n \to \mathbb{R}^n$ is onto.
(ii)′ $T_A: \mathbb{R}^n \to \mathbb{R}^n$ is one-to-one.
(iii) The columns of A span $\mathbb{R}^n$.
(iii)′ The columns of A are linearly independent.
(iv) For every $\mathbf{b} \in \mathbb{R}^n$, the matrix equation $A\mathbf{x} = \mathbf{b}$ is consistent.
(iv)′ The matrix equation $A\mathbf{x} = \mathbf{0}_{n\times 1}$ has only the trivial solution $\mathbf{x} = \mathbf{0}_{n\times 1}$.
(iv)″ For every $\mathbf{b} \in \mathbb{R}^n$, the matrix equation $A\mathbf{x} = \mathbf{b}$ has a unique solution.
(v) A is invertible.
(vi) $GJ(A) = I_n$.

(vii) $A \underset{r}{\sim} I_n$.
(viii) A is expressible as a product of elementary matrices.
(ix) Rank A $= n$.
(x) The rows of A span $\mathbb{R}^n$.
(x)′ The rows of A are linearly independent.

20.2 Exercise

Provide a proof of Theorem 20.1 by combining Exercise 8.5 with Theorems 6.11, 9.6, 19.5, and 12.1.

20.3 Example (Twenty Questions)

Consider the matrix

$$A = \begin{bmatrix} 3 & -1 & 2 & 1 \\ -3 & 2 & -1 & 1 \\ 0 & 4 & 1 & 2 \\ 1 & 0 & 2 & 3 \end{bmatrix}$$

and the associated matrix transformation $T_A: \mathbb{R}^4 \to \mathbb{R}^4$ defined by $T_A(x_1, x_2, x_3, x_4) = (3x_1 - x_2 + 2x_3 + x_4, -3x_1 + 2x_2 - x_3 + x_4, 4x_2 + x_3 + 2x_4, x_1 + 2x_3 + 3x_4)$.

Q1. Is A invertible?
Q2. Is T_A one-to-one?
Q3. Do the rows of A span $\mathbb{R}^4$?
Q4. Is rank (A) = 4?
Q5. Is A row-equivalent to I_4?
Q6. Are the columns of A linearly independent?
Q7. Does the system $A\mathbf{x} = \mathbf{0}$ have infinitely many solutions?
Q8. Is CS(A) $= \mathbb{R}^4$?
Q9. Does GJ(A) have an all-zero row?
Q10. Is $\ker T_A = \{\mathbf{0}\}$?
Q11. Do the rows of A form a basis for $\mathbb{R}^4$?
Q12. Is ran $T_A = \mathbb{R}^4$?
Q13. Does A have a left inverse?
Q14. Are the rows of A linearly independent?
Q15. Does there exist a vector $\mathbf{b} \in \mathbb{R}^4$ such that the system $A\mathbf{x} = \mathbf{b}$ is inconsistent?
Q16. Is dim RS(A) = 4?
Q17. Does T_A have a right inverse?
Q18. Does G(A) have an all-zero row?
Q19. Do the columns of A form a basis for $\mathbb{R}^4$?
Q20. Can A be expressed as a product of elementary matrices?

What we have here, of course, are 20 "different" ways of asking the "same" question. Understanding the first four chapters of this text and knowing the answer to any *one* of these questions allows you to answer the remaining 19. Moreover, any one of these questions, in isolation, would be answered by the same computational procedure, that is, computing G(A). You should verify that

$$G(A) = \begin{bmatrix} 1 & -\frac{1}{3} & \frac{2}{3} & \frac{1}{3} \\ 0 & 1 & 1 & 2 \\ 0 & 0 & 1 & 2 \\ 0 & 0 & 0 & 0 \end{bmatrix}$$

So the answer to Q18 is yes.

20.4 Exercise

Answer the remaining 19 questions from Example 20.3 without doing any additional calculations.

20.5 Example (Example 20.3 Continued)

Consider the following additional questions about A and T_A.

Q21. Find a reduced description of CS(A).
Q22. Find a vector $\mathbf{w} \in \mathbb{R}^4$ such that $\mathbf{w} \notin \operatorname{ran} T_A$.
Q23. Find distinct vectors $\mathbf{v}_1$ and $\mathbf{v}_2$ such that $T_A(\mathbf{v}_1) = T_A(\mathbf{v}_2)$.
Q24. Find a reduced basis for ran T_A.
Q25. Find a reduced basis for the row-space of A.

You should realize why none of these can be answered knowing only G(A), as above.

Questions 21, 22, and 24 can be answered simultaneously by computing G([A ⋮ **b**]). You should verify that

$$G([A \,\vdots\, \mathbf{b}]) = \left[\begin{array}{cccc|c} 1 & -\frac{1}{3} & \frac{2}{3} & \frac{1}{3} & b_1/3 \\ 0 & 1 & 1 & 2 & b_2 + b_1 \\ 0 & 0 & 1 & 2 & \frac{1}{3}(b_3 - 4b_2 - 4b_1) \\ 0 & 0 & 0 & 0 & b_4 + b_3/3 - \frac{5}{3}b_2 - 2b_1 \end{array}\right]$$

Thus $CS(A) = \{(b_1, b_2, b_3, b_4) \in \mathbb{R}^4\colon\ b_4 = 2b_1 + \frac{5}{3}b_2 - \frac{1}{3}b_3\}$. This answers Q21. Since ran T_A = CS(A), we can answer Q22 by choosing **w** to be any vector that does not belong to CS(A), for example, $\mathbf{w} = (0, 0, 0, 1)$. Q24 can also be answered: a reduced basis for ran T_A is $\{(1, 0, 0, 2), (0, 1, 0, \frac{5}{3}), (0, 0, 1, -\frac{1}{3})\}$.

To answer Q23 and Q25, we need to continue the row-reduction of A to Gauss–Jordan form. You should check that

$$\mathrm{GJ(A)} = \begin{bmatrix} 1 & 0 & 0 & -1 \\ 0 & 1 & 0 & 0 \\ 0 & 0 & 1 & 2 \\ 0 & 0 & 0 & 0 \end{bmatrix}$$

Thus a reduced basis for RS(A) is $\{(1,0,0,-1),(0,1,0,0),(0,0,1,2)\}$. We can find distinct vectors whose images under T_{A} are the same by taking $\mathbf{v}_1 = \mathbf{0}$ and taking $\mathbf{v}_2$ to be any nonzero vector in ker T_{A}. From GJ(A) above, we conclude that ker $T_{\mathrm{A}} = \{(x_1, x_2, x_3, x_4) \in \mathbb{R}^4\colon\ x_1 = x_4,\ x_2 = 0,\ x_3 = -2x_4\}$. So, for example, $T_{\mathrm{A}}(\mathbf{0}) = T_{\mathrm{A}}(1,0,-2,1) = \mathbf{0}$.

Problem Set 20

20.1. Go back and do any of the exercises from Section 20 that you may have skipped.

*20.2. Prove that if A and B are $n \times n$ matrices and $\mathrm{A} \underset{r}{\sim} \mathrm{B}$, then A is invertible if and only if B is invertible.

The next four problems should be answered *without* doing any numerical calculations requiring pencil and paper. All answers must be justified.

*20.3. A is a 2×2 matrix such that $T_{\mathrm{A}}(3,-1) = (2,2)$ and $T_{\mathrm{A}}(2,2) = (1,1)$.
(i) Is A invertible?
(ii) Is T_{A} one-to-one?

*20.4. A is a 2×2 matrix such that

$$\mathrm{A}\begin{bmatrix} 1 \\ -2 \end{bmatrix} = \begin{bmatrix} 4 \\ -2 \end{bmatrix} \quad \text{and} \quad \mathrm{A}\begin{bmatrix} 2 \\ -3 \end{bmatrix} = \begin{bmatrix} 3 \\ 6 \end{bmatrix}$$

(i) Is A invertible?
(ii) Does the vector $(7, -39)$ belong to the range of T_{A}?

*20.5. A is a 2×2 matrix such that

$$\mathrm{A}\begin{bmatrix} 1 \\ -2 \end{bmatrix} = \begin{bmatrix} 1 \\ 0 \end{bmatrix} \quad \text{and} \quad \mathrm{A}\begin{bmatrix} 2 \\ -3 \end{bmatrix} = \begin{bmatrix} 0 \\ 1 \end{bmatrix}$$

(i) Prove that A is invertible.
(ii) What is A^{-1}?

*20.6. Find a 2×2 matrix A such that the range of T_{A} is the set $\{(x_1, x_2) \in \mathbb{R}^2\colon\ x_2 = 3x_1 - 4\}$.

21

Determinants (Motivation)

21.1 Observation

Recall that in Problem 3.6 you discovered that the 2×2 matrix $\begin{bmatrix} a_{11} & a_{12} \\ a_{21} & a_{22} \end{bmatrix}$ is invertible if and only if $a_{11}a_{22} - a_{12}a_{21} \neq 0$.

This result, relating the invertibility of A with the fact that a certain algebraic combination of its entries is nonzero, is easy to remember. Nonetheless, we will point out the standard mnemonic device for recalling the particular algebraic expression involved.

21.2 Mnemonic Device

In the 2×2 case, the expression involves two products, each with two entries from A as factors:

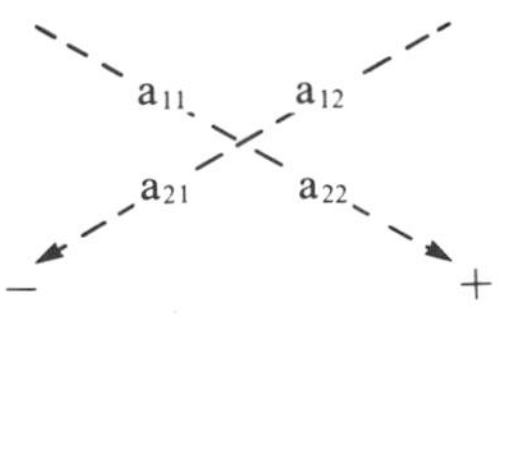

The product $a_{11}a_{22}$ occurs with a $+$, while the product $a_{12}a_{21}$ occurs with a $-$.

21.3 Exercise

By applying the Gaussian algorithm to

$$\begin{bmatrix} a_{11} & a_{12} & a_{13} \\ a_{21} & a_{22} & a_{23} \\ a_{31} & a_{32} & a_{33} \end{bmatrix}$$

and invoking the fact that a matrix is invertible iff it is row-reducible to the identity, conclude that the general 3×3 matrix A is invertible if and only if

$$a_{11}a_{22}a_{33} + a_{12}a_{23}a_{31} + a_{13}a_{21}a_{32} - a_{13}a_{22}a_{31} - a_{12}a_{21}a_{33} - a_{11}a_{23}a_{32} \neq 0.$$

This result, relating the invertibility of A with the fact that a certain algebraic combination of its entries is nonzero, is less easy to recall. However, it is useful in practice to memorize the expression from 21.3 as well. Again, there is a standard pictorial mnemonic.

21.4 Mnemonic Device

Form the 3×5 array in which the fourth and fifth columns are duplicates of the first and second columns, respectively.

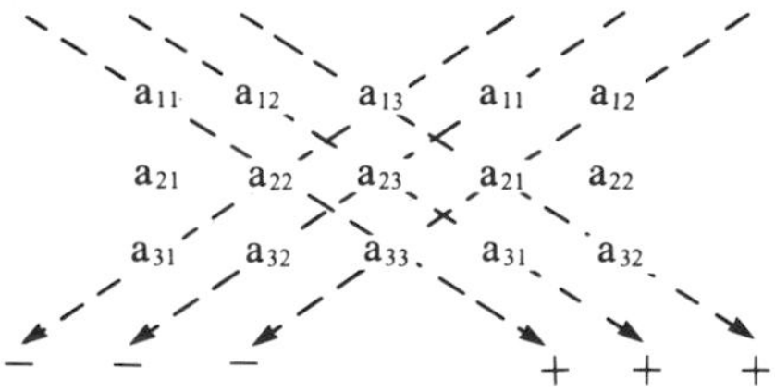

The relevant expression involves six products, each with three entries from A as factors; the products corresponding to the arrows from upper left to lower right occur with $+$; those corresponding to arrows from upper right to lower left occur with $-$.

These observations in low dimensions suggest that something is going on. Even the form that a general statement might take is suggested by our experience so far: an $n \times n$ matrix A is invertible iff some algebraic combination of the entries in A is nonzero. The rule that gives this algebraic combination is not easy to discover, however, even if one were painstakingly to undertake the details for the general 4×4 and 5×5 matrices.

We should also immediately caution the reader that, because the corresponding algebraic expressions for larger matrices are quite complicated, there does *not* exist a mnemonic device analogous to those for 2×2 and 3×3 matrices.

21.5 Exercise

Use 21.2 and 21.4 to decide whether the following matrices are invertible.

$$\begin{bmatrix} 2 & 1 \\ 4 & 2 \end{bmatrix} \quad \begin{bmatrix} \frac{1}{2} & 3 \\ -4 & -12 \end{bmatrix} \quad \begin{bmatrix} \frac{1}{2} & 0 \\ -\frac{1}{2} & -1 \end{bmatrix}$$

$$\begin{bmatrix} 3 & 4 & 2 \\ -1 & 1 & 1 \\ 4 & 1 & -6 \end{bmatrix} \quad \begin{bmatrix} 3 & 2 & 1 \\ -1 & 1 & 3 \\ 4 & -6 & -16 \end{bmatrix} \quad \begin{bmatrix} 3 & 0 & 2 \\ -1 & 0 & 1 \\ 4 & 0 & -6 \end{bmatrix}$$

21.6 Observation (Geometric Interpretation of the 2×2 Case)

Let A be the matrix $\begin{bmatrix} a & b \\ c & d \end{bmatrix}$. The expression $ad - bc$ has an important geometric significance. Consider the parallelogram in $\mathbb{R}^2$ determined by the vectors (a, c) and (b, d). See Figure 21.1.

Let us compute the area of this parallelogram. One way to do this is to subtract the areas of the six regions labeled I to VI from the area of the large outer rectangle. Thus, the area of the parallelogram is

$$\begin{aligned} &(a + b)(c + d) - \frac{1}{2}ac - bc - \frac{1}{2}bd - \frac{1}{2}ac - bc - \frac{1}{2}bd \\ &\quad = ac + bc + ad + bd - ac - 2bc - bd \\ &\quad = ad - bc \end{aligned}$$

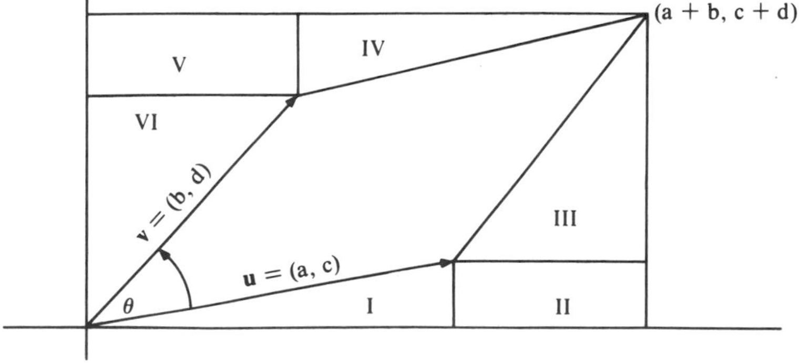

Figure 21.1

21.7 Observation

For any ordered pair of nonzero vectors $(\mathbf{u}, \mathbf{v})$ in $\mathbb{R}^2$, there is a unique angle θ in the interval $(-\pi, \pi]$ through which the first vector, $\mathbf{u}$, can be rotated into the second vector, $\mathbf{v}$. See Figure 21.2.

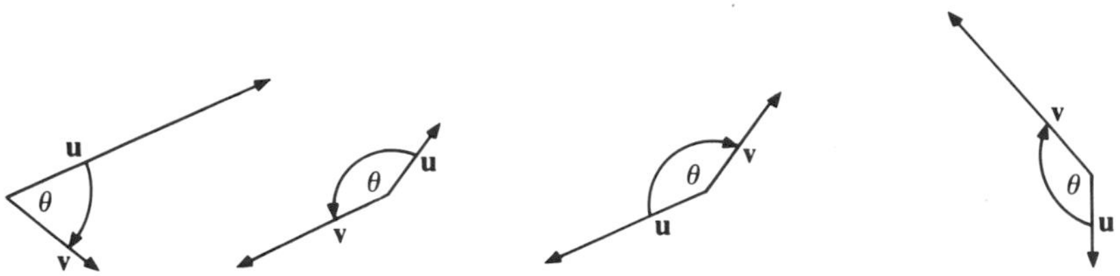

Figure 21.2

Before leaping to a premature conclusion from 21.6, we should note that, in drawing Figure 21.1, we *assumed* that to rotate the vector $\mathbf{u} = (a, c)$ into the vector $\mathbf{v} = (b, d)$ required a counterclockwise rotation (i.e., that the angle θ is positive). Is the same answer obtained when the required rotation is clockwise? See Figure 21.3.

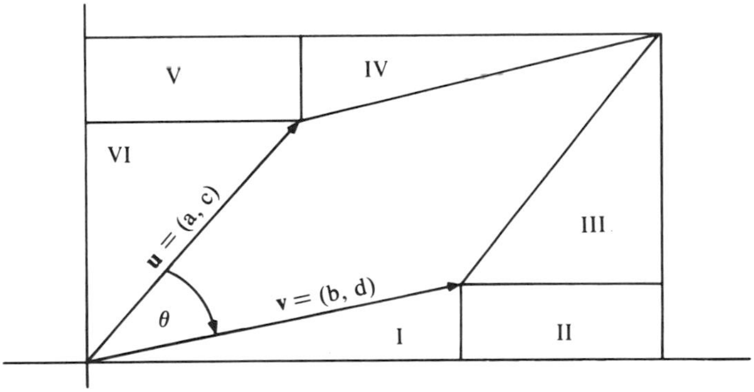

Figure 21.3

21.8 Exercise

Check that, when the argument from 21.6 is repeated for Figure 21.3, the conclusion is that the area of the parallelogram is $-ad + bc$.

21.9 Remark

We are now in position to draw the correct conclusions from 21.6 and 21.8. Let (a, c) and (b, d) be any two vectors in $\mathbb{R}^2$. First, recall that $A = \begin{bmatrix} a & b \\ c & d \end{bmatrix}$

is singular iff $ad - bc = 0$ iff the columns of A are linearly dependent iff the parallelogram determined by (a, c) and (b, d) collapses to a line or to the single point $\{(0, 0)\}$. Thus, we have the following geometric interpretation of the magnitude and sign of $ad - bc$.

(i) The magnitude or absolute value, $|ad - bc|$, of the expression $ad - bc$ is the area of the parallelogram in $\mathbb{R}^2$ determined by the columns (a, c) and (b, d) of A.
(ii) When $\{(a, c), (b, d)\}$ is linearly independent, this area is nonzero and the sign of $ad - bc$ is $+$ or $-$ according as the angle in the interval $(-\pi, \pi)$ required to rotate (a, c) into (b, d) is $+$ or $-$.

21.10 Exercise

(i) Find the area of the parallelogram in $\mathbb{R}^2$ determined by the vectors $(-6, 2)$ and $(3, 5)$.
(ii) When the vector $(2, 5)$ is rotated by an angle $\theta \in (-\pi, \pi]$ into the vector $(-6, -14)$, is θ positive or negative?

21.11 Observation (Geometric interpretation in the 3 × 3 case)

There are analogous results, which we will now state without proof, for ordered sets $(\mathbf{u}, \mathbf{v}, \mathbf{w})$ of three vectors from $\mathbb{R}^3$. We do not bother with the proofs since we only intend to use these results as motivation; nothing in later sections depends logically on the facts that follow.

The analogue of the parallelogram in $\mathbb{R}^2$ determined by two vectors $\mathbf{u}, \mathbf{v} \in \mathbb{R}^2$ is the parallelepiped in $\mathbb{R}^3$ determined by three vectors $\mathbf{u}, \mathbf{v}, \mathbf{w} \in \mathbb{R}^3$. If the set $\{\mathbf{u}, \mathbf{v}, \mathbf{w}\}$ is linearly dependent, this parallelepiped collapses, depending on the dimension of $\mathscr{L}(\mathbf{u}, \mathbf{v}, \mathbf{w})$, to a parallelogram, to a line segment, or to the single point $\{(0, 0, 0)\}$. If $\{\mathbf{u}, \mathbf{v}, \mathbf{w}\}$ is linearly independent, this parallelepiped will have nonzero volume.

Fact: Let

$$A = \begin{bmatrix} a & b & c \\ d & e & f \\ g & h & i \end{bmatrix}$$

The magnitude (i.e., absolute value) of the expression

$$aei + bfg + cdh - ceg - bdi - afh$$

is the volume of the parallelepiped in $\mathbb{R}^3$ determined by the three vectors $\mathbf{u} = (a, d, g)$, $\mathbf{v} = (b, c, h)$, and $\mathbf{w} = (c, f, i)$. To interpret the sign of this expression requires that we introduce the concept of *orientation*. We have already done this implicitly for $\mathbb{R}^2$ earlier in this section.

A line L through the origin in $\mathbb{R}^2$ obviously cuts $\mathbb{R}^2$ into two pieces. Is there any way to distinguish these two pieces from one another without additional information? The answer is no!

However, if we give the line L an orientation by picking a vector $\mathbf{u} \in L$ and viewing L as $\mathscr{L}(\mathbf{u})$, then we can distinguish the resulting two halves of $\mathbb{R}^2$ and conveniently label them the "positive half" and "negative half" in the following way: a vector $\mathbf{v} \in \mathbb{R}^2 - L$ belongs to the *positive* (*negative*) half if the unique angle $\theta \in (-\pi, \pi)$ through which $\mathbf{u}$ is rotated into $\mathbf{v}$ is positive (negative).

It is clear that, if we reverse the orientation of L by viewing L as $\mathscr{L}(-\mathbf{u})$, then the signs of the two halves are interchanged. See Figure 21.4.

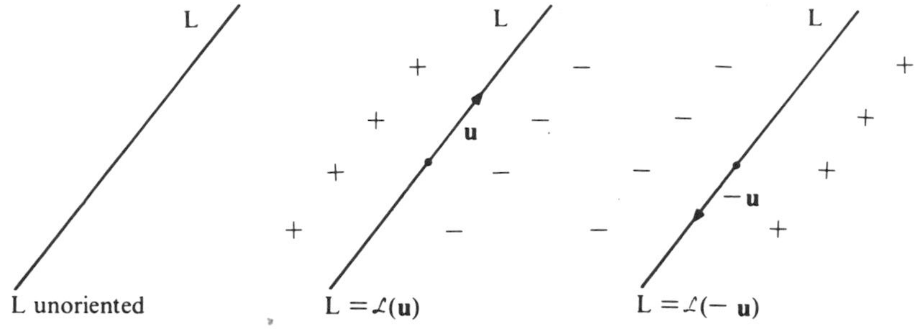

Figure 21.4

We may now rephrase 21.9(ii) as follows: let $\mathbf{u} = (a, c)$ and $\mathbf{v} = (b, d)$; when $ad - bc \neq 0$, the sign of $ad - bc$ is + or − according as the vector $\mathbf{v}$ lies on the positive or negative side of the oriented line $L = \mathscr{L}(\mathbf{u})$.

Similarly, a plane P through the origin in $\mathbb{R}^3$ obviously cuts $\mathbb{R}^3$ into two pieces. Without orienting the plane, we cannot distinguish the two resulting halves of $\mathbb{R}^3$ from one another. But if we orient the plane P by

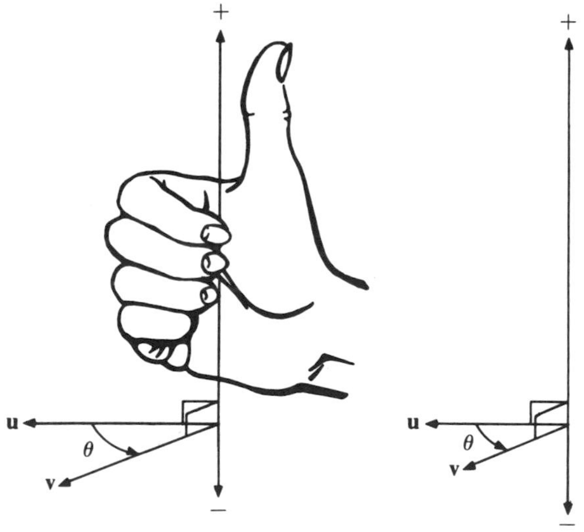

Figure 21.5

viewing it as the linear span of the ordered set $(\mathbf{u}, \mathbf{v})$, then the right-hand screw rule provides a physically intuitive way to distinguish the two sides. Let P be the plane in $\mathbb{R}^3$ determined by the two vectors $\mathbf{u}$ and $\mathbf{v}$, and let $\theta \in (-\pi, \pi]$ be the angle in the plane P through which the first vector, $\mathbf{u}$, is rotated into the second vector, $\mathbf{v}$. By convention, the *positive* side of $\mathbb{R}^3 - P$ is the side toward which a right-threaded screw would move if placed at the origin, perpendicular to P, and rotated through θ. Equivalently, it is the side toward which the thumb points when your right hand is placed on P with the fingers circled in the direction of θ. See Figure 21.5.

It is clear that if we change the orientation of P by viewing P as $\mathscr{L}(-\mathbf{u}, \mathbf{v})$ or $\mathscr{L}(-\mathbf{u}, -\mathbf{v})$ or $\mathscr{L}(\mathbf{u}, -\mathbf{v})$ or $\mathscr{L}(\mathbf{v}, \mathbf{u})$ or ..., then this affects which side of P is the positive side. Further details are unnecessary at this point.

21.12 Definition

The linearly independent ordered triple $(\mathbf{u}, \mathbf{v}, \mathbf{w}) \subset \mathbb{R}^3$ *obeys* (*violates*) *the right-hand screw rule* if $\mathbf{w}$ is on the positive (negative) side of the oriented plane $P = \mathscr{L}(\mathbf{u}, \mathbf{v})$.

We may now summarize the geometric interpretation of the 3×3 case.

21.13 Observation

Let

$$A = \begin{bmatrix} a_{11} & a_{12} & a_{13} \\ a_{21} & a_{22} & a_{23} \\ a_{31} & a_{32} & a_{33} \end{bmatrix}$$

and let $\mathbf{u} = (a_{11}, a_{21}, a_{31})$, $\mathbf{v} = (a_{12}, a_{22}, a_{32})$, and $\mathbf{w} = (a_{13}, a_{23}, a_{33})$ be the columns of A.

(i) The absolute value of the expression

$$a_{11}a_{22}a_{33} + a_{12}a_{23}a_{31} + a_{13}a_{21}a_{32} - a_{13}a_{22}a_{31} - a_{12}a_{21}a_{33} - a_{11}a_{23}a_{32}$$

is the volume of the parallelepiped in $\mathbb{R}^3$ determined by the three vectors $\mathbf{u}$, $\mathbf{v}$, and $\mathbf{w}$. This value is 0 if and only if $(\mathbf{u}, \mathbf{v}, \mathbf{w})$ is linearly dependent.

(ii) When $(\mathbf{u}, \mathbf{v}, \mathbf{w})$ is linearly independent, the sign of the expression

$$a_{11}a_{22}a_{33} + a_{12}a_{23}a_{31} + a_{13}a_{21}a_{32} - a_{13}a_{22}a_{31} - a_{12}a_{21}a_{33} - a_{11}a_{23}a_{32}$$

is + or − according as $(\mathbf{u}, \mathbf{v}, \mathbf{w})$ obeys or violates the right-hand screw rule.

To conclude this motivating section, we recall, from Section 4, the discussion of hyperplanes in $\mathbb{R}^n$. As do a line in $\mathbb{R}^2$ and a plane in $\mathbb{R}^3$, a hyperplane in $\mathbb{R}^n$ cuts $\mathbb{R}^n$ into two indistinguishable halves. If we take a hyperplane, W, through the origin and give it an orientation by viewing W as the span of a linearly independent ordered set $(\mathbf{u}_1, \mathbf{u}_2, \ldots, \mathbf{u}_{n-1}) \subset \mathbb{R}^n$, then we will be able to tell the two sides of $\mathbb{R}^n - W$ apart and to label them positive and negative according to some convention.

Obviously, some sort of convention will be needed in this general context, that is, a rule for $\mathbb{R}^n$ that is an analogue of the rules

(i) "the angle $\theta \in (-\pi, \pi]$ for rotating $\mathbf{u}$ into $\mathbf{v}$," used for $\mathbb{R}^2$, and
(ii) "the right-hand screw rule," used for $\mathbb{R}^3$

for deciding which side to label + and which to label −. Here our physical intuition lets us down (inhabitants of hyperspace are known to have more than two five-fingered hands), and we are forced to invent a "mathematical rule" for accomplishing this goal. This, and many other considerations, leads to the subject of the next section: permutations!

To summarize, the general result we seek will have the following form and geometric interpretation: let A be the $n \times n$ matrix having the vectors $\mathbf{u}_1, \mathbf{u}_2, \ldots, \mathbf{u}_n$ as its columns; then

(i) A is invertible iff a certain algebraic combination of its entries (to be called the *determinant* of A) is nonzero.
(ii) The determinant of A will be nonzero iff the set $\{\mathbf{u}_1, \mathbf{u}_2, \ldots, \mathbf{u}_n\}$ is linearly independent; in this case, the absolute value of the determinant is the n-dimensional volume of the parallelepiped in $\mathbb{R}^n$ determined by the vectors $\mathbf{u}_1, \mathbf{u}_2, \ldots, \mathbf{u}_n$, while the sign of the determinant is + or − according as $\mathbf{u}_n$ lies on the positive or negative side of the oriented hyperplane $\mathscr{L}(\mathbf{u}_1, \mathbf{u}_2, \ldots, \mathbf{u}_{n-1})$ in $\mathbb{R}^n$.

Problem Set 21

21.1. Go back and do any of the exercises from Section 21 that you may have skipped.

*21.2. Does the point $(3, -1)$ lie on the positive or negative side of the oriented line $L = \mathscr{L}(-2, 5)$?

*21.3. Does the point $(5, -1, 3)$ lie on the positive or negative side of the oriented plane $P = \mathscr{L}((2, 1, -3), (7, -2, 0))$?

*21.4. Find the area of the parallelogram determined in $\mathbb{R}^2$ by the vectors $(3, -4)$ and $(5, -6)$.

22

Permutations

The definition of *determinant* to be given in the next section requires a collection of auxiliary concepts, which are very important in their own right.

22.1 Definition

Let n be a fixed positive integer. A *permutation of* n is an ordered set $(j_1, j_2, \ldots, j_n)$ whose elements consist of the integers $1, 2, \ldots, n$.

22.2 Examples

(i) There is only one permutation of 1, that is, (1).
(ii) There are two permutations of 2, that is, $(1, 2)$ and $(2, 1)$.
(iii) There are six permutations of 3, that is, $(1, 2, 3)$, $(1, 3, 2)$, $(2, 1, 3)$, $(2, 3, 1)$, $(3, 1, 2)$, and $(3, 2, 1)$.
(iv) There are 24 permutations of 4. We leave it as an exercise for you to list them all.
(v) It is easy to prove that there are, in general, $n!$-many permutations of n. To see this, note that there are n-many ways to choose the first element, j_1, of the permutation $(j_1, j_2, \ldots, j_n)$; there are then $(n - 1)$-many ways to choose the second element, j_2; and so on.

22.3 Definition

An *inversion* in a permutation of n is an occurrence of a smaller number following a larger one.

This concept is best explained with examples.

22.4 Example

The permutation $(3, 5, 1, 4, 2)$ of 5 contains many inversions: the occurrence of 1 following 3 is an inversion; the occurrence of 4 following 5 is an inversion; so are the occurrences of 1 and 2 following 5; there are other inversions as well, such as the occurrence of 2 following 4.

The number of inversions in a permutation $(j_1, j_2, \ldots, j_n)$ of n is, vaguely speaking, a measure of how far the permutation is from the natural order $(1, 2, \ldots, n)$.

22.5 A Systematic Way to Count the Number of Inversions in a Permutation

The number of inversions in a permutation $(j_1, j_2, \ldots, j_n)$ is the sum of the number of integers smaller than j_1 that follow j_1, plus the number of integers smaller than j_2 that follow j_2, plus, $\ldots$, plus the number of integers smaller than j_{n-1} that follow j_{n-1}.

Let us illustrate this procedure with the permutation from Example 22.4.

22.6 Example

The number of inversions in the permutation $(3, 5, 1, 4, 2)$ is

2 (because there are two numbers smaller than 3 following 3)

+3 (because there are three numbers smaller than 5 following 5)

+0 (because there are no numbers smaller than 1 following 1)

+1 (because there is one number smaller than 4 following 4)

$= 2 + 3 + 0 + 1 = 6$

Similarly, the number of inversions in the permutation $(3, 5, 6, 2, 1, 4)$ of 6 is $2 + 3 + 3 + 1 + 0 = 9$.

22.7 Definition

(i) A permutation is called *even* or *odd* according as the number of inversions it contains is even or odd.
(ii) The evenness or oddness of a permutation is called its *parity*.

Thus the permutation (3, 5, 1, 4, 2) is even; the permutation (3, 5, 6, 2, 1, 4) is odd.

We conclude this section with an important observation concerning parity that is the key to the proofs of most theorems about determinants.

22.8 Definition

If σ and τ are two permutations of n, we say that σ and τ *differ by a transposition* if they can be obtained from one another by interchanging entries in exactly two positions.

22.9 Examples

The following pairs of permutations differ by a transposition.

(i)	(3, 5, 2, 4, 1)	and	(3, 2, 5, 4, 1)
(ii)	(3, 5, 2, 4, 1)	and	(3, 1, 2, 4, 5)
(iii)	(1, 2, 3)	and	(1, 3, 2)
(iv)	(1, 2, 3, 4, 5)	and	(3, 2, 1, 4, 5)
(v)	(3, 4, 5, 1, 2, 6)	and	(6, 4, 5, 1, 2, 3)

22.10 Exercise

Compute the parities of each of the permutations from Example 22.9.

22.11 Theorem

If σ and τ are permutations of n that differ by a transposition, then σ and τ have opposite parities.

Proof: Suppose that σ and τ are obtained from one another by interchanging the entries in the pth and q^{th} positions. Consider first the special case in which these positions are adjacent (i.e., $q = p + 1$). So

$$\sigma = \left(j_1, \ldots, j_{p-1}, j_p, j_{p+1}, j_{p+2}, \ldots, j_n\right)$$

and

$$\tau = (j_1, \ldots, j_{p-1}, j_{p+1}, j_p, j_{p+2}, \ldots, j_n)$$

We will use 22.5 to compare the number of inversions in σ and τ. For $i = 1, \ldots, p-1$ and for $i = p+2, \ldots, n$ the number of integers smaller than j_i that occur to the right of j_i is the same in both σ and τ. It remains to compare the number of inversions in σ and τ attributable to j_p and j_{p+1}.

Case (i): Suppose $j_p \lneqq j_{p+1}$. Then the number of integers smaller than j_p that follow j_p is the same in both σ and τ. The number of integers smaller than j_{p+1} is exactly one more in τ than in σ, since j_p follows j_{p+1} in τ but not in σ.

Case (ii): $j_{p+1} \lneqq j_p$. Then the number of integers smaller than j_{p+1} that follow j_{p+1} is the same in both σ and τ. The number of integers smaller than j_p that follow j_p is one less in τ than in σ, since j_{p+1} follows j_p in σ but not in τ.

The number of inversions in τ is therefore exactly one more (if $j_p \lneqq j_{p+1}$) or exactly one less (if $j_{p+1} \lneqq j_p$) than in σ, so σ and τ have opposite parity.

The proof for the general case, in which the pth and q^{th} positions are not adjacent, makes use of the result just proved for the special case of transpositions involving adjacent entries.

If σ and τ differ by a transposition, they have the general form

$$\sigma = (j_1, \ldots, j_{p-1}, j_p, j_{p+1}, \ldots, j_{q-1}, j_q, j_{q+1}, \ldots, j_n)$$
$$\tau = (j_1, \ldots, j_{p-1}, j_q, j_{p+1}, \ldots, j_{q-1}, j_p, j_{q+1}, \ldots, j_n)$$

where the entries in the pth and q^{th} positions are the ones interchanged. Note that there are $(q-p-1)$-many positions strictly between the pth and q^{th} positions.

We now observe that, if σ and τ differ by an arbitrary transposition, then σ and τ can be obtained from one another by an *odd* number of transpositions involving adjacent entries. To see this, let

$$\sigma = (j_1, \ldots, j_{p-1}, j_p, \underbrace{j_{p+1}, \ldots, j_{q-1}}_{\substack{\text{there are } (q-p-1)\text{-many} \\ \text{positions in here}}}, j_q, j_{q+1}, \ldots, j_n).$$

To move j_q to the pth position by repeatedly transposing adjacent entries, we first transpose j_{q-1} and j_q, then j_{q-2} and $j_q, \ldots,$ and finally j_p and j_q. Thus, after $(q-p-1)+1$ successive transpositions of adjacent entries, the situation is

$$(j_1, \ldots, j_{p-1}, j_q, j_p, \overbrace{j_{p+1}, \ldots, j_{q-1}}^{\substack{\text{there are } (q-p-1)\text{-many} \\ \text{positions in here}}}, j_{q+1}, \ldots, j_n)$$

To move j_p [which is now in $(p+1)^{\text{st}}$ position] to the qth position [now occupied by j_{q-1}], we perform an additional $(q-p-1)$-many transpositions involving adjacent entries.

Thus σ and τ can be obtained from one another using $2(q-p-1)+1$ transpositions involving adjacent entries. We proved earlier that each such transposition changes the parity of the permutation. Since $2(q-p-1)+1$ is odd, σ and τ must have opposite parity. ■

Problem Set 22

22.1. Go back and do any of the exercises from Section 22 that you may have skipped.

*22.2. Classify the following permutations as even or odd.

(i) $(3, 5, 2, 7, 1, 4, 6)$

(ii) $(3, 6, 4, 1, 2, 5)$

(iii) $(3, 5, 2, 1, 4)$

23

Definition of the Determinant

23.1 Definition

Let A be an $n \times n$ matrix. An *elementary product from A* is a product of n-many entries from A of the form $a_{1j_1}a_{2j_2} \cdots a_{nj_n}$ where $(j_1, j_2, \ldots, j_n)$ is a permutation of n.

Thus an elementary product from A is the product of n-many entries from A consisting of one entry from each row and column, with the factors ordered so that the entry from row 1 appears first, the entry from row 2 appears second, and so on.

There are obviously $n!$-many elementary products from an $n \times n$ matrix A, one corresponding to each permutation of n.

23.2 Examples

For the sake of legibility, we will place parentheses around each factor when writing the following elementary products:

(i) Let

$$A = \begin{bmatrix} -3 & 5 & 9 \\ 0 & 6 & -2 \\ -8 & \frac{1}{2} & 5 \end{bmatrix}$$

Here is a complete list of the six elementary products from A, together with their corresponding permutations.

Elementary Products	*Corresponding Permutation*
$(-3)(6)(5)$	(1, 2, 3,)
$(5)(-2)(-8)$	(2, 3, 1)
$(9)(0)(\frac{1}{2})$	(3, 1, 2)
$(9)(6)(-8)$	(3, 2, 1)
$(5)(0)(5)$	(2, 1, 3)
$(-3)(-2)(\frac{1}{2})$	(1, 3, 2)

Note, for instance, that $(-3)(-2)(-8)$ is *not* an elementary product since it involves two factors from the same column; $(9)(-8)(6)$ is *not* an elementary product since the factor (-8) from the third row appears before the factor (6) from the second row.

(ii) Let

$$A = \begin{bmatrix} a_{11} & a_{12} \\ a_{21} & a_{22} \end{bmatrix}$$

There are only two elementary products from A.

Elementary Products	*Corresponding Permutation*
$(a_{11})(a_{22})$	(1, 2)
$(a_{12})(a_{21})$	(2, 1)

(iii) Let

$$A = \begin{bmatrix} 3 & 4 & 7 & 2 & 3 \\ -2 & -1 & -3 & -2 & -2 \\ 1 & 1 & 1 & -1 & 0 \\ 1 & 2 & 5 & 2 & 2 \\ 2 & 3 & -3 & -2 & -1 \end{bmatrix}$$

Of the 120 elementary products from A, we will only list three, for the sake of example.

Elementary Products	*Corresponding Permutation*
$(4)(-2)(1)(2)(-2)$	(2, 1, 3, 5, 4)
$(2)(-1)(1)(1)(-1)$	(4, 2, 3, 1, 5)
$(3)(-3)(1)(2)(3)$	(5, 3, 1, 4, 2)
and so on	and so on

23.3 Definition

A *signed elementary product from A* is an elementary product $a_{1j_1}a_{2j_2} \cdots a_{nj_n}$ preceded by + or − according as the parity of the permutation $(j_1, j_2, \ldots, j_n)$ is even or odd.

23.4 Examples

It is instructive to return to the examples from 23.2. This time we will list the elementary product, the corresponding permutation, the number of inversions in the permutation, the parity of the permutation, and the signed elementary product.

(i) Let

$$A = \begin{bmatrix} -3 & 5 & 9 \\ 0 & 6 & -2 \\ -8 & \frac{1}{2} & 5 \end{bmatrix}$$

Elementary Product	*Corresponding Permutation*	*Number of Inversions*	*Parity*	*Signed Elementary Product*
$(-3)(6)(5)$	$(1, 2, 3)$	0	Even	$+(-3)(6)(5)$
$(5)(-2)(-8)$	$(2, 3, 1)$	2	Even	$+(5)(-2)(-8)$
$(9)(0)(\frac{1}{2})$	$(3, 1, 2)$	2	Even	$+(9)(0)(\frac{1}{2})$
$(9)(6)(-8)$	$(3, 2, 1)$	3	Odd	$-(9)(6)(-8)$
$(5)(0)(5)$	$(2, 1, 3)$	1	Odd	$-(5)(0)(5)$
$(-3)(-2)(\frac{1}{2})$	$(1, 3, 2)$	1	Odd	$-(-3)(-2)(\frac{1}{2})$

(ii) Let

$$A = \begin{bmatrix} a_{11} & a_{12} \\ a_{21} & a_{22} \end{bmatrix}$$

Elementary Product	*Corresponding Permutation*	*Number of Inversions*	*Parity*	*Signed Elementary Product*
$(a_{11})(a_{22})$	$(1, 2)$	0	Even	$+a_{11}a_{22}$
$(a_{12})(a_{21})$	$(2, 1)$	1	Odd	$-a_{12}a_{21}$

(iii) Let

$$A = \begin{bmatrix} 3 & 4 & 7 & 2 & 3 \\ -2 & -1 & -3 & -2 & -2 \\ 1 & 1 & 1 & -1 & 0 \\ 1 & 2 & 5 & 2 & 2 \\ 2 & 3 & -3 & -2 & -1 \end{bmatrix}$$

Elementary Product	*Corresponding Permutation*	*Number of Inversions*	*Parity*	*Signed Elementary Product*
$(4)(-2)(1)(2)(-2)$	$(2, 1, 3, 5, 4)$	2	Even	$+(4)(-2)(1)(2)(-2)$
$(2)(-1)(1)(1)(-1)$	$(4, 2, 3, 1, 5)$	5	Odd	$-(2)(-1)(1)(1)(-1)$
$(3)(-3)(1)(2)(3)$	$(5, 3, 1, 4, 2)$	7	Odd	$-(3)(-3)(1)(2)(3)$
and so on				

23.5 Definition

The *determinant* of an $n \times n$ matrix A is the sum of all the signed elementary products from A. The determinant of A is usually denoted by $|A|$ or by det A.

23.6 Remark

Observe that for 2×2 and 3×3 matrices, this definition gives the results with which we are already familiar. We also realize why there is no mnemonic device for remembering the expression for the determinant of a 4×4 matrix: the expression involves 24 terms! The expression for the determinant of a 5×5 matrix involves 120 terms! It is clear therefore that, for the practical computation of determinants, we will require some theorems that free us from reliance on the definition.

Except for 2×2 and 3×3 matrices, it is only when most of the entries in the matrix are zero that it is feasible to compute the determinant directly from the definition, for there will then be very few (in some cases only one) nonzero elementary products.

We conclude this section with some exercises and examples that exploit this observation.

23.7 Exercise

Prove the following easy consequences of the definition.

(i) The determinant of a diagonal matrix D is the product $d_{11}d_{22} \cdots d_{nn}$ of its entries on the main diagonal.
(ii) If a matrix A has either an all-zero row or an all-zero column, then $|A| = 0$.
(iii) If A is either upper or lower triangular, then $|A| = a_{11}a_{22} \cdots a_{nn}$.

23.8 Example

Compute

$$\begin{vmatrix} 0 & 3 & 0 & \frac{4}{3} \\ 5 & 0 & 0 & 0 \\ 0 & 1 & 0 & 6 \\ 0 & 0 & \frac{1}{2} & 0 \end{vmatrix}$$

directly from the definition of *determinant*.

From each of the second and fourth rows, there is only one possible choice of a nonzero factor; thus the only nonzero elementary products are

$(3)(5)(6)(\frac{1}{2})$ and $(\frac{4}{3})(5)(1)(\frac{1}{2})$. We can therefore calculate the determinant by noting the parity of the permutations associated with these elementary products and adding up the signed elementary products

Elementary Product	*Corresponding Permutation*	*Number of Inversions*	*Parity*	*Signed Elementary Product*
$(3)(5)(6)(\frac{1}{2})$	$(2, 1, 4, 3)$	2	Even	$+(3)(5)(6)(\frac{1}{2})$
$(\frac{4}{3})(5)(1)(\frac{1}{2})$	$(4, 1, 2, 3)$	3	Odd	$-(\frac{4}{3})(5)(1)(\frac{1}{2})$

So the determinant is $+45 - \frac{10}{3} = 41\frac{2}{3}$.

Problem Set 23

23.1. Go back and do any of the exercises from Section 23 that you may have skipped.

*23.2. The entries a_{32}, a_{15}, a_{53}, a_{41}, a_{24}, a_{66} are the only nonzero entries in a 6×6 matrix A. What is $|A|$?

*23.3. A is a $k \times k$ matrix whose only nonzero entries are $a_{i\,k-(i-1)}$ for $i = 1, 2, \ldots, k$.
(i) What is $|A|$ if $k = 6$?
(ii) What is $|A|$ if $k = 99$?
(iii) What is $|A|$ as a function of k?

*23.4. If A is an $n \times n$ matrix and if t is any scalar, then $|tA| = t^n|A|$.

23.5. It is clear that if every elementary product from A is 0, then $|A| = 0$; of course, this condition is not necessary for $|A| = 0$. It is interesting to characterize those $n \times n$ matrices for which *all* the elementary products are 0. Vaguely speaking, the question is, How many zeros must A contain, and how must they be placed, so that every elementary product from A is 0?

A *submatrix of A* is a matrix obtained from A by deleting any number of rows and/or columns from A.
(i) If A is 2×2, show that every elementary product from A is 0 iff either $0_{2\times1}$ or $0_{1\times2}$ is a submatrix of A.
(ii) If A is 3×3, show that every elementary product from A is 0 iff either $0_{3\times1}$ or $0_{2\times2}$ or $0_{1\times3}$ is a submatrix of A.
(iii) If A is 4×4, show that every elementary product from A is 0 iff either $0_{4\times1}$ or $0_{3\times2}$ or $0_{2\times3}$ or $0_{1\times4}$ is a submatrix of A.

We conclude this problem with a statement of the general result; its proof is beyond the scope of our text.
(iv) *Theorem*: (Frobenius) Every elementary product from an $n \times n$ matrix A is 0 iff for some k such that $1 \le k \le n$, $0_{k\times(n-k+1)}$ is a submatrix of A.

24

Computation of Determinants by Row Reduction

When the Gauss–Jordan algorithm was first introduced, we mentioned that it would be useful for other purposes. After analyzing the effects that the three types of elementary row-operations have on the determinant of a matrix, we will be able to describe a slight variant of the Gaussian algorithm that can be used to compute determinants. Recall that performing an elementary row-operation on a matrix has the same effect as multiplication on the left by an appropriate elementary matrix.

So, first, we need to examine the determinants of elementary matrices themselves.

24.1 Theorem

(i) If $E = \mathcal{O}(I_n)$, where $\mathcal{O}$ is sR_p and $s \neq 0$, then $|E| = s$.
(ii) If $E = \mathcal{O}(I_n)$, where $\mathcal{O}$ is $R_p \leftrightarrow R_q$, then $|E| = -1$.
(iii) If $E = \mathcal{O}(I_n)$, where $\mathcal{O}$ is $R_p + tR_q$, then $|E| = 1$.

Proof:

Case (i):

$$E = \begin{bmatrix} 1 & & & & & & \\ & \ddots & & & & & \\ & & 1 & & & & \\ & & & s & & & \\ & & & & 1 & & \\ & & & & & \ddots & \\ & & & & & & 1 \end{bmatrix} \begin{matrix} \\ \\ \\ \leftarrow p^{\text{th}} \text{ row} \\ \\ \\ \\ \end{matrix}$$

All other entries in E are 0. Since E is a diagonal matrix, $|E| = (1)\cdots(1)(s)(1)\cdots(1) = s$ by 23.7(i).

Case (ii): Assume, without loss of generality, that $p \lneq q$. Thus

$$E = \begin{bmatrix} 1 & & & & & & & & & & & & & \\ & \cdot & & & & & & & & & & & & \\ & & \cdot & & & & & & & & & & & \\ & & & \cdot & & & & & & & & & & \\ 0 & \cdot & \cdot & \cdot & 0 & \cdot & \cdot & \cdot & 0 & 1 & 0 & \cdot & 0 \\ & & & & & \cdot & & & & & & & \\ & & & & & & \cdot & & & & & & \\ & & & & & & & \cdot & & & & & \\ 0 & \cdot & 0 & 1 & 0 & \cdot & \cdot & \cdot & 0 & \cdot & \cdot & \cdot & 0 \\ & & & & & & & & & \cdot & & & \\ & & & & & & & & & & \cdot & & \\ & & & & & & & & & & & \cdot & \\ & & & & & & & & & & & & 1 \end{bmatrix} \begin{matrix} \\ \\ \\ \\ \leftarrow p^{\text{th}} \text{ row} \\ \\ \\ \\ \leftarrow q^{\text{th}} \text{ row} \\ \\ \\ \\ \\ \end{matrix}$$

$\qquad\qquad\overbrace{\qquad\qquad}^{}\qquad\qquad\qquad\overbrace{\qquad\qquad}^{}$

$\qquad\qquad p^{\text{th}}$ column $\qquad\qquad\qquad q^{\text{th}}$ column

E has 1's down the main diagonal, except as shown, and all entries not shown are 0. There is clearly only one nonzero elementary product, $(1)\cdots(1)$; the corresponding permutation is $\sigma = (1, 2, \ldots, q, \ldots, p, \ldots, n)$. The only difference between this permutation and the natural order $\rho = (1, 2, \ldots, p, \ldots, q, \ldots, n)$ is that the two integers p and q have exchanged their positions: q now occupies the pth position, while p occupies the qth position. Since ρ is even, σ is odd by Theorem 22.11; hence $|E| = -(1)\cdots(1) = -1$.

Case (iii): E has 1's down the main diagonal and is either upper triangular, if $p \lneq q$, or lower triangular, if $q \lneq p$. Thus $|E| = 1$ by 23.7(iii). ■

24.2 Exercise

Repeat Exercise 13.2, computing the determinants of those matrices that are elementary.

The next result is one that we will not need until Section 27, but is more appropriately placed here since it is an easy consequence of 24.1.

24.3 Theorem

For any elementary matrix E, $|E| = |E^T|$.

Proof: If $E = \mathcal{O}(I_n)$ and $\mathcal{O}$ is either type I or II, then $E^T = E$ (recall problem 15.5). If $\mathcal{O}$ is $R_p + tR_q$, then E^T is also an elementary matrix of type III (specifically, E^T is obtained by performing $R_q + tR_p$ on I_n) and so $|E| = 1 = |E^T|$ by 24.1(iii). ■

We will eventually prove in Section 26 a very important theorem for $n \times n$ matrices A and B; specifically, that $|AB| = |A|\,|B|$; in other words, that the determinant of a product is the product of the determinants. The next theorem is an essential first step toward that goal: it is the special case of this result when the left factor is an elementary matrix.

24.4 Theorem

For every $n \times n$ matrix A and every elementary $n \times n$ matrix E, $|EA| = |E|\,|A|$.

Proof: We need separate proofs for each of the three types of elementary matrices. Throughout, we will make use of the facts, proved in 13.3, concerning multiplication on the left by elementary matrices: if $E = \mathcal{O}(I_n)$, then $EA = \mathcal{O}(A)$.

Case (i): Suppose $E = \mathcal{O}(I_n)$, where $\mathcal{O}$ is sR_p with $s \neq 0$. Thus

$$EA = \begin{bmatrix} a_{11} & a_{12} & \cdots & a_{1n} \\ \vdots & \vdots & & \vdots \\ sa_{p1} & sa_{p2} & \cdots & sa_{pn} \\ \vdots & \vdots & & \vdots \\ a_{n1} & a_{n2} & \cdots & a_{nn} \end{bmatrix} \leftarrow p^{\text{th}} \text{ row}$$

It is clear that $a_{1j_1}a_{2j_2} \cdots a_{pj_p} \cdots a_{nj_n}$ is an elementary product from A iff $a_{1j_1}a_{2j_2} \cdots sa_{pj_p} \cdots a_{nj_n}$ is an elementary product from EA.

So the signed elementary products from EA are precisely the signed elementary products from A with each one multiplied by s. Hence $|EA| = s|A|$. But $s|A| = |E|\,|A|$ by 24.1(i), so we are done.

Case (ii): Suppose $E = \mathcal{O}(I_n)$, where $\mathcal{O}$ is $R_p \leftrightarrow R_q$.

$$A = \begin{bmatrix} a_{11} & \cdots & a_{1r} & \cdots & a_{1s} & \cdots & a_{1n} \\ \vdots & & \vdots & & \vdots & & \vdots \\ a_{p1} & \cdots & (a_{pr}) & \cdots & a_{ps} & \cdots & a_{pn} \\ \vdots & & \vdots & & \vdots & & \vdots \\ a_{q1} & \cdots & a_{qr} & \cdots & (a_{qs}) & \cdots & a_{qm} \\ \vdots & & \vdots & & \vdots & & \vdots \\ a_{n1} & \cdots & a_{nr} & \cdots & a_{ns} & \cdots & a_{nn} \end{bmatrix} \begin{matrix} \\ \\ \leftarrow p^{th} \text{ row} \rightarrow \\ \\ \leftarrow q^{th} \text{ row} \rightarrow \\ \\ \\ \end{matrix} \begin{bmatrix} a_{11} & \cdots & a_{1r} & \cdots & a_{1s} & \cdots & a_{1n} \\ \vdots & & \vdots & & \vdots & & \vdots \\ a_{q1} & \cdots & a_{qr} & \cdots & (a_{qs}) & \cdots & a_{qn} \\ \vdots & & \vdots & & \vdots & & \vdots \\ a_{p1} & \cdots & (a_{pr}) & \cdots & a_{ps} & \cdots & a_{pn} \\ \vdots & & \vdots & & \vdots & & \vdots \\ a_{n1} & \cdots & a_{nr} & \cdots & a_{ns} & \cdots & a_{nn} \end{bmatrix} = EA$$

Because an elementary product involves one entry from each row and column, it is clear that the values of the elementary products from A and from EA are the same: when the entries $a_{1j_1}, \ldots, a_{pr}, \ldots, a_{qs}, \ldots, a_{nj_n}$ are chosen to form an elementary product either from EA or from A, the resulting elementary products are

$$a_{1j_1} \cdots \underbrace{a_{qs}}_{\text{This is the factor from the } p^{th} \text{ row}} \cdots \underbrace{a_{pr}}_{\text{This is the factor from the } q^{th} \text{ row}} \cdots a_{nj_n} \qquad \text{[from EA]}$$

$$a_{1j_1} \cdots a_{pr} \cdots a_{qs} \cdots a_{nj_n} \qquad \text{[from A]}$$

(In the preceding matrices we have circled the entries from rows p and q that were chosen to form these elementary products.)

Since ordinary multiplication is commutative, these elementary products have the same value. However, the sign associated with $a_{1j_1} \cdots a_{qs} \cdots a_{pr} \cdots a_{nj_n}$, whose corresponding permutation is $\sigma = (1, \ldots, s, \ldots, r, \ldots, n)$, will be opposite to the sign associated with $a_{1j_1} \cdots a_{pr} \cdots a_{qs} \cdots a_{nj_n}$, whose corresponding permutation is $\tau = (1, \ldots, r, \ldots, s, \ldots, n)$, because σ and τ have opposite parity by Theorem 22.11.

The observation just made applies to each of the $n!$-many ways of forming elementary products from EA and from A.

Thus $|EA| = -|A|$. But $-|A| = |E|\,|A|$ by 24.1(ii), so we are done.

Case (iii): Suppose $E = \mathcal{O}(I_n)$, where $\mathcal{O}$ is $R_p + tR_q$. Thus

$$EA = \begin{bmatrix} a_{11} & a_{12} & \cdots & a_{1n} \\ \vdots & \vdots & & \vdots \\ a_{p1} + ta_{q1} & a_{p2} + ta_{q2} & \cdots & a_{pn} + ta_{qn} \\ \vdots & \vdots & & \vdots \\ a_{n1} & a_{n2} & \cdots & a_{nn} \end{bmatrix} \leftarrow p^{th} \text{ row}$$

A typical elementary product from EA has the form

$$a_{1j_1}a_{2j_2}\cdots\left(a_{pj_p}+ta_{qj_p}\right)\cdots a_{nj_n}$$

which can be rewritten as the sum of two other elementary products:

$$a_{1j_1}a_{2j_2}\cdots a_{pj_p}\cdots a_{nj_n}+a_{1j_1}a_{2j_2}\cdots ta_{qj_p}\cdots a_{nj_n}$$

where the first term in this sum is an elementary product from

$$A=\begin{bmatrix} a_{11} & a_{12} & \cdots & a_{1n} \\ \vdots & \vdots & & \vdots \\ a_{p1} & a_{p2} & \cdots & a_{pn} \\ \vdots & \vdots & & \vdots \\ a_{n1} & a_{n2} & \cdots & a_{nn} \end{bmatrix}$$

and the second term in the sum is an elementary product from

$$B=\begin{bmatrix} a_{11} & a_{12} & \cdots & a_{1n} \\ \vdots & \vdots & & \vdots \\ ta_{q1} & ta_{q2} & \cdots & ta_{qn} \\ \vdots & \vdots & & \vdots \\ a_{n1} & a_{n2} & \cdots & a_{nn} \end{bmatrix} \leftarrow p^{\text{th}} \text{ row}$$

The same permutation $(j_1, j_2, \ldots, j_p, \ldots, j_n)$ is associated with all three of these elementary products.

These observations establish that $|EA| = |A| + |B|$. If we can show that $|B| = 0$, we are done, for then $|EA| = |A| = 1|A| = |E|\,|A|$, using 24.1(iii).

To see that $|B| = 0$, we use the two parts of 24.4 that have been proved so far.

By 24.4(i), $|B| = t|C|$, where

$$C=\begin{bmatrix} a_{11} & a_{12} & \cdots & a_{1n} \\ \vdots & \vdots & & \vdots \\ a_{q1} & a_{q2} & \cdots & a_{qn} \\ \vdots & \vdots & & \vdots \\ a_{q1} & a_{q2} & \cdots & a_{qn} \\ \vdots & \vdots & & \vdots \\ a_{n1} & a_{n2} & \cdots & a_{nn} \end{bmatrix} \begin{matrix} \\ \\ \leftarrow p^{\text{th}} \text{ row} \\ \\ \leftarrow q^{\text{th}} \text{ row} \\ \\ \\ \end{matrix}$$

By 24.4(ii), $|C| = -|\mathcal{O}(C)|$, where $\mathcal{O}$ is $R_p \leftrightarrow R_q$. But since the pth and q^{th} rows of C are equal, $\mathcal{O}(C) = C$. Thus $|C| = -|C|$, so $|C| = 0$. ■

We can rephrase the results from 24.4 as facts concerning the effects that performing elementary row-operations have on the determinant of an $n \times n$ matrix A.

24.5 Corollary

(i) Multiplying one row of an $n \times n$ matrix A by a nonzero constant s produces a new matrix whose determinant is $s|A|$.
(ii) Interchanging two rows of an $n \times n$ matrix changes the sign of the determinant.
(iii) A type III elementary operation on an $n \times n$ matrix has no effect on its determinant. ■

This corollary, in conjunction with the Gaussian algorithm, provides a practical method for computing the determinant of any size square matrix: one simply applies the algorithm and simultaneously keeps track of the effects on the determinant of any operations of types I or II; eventually, one arrives at an upper-triangular matrix, whose determinant, by 23.7(iii), is simply the product of its main diagonal entries. This is best illustrated with examples.

24.6 Examples of Computation of Determinants by Row-Reduction

(i) Find $\begin{vmatrix} 2 & -2 & 3 \\ 2 & 1 & -2 \\ -1 & 3 & 1 \end{vmatrix}$.

$$\begin{vmatrix} 2 & -2 & 3 \\ 2 & 1 & -2 \\ -1 & 3 & 1 \end{vmatrix} = (2)\begin{vmatrix} 1 & -1 & \frac{3}{2} \\ 2 & 1 & -2 \\ -1 & 3 & 1 \end{vmatrix}$$

$$= (2)\begin{vmatrix} 1 & -1 & \frac{3}{2} \\ 0 & 3 & -5 \\ -1 & 3 & 1 \end{vmatrix} = (2)\begin{vmatrix} 1 & -1 & \frac{3}{2} \\ 0 & 3 & -5 \\ 0 & 2 & \frac{5}{2} \end{vmatrix}$$

$$= (2)(3)\begin{vmatrix} 1 & -1 & \frac{3}{2} \\ 0 & 1 & -\frac{5}{3} \\ 0 & 2 & \frac{5}{2} \end{vmatrix} = (2)(3)\begin{vmatrix} 1 & -1 & \frac{3}{2} \\ 0 & 1 & -\frac{5}{3} \\ 0 & 0 & \frac{35}{6} \end{vmatrix}$$

$$= (2)(3)(1)(1)\left(\frac{35}{6}\right) = 35$$

(ii) Find $\begin{vmatrix} \frac{1}{2} & 3 & 2 \\ 2 & 12 & -1 \\ -4 & 0 & -6 \end{vmatrix}$.

$$\begin{vmatrix} \frac{1}{2} & 3 & 2 \\ 2 & 12 & -1 \\ -4 & 0 & -6 \end{vmatrix} = \left(\frac{1}{2}\right)\begin{vmatrix} 1 & 6 & 4 \\ 2 & 12 & -1 \\ -4 & 0 & -6 \end{vmatrix}$$

$$= \left(\frac{1}{2}\right)\begin{vmatrix} 1 & 6 & 4 \\ 0 & 0 & -9 \\ 0 & 24 & 10 \end{vmatrix} = \left(\frac{1}{2}\right)(-1)\begin{vmatrix} 1 & 6 & 4 \\ 0 & 24 & 10 \\ 0 & 0 & -9 \end{vmatrix}$$

$$= \left(\frac{1}{2}\right)(-1)(1)(24)(-9) = 108$$

(iii) Find $\begin{vmatrix} 2 & -8 & -6 & 4 \\ 2 & -1 & 1 & 8 \\ 5 & 1 & 6 & 0 \\ 3 & 2 & 5 & 4 \end{vmatrix}$.

$$\begin{vmatrix} 2 & -8 & -6 & 4 \\ 2 & -1 & 1 & 8 \\ 5 & 1 & 6 & 0 \\ 3 & 2 & 5 & 4 \end{vmatrix} = (2)\begin{vmatrix} 1 & -4 & -3 & 2 \\ 2 & -1 & 1 & 8 \\ 5 & 1 & 6 & 0 \\ 3 & 2 & 5 & 4 \end{vmatrix}$$

$$= (2)\begin{vmatrix} 1 & -4 & -3 & 2 \\ 0 & 7 & 7 & 4 \\ 0 & 21 & 21 & -10 \\ 0 & 14 & 14 & -2 \end{vmatrix} = (2)(7)\begin{vmatrix} 1 & -4 & -3 & 2 \\ 0 & 1 & 1 & \frac{4}{7} \\ 0 & 21 & 21 & -10 \\ 0 & 14 & 14 & -2 \end{vmatrix}$$

$$= (2)(7)\begin{vmatrix} 1 & -4 & -3 & 2 \\ 0 & 1 & 1 & \frac{4}{7} \\ 0 & 0 & 0 & -94 \\ 0 & 0 & 0 & -58 \end{vmatrix} = (2)(7)(1)(1)(0)(-58) = 0$$

Problem Set 24

24.1. Go back and do any of the exercises from Section 24 that you may have skipped.

*24.2. Prove that if $A \in \mathbb{R}_{n\times n}$ has two identical rows then $|A| = 0$.

*24.3. Compute the determinants of the following matrices:

(i) $\begin{bmatrix} 1 & 2 & 0 \\ 6 & 4 & 0 \\ -3 & 5 & -1 \end{bmatrix}$

(ii) $\begin{bmatrix} 2 & 0 & 3 \\ 5 & 2 & -2 \\ 0 & 0 & -1 \end{bmatrix}$

(iii) $\begin{bmatrix} 3 & 1 & 2 & 3 \\ -1 & 2 & 4 & 0 \\ 2 & -1 & 1 & 2 \\ 1 & 1 & 2 & 3 \end{bmatrix}$

(iv) $\begin{bmatrix} 0 & 3 & -3 & 6 \\ -2 & 5 & -1 & 2 \\ 2 & -4 & 3 & -3 \\ 3 & 3 & 0 & 9 \end{bmatrix}$

(v) $\begin{bmatrix} 2 & 3 & -2 & 7 & 1 \\ 1 & 2 & -3 & 3 & 1 \\ 2 & 1 & 2 & 1 & 1 \\ -1 & -1 & 1 & -1 & 2 \\ 3 & -1 & 1 & -1 & -2 \end{bmatrix}$

(vi) $\begin{bmatrix} 2 & -7 & 5 & -1 & 1 \\ 3 & -15 & 11 & -4 & 2 \\ 1 & 4 & 5 & 2 & 3 \\ -1 & 8 & -6 & 3 & 1 \\ -1 & 6 & 0 & 1 & 1 \end{bmatrix}$

24.4 The result in this problem will be needed only in Section 30; it can therefore be omitted by readers who plan to skip Section 30.
Suppose the $n \times n$ matrices A and B satisfy $a_{ij} = b_{ij}$ for all $i \neq p$. For such matrices, define $A +_p B$ as follows:

$$(i, j)^{\text{th}} \text{ entry in } A +_p B = \begin{cases} a_{ij}(= b_{ij}) & \text{if } i \neq p \\ a_{pj} + b_{pj} & \text{if } i = p \end{cases}$$

Prove that if $C = A +_p B$, then $|C| = |A| + |B|$.
Hint: In the proof of Theorem 24.4, case (iii), we had $EA = A +_p B$; imitate this proof.

25

A Is Invertible If and Only If det A $\neq 0$

The title of this section says it all!

25.1 Theorem

An $n \times n$ matrix A is invertible if and only if $|A| \neq 0$.

Proof: ($\rightarrow$) If A is invertible then, by Theorem 16.1, there exist elementary matrices $E_1, E_2, \ldots, E_k$ such that $A = E_1E_2 \cdots E_k$. Thus

$$\begin{aligned} |A| &= |E_1E_2 \cdots E_k| \\ &= |E_1|\,|E_2| \cdots |E_k| && \text{[by repeated use of Theorem 24.3]} \\ &\neq 0 && \text{[by Theorem 24.1]} \end{aligned}$$

($\leftarrow$) The results from Sections 14 and 15 concerning the Gauss–Jordan algorithm and the row-reduction process guarantee the existence of elementary matrices $E_1, E_2, \ldots, E_p$ such that $E_p \cdots E_2E_1A = GJ(A)$. Thus

$$\begin{aligned} |GJ(A)| &= |E_p \cdots E_2E_1A| \\ &= |E_p| \cdots |E_2|\,|E_1|\,|A| && \text{[by repeated use of Theorem 24.3]} \\ &\neq 0 && \text{[by Theorem 24.1 plus the assumption that } |A| \neq 0] \end{aligned}$$

Therefore, by 23.7(ii), GJ(A) does not have an all-zero row and so, by Problem 12.4, $GJ(A) = I_n$. Thus A is invertible since it is row-equivalent to the identity. ■

Problem Set 25

25.1. For what values of t is the matrix

$$\begin{bmatrix} t^3 & 0 & t^2 \\ 17 & 0 & 1 \\ t^4 & 2 & -17 \end{bmatrix}$$

invertible?

*25.2. For what values of t is the matrix

$$\begin{bmatrix} 5t+1 & 6 & -1 & 0 \\ 0 & 3 & 0 & 4 \\ 0 & 6 & 4t & 0 \\ 0 & 9 & 0 & 3t^2 \end{bmatrix}$$

singular?

25.3. If A is an $m \times n$ matrix, let $A[i_1, i_2, \ldots, i_p; j_1, j_2, \ldots, j_q]$ denote the $p \times q$ submatrix comprising rows $i_1, \ldots, i_p$ and columns $j_1, \ldots, j_q$ from A. For example, if A is a 5×6 matrix, then $A[2, 3, 5; 3, 6]$ is the 3×2 matrix consisting of rows 2, 3, and 5 and columns 3 and 6 from A.

Definition: The *determinant rank* of an $m \times n$ matrix A is the order of the largest square submatrix of A having nonzero determinant.

Theorem: For every $m \times n$ matrix A, the determinant rank of A is equal to the rank of A.

Provide all the details of the proof that will now be sketched.

Proof: Suppose rank $A = r$. Choose $j_1, j_2, \ldots, j_r$ so that the set consisting of columns $j_1, j_2, \ldots, j_r$ is linearly independent. The $m \times r$ submatrix $A[1, 2, \ldots, m; j_1, j_2, \ldots, j_r]$ also has rank r, so we may choose $i_1, i_2, \ldots, i_r$ such that $|A[i_1, i_2, \ldots, i_r; j_1, j_2, \ldots, j_r]| \neq 0$. Conversely, suppose A has a $p \times p$ submatrix $A[i_1, \ldots, i_p; j_1, \ldots, j_p]$ with nonzero determinant. There is a sequence of row interchanges, followed by another sequence of column interchanges, that we can perform on A to obtain the matrix

$$M = \left[\begin{array}{c|c} A[i_1, \ldots, i_p; j_1, \ldots, j_p] & B \\ \hline C & D \end{array}\right]$$

Then rank A = rank M $\geq p$. ■

*25.4. Use determinants to prove that if a 2×2 matrix A can be expressed as a product, A = BC, of a 2×1 matrix B and a 1×2 matrix C, then A is singular.

*25.5. Consider the following question: Is $(-2, 2, 3, -7) \in \mathscr{L}\{(1, 0, -1, 3), (2, 5, 3, -4), (1, 1, 0, 1)\}$? A student claims to answer this question by computing the determinant of the matrix

$$A = \begin{bmatrix} 1 & 2 & 1 & -2 \\ 0 & 5 & 1 & 2 \\ -1 & 3 & 0 & 3 \\ 3 & -4 & 1 & -7 \end{bmatrix},$$

finding that $|A| = 0$, and concluding from this that the answer to the question is yes. What is the correct answer to the question and what is wrong with the student's argument?

26

det(AB) = det(A)det(B)

Again, the content of this section is essentially contained in the title.

The determinant of a product of two $n \times n$ matrices is the product of their determinants.

Before giving the proof of this result, let us examine intuitively why it should be true. We will treat separately the questions of magnitude and sign of the determinant.

Recall from Section 21 that the magnitude of $\det A$ is the n-dimensional volume of the parallelepiped in $\mathbb{R}^n$ determined by the columns of A.

Also recall Observation 17.8: the columns of an $n \times n$ matrix A are just the images of the standard basis vectors $\mathbf{e}_1^n, \mathbf{e}_2^n, \ldots, \mathbf{e}_n^n$ under the linear transformation $T_A : \mathbb{R}^n \to \mathbb{R}^n$.

The parallelepiped in $\mathbb{R}^n$ determined by $(\mathbf{e}_1^n, \mathbf{e}_2^n, \ldots, \mathbf{e}_n^n)$ is the *unit hypercube*, whose n-dimensional volume is 1. (Note that for $n = 1$, this is the familiar unit interval, for $n = 2$, the unit square, and for $n = 3$, the unit cube.) Thus T_A takes the unit hypercube into a parallelepiped whose volume is $\det A$; in other words, the absolute value of $\det A$ is the factor by which the linear transformation T_A affects volumes.

One final point to recall is that matrix multiplication "represents" composition of linear transformations, as in Theorem 18.3; so when A and B are $n \times n$ matrices,

$$R^n \xrightarrow{T_B} R^n \xrightarrow{T_A} R^n$$

$$T_A \circ T_B = T_{AB}$$

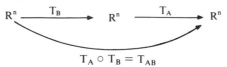

It is to be expected therefore that if the transformation T_B affects volumes by a factor of 5, for example, and the transformation T_A affects volumes by a factor of, say, $\frac{3}{4}$, then performing first T_B followed by T_A (i.e., performing T_{AB}) should affect volumes by a factor of $\frac{15}{4}$.

This explains why det(AB) and the product, det A det B, should have the same magnitude.

The reason they have the same sign has to do, vaguely speaking, with orientation. If T_A and T_B both "preserve" orientation (i.e., if $|A|$ and $|B|$ are both positive), then so does $T_A \circ T_B$. If exactly one of T_A or T_B "reverses" orientation, then so does $T_A \circ T_B$. If T_A and T_B both "reverse" orientation, then $T_A \circ T_B$ will "preserve" orientation. The preceding remarks are vague in that we have not given a precise definition of orientation (for $n > 3$), but they do serve a purpose in motivating the theorem.

With these preliminaries aside, we turn to an exact statement and proof of the theorem.

26.1 Theorem

For all $n \times n$ matrices A and B, $|AB| = |A|\,|B|$.

Proof:

Case (i): Assume A is not invertible. Then neither is AB by Problem 19.5. So $|AB| = 0 = 0|B| = |A|\,|B|$ by Theorem 25.1.

Case (ii): Assume A is invertible. Then A is expressible as a product of elementary matrices $E_1E_2 \cdots E_k$, and so

$$
\begin{aligned}
|AB| &= |E_1E_2 \cdots E_kB| \\
&= |E_1|\,|E_2 \cdots E_kB| \\
&= \cdots && \text{[by repeated use of 24.3]} \\
&= |E_1| \cdots |E_k|\,|B| \\
&= |E_1| \cdots |E_{k-1}E_k|\,|B| \\
&= \cdots && \text{[by repeated use of 24.3]} \\
&= |E_1E_2 \cdots E_k|\,|B| \\
&= |A|\,|B|
\end{aligned}
$$

■

Problem Set 26

26.1. Illustrate Theorem 26.1 when A and B are the matrices from Problem 24.3(i) and (ii), respectively.

*26.2. (i) Prove for $A, B \in \mathbb{R}_{n\times n}$ that, if n is odd and $AB = -BA$, then at least one of A or B is singular.
(ii) Find a pair of invertible 2×2 matrices A and B satisfying $AB = -BA$.

27

$\det A^T = \det A$

Here is another brief section with just a single theorem, asserting that an $n \times n$ matrix and its transpose have the same determinant.

27.1 Theorem

For any $n \times n$ matrix A, $|A^T| = |A|$.

Proof: Since A and A^T have the same rank (recall Problem 9.4), A and A^T are either both invertible or both singular by Theorem 19.5. If they are both singular, then $|A^T| = 0 = |A|$ by Theorem 25.1. If A and A^T are invertible, then there exist elementary matrices $E_1, E_2, \ldots, E_k$ such that $A = E_1E_2 \cdots E_k$, and hence $A^T = E_k^T E_{k-1}^T \cdots E_1^T$. So

$$\begin{aligned}
|A| &= |E_1E_2 \cdots E_k| \\
&= |E_1|\,|E_2| \cdots |E_k| && \text{[by repeated use of 24.4]} \\
&= |E_k|\,|E_{k-1}| \cdots |E_1| && \text{[multiplication of real numbers is commutative]} \\
&= |E_k^T|\,|E_{k-1}^T| \cdots |E_1^T| && \text{[by repeated use of 24.3]} \\
&= |E_k^T E_{k-1}^T \cdots E_1^T| && \text{[by repeated use of 24.4]} \\
&= |A^T|
\end{aligned}$$

■

As a consequence of this theorem, we obtain analogues for columns of all the results proved earlier for rows.

27.2 Corollary (Analogue of 24.5)

(i) Multiplying one column of an $n \times n$ matrix A by a nonzero constant s produces a new matrix whose determinant is $s|A|$.
(ii) Interchanging two columns of an $n \times n$ matrix changes the sign of the determinant.
(iii) A type III elementary column operation on an $n \times n$ matrix has no effect on its determinant.

Proof: Combine 24.5 and 27.1 in the obvious way. ■

Problem Set 27

*27.1. A is a 3×3 matrix and $|A| = 10$. Find the following: $|\frac{1}{2}A|$, $|3A^T|$, $|-A^{-1}|$, $|3A^2|$, $(|3A|)^2$, $|2(A^T)^{-2}|$.

*27.2. Let $A = \begin{bmatrix} 2 & 4 & 1 \\ 3 & 0 & 1 \\ 1 & 2 & 0 \end{bmatrix}$. Find the following: $|A|$, $|5A^T|$, $|A - 2I_3|$, $|(2A)^2|$, $(|2A|)^2$, $|I_3 - A|$, $|I_3 - A^T|$.

27.3. Try to find a geometric interpretation in $\mathbb{R}^2$ of the fact that $|A| = |A^T|$.

27.4. Prove that, if A has two identical columns, then $|A| = 0$.

28

The Cofactor Expansion Theorems

We have already described in Section 24 a systematic method for computing determinants using a slightly modified Gaussian algorithm. In this section, we present another method that is often convenient for hand calculation. It is also very important for computerized calculations when the size of the computer memory is a factor; this is because the method is inductive: it reduces the problem of computing the determinant of a single matrix of size $n \times n$ to that of computing the determinants of n-many matrices each of size $(n - 1) \times (n - 1)$.

If A is an $n \times n$ matrix, let $\hat{A}_{ij}$ denote the $(n - 1) \times (n - 1)$ matrix obtained by deleting the ith row and j^{th} column from A.

28.1 Definition

Let A be an $n \times n$ matrix.

(i) The determinant, $|\hat{A}_{ij}|$, of $\hat{A}_{ij}$ is called the *minor* of a_{ij} and is denoted by a_{ij}^m.

(ii) The number $(-1)^{i+j}|\hat{A}_{ij}|$ is called the *cofactor* of a_{ij} and is denoted by a_{ij}^c. Thus $a_{ij}^c = (-1)^{i+j}a_{ij}^m$.

(i) Let $A = \begin{bmatrix} 2 & 3 \\ -1 & 5 \end{bmatrix}$. Then

$$a_{11}^c = (-1)^{1+1} a_{11}^m = (-1)^{1+1} 5 = 5$$

$$a_{12}^c = (-1)^{1+2} a_{12}^m = (-1)^{1+2}(-1) = 1$$

$$a_{21}^c = (-1)^{2+1} a_{21}^m = (-1)^{2+1} 3 = -3$$

$$a_{22}^c = (-1)^{2+2} a_{22}^m = (-1)^{2+2} 2 = 2$$

(ii) Let $B = \begin{bmatrix} 2 & -2 & 3 \\ 2 & 1 & -2 \\ -1 & 3 & 1 \end{bmatrix}$. Then

$$b_{11}^c = (-1)^{1+1} \begin{vmatrix} 1 & -2 \\ 3 & 1 \end{vmatrix} = 1 + 6 = 7$$

$$b_{12}^c = (-1)^{1+2} \begin{vmatrix} 2 & -2 \\ -1 & 1 \end{vmatrix} = -(2 - 2) = 0$$

$$b_{13}^c = (-1)^{1+3} \begin{vmatrix} 2 & 1 \\ -1 & 3 \end{vmatrix} = 6 + 1 = 7$$

$$b_{21}^c = (-1)^{2+1} \begin{vmatrix} -2 & 3 \\ 3 & 1 \end{vmatrix} = -(-2 - 9) = 11$$

$$b_{22}^c = (-1)^{2+2} \begin{vmatrix} 2 & 3 \\ -1 & 1 \end{vmatrix} = 2 + 3 = 5.$$

$$b_{23}^c = (-1)^{2+3} \begin{vmatrix} 2 & -2 \\ -1 & 3 \end{vmatrix} = -(6 - 2) = -4$$

$$b_{31}^c = (-1)^{3+1} \begin{vmatrix} -2 & 3 \\ 1 & -2 \end{vmatrix} = 4 - 3 = 1$$

$$b_{32}^c = (-1)^{3+2} \begin{vmatrix} 2 & 3 \\ 2 & -2 \end{vmatrix} = -(-4 - 6) = 10$$

$$b_{33}^c = (-1)^{3+3} \begin{vmatrix} 2 & -2 \\ 2 & 1 \end{vmatrix} = 2 + 4 = 6$$

(iii) Let $C = \begin{bmatrix} 2 & -8 & -6 & 4 \\ 2 & -1 & 1 & 8 \\ 5 & 1 & 6 & 0 \\ 3 & 2 & 5 & 4 \end{bmatrix}$. We will not compute the cofactors of each of the 16 entries. For the sake of example, let us merely compute c_{33}^c.

$$c_{33}^c = (-1)^{3+3}\begin{vmatrix} 2 & -8 & 4 \\ 2 & -1 & 8 \\ 3 & 2 & 4 \end{vmatrix} = (2)\begin{vmatrix} 1 & -4 & 2 \\ 2 & -1 & 8 \\ 3 & 2 & 4 \end{vmatrix}$$

$$= (2)\begin{vmatrix} 1 & -4 & 2 \\ 0 & 7 & 4 \\ 0 & 14 & -2 \end{vmatrix} = (2)(7)\begin{vmatrix} 1 & -4 & 2 \\ 0 & 1 & \frac{4}{7} \\ 0 & 14 & -2 \end{vmatrix}$$

$$= (2)(7)\begin{vmatrix} 1 & -4 & 2 \\ 0 & 1 & \frac{4}{7} \\ 0 & 0 & -10 \end{vmatrix} = (2)(7)(-10) = -140$$

The Cofactor Expansion Theorems have two forms, one for rows, another for columns. When the entries from a given row of A are paired with their respective cofactors, then the sum of the products of the pairs is det A: this is the content, in words, of 28.3(i). Theorem 28.3(ii) asserts that, when the entries from a given row are paired with the cofactors of the corresponding entries from a *different* row, then the sum of the products of the pairs is 0. Parts (iii) and (iv) of Theorem 28.3 are the analogues for columns of parts (i) and (ii). Here is a precise statement.

28.3 Theorem (Cofactor Expansion Theorems)

Let A be an $n \times n$ matrix.

(i) For each integer i between 1 and n,

$$a_{i1}a_{i1}^c + a_{i2}a_{i2}^c + \cdots + a_{in}a_{in}^c = \det(\mathrm{A})$$

(ii) For each pair of distinct integers $i \neq p$ between 1 and n,

$$a_{i1}a_{p1}^c + a_{i2}a_{p2}^c + \cdots + a_{in}a_{pn}^c = 0$$

(iii) For each integer j between 1 and n,

$$a_{1j}a_{1j}^c + a_{2j}a_{2j}^c + \cdots + a_{nj}a_{nj}^c = \det(\mathrm{A})$$

(iv) For each pair of distinct integers $j \neq k$ between 1 and n,

$$a_{1j}a_{1k}^c + a_{2j}a_{2k}^c + \cdots + a_{nj}a_{nk}^c = 0$$

To compute $|\mathrm{A}|$ using (i) is called *computing the determinant by cofactor expansion along the* i^{th} *row*. To compute $|\mathrm{A}|$ using (iii) is called *computing the determinant by cofactor expansion along the* j^{th} *column*.

The proofs of these theorems, although not difficult to follow step by step, involve some painstaking details. We have relegated these proofs to a separate, optional section at the end of this chapter.

28.4 Examples

Since we have already computed all the cofactors of the matrix B from Example 28.2 and have computed $|B| = 35$ in Example 24.6(ii), let us use this matrix to illustrate the Cofactor Expansion Theorems.

(i) When the entries from the third row of B are paired with their respective cofactors, the sum of the products of the pairs is $(-1)(1) + (3)(10) + (1)(6) = -1 + 30 + 6 = 35$, as expected.
(ii) When the entries from the third row are paired with the cofactors of the corresponding entries from, say, the second row, the sum of the products of the pairs is $(-1)(11) + (3)(5) + (1)(-4) = -11 + 15 - 4 = 0$, as expected.

You may verify the remaining instances on your own.

The Cofactor Expansion Theorems can be useful for calculating determinants by hand, but their use in this context is *ad hoc*, depending on the specific problem. The idea is to begin with some combination of elementary row-operations and/or elementary column-operations (the latter are permitted by Corollary 27.2) that will result in a row or column all but one of whose entries are zero, and then to expand the determinant by cofactor expansion along that row or column.

28.5 Example (Example 24.6(iii) Revisited)

$$\begin{vmatrix} 2 & -8 & -6 & 4 \\ 2 & -1 & 1 & 8 \\ 5 & 1 & 6 & 0 \\ 3 & 2 & 5 & 4 \end{vmatrix} = \begin{vmatrix} 2 & -8 & -6 & 4 \\ -2 & 15 & 13 & 0 \\ 5 & 1 & 6 & 0 \\ 1 & 10 & 11 & 0 \end{vmatrix} \quad \text{[using } R_2 - 2R_1 \text{ and } R_4 - R_1\text{]}$$

$$= (4)(-1)^{1+4} \begin{vmatrix} -2 & 15 & 13 \\ 5 & 1 & 6 \\ 1 & 10 & 11 \end{vmatrix} \quad \text{[by cofactor expansion along the fourth column]}$$

$$= (-4) \begin{vmatrix} 13 & 15 & 13 \\ 6 & 1 & 6 \\ 11 & 10 & 11 \end{vmatrix} \quad \text{[using } C_1 + C_2\text{]}$$

$$= 0 \quad \text{[since this last matrix has two identical columns]}$$

28.6 Caution

When using the cofactor expansion theorems, beginning students invariably forget to include the expressions of the form $(-1)^{i+j}$ that are involved in the definition of cofactor.

Problem Set 28

*28.1. The Cofactor Expansion Theorems can be used to prove the following important fact about determinants of certain partitioned matrices.

Theorem: Where A is a $k \times k$ matrix, the determinants of the partitioned $n \times n$ matrices

$$M = \left[\begin{array}{c|c} A & B \\ \hline 0_{(n-k)\times k} & D \end{array}\right] \quad \text{and} \quad N = \left[\begin{array}{c|c} A & 0_{k\times(n-k)} \\ \hline C & D \end{array}\right]$$

are both equal to $|A|\,|D|$.

Hint: Prove this by induction on k. Use cofactor expansion along the kth column (row) to evaluate $|M|$ ($|N|$). *Caution*: you must distinguish the cofactor of a_{ij} in A from the cofactor of a_{ij} in M!

*28.2. Use the preceding theorem to compute the determinant of the following matrix.

$$\begin{bmatrix} 2 & 5 & -3 & 77 & 0 \\ -3 & 2 & 2 & 5 & 55 \\ 4 & 0 & -4 & -8 & 8 \\ 0 & 0 & 0 & -5 & 6 \\ 0 & 0 & 0 & \frac{1}{2} & 1 \end{bmatrix}$$

28.3. Recompute the determinants of the matrices from Problem 24.3 using the Cofactor Expansion Theorems.

29

$A^{-1} = (1/|A|)\text{adj } A$

In this section, we present a formula for the inverse of an invertible matrix. This result is included for the sake of completeness; it will not be used subsequently.

We should stress from the beginning that the importance of this formula is theoretical rather than practical. For $n = 2$, this formula reduces to the familiar one obtained in Problem 3.6. For *computing* the inverse of a given matrix of size 3×3 or larger, the method from Section 14 should be used, as it is far more efficient.

Recall from Section 28 that the cofactor of a_{ij} is denoted by a_{ij}^c and is defined by $a_{ij}^c = (-1)^{i+j}|\hat{A}_{ij}|$.

29.1 Definition

Let A be an $n \times n$ matrix.

(i) The $n \times n$ matrix whose $(i, j)^{\text{th}}$ entry is a_{ij}^c is called the *matrix of cofactors of A* and is denoted by cof A or by A^c.

(ii) The transpose of the matrix of cofactors of A is called the *adjugate of A* and is denoted by adj A.

Thus $\text{adj } A = (A^c)^T$.

29.2 Remark on Terminology

The matrix $(A^c)^T$ is often called the *adjoint of A*, and the formula in the title of this section is then known as the *adjoint formula for the inverse*. But the word adjoint is also in common use for another (unrelated and more important) concept (see 50.11 and 50.12). Using this word for both concepts is confusing. Some authors distinguish them by calling $(A^c)^T$ the *classical adjoint of A*. A few recent textbooks avoid this confusion altogether by using some other name for $(A^c)^T$, beginning with the three letters "adj. . . ." It is not too late, historically, to attempt a change and it is hoped the name *adjugate* will stick. (I am following Strang, Whitelaw, and perhaps others in this usage.)

The theorem in the title gives a formula for the inverse of an invertible matrix. Specifically, if A is invertible, then its inverse is a certain scalar multiple of its adjugate; the scalar multiple involved is the reciprocal of $|A|$.

29.3 Theorem

If A is an invertible $n \times n$ matrix, then $A^{-1} = (1/|A|)\operatorname{adj} A$.

Proof: It suffices by Section 19 to prove that the product A adj A is equal to $|A|I_n$. But this follows immediately from the Cofactor Expansion Theorems:

$$(i, j)^{\text{th}} \text{ entry of } A \operatorname{adj} A = [i^{\text{th}} \text{ row of } A]\begin{bmatrix} j^{\text{th}} \\ \text{column} \\ \text{of adj } A \end{bmatrix}$$

$$= [a_{i1}a_{i2} \cdots a_{in}]\begin{bmatrix} a^c_{j1} \\ a^c_{j2} \\ \vdots \\ a^c_{jn} \end{bmatrix} \quad \text{[by Definition 29.1]}$$

$$= a_{i1}a^c_{j1} + a_{i2}a^c_{j2} + \cdots + a_{in}a^c_{jn}$$

$$= \begin{cases} |A|, & \text{if } i = j \\ 0, & \text{if } i \neq j \end{cases} \quad \text{[by Theorem 28.3]}$$

■

29.4 Examples

(i) Let $A = \begin{bmatrix} 2 & 3 \\ -1 & 5 \end{bmatrix}$. As we observed in Example 28.2(i),

$$A^c = \begin{bmatrix} 5 & -3 \\ 1 & 2 \end{bmatrix} \quad \text{so} \quad \operatorname{adj} A = (A^c)^T = \begin{bmatrix} 5 & 1 \\ -3 & 2 \end{bmatrix}$$

Now $|A| = (2)(5) - (3)(-1) = 10 + 3 = 13$, so, by Theorem 29.3,

$$A^{-1} = \frac{1}{13}\begin{bmatrix} 5 & -3 \\ 1 & 2 \end{bmatrix} = \begin{bmatrix} \frac{5}{13} & -\frac{3}{13} \\ \frac{1}{13} & \frac{2}{13} \end{bmatrix}$$

(ii) Let $B = \begin{bmatrix} 2 & -2 & 3 \\ 2 & 1 & -2 \\ -1 & 3 & 1 \end{bmatrix}$. In Example 28.2(ii), we computed the cofactors of the entries of B:

$$B^c = \begin{bmatrix} 7 & 0 & 7 \\ 11 & 5 & -4 \\ 1 & 10 & 6 \end{bmatrix} \quad \text{so} \quad \text{adj}\, B = (B^c)^T = \begin{bmatrix} 7 & 11 & 1 \\ 0 & 5 & 10 \\ 7 & -4 & 6 \end{bmatrix}$$

In Example 24.6(iii), we computed $|B| = 35$. So, by Theorem 29.3,

$$B^{-1} = \tfrac{1}{35}\begin{bmatrix} 7 & 11 & 1 \\ 0 & 5 & 10 \\ 7 & -4 & 6 \end{bmatrix} = \begin{bmatrix} \frac{1}{5} & \frac{11}{35} & \frac{1}{35} \\ 0 & \frac{1}{7} & \frac{2}{7} \\ \frac{1}{5} & -\frac{4}{35} & \frac{6}{35} \end{bmatrix}$$

These same matrices A and B were used in Examples 14.4 and 14.5, where A^{-1} and B^{-1} were computed by conventional means. You may wish to compare the relative amounts of work required.

Problem Set 29

29.1. **Cramer's Rule:** Let $A[j;\mathbf{b}]$ denote the matrix obtained from A by replacing the jth column of A by the column-vector $\mathbf{b}$. Prove that if A is an invertible $n \times n$ matrix then for every $\mathbf{b} \in \mathbb{R}^n$, the unique solution to the system $A\mathbf{x} = \mathbf{b}$ is the vector $\mathbf{x} = (x_1, x_2, \ldots, x_n)$ whose j^{th} coordinate is $\dfrac{|A[j;\mathbf{b}]|}{|A|}$.
Hint: Theorem 6.11 guarantees that the unique solution is $\mathbf{x} = A^{-1}\mathbf{b}$. The result follows when you evaluate the determinant $|A[j;\mathbf{b}]|$ by cofactor expansion along the jth column and invoke the fact that $A^{-1} = \dfrac{1}{|A|}\,\text{adj}\, A$.

29.2. Illustrate Cramer's rule by using it to redo Problems 12.6, 12.10, and 12.18.

*29.3. To solve a system of n equations in n unknowns when $n = 2$ or 3, Cramer's rule requires approximately the same amount of work as the Gauss–Jordan algorithm, although its use is still inferior because it is limited to the case when the coefficient matrix is invertible. When $n \geq 4$, it is hopelessly inefficient. Convince yourself of this once and for all by solving the following system twice, first using Cramer's rule and then using the Gauss–Jordan algorithm.

$$\begin{aligned} 2x_1 - x_2 + x_3 + x_4 &= 5 \\ 5x_1 + 3x_2 + 3x_3 + x_4 &= 3 \\ 3x_1 + x_2 + 3x_3 + x_4 &= 1 \\ x_1 - x_2 - 2x_3 - 2x_4 &= -2 \end{aligned}$$

*29.4. If A is an $n \times n$ matrix, then $|\text{adj A}| = (|\text{A}|)^{n-1}$.

*29.5. Let $\text{A} = \begin{bmatrix} 2 & 1 & -1 \\ -1 & 2 & 1 \\ 4 & 1 & 0 \end{bmatrix}$. Find adj A, A^{-1}, $|\text{A}|$, and $|2\text{A}^2 \text{ adj A}|$.

*29.6. If A is an invertible matrix all of whose entries are integers and if $|\text{A}| = 1$, then all the entries in A^{-1} are integers.

*29.7. Let $\text{A} \circ \text{B}$ denote the Hadamard product (defined in 2.4) of the $n \times n$ matrices A and B. Suppose that A is invertible. Show that the entries from any fixed row or column of $\text{A} \circ (\text{A}^{-1})^T$ add up to 1. *Hint*: Use Theorems 28.3 and 29.3.

30

Proofs of the Uniqueness of the Determinant Function and of the Cofactor Expansion Theorems

(Note that this entire section is optional.)

Because this section is the most difficult one encountered thus far, we begin with a brief survey of its contents. Let us denote by $\mathbb{R}_{n \times n}$ the set of all $n \times n$ matrices with entries from $\mathbb{R}$. The determinant, as defined in Section 23, should be viewed as a function, $\det:\mathbb{R}_{n \times n} \rightarrow \mathbb{R}$. We have proved that this function has several important properties, among which are the four that follow:

(i) $\det(\mathrm{EA}) = s \det(\mathrm{A})$, where E is the elementary matrix corresponding to the elementary row-operation sR_p, $s \neq 0$. [24.4]
(ii) $\det(\mathrm{C}) = \det(\mathrm{A}) + \det(\mathrm{B})$ if $\mathrm{C} = \mathrm{A} +_p \mathrm{B}$. [P24.4]
(iii) $\det(\mathrm{A}) = 0$ if A has two equal rows. [P24.2]
(iv) $\det(\mathrm{I}_n) = 1$. [23.7(i)]

Our proofs of the cofactor expansion theorems will proceed by first characterizing det, in Theorem 30.12, as the *unique* function from $\mathbb{R}_{n \times n} \rightarrow \mathbb{R}$ having the four properties just listed. Specifically, we will show that if $\Delta:\mathbb{R}_{n \times n} \rightarrow \mathbb{R}$ is *any* function satisfying conditions (i) to (iv), then $\Delta = \det$, that is, $\Delta(\mathrm{A}) = \det(\mathrm{A})$ for all $\mathrm{A} \in \mathbb{R}_{n \times n}$.

This conclusion will be arrived at gradually by showing first that Δ and det must agree on elementary matrices; we simultaneously show that Δ must also share with det the property from Theorem 24.4 that $\Delta(\mathrm{EA}) = \Delta(\mathrm{E})\Delta(\mathrm{A})$ for all $\mathrm{A} \in \mathbb{R}_{n \times n}$ and all elementary $\mathrm{E} \in \mathbb{R}_{n \times n}$; eventually, we will conclude that Δ and det agree on all matrices.

This characterization of the determinant function, which is important in its own right, will then be applied to prove the cofactor expansion theorems. We will view the various cofactor expansions as functions from $\mathbb{R}_{n\times n} \to \mathbb{R}$ and prove, in 30.19, that they all satisfy conditions (i) to (iv), and so conclude from Theorem 30.12 that they are all equal to det.

This finishes the survey and we continue, with a succession of short theorems and corollaries, to treat separately the three types of elementary matrices. Since these theorems all share the same hypothesis, which is fairly lengthy to state, we introduce the following definition as a convenient abbreviation.

30.1 Definition

A function $\Delta: \mathbb{R}_{n\times n} \to \mathbb{R}$ will be called a *pseudodeterminant* if it satisfies the following four conditions:

(i) $\Delta(\mathrm{EA}) = s\,\Delta(\mathrm{A})$, where E is the elementary matrix corresponding to the elementary row-operation sR_p, $s \neq 0$.
(ii) $\Delta(\mathrm{C}) = \Delta(\mathrm{A}) + \Delta(\mathrm{B})$ if $\mathrm{C} = \mathrm{A} +_p \mathrm{B}$.
(iii) $\Delta(\mathrm{A}) = 0$ if A has two equal rows.
(iv) $\Delta(\mathrm{I}_n) = 1$.

30.2 Theorem

If Δ is a pseudodeterminant, then $\Delta(\mathrm{E}) = s$ if E is the elementary matrix corresponding to the elementary row-operation sR_p, $s \neq 0$.

Proof:

$$\begin{aligned} \Delta(\mathrm{E}) &= \Delta(\mathrm{EI}_n) && [\text{property of } \mathrm{I}_n] \\ &= s\,\Delta(\mathrm{I}_n) && [\Delta \text{ satisfies (i)}] \\ &= (s)(1) && [\Delta \text{ satisfies (iv)}] \\ &= s && \blacksquare \end{aligned}$$

30.3 Corollary

If Δ is a pseudodeterminant, then $\Delta(\mathrm{EA}) = \Delta(\mathrm{E})\Delta(\mathrm{A})$ for any $\mathrm{A} \in \mathbb{R}_{n\times n}$ and any type I elementary matrix $\mathrm{E} \in \mathbb{R}_{n\times n}$.

Proof: Suppose E corresponds to the operation sR_p, $s \neq 0$. Then

$$\begin{aligned} \Delta(\mathrm{EA}) &= s\,\Delta(\mathrm{A}) && [\Delta \text{ satisfies (i)}] \\ &= \Delta(\mathrm{E})\Delta(\mathrm{A}) && [\text{Theorem 30.2}] \quad \blacksquare \end{aligned}$$

30.4 Theorem

If Δ is a pseudodeterminant, then $\Delta(\mathrm{EA}) = -\Delta(\mathrm{A})$ for any $\mathrm{A} \in \mathbb{R}_{n\times n}$ and any type II elementary matrix, $\mathrm{E} \in \mathbb{R}_{n\times n}$.

Proof: Suppose E is the matrix corresponding to the operation $R_p \leftrightarrow R_q$. Throughout the argument that follows, we will display only the pth and q^{th} rows of the matrices involved; for $i \neq p, q$, the ith row of all these matrices is the same as the ith row of A. Thus we have

$$A = \begin{bmatrix} a_{p1} & a_{p2} & \cdots & a_{pn} \\ a_{q1} & a_{q2} & \cdots & a_{qn} \end{bmatrix} \begin{matrix} \leftarrow p^{\text{th}} \text{ row} \rightarrow \\ \leftarrow q^{\text{th}} \text{ row} \rightarrow \end{matrix} \begin{bmatrix} a_{q1} & a_{q2} & \cdots & a_{qn} \\ a_{p1} & a_{p2} & \cdots & a_{pn} \end{bmatrix} = EA.$$

$$0 = \Delta\left(\begin{bmatrix} a_{p1} + a_{q1} & a_{p2} + a_{q2} & \cdots & a_{pn} + a_{qn} \\ a_{q1} + a_{p1} & a_{q2} + a_{p2} & \cdots & a_{qn} + a_{pn} \end{bmatrix}\right) \qquad [\Delta \text{ satisfies (iii)}]$$

$$= \Delta\left(\begin{bmatrix} a_{p1} & \cdots & a_{pn} \\ a_{q1} + a_{p1} & \cdots & a_{qn} + a_{pn} \end{bmatrix} +_p \begin{bmatrix} a_{q1} & \cdots & a_{qn} \\ a_{q1} + a_{p1} & \cdots & a_{qn} + a_{pn} \end{bmatrix}\right) \qquad [\text{definition of } +_p]$$

$$= \Delta\left(\begin{bmatrix} a_{p1} & \cdots & a_{pn} \\ a_{q1} + a_{p1} & \cdots & a_{qn} + a_{pn} \end{bmatrix}\right) + \Delta\left(\begin{bmatrix} a_{q1} & \cdots & a_{qn} \\ a_{q1} + a_{p1} & \cdots & a_{qn} + a_{pn} \end{bmatrix}\right) \qquad [\Delta \text{ satisfies (ii)}]$$

$$= \Delta\left(A +_q \begin{bmatrix} a_{p1} & \cdots & a_{pn} \\ a_{p1} & \cdots & a_{pn} \end{bmatrix}\right) + \Delta\left(\begin{bmatrix} a_{q1} & \cdots & a_{qn} \\ a_{q1} & \cdots & a_{qn} \end{bmatrix} +_q EA\right)$$

$[\text{definition of } +_q]$

$$= \Delta(A) + 0 + 0 + \Delta(EA) \qquad [\Delta \text{ satisfies (ii) and (iii)}] \quad \blacksquare$$

30.5 Corollary

If Δ is a pseudodeterminant, then $\Delta(\mathrm{E}) = -1$ for any type II elementary matrix $\mathrm{E} \in \mathbb{R}_{n\times n}$.

Proof:

$$\begin{aligned} \Delta(\mathrm{E}) &= \Delta(\mathrm{EI}_n) && \text{[property of } \mathrm{I}_n\text{]} \\ &= -\Delta(\mathrm{I}_n) && \text{[Theorem 30.4]} \\ &= -1 && [\Delta \text{ satisfies (iv)}] \quad \blacksquare \end{aligned}$$

30.6 Corollary

If Δ is a pseudodeterminant, then $\Delta(\mathrm{EA}) = \Delta(\mathrm{E})\Delta(\mathrm{A})$ for any $\mathrm{A} \in \mathbb{R}_{n\times n}$ and any type II elementary matrix $\mathrm{E} \in \mathbb{R}_{n\times n}$.

Proof:

$$\begin{aligned} \Delta(\mathrm{EA}) &= -\Delta(\mathrm{A}) && \text{[Theorem 30.4]} \\ &= \Delta(\mathrm{E})\Delta(\mathrm{A}) && \text{[Corollary 30.5]} \quad \blacksquare \end{aligned}$$

30.7 Theorem

If Δ is a pseudodeterminant, then $\Delta(\mathrm{EA}) = \Delta(\mathrm{A})$ for any $\mathrm{A} \in \mathbb{R}_{n\times n}$ and any type III elementary matrix $\mathrm{E} \in \mathbb{R}_{n\times n}$.

Proof: Suppose E corresponds to the operation $R_p + tR_q$. Define B and C to be the following matrices, which are equal to A except in the pth row.

$$\mathrm{B} = \begin{bmatrix} ta_{q1} & ta_{q2} & \cdots & ta_{qn} \\ a_{q1} & a_{q2} & \cdots & a_{qn} \end{bmatrix} \begin{matrix} \leftarrow p^{\text{th}} \text{ row} \rightarrow \\ \leftarrow q^{\text{th}} \text{ row} \rightarrow \end{matrix} \begin{bmatrix} a_{q1} & a_{q2} & \cdots & a_{qn} \\ a_{q1} & a_{q2} & \cdots & a_{qn} \end{bmatrix} = \mathrm{C}$$

Then $\mathrm{EA} = \mathrm{A} +_p \mathrm{B}$. Hence

$$\begin{aligned} \Delta(\mathrm{EA}) &= \Delta(\mathrm{A} +_p \mathrm{B}) \\ &= \Delta(\mathrm{A}) + \Delta(\mathrm{B}) && [\Delta \text{ satisfies (ii)}] \\ &= \Delta(\mathrm{A}) + t\,\Delta(\mathrm{C}) && \text{[Corollary 30.3]} \\ &= \Delta(\mathrm{A}) + (t)(0) && [\Delta \text{ satisfies (iii)}] \\ &= \Delta(\mathrm{A}) && \blacksquare \end{aligned}$$

30.8 Corollary

If Δ is a pseudodeterminant, then $\Delta(E) = 1$ for any type III elementary matrix $E \in \mathbb{R}_{n \times n}$.

Proof:

$$\begin{aligned} \Delta(E) &= \Delta(EI_n) && \text{[property of } I_n\text{]} \\ &= \Delta(I_n) && \text{[Theorem 30.7]} \\ &= 1 && [\Delta \text{ satisfies (iv)}] \quad \blacksquare \end{aligned}$$

30.9 Corollary

If Δ is a pseudodeterminant, then $\Delta(EA) = \Delta(E)\Delta(A)$ for any $A \in \mathbb{R}_{n \times n}$ and any type III elementary matrix $E \in \mathbb{R}_{n \times n}$.

Proof:

$$\begin{aligned} \Delta(EA) &= \Delta(A) && \text{[Theorem 30.7]} \\ &= \Delta(E)\Delta(A) && \text{[Corollary 30.8]} \quad \blacksquare \end{aligned}$$

For convenience, we consolidate the preceding chain of results into the following theorem.

30.10 Theorem

If Δ is a pseudodeterminant, then

(i) $\Delta(E) = \det(E)$ for any elementary matrix $E \in \mathbb{R}_{n \times n}$.
(ii) $\Delta(EA) = \Delta(E)\Delta(A)$ for any $A \in \mathbb{R}_{n \times n}$ and any elementary matrix $E \in \mathbb{R}_{n \times n}$. ■

We are now almost in position to prove the main result that if Δ is a pseudodeterminant then $\Delta = \det$. Before proceeding, we need one more property of pseudodeterminants.

30.11 Theorem

If Δ is a pseudodeterminant and if $A \in \mathbb{R}_{n \times n}$ has an all-zero row, then $\Delta(A) = 0$.

Proof: Suppose the pth row of A is all-zero. Let $s \neq 0$; then $A = A +_p EA$, where E is the elementary matrix corresponding to the elementary row-operation sR_p. Thus

$$\begin{aligned}\Delta(A) &= \Delta(A +_p EA) \\ &= \Delta(A) + \Delta(EA) && [\Delta \text{ satisfies (ii)}] \\ &= \Delta(A) + s\Delta(A) && [\Delta \text{ satisfies (i)}]\end{aligned}$$

Hence, $0 = s\Delta(A)$, and since $s \neq 0$, $\Delta(A) = 0$. ■

At last, we obtain the theorem which shows that there is really only one pseudodeterminant, that is, the function det itself.

30.12 Theorem (Uniqueness of the Determinant Function)

If Δ is a pseudodeterminant, then $\Delta(A) = \det(A)$ for all $A \in \mathbb{R}_{n\times n}$.

Proof:

Case 1: If A is invertible, then A is expressible as a product of elementary matrices. If $A = E_1E_2 \cdots E_k$, then

$$\begin{aligned}\Delta(A) &= \Delta(E_1E_2 \cdots E_k) \\ &= \Delta(E_1)\Delta(E_2) \cdots \Delta(E_k) && [\text{repeated application of Theorem 30.10(ii)}] \\ &= \det(E_1)\det(E_2) \cdots \det(E_k) && [\text{repeated application of Theorem 30.10(i)}] \\ &= \det(E_1E_2 \cdots E_k) && [\text{repeated application of Theorem 24.4}] \\ &= \det(A)\end{aligned}$$

Case 2: If A is singular, then, as in the proof of 14.1, GJ(A) has an all-zero row. Let $E_1, E_2, \ldots, E_k$ be elementary matrices satisfying $E_k \cdots E_2E_1A = GJ(A)$. Then

$$\begin{aligned}0 &= \Delta(GJ(A)) && [\text{Theorem 30.11}] \\ &= \Delta(E_k \cdots E_2E_1A) \\ &= \Delta(E_k) \cdots \Delta(E_1)\Delta(A) && [\text{repeated application of Theorem 30.10(ii)}]\end{aligned}$$

Hence

$$\begin{aligned}\Delta(A) &= 0 && [\Delta(E_i) = \det(E_i) \neq 0 \text{ by 30.10(i) and 24.1}] \\ &= \det(A) && [\text{Theorem 25.1}]\end{aligned}$$ ■

See [F]30.1.

We turn now to the proofs of the cofactor expansion theorems, which we will restate here.

30.13 Theorem (Cofactor Expansion Theorems)

Assume $A \in \mathbb{R}_{n\times n}$.

(i) For each integer i between 1 and n,

$$a_{i1}a^c_{i1} + a_{i2}a^c_{i2} + \cdots + a_{in}a^c_{in} = \det(A)$$

(ii) For each pair of distinct integers $i \neq p$ between 1 and n,

$$a_{i1}a^c_{p1} + a_{i2}a^c_{p2} + \cdots + a_{in}a^c_{pn} = 0$$

(iii) For each integer j between 1 and n,

$$a_{1j}a^c_{1j} + a_{2j}a^c_{2j} + \cdots + a_{nj}a^c_{nj} = \det(A)$$

(iv) For each pair of distinct integers $j \neq k$ between 1 and n,

$$a_{1j}a^c_{1k} + a_{2j}a^c_{2k} + \cdots + a_{nj}a^c_{nk} = 0$$

We will begin with the proof of part (iii), cofactor expansion along the jth column (this will be lengthy), and then use (iii) to provide short proofs of (i), (ii), and (iv). To prove (iii), it suffices, by Theorem 30.12, to prove for arbitrary j that the function Δ_j is a pseudodeterminant, where $\Delta_j: \mathbb{R}_{n\times n} \to \mathbb{R}$ is defined by $\Delta_j(A) = a_{1j}a^c_{1j} + a_{2j}a^c_{2j} + \cdots + a_{nj}a^c_{nj}$.

30.14 Proof That $\Delta_j(EA) = s\Delta_j(A)$

Suppose E is the elementary matrix corresponding to the elementary row operation sR_p, $s \neq 0$. Let

$$B = EA = \begin{bmatrix} a_{11} & a_{12} & \cdots & a_{1n} \\ \vdots & \vdots & \vdots & \vdots \\ sa_{p1} & sa_{p2} & \cdots & sa_{pn} \\ \vdots & \vdots & \vdots & \vdots \\ a_{n1} & a_{n2} & \cdots & a_{nn} \end{bmatrix}$$

Thus, by definition, for $i = 1, 2, \ldots, n$,

$$b_{ij} = \begin{cases} a_{ij}, & \text{if } i \neq p \\ sa_{pj}, & \text{if } i = p \end{cases}$$

Moreover, because cofactors are *defined* in terms of the function det, we know from Theorem 24.4 that for $i = 1, 2, \ldots, n$,

$$b_{ij}^c = \begin{cases} sa_{ij}^c, & \text{if } i \neq p \\ a_{pj}^c, & \text{if } i = p \end{cases}$$

Thus

$$\begin{aligned} \Delta_j(\mathrm{EA}) &= b_{1j}b_{1j}^c + \cdots + b_{pj}b_{pj}^c + \cdots + b_{nj}b_{nj}^c \\ &= a_{1j}sa_{1j}^c + \cdots + sa_{pj}a_{pj}^c + \cdots + a_{nj}sa_{nj}^c \\ &= s\,\Delta_j(\mathrm{A}) \end{aligned}$$

■

30.15 Proof That $\Delta_j(C) = \Delta_j(A) + \Delta_j(B)$ if $C = A +_p B$

This time, for $i = 1, 2, \ldots, n$, we have, by the definition of $+_p$, that

$$c_{ij} = \begin{cases} a_{ij} = b_{ij}, & \text{if } i \neq p \\ a_{pj} + b_{pj}, & \text{if } i = p \end{cases}$$

Moreover, by Theorem 24.4, we have that

$$c_{ij}^c = \begin{cases} a_{ij}^c + b_{ij}^c, & \text{if } i \neq p \\ a_{pj}^c = b_{pj}^c, & \text{if } i = p \end{cases}$$

Thus

$$\begin{aligned} \Delta_j(C) &= c_{1j}c_{1j}^c + \cdots + c_{pj}c_{pj}^c + \cdots + c_{nj}c_{nj}^c \\ &= a_{1j}(a_{1j}^c + b_{1j}^c) + \cdots + (a_{pj} + b_{pj})a_{pj}^c + \cdots + a_{nj}(a_{nj}^c + b_{nj}^c) \\ &= (a_{1j}a_{1j}^c + \cdots + a_{pj}a_{pj}^c + \cdots + a_{nj}a_{nj}^c) \\ &\quad + (b_{1j}b_{1j}^c + \cdots + b_{pj}b_{pj}^c + \cdots + b_{nj}b_{nj}^c) \\ &= \Delta_j(\mathrm{A}) + \Delta_j(\mathrm{B}) \end{aligned}$$

■

Before showing that Δ_j has the third property in the definition of a pseudodeterminant, we pause to present a lemma concerning minors that will be needed in that argument.

30.16 Lemma

Suppose that the matrix A has two identical rows, row i and row p, where $i \lneq p$; then $a_{ij}^m = (-1)^{p-i-1} a_{pj}^m$.

Proof: Recall that $a_{ij}^m = \det(\hat{A}_{ij})$ and that $a_{pj}^m = \det(\hat{A}_{pj})$.

$$A = \begin{bmatrix} a_{i1} & a_{i2} & \cdots & a_{in} \\ & & & \\ & & & \\ a_{i1} & a_{i2} & \cdots & a_{in} \end{bmatrix} \begin{matrix} \leftarrow \text{row } i \\ \text{there are } (p-i-1)\text{-many rows} \\ \text{separating row } i \text{ and row } p \\ \leftarrow \text{row } p \end{matrix}$$

The conclusion of the lemma follows by observing how the matrices $\hat{A}_{ij}$ and $\hat{A}_{pj}$ are related: $\hat{A}_{pj}$ can be obtained from $\hat{A}_{ij}$ by a succession of $(p-i-1)$-many interchanges of adjacent rows. Thus, $a_{ij}^m = (-1)^{p-i-1} a_{pj}^m$ by repeated application of Theorem 24.5(ii). ■

30.17 Proof That $\Delta_j(A) = 0$ if A Has Two Equal Rows

Suppose that row i and row p are equal, where $i \lneq p$. Note that $a_{qj}^c = 0$ for $q \neq i, p$, by Problem 24.2, because the $(n-1) \times (n-1)$ matrix $\hat{A}_{qj}$ has two identical rows. Thus

$$\begin{aligned}
\Delta_j(A) &= a_{1j}a_{1j}^c + \cdots + a_{ij}a_{1j}^c + \cdots + a_{pj}a_{pj}^c + \cdots + a_{nj}a_{nj}^c \\
&= a_{ij}a_{ij}^c + a_{pj}a_{pj}^c \\
&= a_{ij}\left(a_{ij}^c + a_{pj}^c\right) && [\text{because } a_{ij} = a_{pj}] \\
&= a_{ij}\left[(-1)^{i+j}a_{ij}^m + (-1)^{p+j}a_{pj}^m\right] && [\text{definition of minors and cofactors}] \\
&= a_{ij}\left[(-1)^{i+j}(-1)^{p-i-1}a_{pj}^m + (-1)^{p+j}a_{pj}^m\right] && [\text{Lemma 30.16}] \\
&= a_{ij}\left[(-1)^{p+j-1}a_{pj}^m + (-1)^{p+j}a_{pj}^m\right] \\
&= 0 && [\text{because } p+j-1 \text{ and } p+j \text{ have opposite parity}]
\end{aligned}$$ ■

30.18 Proof That $\Delta_j(\mathrm{A}) = 1$ if $\mathrm{A} = \mathrm{I}_n$

$$\begin{aligned}\Delta_j(\mathrm{A}) &= a_{1j}a^c_{1j} + \cdots + a_{jj}a^c_{jj} + \cdots + a_{nj}a^c_{nj} \\ &= (0)(a^c_{1j}) + \cdots + (1)(-1)^{j+j}\det(\mathrm{I}_{n-1}) + \cdots + (0)(a^c_{nj}) \\ &= 1\end{aligned}$$ ■

30.19 Proof of Theorem 30.13(iii)

Combining items 30.14 through 30.18 provides a proof that Δ_j is a pseudodeterminant and hence, by the uniqueness of the determinant, that $\Delta_j(\mathrm{A}) = \det(\mathrm{A})$ for all $\mathrm{A} \in \mathbb{R}_{n\times n}$. ■

After this considerable effort, it is a relief to be able to exploit what we have just proved to polish off the remaining cofactor expansion theorems rather easily.

30.20 Proof of Theorem 30.13(iv)

We wish to show that the sum of the products of the pairs obtained by pairing the entries in a given column with the corresponding cofactors from a different column is equal to zero. So suppose $j \neq k$, and let B be the matrix obtained from A by deleting column k and replacing it with a duplicate of column j. Thus

$$\mathrm{A} = \begin{bmatrix} \cdots & a_{1j} & \cdots & a_{1k} & \cdots \\ \cdots & a_{2j} & \cdots & a_{2k} & \cdots \\ & \vdots & & \vdots & \\ \cdots & a_{nj} & \cdots & a_{nk} & \cdots \end{bmatrix},$$

$$\mathrm{B} = \begin{bmatrix} \cdots & a_{1j} & \cdots & a_{1j} & \cdots \\ \cdots & a_{2j} & \cdots & a_{2j} & \cdots \\ & \vdots & & \vdots & \\ \cdots & a_{nj} & \cdots & a_{nj} & \cdots \end{bmatrix}$$

Note that for $p = 1, 2, \ldots, n$, $a_{pj} = b_{pk}$ and that $a^c_{pk} = b^c_{pk}$. Thus

$$\begin{aligned}&a_{1j}a^c_{1k} + a_{2j}a^c_{2k} + \cdots + a_{nj}a^c_{nk} \\ &\quad = b_{1k}b^c_{1k} + b_{2k}b^c_{2k} + \cdots + b_{nk}b^c_{nk} \\ &\quad = \det(\mathrm{B}) \qquad \text{[Theorem 30.13(iii)]} \\ &\quad = 0 \qquad \text{[Problem 27.4]}\end{aligned}$$ ■

30.21 Exercise

Prove the cofactor expansion theorems for rows [Theorems 30.13(i) and (ii)] by applying the cofactor expansion theorems for columns [Theorems 30.13(iii) and (iv)] to the matrix A^T.

Problem Set 30

30.1. In more advanced texts, the theorem on the uniqueness of the determinant function is usually cast in different language: "det is the unique alternating n-linear functional on $\mathbb{R}_{n \times n}$ whose value on the identity matrix is 1." The proof that we have given can be reexpressed in these terms. Consult such a treatment (e.g., in [HK] or [FIS]) and redo the proof from this point of view.

31

Representing Linear Transformations from $\mathbb{R}^n$ to $\mathbb{R}^m$ by Matrices

The time has finally arrived to explain the remark that preceded Example 17.8: "The fact is that *every* linear transformation $T:\mathbb{R}^n \to \mathbb{R}^m$ can be 'represented' as a matrix transformation T_A for some $m \times n$ matrix A. The problem is to explain what is meant by the word 'represent' in this context. It turns out that a given linear transformation $T:\mathbb{R}^n \to \mathbb{R}^m$ can be 'represented' by infinitely many 'different' matrix transformations."

A linear transformation, as we have seen, is just a special kind of function. In general, a function $f: X \to Y$ should be thought of as a machine which, when presented with an arbitrary element, x, from its domain, X, outputs a determined element, $f(x)$, of its codomain, Y. It is natural to ask whether the behavior of a given function (machine) can be described in numerical terms (i.e., using "numbers" in some way). (The word "numbers" is in quotes because we wish deliberately, in the discussion that follows, to keep its meaning vague.) If a numerical description is to be possible, there are three conditions to meet:

(i) We need a way to associate "numbers" to points $x \in X$.
(ii) We need a way to associate "numbers" to points $y \in Y$.
(iii) We need a "numerical way" to explain the general relationship between the "numbers" that are associated to x by (i) and to $f(x)$ by (ii).

[Ideally, the associations in (i) and (ii) are one-to-one and onto; that is, each point is associated with a unique number, and vice versa.]

Throughout high school and in freshman calculus, we naturally overlook conditions (i) and (ii) because the functions we deal with have sets of real numbers for their domains and codomains. Conditions (i) and (ii) are already met in an obvious and natural way.

The mathematics student's first encounter with condition (i) is usually provided by the subject of polar coordinates. (The next four numbered items may be omitted by readers unfamiliar with this subject.)

31.1 Example

Consider the function $f\colon \mathbb{R}^2 \to \mathbb{R}$ whose standard description, using Cartesian coordinates (x, y) to label the points of the domain $\mathbb{R}^2$, is $f(x, y) = (x^2 + y^2)^{1/2}$.

An alternate way of associating "numbers" to points in $\mathbb{R}^2$ is to use polar coordinates. The problem of uniqueness does arise with this labeling since (r, θ), $(r, \theta + 2\pi)$, and $(-r, \theta + \pi)$ denote the same point; there are in fact infinitely many labels

$$\{(r, \theta \pm 2k\pi) : k = 0, 1, 2, \ldots\}$$

and

$$\{(-r, \theta \pm (2k+1)\pi) : k = 0, 1, 2, \ldots\}$$

all associated with the same point. This aspect of the problem can be essentially overcome by restricting r and θ to certain ranges, for example, $r \geqq 0$ and $0 \leqq \theta \lneqq 2\pi$.

Then the expression $g(r, \theta) = r$ is simply an alternate description of the function f described previously.

31.2 Example

The function $f\colon \mathbb{R}^2 \to \mathbb{R}$ whose description, using Cartesian coordinates to label the points of $\mathbb{R}^2$, is $f(x, y) = x^2 - 8xy + y^2$ can also be described, using polar coordinates to label the points of $\mathbb{R}^2$, by $g(r, \theta) = r^2(1 - 4\sin 2\theta)$.

31.3 Exercise

The claim was just made in Example 31.2 that the two expressions $f(x, y) = x^2 - 8xy + y^2$ and $g(r, \theta) = r^2(1 - 4\sin 2\theta)$ represent the same function from $\mathbb{R}^2$ into $\mathbb{R}$.

(i) Verify that the point in $\mathbb{R}^2$ with Cartesian coordinates $(-4, 3)$ has polar coordinates $\left(5, \arctan\left(-\frac{3}{4}\right)\right)$. Illustrate the claim by computing $f(-4, 3)$ and $g\left(5, \arctan\left(-\frac{3}{4}\right)\right)$.

(ii) Verify that the point in $\mathbb{R}^2$ with polar coordinates $(4, \pi/3)$ has Cartesian coordinates $(2, 2\sqrt{3})$. Illustrate the claim by computing $f(2, 2\sqrt{3})$ and $g(4, \pi/3)$.

31.4 Exercise

Prove the claim made in 31.2.

The purpose of these examples is to convince you that a given function $f: X \to Y$ can be "represented" in "different ways" and that the differences among the descriptions arise simply from the use of different ways of labeling the points of X and Y with "numbers."

Let us consider, at last, the case of a linear transformation $T:\mathbb{R}^n \to \mathbb{R}^m$. Because there is already an obvious and natural way to associate "numbers" to points in $\mathbb{R}^n$ and $\mathbb{R}^m$, we do not initially give much thought to conditions (i) and (ii) raised at the beginning of this section. The natural way is this: to the point $\mathbf{x} \in \mathbb{R}^n$, we associate the n-tuple of real numbers $(x_1, x_2, \ldots, x_n) = (\mathbf{x})_{\alpha^n}$ consisting of the coordinates of $\mathbf{x}$ with respect to the standard basis, α^n, for $\mathbb{R}^n$. When we do this for $\mathbb{R}^n$ and the analogous thing for $\mathbb{R}^m$, we can *then* ask whether the linear transformation $T:\mathbb{R}^n \to \mathbb{R}^m$ can be represented as a matrix transformation. We interpret the word represent in this context as follows: is there an $m \times n$ matrix A such that for all $\mathbf{x} \in \mathbb{R}^n$, $[T(\mathbf{x})]_{\alpha^m} = \mathrm{A}[\mathbf{x}]_{\alpha^n}$, that is, such that whenever the standard coordinate matrix, $[\mathbf{x}]_{\alpha^n}$, of a vector in the domain of T is multiplied on the left by the $m \times n$ matrix A, the result is the standard coordinate matrix of the image, $T(\mathbf{x})$, of $\mathbf{x}$ under T?

Surely, if this happens, it would be legitimate to say that the matrix transformation $T_{\mathrm{A}}:\mathbb{R}^n \to \mathbb{R}^m$ is an accurate representation of the linear transformation $T:\mathbb{R}^n \to \mathbb{R}^m$. What needs to be stressed is that this representation depends on how points in $\mathbb{R}^n$ and $\mathbb{R}^m$ are labeled.

31.5 Example

When we encounter a phrase such as

"let $T: \mathbb{R}^5 \to \mathbb{R}^3$ be the linear transformation defined by
$T(x_1, x_2, x_3, x_4, x_5) = (x_1 - x_2 + 2x_3, x_2 + x_3, x_1 - x_4 + 3x_5)$,"

it is implicit that points are being labeled with their standard coordinates, that is, that $(x_1, x_2, x_3, x_4, x_5) = (\mathbf{x})_{\alpha^5}$ and that $(x_1 - x_2 + 2x_3, x_2 + x_3, x_1 - x_4 + 3x_5) = (T(\mathbf{x}))_{\alpha^3}$.

Observe, for this example, that the 3×5 matrix

$$\mathrm{A} = \begin{bmatrix} 1 & -1 & 2 & 0 & 0 \\ 0 & 1 & 1 & 0 & 0 \\ 1 & 0 & 0 & -1 & 3 \end{bmatrix}$$

represents T with respect to the standard bases α^5 for the domain and α^3 for the codomain, since for arbitrary $\mathbf{x} \in \mathbb{R}^5$

$$\mathrm{A}[\mathbf{x}]_{\alpha^5} = \begin{bmatrix} 1 & -1 & 2 & 0 & 0 \\ 0 & 1 & 1 & 0 & 0 \\ 1 & 0 & 0 & -1 & 3 \end{bmatrix} \begin{bmatrix} x_1 \\ x_2 \\ x_3 \\ x_4 \\ x_5 \end{bmatrix} = \begin{bmatrix} x_1 - x_2 + 2x_3 \\ x_2 + x_3 \\ x_1 - x_4 + 3x_5 \end{bmatrix} = [T(\mathbf{x})]_{\alpha^3}$$

The general procedure for finding such a matrix will be explained in Computational Procedure 31.9.

Let us return to the general case of a linear transformation $T:\mathbb{R}^n \to \mathbb{R}^m$. There are other ways, in addition to the standard one, of associating numbers to points of the domain and codomain of T. In addition to α^n, there are many other ordered bases (infinitely many, in fact) for $\mathbb{R}^n$. Having chosen an ordered basis, β, for $\mathbb{R}^n$, we can associate to each point $\mathbf{x} \in \mathbb{R}^n$ the n-tuple of real numbers $(t_1, t_2, \ldots, t_n) = (\mathbf{x})_\beta$ consisting of the β-coordinates of $\mathbf{x}$. This way of associating numbers to points does not suffer from the lack of uniqueness that we encountered with polar coordinates. Recall Theorem 8.11: if $\beta = (\mathbf{u}_1, \mathbf{u}_2, \ldots, \mathbf{u}_n)$ is an ordered basis for $\mathbb{R}^n$, then every vector $\mathbf{x} \in \mathbb{R}^n$ can be uniquely expressed in the form $\mathbf{x} = t_1\mathbf{u}_1 + t_2\mathbf{u}_2 + \cdots + t_n\mathbf{u}_n$. We will defer to Section 32 the question of *why* we would even consider using coordinate systems other than the standard ones.

31.6 Definition

An $m \times n$ matrix A represents the linear transformation T: $\mathbb{R}^n \to \mathbb{R}^m$ with respect to the ordered bases β for $\mathbb{R}^n$ and γ for $\mathbb{R}^m$ if and only if

$$\text{for all } \mathbf{x} \in \mathbb{R}^n, \quad [T(\mathbf{x})]_\gamma = \mathrm{A}[\mathbf{x}]_\beta \qquad (*)$$

This condition says essentially that T *is* T_{A}, provided points in the domain, $\mathbb{R}^n$, are labeled by their β-coordinates and points in the codomain, $\mathbb{R}^m$, are labeled by their γ-coordinates.

31.7 Theorem

For every linear transformation T: $\mathbb{R}^n \to \mathbb{R}^m$ and for every pair of ordered bases $\beta = (\mathbf{u}_1, \mathbf{u}_2, \ldots, \mathbf{u}_n)$ for $\mathbb{R}^n$ and $\gamma = (\mathbf{v}_1, \mathbf{v}_2, \ldots, \mathbf{v}_m)$ for $\mathbb{R}^m$, there is a unique $m \times n$ matrix A satisfying

$$\text{for all } \mathbf{x} \in \mathbb{R}^n, \quad [T(\mathbf{x})]_\gamma = \mathrm{A}[\mathbf{x}]_\beta \qquad (*)$$

Proof: $(*)$ is an assertion about all $\mathbf{x} \in \mathbb{R}^n$. If it is to hold, it must hold in particular for each of the vectors $\mathbf{u}_1, \mathbf{u}_2, \ldots, \mathbf{u}_n$. Specifically, if $(*)$ is to hold for $\mathbf{u}_j$, we must have

$$[T(\mathbf{u}_j)]_\gamma = \mathrm{A}[\mathbf{u}_j]_\beta = \mathrm{A}\begin{bmatrix} 0 \\ \vdots \\ 0 \\ 1 \\ 0 \\ \vdots \\ 0 \end{bmatrix} = \begin{bmatrix} j^{\text{th}} \\ \text{column} \\ \text{of A} \end{bmatrix}$$

since $[\mathbf{u}_j]_\beta = [\mathbf{e}_j^n]$. The proof given thus far establishes uniqueness: it shows that the only possible matrix that could satisfy $(*)$ is the $m \times n$ matrix

$$\mathrm{A} = \left[[T(\mathbf{u}_1)]_\gamma \;\vdots\; \cdots \;\vdots\; [T(\mathbf{u}_j)]_\gamma \;\vdots\; \cdots \;\vdots\; [T(\mathbf{u}_m)]_\gamma \right]$$

having $[T(\mathbf{u}_j)]_\gamma$ as its jth column for $j = 1, 2, \ldots, n$.

To conclude the proof, we must verify that this matrix does in fact satisfy $(*)$. So let $\mathbf{x} \in \mathbb{R}^n$ be given and suppose $(\mathbf{x})_\beta = (t_1, t_2, \ldots, t_n)$. Then

$$\mathrm{A}[\mathbf{x}]_\beta = \begin{bmatrix} a_{11} & \cdots & a_{1j} & \cdots & a_{1n} \\ a_{21} & \cdots & a_{2j} & \cdots & a_{2n} \\ \vdots & & \vdots & & \vdots \\ a_{m1} & \cdots & a_{mj} & \cdots & a_{mn} \end{bmatrix} \begin{bmatrix} t_1 \\ t_2 \\ \vdots \\ t_n \end{bmatrix} = \begin{bmatrix} a_{11}t_1 + a_{12}t_2 + \cdots + a_{1n}t_n \\ a_{21}t_1 + a_{22}t_2 + \cdots + a_{2n}t_n \\ \vdots \\ a_{m1}t_1 + a_{m2}t_2 + \cdots + a_{mn}t_n \end{bmatrix}$$

Is this result equal to $[T(\mathbf{x})]_\gamma$? Yes, since

$$\begin{aligned} T(\mathbf{x}) &= T(t_1\mathbf{u}_1 + t_2\mathbf{u}_2 + \cdots + t_n\mathbf{u}_n) \\ &= t_1T(\mathbf{u}_1) + t_2T(\mathbf{u}_2) + \cdots + t_nT(\mathbf{u}_n) && \text{[by linearity of } T\text{]} \\ &= t_1(a_{11}\mathbf{v}_1 + a_{21}\mathbf{v}_2 + \cdots + a_{m1}\mathbf{v}_m) \\ &\quad \vdots \\ &\quad + t_j(a_{1j}\mathbf{v}_1 + a_{2j}\mathbf{v}_2 + \cdots + a_{mj}\mathbf{v}_m) && \left[\text{because the } j^{\text{th}} \text{ column of A is } [T(\mathbf{u}_j)]_\gamma\right] \\ &\quad \vdots \\ &\quad + t_n(a_{1n}\mathbf{v}_1 + a_{2n}\mathbf{v}_2 + \cdots + a_{mn}\mathbf{v}_m) \\ &= (a_{11}t_1 + a_{12}t_2 + \cdots + a_{1n}t_n)\mathbf{v}_1 && \text{[by regrouping the terms]} \\ &\quad + (a_{21}t_1 + a_{22}t_2 + \cdots + a_{2n}t_n)\mathbf{v}_2 \\ &\quad \vdots \\ &\quad + (a_{m1}t_1 + a_{m2}t_2 + \cdots + a_{mn}t_n)\mathbf{v}_m \end{aligned}$$

■

It is Theorem 31.7 which guarantees that the following definition makes sense.

31.8 Definition

For a linear transformation $T: \mathbb{R}^n \to \mathbb{R}^m$ and ordered bases β for $\mathbb{R}^n$ and γ for $\mathbb{R}^m$, the unique $m \times n$ matrix satisfying $(*)$ from Definition 31.6 is called the *matrix that represents T with respect to β and γ* and is denoted by ${}_\gamma[T]_\beta$ (this symbol is to be read as T-beta-gamma). The matrix ${}_{\alpha^m}[T]_{\alpha^n}$ is called the *standard matrix for T*.

The proof of Theorem 31.7 reveals what the columns of ${}_\gamma[T]_\beta$ *must* be; this justifies the following computational procedure.

31.9 Computational Procedure

Given a linear transformation $T: \mathbb{R}^n \to \mathbb{R}^m$ and ordered bases $\beta = (\mathbf{u}_1, \mathbf{u}_2, \ldots, \mathbf{u}_n)$ for $\mathbb{R}^n$ and $\gamma = (\mathbf{v}_1, \mathbf{v}_2, \ldots, \mathbf{v}_m)$ for $\mathbb{R}^m$, to compute ${}_\gamma[T]_\beta$: form the $m \times n$ matrix whose j^{th} column, for $j = 1, 2, \ldots, n$, is $[T(\mathbf{u}_j)]_\gamma$.

31.10 Example

Find the standard matrix for the linear transformation $T: \mathbb{R}^2 \to \mathbb{R}^3$ defined by $T(x_1, x_2) = (3x_1, 2x_1 - x_2, 3x_1 + 2x_2)$.

Solution: ${}_{\alpha^3}[T]_{\alpha^2}$ is the matrix whose columns are the standard coordinates of the images under T of the standard basis vectors in α^2. Since $T(1,0) = (3,2,3)$ and $T(0,1) = (0,-1,2)$, the desired matrix is ${}_{\alpha^3}[T]_{\alpha^2}$ $= \begin{bmatrix} 3 & 0 \\ 2 & -1 \\ 3 & 2 \end{bmatrix}$.

31.11 Example

Where T is the same linear transformation as in the previous example, find ${}_{\alpha^3}[T]_\beta$ when $\beta = ((2,1),(-1,4))$.

Solution: ${}_{\alpha^3}[T]_\beta$ is the matrix whose columns are the α^3-coordinates of the images under T of the basis vectors $(2,1)$ and $(-1,4)$ from β. Since $T(2,1) = (6,3,8)$ and $T(-1,4) = (-3,-6,5)$, ${}_{\alpha^3}[T]_\beta = \begin{bmatrix} 6 & -3 \\ 3 & -6 \\ 8 & 5 \end{bmatrix}$.

31.12 Example

Consider once more the linear transformation T given by $T(x_1, x_2) = (3x_1, 2x_1 - x_2, 3x_1 + 2x_2)$. Let $\beta = ((1,0),(1,1))$ and let $\gamma = ((0,0,1),(1,0,1),(0,2,1))$. Find ${}_{\gamma}[T]_{\beta}$.

Solution: ${}_{\gamma}[T]_{\beta}$ is the matrix whose columns are the γ-coordinates of the images under T of the vectors in β. So we begin by computing $T(1,0) = (3,2,3)$ and $T(1,1) = (3,1,5)$. It is important to realize that $\begin{bmatrix} 3 & 3 \\ 2 & 1 \\ 3 & 5 \end{bmatrix}$ is *not* the answer to the question. This is the matrix whose columns are the *standard* coordinates of $T(1,0)$ and $T(1,1)$; thus $\begin{bmatrix} 3 & 3 \\ 2 & 1 \\ 3 & 5 \end{bmatrix} = {}_{\alpha^3}[T]_{\beta}$. To find ${}_{\gamma}[T]_{\beta}$, we must compute the γ-coordinates of the vectors $(3,2,3)$ and $(3,1,5)$. This could be done by brute-force methods, which consist in setting up and solving an appropriate system of equations. But we have, virtually at hand, a very efficient approach to this latter problem. So let us postpone the answer to 31.12 in order to present this approach.

There is a very important special case of the ideas presented thus far in Section 31; this concerns the identity transformation $I: \mathbb{R}^n \to \mathbb{R}^n$ defined, for all $\mathbf{x} \in \mathbb{R}^n$, by $I(\mathbf{x}) = \mathbf{x}$. When we use 31.9 to compute the standard matrix for I, the result is, not surprisingly, the $n \times n$ identity matrix, I_n.

Although it may not seem natural when $T: \mathbb{R}^n \to \mathbb{R}^n$ (i.e., when T is a linear transformation whose codomain is the *same* as its domain), to represent T using *different* bases β for the domain of T and γ for the codomain of T, the concept of ${}_{\gamma}[T]_{\beta}$ is meaningful. Indeed, we will write ${}_{\beta}[T]_{\beta}$, with the subscript β in *both* positions, to denote the $n \times n$ matrix that represents T when the same basis, β, is used for both the domain and codomain.

31.13 Exercise

Verify that if β is *any* ordered basis for $\mathbb{R}^n$, then ${}_{\beta}[I]_{\beta} = \mathrm{I}_n$.

Given two bases β and γ for $\mathbb{R}^n$, ${}_{\gamma}[I]_{\beta}$ is, by Definition 31.8, the unique $n \times n$ matrix such that

$$\text{for all } \mathbf{x} \in \mathbb{R}^n, \quad [\mathbf{x}]_{\gamma} = [I(\mathbf{x})]_{\gamma} = {}_{\gamma}[I]_{\beta}\,[\mathbf{x}]_{\beta}$$

In other words, whenever the β-coordinate matrix of a vector $\mathbf{x}$ is multiplied on the left by ${}_{\gamma}[I]_{\beta}$, the result is the γ-coordinate matrix of $\mathbf{x}$.

For this reason, ${}_{\gamma}[I]_{\beta}$ is usually called the *change-of-basis matrix* (or *the transition matrix*) from the basis β to the basis γ.

The formula for computing ${}_{\gamma}[I]_{\beta}$ is given by 31.9, specialized to the case when $T = I$. If $\beta = (\mathbf{u}_1, \mathbf{u}_2, \ldots, \mathbf{u}_n)$, then

$$ {}_{\gamma}[I]_{\beta} = \left[[\mathbf{u}_1]_{\gamma} \mid [\mathbf{u}_2]_{\gamma} \mid \cdots \mid [\mathbf{u}_n]_{\gamma} \right] $$

When discussing the change-of-basis matrix from β to γ, it is often convenient to think of β as the "old" basis and of γ as the "new" basis. Using this language, we can described ${}_{\gamma}[I]_{\beta}$ as "the matrix whose columns are the new coordinates of the old basis vectors."

The following theorem will enormously simplify the computational treatment of these ideas, so we present it first and then proceed to some examples.

31.14 Theorem

For any pair of ordered bases $\beta = (\mathbf{u}_1, \mathbf{u}_2, \ldots, \mathbf{u}_n)$ and γ for $\mathbb{R}^n$, the transition matrix ${}_{\gamma}[I]_{\beta}$ is invertible, and its inverse is the transition matrix ${}_{\beta}[I]_{\gamma}$.

Proof: It suffices, by Corollary 19.6, to show that ${}_{\beta}[I]_{\gamma}\, {}_{\gamma}[I]_{\beta} = \mathrm{I}_n$. We do this by showing, for arbitrary j, that the jth column of the product ${}_{\beta}[I]_{\gamma}\, {}_{\gamma}[I]_{\beta}$ is equal to the jth column, $[\mathbf{e}_j^n]$, of the identity matrix I_n.

$$
\begin{aligned}
j^{\text{th}} \text{ column of } {}_{\beta}[I]_{\gamma}\, {}_{\gamma}[I]_{\beta} &= {}_{\beta}[I]_{\gamma} \begin{bmatrix} j^{\text{th}} \\ \text{column} \\ \text{of} \\ {}_{\gamma}[I]_{\beta} \end{bmatrix} && \text{[by Observation 1.12(i)]} \\
&= {}_{\beta}[I]_{\gamma} \left[I(\mathbf{u}_j) \right]_{\gamma} && \text{[by the proof of Theorem 31.7]} \\
&= {}_{\beta}[I]_{\gamma} [\mathbf{u}_j]_{\gamma} && \text{[by definition of } I\text{]} \\
&= \left[I(\mathbf{u}_j) \right]_{\beta} && \text{[by definition of } {}_{\beta}[I]_{\gamma}\text{]} \\
&= [\mathbf{u}_j]_{\beta} && \text{[by definition of } I\text{]} \\
&= [\mathbf{e}_j^n] && \text{[by Definition 8.12]}
\end{aligned}
$$

■

31.15 Example

Find the transition matrix from the basis $\beta = ((2,1), (-1,4))$ to the standard basis α^2 for $\mathbb{R}^2$.

Solution: ${}_{\alpha^2}[I]_\beta$ is the matrix whose columns are the α^2-coordinates of the images under I of the vectors from β. Since $I(2,1)=(2,1)$ and $I(-1,4)=(-1,4)$, ${}_{\alpha^2}[I]_\beta = \begin{bmatrix} 2 & -1 \\ 1 & 4 \end{bmatrix}$.

This deceptively simple example illustrates an important special case: the transition matrix *from* an arbitrary basis *to* the standard basis can always be obtained by inspection. Here is another illustration.

31.16 Example

Find the transition matrix from the basis $\gamma = ((0,0,1),(1,0,1),(0,2,1))$ to the standard basis α^3 for $\mathbb{R}^3$.

Solution: $${}_{\alpha^3}[I]_\gamma = \begin{bmatrix} 0 & 1 & 0 \\ 0 & 0 & 2 \\ 1 & 1 & 1 \end{bmatrix}.$$

31.17 Computational Procedure

If $\beta = (\mathbf{u}_1, \mathbf{u}_2, \ldots, \mathbf{u}_n)$ is a basis for $\mathbb{R}^n$, then ${}_{\alpha^n}[I]_\beta$ is the matrix having the vectors from β as its columns; that is,

$${}_{\alpha^n}[I]_\beta = \begin{bmatrix} u_{11} & u_{21} & \cdots & u_{n1} \\ \vdots & \vdots & \cdots & \vdots \\ u_{1n} & u_{2n} & \cdots & u_{nn} \end{bmatrix}$$

Justification: By 31.9, ${}_{\alpha^n}[I]_\beta$ is the matrix whose j^{th} column consists of the standard coordinates of the image under I of the jth vector in β; since $I(\mathbf{u}_j) = \mathbf{u}_j$, we are done. ■

31.18 Example

Find the transition matrix from α^2 to the basis $\beta = ((2,1),(-1,4))$.

Solution:

$$\begin{aligned} {}_\beta[I]_{\alpha^2} &= \left({}_{\alpha^2}[I]_\beta\right)^{-1} && \text{[Theorem 31.14]} \\ &= \begin{bmatrix} 2 & -1 \\ 1 & 4 \end{bmatrix}^{-1} && \text{[31.17]} \\ &= \frac{1}{9}\begin{bmatrix} 4 & 1 \\ -1 & 2 \end{bmatrix} = \begin{bmatrix} \frac{4}{9} & \frac{1}{9} \\ -\frac{1}{9} & \frac{2}{9} \end{bmatrix} \end{aligned}$$

This example illustrates another special case: the transition matrix *from* the standard basis *to* a given arbitrary basis is obtained by inverting the matrix that has the vectors from the given basis as its columns. Here is another example.

31.19 Example

Find the transition matrix from α^3 to the basis $\gamma = ((0, 0, 1), (1, 0, 1), (0, 2, 1))$.

Solution:

$$_{\gamma}[I]_{\alpha^3} = \left(_{\alpha^3}[I]_{\gamma}\right)^{-1} = \begin{bmatrix} 0 & 1 & 0 \\ 0 & 0 & 2 \\ 1 & 1 & 1 \end{bmatrix}^{-1} = \begin{bmatrix} -1 & -\frac{1}{2} & 1 \\ 1 & 0 & 0 \\ 0 & \frac{1}{2} & 0 \end{bmatrix}$$

In the last step, the inverse was found using the method from Section 14.

31.20 Computational Procedure

If $\beta = (\mathbf{u}_1, \mathbf{u}_2, \ldots, \mathbf{u}_n)$ is a basis for $\mathbb{R}^n$, then

$$_{\beta}[I]_{\alpha^n} = \begin{bmatrix} u_{11} & u_{21} & \cdots & u_{n1} \\ \vdots & \vdots & \cdots & \vdots \\ u_{1n} & u_{2n} & \cdots & u_{nn} \end{bmatrix}^{-1}$$

Justification: Theorem 31.14 plus 31.17. ■

Thus far we have considered the special cases of a transition matrix $_{\gamma}[I]_{\beta}$ in which one or the other of β or γ is the standard basis. The following theorem will allow us to treat the general case, for arbitrary β and γ, by combining these two special cases.

31.21 Theorem

Let β and γ be bases for $\mathbb{R}^n$; then $_{\gamma}[I]_{\beta} = {_{\gamma}[I]_{\alpha^n}}\, {_{\alpha^n}[I]_{\beta}}$.

Proof: Let $\mathbf{x} \in \mathbb{R}^n$; then

$$\begin{aligned} {_{\gamma}[I]_{\alpha^n}}\, {_{\alpha^n}[I]_{\beta}}\, [\mathbf{x}]_{\beta} &= {_{\gamma}[I]_{\alpha^n}}\, [\mathbf{x}]_{\alpha^n} && \left[\text{defining property of } {_{\alpha^n}[I]_{\beta}}\right] \\ &= [\mathbf{x}]_{\gamma} && \left[\text{defining property of } {_{\gamma}[I]_{\alpha^n}}\right] \end{aligned}$$

This shows that the product $_{\gamma}[I]_{\alpha^n}\, {_{\alpha^n}[I]_{\beta}}$ satisfies the property that uniquely defines $_{\gamma}[I]_{\beta}$, so it is therefore equal to $_{\gamma}[I]_{\beta}$. ■

31.22 Example

Find ${}_{\gamma}[I]_{\beta}$, where $\beta = ((2,1),(-1,4))$ and $\gamma = ((0,2),(1,3))$.

Solution:

$$
\begin{aligned}
{}_{\gamma}[I]_{\beta} &= \begin{bmatrix} 0 & 1 \\ 2 & 3 \end{bmatrix}^{-1} \begin{bmatrix} 2 & -1 \\ 1 & 4 \end{bmatrix} \qquad \text{[31.21, 31.20, and 31.17]} \\
&= \frac{1}{-2} \begin{bmatrix} 3 & -1 \\ -2 & 0 \end{bmatrix} \begin{bmatrix} 2 & -1 \\ 1 & 4 \end{bmatrix} = \begin{bmatrix} -\frac{3}{2} & \frac{1}{2} \\ 1 & 0 \end{bmatrix} \begin{bmatrix} 2 & -1 \\ 1 & 4 \end{bmatrix} \\
&= \begin{bmatrix} -\frac{5}{2} & \frac{7}{2} \\ 2 & -1 \end{bmatrix}
\end{aligned}
$$

31.23 Example

Find the transition matrix from the basis $\beta = ((-1, 0, -1), (2,0,1),(1,1,0))$ to the basis $\gamma = ((0,0,1),(1,0,1),(0,2,1))$.

Solution:

$$
\begin{aligned}
{}_{\gamma}[I]_{\beta} &= \begin{bmatrix} 0 & 1 & 0 \\ 0 & 0 & 2 \\ 1 & 1 & 1 \end{bmatrix}^{-1} \begin{bmatrix} -1 & 2 & 1 \\ 0 & 0 & 1 \\ -1 & 1 & 0 \end{bmatrix} \qquad \text{[from 31.21]} \\
&= \begin{bmatrix} -1 & -\frac{1}{2} & 1 \\ 1 & 0 & 0 \\ 0 & \frac{1}{2} & 0 \end{bmatrix} \begin{bmatrix} -1 & 2 & 1 \\ 0 & 0 & 1 \\ -1 & 1 & 0 \end{bmatrix} \\
&= \begin{bmatrix} 0 & -1 & -\frac{3}{2} \\ -1 & 2 & 1 \\ 0 & 0 & \frac{1}{2} \end{bmatrix}
\end{aligned}
$$

With 31.17, 31.20, and 31.21, we have effectively automated the initial step of setting up, by inspection, any calculation involving transition matrices. Although you may make numerical errors while inverting or multiplying matrices by hand, you should *never* mistakenly set up the wrong calculation.

Here are some examples of transition matrices at work.

31.24 Example

Let $\beta = ((2,1),(-1,4))$ and let $\mathbf{u} = (9, -18)$. Find the β-coordinates of $\mathbf{u}$. That is, find scalars s_1 and s_2 such that $(9, -18) = s_1(2,1) + s_2(-1,4)$.

Solution:

$$\begin{bmatrix} s_1 \\ s_2 \end{bmatrix} = [\mathbf{u}]_\beta$$

$$= {}_\beta[I]_{\alpha^2}\,[\mathbf{u}]_{\alpha^2} \qquad \text{[Theorem 31.7]}$$

$$= \begin{bmatrix} 2 & -1 \\ 1 & 4 \end{bmatrix}^{-1} \begin{bmatrix} 9 \\ -18 \end{bmatrix} \qquad \text{[Theorem 31.20]}$$

$$= \begin{bmatrix} \frac{4}{9} & \frac{1}{9} \\ -\frac{1}{9} & \frac{2}{9} \end{bmatrix} \begin{bmatrix} 9 \\ -18 \end{bmatrix} = \begin{bmatrix} 2 \\ -5 \end{bmatrix}$$

31.25 Example (Example 31.12 Revisited)

We delayed finding ${}_\gamma[T]_\beta$ in 31.12 in order to discuss transition matrices. In that example, we need to compute the γ-coordinates of the vectors (3, 2, 3) and (3, 1, 5), where $\gamma = ((0,0,1),(1,0,1),(0,2,1))$. We now know how to do this efficiently.

$$[(3,2,3)]_\gamma = {}_\gamma[I]_{\alpha^3} \begin{bmatrix} 3 \\ 2 \\ 3 \end{bmatrix} \qquad \text{[Theorem 31.7]}$$

$$= \begin{bmatrix} 0 & 1 & 0 \\ 0 & 0 & 2 \\ 1 & 1 & 1 \end{bmatrix}^{-1} \begin{bmatrix} 3 \\ 2 \\ 3 \end{bmatrix} \qquad \text{[by inspection, using 31.20]}$$

$$= \begin{bmatrix} -1 & -\frac{1}{2} & 1 \\ 1 & 0 & 0 \\ 0 & \frac{1}{2} & 0 \end{bmatrix} \begin{bmatrix} 3 \\ 2 \\ 3 \end{bmatrix} \qquad \text{[inverse computed using 14.3]}$$

$$= \begin{bmatrix} -1 \\ 3 \\ 1 \end{bmatrix}$$

Similarly,

$$[(3,1,5)]_\gamma = \begin{bmatrix} -1 & -\frac{1}{2} & 1 \\ 1 & 0 & 0 \\ 0 & \frac{1}{2} & 0 \end{bmatrix} \begin{bmatrix} 3 \\ 1 \\ 5 \end{bmatrix} = \begin{bmatrix} \frac{3}{2} \\ 3 \\ \frac{1}{2} \end{bmatrix}$$

Thus

$$ {}_{\gamma}[T]_{\beta} = \begin{bmatrix} -1 & \frac{3}{2} \\ 3 & 3 \\ 1 & \frac{1}{2} \end{bmatrix} $$

is the answer to 31.12.

In the final example of this type, we will express the answers only as products of matrices and/or their inverses. The subsequent calculations are left as exercises.

31.26 Example

Let the basis $\gamma = ((0,0,1),(1,0,1),(0,2,1))$ and let the basis $\delta = ((-1,3,1),(2,4,-1),(6,2,0))$.

(i) Find the standard coordinates of the vector **u** whose δ-coordinates are $(3, 1, -9)$?

(ii) Find the γ-coordinates of the vector **u** whose δ-coordinates are $(3, 1, -9)$?

(iii) Find the δ-coordinates of the vector **u** whose standard coordinates are $(3, 1, -9)$?

(iv) Find the δ-coordinates of the vector **u** whose γ-coordinates are $(3, 1, -9)$?

Solutions:

(i) $$[\mathbf{u}]_{\alpha^3} = \begin{bmatrix} -1 & 2 & 6 \\ 3 & 4 & 2 \\ 1 & -1 & 0 \end{bmatrix} \begin{bmatrix} 3 \\ 1 \\ -9 \end{bmatrix}$$

(ii) $$[\mathbf{u}]_{\gamma} = \begin{bmatrix} 0 & 1 & 0 \\ 0 & 0 & 2 \\ 1 & 1 & 1 \end{bmatrix}^{-1} \begin{bmatrix} -1 & 2 & 6 \\ 3 & 4 & 2 \\ 1 & -1 & 0 \end{bmatrix} \begin{bmatrix} 3 \\ 1 \\ -9 \end{bmatrix}$$

(iii) $$[\mathbf{u}]_{\delta} = \begin{bmatrix} -1 & 2 & 6 \\ 3 & 4 & 2 \\ 1 & -1 & 0 \end{bmatrix}^{-1} \begin{bmatrix} 3 \\ 1 \\ -9 \end{bmatrix}$$

(iv) $$[\mathbf{u}]_{\delta} = \begin{bmatrix} -1 & 2 & 6 \\ 3 & 4 & 2 \\ 1 & -1 & 0 \end{bmatrix}^{-1} \begin{bmatrix} 0 & 1 & 0 \\ 0 & 0 & 2 \\ 1 & 1 & 1 \end{bmatrix} \begin{bmatrix} 3 \\ 1 \\ -9 \end{bmatrix}$$

We conclude this section with an exercise that is a partial converse to Theorem 31.14. It shows that every invertible matrix can be thought of, in infinitely many different ways, as a transition matrix.

31.27 Exercise

If P is an invertible $n \times n$ matrix, then for every ordered basis $\gamma = (\mathbf{v}_1, \mathbf{v}_2, \ldots, \mathbf{v}_n)$, there is a unique ordered basis $\beta = (\mathbf{u}_1, \mathbf{u}_2, \ldots, \mathbf{u}_n)$ such that $P =_\gamma [I]_\beta$.

Hint: Let $\mathbf{u}_j = p_{1j}\mathbf{v}_1 + p_{2j}\mathbf{v}_2 + \cdots + p_{nj}\mathbf{v}_n$.

Problem Set 31

31.1. Go back and do any of the exercises from Section 31 that you may have skipped.

*31.2. Let $T:\mathbb{R}^2 \to \mathbb{R}^3$ be a linear transformation satisfying $T(1,0) = (1,1,-2)$ and $T(0,1) = (2,3,8)$.
 (i) Find $T(3,1)$.
 (ii) Find the general expression for $T(x_1, x_2)$.
 (iii) What is the rank of T?

*31.3. Suppose $T:\mathbb{R}^2 \to \mathbb{R}^2$ is an invertible linear transformation satisfying $T^{-1}(1,0) = (2,-1)$ and $T^{-1}(0,1) = (4,-3)$.
 (i) Find the general expression for $T(x_1, x_2)$.
 (ii) Find $T(4,-1)$.

*31.4. Let $T:\mathbb{R}^2 \to \mathbb{R}^3$ be a linear transformation satisfying $T(1,0) = (2,-3,6)$ and $T(0,1) = (-1,4,-2)$.
 (i) Find $T(2,4)$.
 (ii) Find the kernel of T.
 (iii) Find the rank of T.
 (iv) Find the general expression for $T(x_1, x_2)$.
 (v) Find a reduced description of the range of T.
 (vi) Does $(2,-2,6)$ belong to the range of T?

*31.5. $T:\mathbb{R}^4 \to \mathbb{R}^3$ is the linear transformation defined by

$$T(x_1, x_2, x_3, x_4) = (x_1 + x_3 - x_4, -x_2 - 2x_3 + 4x_4, 2x_1 - x_2 + 2x_4)$$

 (i) Find the standard matrix representation of T.
 (ii) Find bases for the kernel and range of T.
 (iii) Find the rank and nullity of T.

*31.6. $\beta = ((1,-2,0), (0,2,5), (1,-1,2))$ is a basis for $\mathbb{R}^3$.
 (i) Find the transition matrix from the standard basis α^3 to the basis β.
 (ii) Find the transition matrix from the basis β to the standard basis α^3.
 (iii) Find the β-coordinates of the vector $(1,-1,1)$.

*31.7. $\beta = ((2,-1,2), (1,3,5), (1,-1,-5))$ and $\gamma = ((1,1,0), (0,1,1), (0,0,1))$ are bases for $\mathbb{R}^3$.
 (i) Find the transition matrix from β to γ.
 (ii) Find the γ-coordinates of the vector whose standard coordinates are $(1,-1,-1)$.

(iii) Find the γ-coordinates of the vector whose β-coordinates are $(1, -1, -1)$.

*31.8. $T:\mathbb{R}^3 \to \mathbb{R}^4$ is a linear transformation satisfying $T(1,0,0) = (2, -3, -6, -1)$, $T(0,1,0) = (1,2,0,-2)$, and $T(0,0,1) = (3, -1, -6, -3)$.

(i) Find the general expression for $T(x_1, x_2, x_3)$.

(ii) Find bases for the kernel and range of T.

(iii) What are the rank and nullity of T?

*31.9. Let $T:\mathbb{R}^4 \to \mathbb{R}^3$ be multiplication by the matrix

$$\begin{bmatrix} 1 & 3 & 4 & -1 \\ 2 & 4 & 1 & 3 \\ 5 & 9 & 2 & 1 \end{bmatrix}$$

(i) Which of the following are in the range of T: $(0,0,2)$, $(1,6,3)$, and/or $(2,1,-1)$?

(ii) Which of the following are in the kernel of T: $(2,1,0,4)$, $(5,9,2,1)$, and/or $(39, -24, 9, 3)$?

32

Similarity

We saw in Section 31 that if $T: \mathbb{R}^n \to \mathbb{R}^m$ is a linear transformation then, for each pair of bases β for $\mathbb{R}^n$ and γ for $\mathbb{R}^m$, there is a unique $m \times n$ matrix ${}_\gamma[T]_\beta$ such that

$$\text{for all } \mathbf{x} \in \mathbb{R}^n, \quad [T(\mathbf{x})]_\gamma = {}_\gamma[T]_\beta\,[\mathbf{x}]_\beta$$

In this section, we answer the following question: if α and β are bases for $\mathbb{R}^n$ and γ and δ are bases for $\mathbb{R}^m$, how are the two $m \times n$ matrices ${}_\gamma[T]_\beta$ and ${}_\delta[T]_\alpha$ related?

The answer is quite straightforward and the proof only uses ideas from Section 31.

32.1 Theorem

Let $T: \mathbb{R}^n \to \mathbb{R}^m$ be a linear transformation, let α and β be bases for $\mathbb{R}^n$, and let γ and δ be bases for $\mathbb{R}^m$; then

$$_\delta[T]_\alpha = {}_\delta[I]_\gamma\,{}_\gamma[T]_\beta\,{}_\beta[I]_\alpha \qquad (\dagger)$$

Before giving the proof, let us paraphrase the statement. ${}_\gamma[T]_\beta$ is an $m \times n$ matrix that accurately describes the behavior of T provided the points of $\mathbb{R}^n$ and $\mathbb{R}^m$ are labeled with their β-coordinates and γ-coordinates, respectively. Suppose this description of T is known, but that we also want another matrix that describes T when α-coordinates and δ-coordinates are used to label, respectively, the points in $\mathbb{R}^n$ and $\mathbb{R}^m$. It is instinctively clear that this alternate description, ${}_\delta[T]_\alpha$, can be obtained by properly combining the known description, ${}_\gamma[T]_\beta$, with the appropriate change-of-basis matrices ${}_\beta[I]_\alpha$ and ${}_\delta[I]_\gamma$. The proper combination is the product of these three matrices in the order indicated in equation (†).

Proof: By 31.7 and 31.8, ${}_\delta[T]_\alpha$ is the unique $m \times n$ matrix A satisfying

$$\text{for all } \mathbf{x} \in \mathbb{R}^n, \quad [T(\mathbf{x})]_\delta = \mathrm{A}[\mathbf{x}]_\alpha \tag{$*$}$$

So to conclude (†), we will show that the product ${}_\delta[I]_\gamma\, {}_\gamma[T]_\beta\, {}_\beta[I]_\alpha$ satisfies ($*$). Let $\mathbf{x} \in \mathbb{R}^n$; then

$$\begin{aligned} {}_\delta[I]_\gamma\, {}_\gamma[T]_\beta\, {}_\beta[I]_\alpha\, [\mathbf{x}]_\alpha &= {}_\delta[I]_\gamma\, {}_\gamma[T]_\beta\, [\mathbf{x}]_\beta \\ &= {}_\delta[I]_\gamma\, [T(\mathbf{x})]_\gamma \\ &= [T(\mathbf{x})]_\delta \end{aligned}$$

■

The proof shows what is happening: ${}_\beta[I]_\alpha$ converts the α-coordinates of $\mathbf{x}$ to the β-coordinates of $\mathbf{x}$ so that they can be acted on by ${}_\gamma[T]_\beta$; this action yields the γ-coordinates of $T(\mathbf{x})$, which are then converted by ${}_\delta[I]_\gamma$ to the δ-coordinates of $T(\mathbf{x})$.

To illustrate this Theorem, let us reuse some of the examples from Section 31.

32.2 Example

Define $T: \mathbb{R}^2 \to \mathbb{R}^3$ by $T(x_1, x_2) = (3x_1, 2x_1 - x_2, 3x_1 + 2x_2)$. In 31.10, we found the standard matrix for T:

$${}_{\alpha^3}[T]_{\alpha^2} = \begin{bmatrix} 3 & 0 \\ 2 & -1 \\ 3 & 2 \end{bmatrix}$$

Let $\beta = ((2,1),(-1,4))$ and let $\gamma = ((0,0,1),(1,0,1),(0,2,1))$. The transition matrices ${}_{\alpha^2}[I]_\beta$ and ${}_\gamma[I]_{\alpha^3}$ were computed in 31.15 and 31.19, respectively. Here is the computation of ${}_\gamma[T]_\beta$ using Theorem 32.1.

$$\begin{aligned} {}_\gamma[T]_\beta &= {}_\gamma[I]_{\alpha^3}\, {}_{\alpha^3}[T]_{\alpha^2}\, {}_{\alpha^2}[I]_\beta \\ &= \begin{bmatrix} -1 & -\frac{1}{2} & 1 \\ 1 & 0 & 0 \\ 0 & \frac{1}{2} & 0 \end{bmatrix} \begin{bmatrix} 3 & 0 \\ 2 & -1 \\ 3 & 2 \end{bmatrix} \begin{bmatrix} 2 & -1 \\ 1 & 4 \end{bmatrix} = \begin{bmatrix} \frac{1}{2} & 11 \\ 6 & -3 \\ \frac{3}{2} & -3 \end{bmatrix} \end{aligned}$$

As a check of our answer, and as a further illustration of these ideas, we will recalculate ${}_{\gamma}[T]_{\beta}$ directly from the definition using 31.9.

32.3 Example

Let T, β, and γ be as in the previous example. ${}_{\gamma}[T]_{\beta}$ is the matrix whose columns are the γ-coordinates of the images under T of the basis vectors from β. To compute this, we first find $T(2,1) = (6,3,8)$ and $T(-1,4) = (-3,-6,5)$. Next we find the γ-coordinates of $(6,3,8)$ and $(-3,-6,5)$ using the transition matrix ${}_{\gamma}[I]_{\alpha^3}$, which was calculated in 31.19. The first column of ${}_{\gamma}[T]_{\beta}$ is

$$
{}_{\gamma}[I]_{\alpha^3}\begin{bmatrix} 6 \\ 3 \\ 8 \end{bmatrix} = \begin{bmatrix} -1 & -\frac{1}{2} & 1 \\ 1 & 0 & 0 \\ 0 & \frac{1}{2} & 0 \end{bmatrix}\begin{bmatrix} 6 \\ 3 \\ 8 \end{bmatrix} = \begin{bmatrix} \frac{1}{2} \\ 6 \\ \frac{3}{2} \end{bmatrix}
$$

The second column of ${}_{\gamma}[T]_{\beta}$ is

$$
{}_{\gamma}[I]_{\alpha^3}\begin{bmatrix} -3 \\ -6 \\ 5 \end{bmatrix} = \begin{bmatrix} -1 & -\frac{1}{2} & 1 \\ 1 & 0 & 0 \\ 0 & \frac{1}{2} & 0 \end{bmatrix}\begin{bmatrix} -3 \\ -6 \\ 5 \end{bmatrix} = \begin{bmatrix} 11 \\ -3 \\ -3 \end{bmatrix}
$$

It is instructive to compare 32.2 and 32.3 to realize that the *same* calculations are being done; only the perspectives from which they are viewed are different.

The special case of the setup in Theorem 32.1 in which $m = n$, $\alpha = \delta$, and $\beta = \gamma$ leads to the concept of similarity. Suppose that $T\colon \mathbb{R}^n \to \mathbb{R}^n$ is a linear transformation and that α and β are bases for $\mathbb{R}^n$. How are the two $n \times n$ matrices ${}_{\alpha}[T]_{\alpha}$ and ${}_{\beta}[T]_{\beta}$ related?

32.4 Corollary to Theorem 32.1

Where T, α, and β are as just described,

$$
{}_{\beta}[T]_{\beta} = {}_{\beta}[I]_{\alpha}\,{}_{\alpha}[T]_{\alpha}\,{}_{\alpha}[I]_{\beta} \qquad \blacksquare
$$

Because ${}_{\beta}[I]_{\alpha} = ({}_{\alpha}[I]_{\beta})^{-1}$, a pair of $n \times n$ matrices A and B that represent the same linear transformation $T\colon \mathbb{R}^n \to \mathbb{R}^n$ with respect to different bases will, by 32.4, satisfy the condition

There exists an invertible $n \times n$ matrix P such that $\mathrm{B} = \mathrm{P}^{-1}\mathrm{AP}$ (#)

Conversely, because any invertible matrix can be interpreted as a change-of-basis matrix (recall Exercise 31.27), we also know that, if a pair of matrices A and B satisfies condition (#), then A and B represent the same linear transformation with respect to different bases. More precisely, if $B = P^{-1}AP$ and if A represents T with respect to the basis α, then B represents T with respect to the basis β, where β is the unique basis such that $P = {}_{\alpha}[I]_{\beta}$, as explained in 31.27.

This concept is obviously important, so we give it a name.

32.5 Definition

The $n \times n$ matrix B is *similar* to the $n \times n$ matrix A if and only if there exists an invertible $n \times n$ matrix P such that $B = P^{-1}AP$.

32.6 Analogy

Consider the formal relationship in 32.4 between the similar matrices ${}_{\alpha}[T]_{\alpha}$ and ${}_{\beta}[T]_{\beta}$:

$$ {}_{\beta}[T]_{\beta} = {}_{\beta}[I]_{\alpha}\, {}_{\alpha}[T]_{\alpha}\, {}_{\alpha}[I]_{\beta} $$

This relationship is natural if we think of T as a machine, of ${}_{\alpha}[T]_{\alpha}$ as the English-language instruction manual for T, of ${}_{\beta}[T]_{\beta}$ as the French-language instruction manual for T, of ${}_{\beta}[I]_{\alpha}$ as the English-to-French dictionary, and of ${}_{\alpha}[I]_{\beta}$ as the French-to-English dictionary.

What is the relationship between the French and English manuals? To obtain the French manual, first apply the French-to-English dictionary, followed by the English instruction manual, followed in turn by the English-to-French dictionary.

32.7 Exercise

Prove that similarity is reflexive (A is similar to itself), symmetric (if B is similar to A, then A is similar to B), and transitive (if B is similar to A and C is similar to B, then C is similar to A).

The problem of deciding whether two given $n \times n$ matrices are similar is rather difficult. An answer to this question is provided by the Jordan canonical form, to be treated in Chapter IX.

Problem Set 32

32.1. Go back and do any of the exercises in Section 32 that you may have skipped.

*32.2. $T:\mathbb{R}^2 \to \mathbb{R}^2$ is defined by $T(x_1, x_2) = (2x_1 + 3x_2, x_1 - x_2)$. $\beta = ((-5, 4), (4, -3))$ is a basis for $\mathbb{R}^2$.

(i) Find the standard matrix that represents T.

(ii) Find the matrix that represents T with respect to the basis β.
(iii) Find $T(1,1)$.
(iv) Find the β-coordinates of $T(1,1)$.
(v) Find the β-coordinates of $T(\mathbf{w})$, where $\mathbf{w}$ is the vector whose β-coordinates are $(1,1)$.

*32.3. Let $\beta = ((1,-1),(1,1))$ and $\gamma = ((-1,1),(0,-1))$. $T:\mathbb{R}^2 \to \mathbb{R}^2$ is the linear transformation whose matrix representation with respect to the basis β is $\begin{bmatrix} 3 & \frac{3}{2} \\ -4 & -1 \end{bmatrix}$. Find the matrix that represents T with respect to the basis γ.

*32.4. Can

$$\begin{bmatrix} 1 & 2 & 3 & -7 \\ 0 & -9 & 7 & -6 \\ 0 & 6 & -8 & 4 \\ 0 & 0 & 3 & 0 \end{bmatrix}$$

be a transition matrix between two bases for $\mathbb{R}^4$?

*32.5 If A and B are similar $n \times n$ matrices then det(A) = det(B).

*32.6 If A and B are similar $n \times n$ matrices then trace(A) = trace(B). (See P3.18 for the definition and properties of the trace function).

33

Diagonalizability (Motivation)

As was suggested at the end of Section 31, the object under study in this context is a fixed linear operator $T\colon \mathbb{R}^n \to \mathbb{R}^n$. We will illustrate the ideas involved with a specific example. **In this section, we will ask questions and provide the answers without explaining how these answers are obtained. The explanations will be forthcoming in subsequent sections. For the moment, just "watch the demo."**

33.1 Example

Let $T\colon \mathbb{R}^2 \to \mathbb{R}^2$ be defined by $T(x_1, x_2) = (-2x_1 - 2x_2, -5x_1 + x_2)$. We may obtain by inspection the matrix that represents T with respect to the standard basis α^2 for $\mathbb{R}^2$.

$$ {}_{\alpha^2}[T]_{\alpha^2} = \begin{bmatrix} -2 & -2 \\ -5 & 1 \end{bmatrix} $$

For the sake of example, let $\mathbf{u} = (\frac{1}{5}, 3)$; then $T(\mathbf{u}) = (-\frac{32}{5}, 2)$. Similarly, if $\mathbf{v} = (-1, -2)$, then $T(\mathbf{v}) = (6, 3)$.

Because linear transformations preserve scalar multiplication, any vector in $\mathscr{L}(\mathbf{u})$ (i.e., any vector on the line determined by $\mathbf{u}$) will be sent by

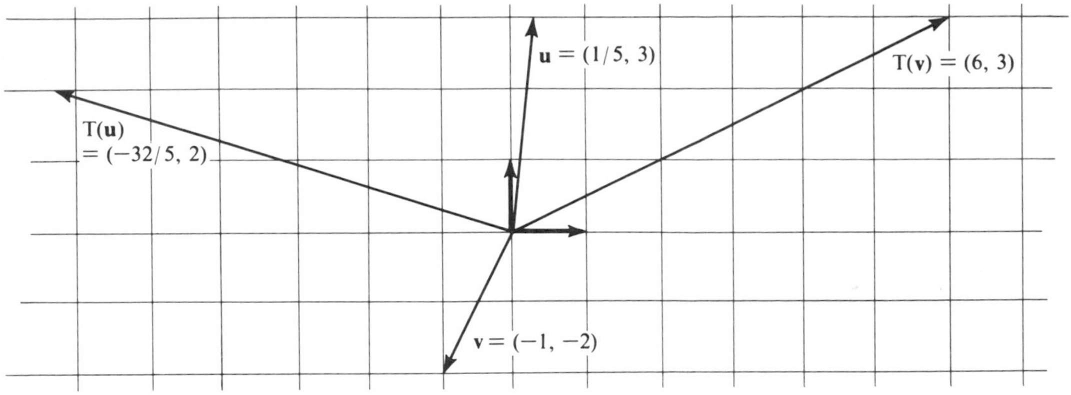

Figure 33.1

T into a vector on the line determined by $T(\mathbf{u})$. The analogue holds for $\mathbf{v}$ and $T(\mathbf{v})$. In Figure 33.1, we sketch the behavior of T at the points $\mathbf{u}$ and $\mathbf{v}$.

There is nothing particularly simple about the standard description of T. The first coordinate of $T(\mathbf{x})$, $-2x_1 - 2x_2$, depends on both x_1 and x_2. So does the second coordinate, $-5x_1 + x_2$. The general form of this description is the following:

$$\text{If } (\mathbf{x})_{\alpha^2} = (x_1, x_2), \text{ then } (T(x))_{\alpha^2} = (s_1x_1 + s_2x_2, t_1x_1 + t_2x_2).$$

33.2 Exercise

Use the methods of Section 32 to find the matrix that represents T with respect to the basis $\beta = ((1,1), (-\frac{2}{5}, 1))$. You will find that

$$_{\beta}[T]_{\beta} = \begin{bmatrix} -4 & 0 \\ 0 & 3 \end{bmatrix}.$$

What has been gained? At the cost of using the different coordinate system, β, to label the points of $\mathbb{R}^2$, the description of the behavior of T is greatly simplified. This description has the form:

$$\text{if } (\mathbf{x})_\beta = (x_1, x_2) \qquad \text{then } (T(\mathbf{x}))_\beta = (-4x_1, 3x_2).$$

In other words, T simply "multiplies each coordinate by a constant factor." In Figure 33.2, we reproduce Figure 33.1 as it would be seen by someone working in the β-coordinate system. The vectors $\mathbf{u}$ and $\mathbf{v}$ are the same. You should verify that $(\mathbf{u})_\beta = (1, 2)$, so $(T(\mathbf{u}))_\beta = ((-4)(1), (3)(2)) = (-4, 6)$. Similarly, $(\mathbf{v})_\beta = (-\frac{9}{7}, -\frac{5}{7})$, so $(T(\mathbf{v}))_\beta = ((-4)(-\frac{9}{7}), (3)(-\frac{5}{7})) = (\frac{36}{7}, -\frac{15}{7})$.

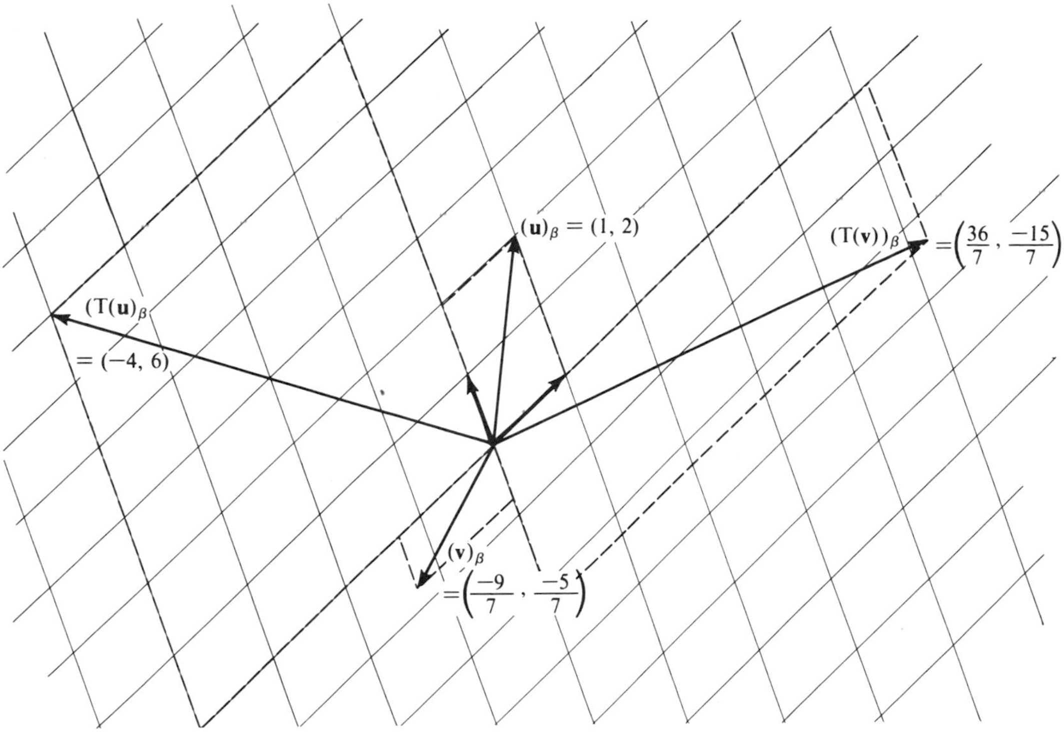

Figure 33.2

In the n-dimensional case, an arbitrary linear operator $T:\mathbb{R}^n \to \mathbb{R}^n$ has a standard description of the form

$$\begin{aligned} T(\mathbf{x}) &= T(x_1, x_2, \ldots, x_n) \\ &= (a_{11}x_1 + a_{12}x_2 + \cdots + a_{1n}x_n, \ldots, a_{n1}x_1 + a_{n2}x_2 + \cdots + a_{nn}x_n). \end{aligned}$$

Each coordinate of the output, $T(\mathbf{x})$, depends in general on *each* coordinate of the input, $\mathbf{x}$. We seek, if possible, a simpler description of the behavior of T. Specifically, we would like to have a basis, β, for $\mathbb{R}^n$ such that, if $(\mathbf{x})_\beta = (x_1, x_2, \ldots, x_n)$, then

$$(T(\mathbf{x}))_\beta = (t_1x_1, t_2x_2, \ldots, t_nx_n).$$

By labeling points in $\mathbb{R}^n$ with their β-coordinates, the resulting description of T is such that for $i = 1, 2, \ldots, n$ the ith coordinate of the image $T(\mathbf{x})$ is simply a constant times the ith coordinate of the input.

Let us reconsider some of these ideas from the geometric point of view.

33.3 Exercise

Let $T:\mathbb{R}^n \to \mathbb{R}^n$ be a linear operator, let W be a linear subspace of $\mathbb{R}^n$, and let H be an affine subspace of $\mathbb{R}^n$.

(i) Prove that $T(W) = \{T(\mathbf{w}): \mathbf{w} \in W\}$ is a linear subspace of $\mathbb{R}^n$.
(ii) Prove that $T(H) = \{T(\mathbf{v}): \mathbf{v} \in H\}$ is an affine subspace of $\mathbb{R}^n$.
(iii) Prove, when T is an invertible linear transformation, that if $\dim W = p$, then $\dim T(W) = p$.

In other words, under an invertible linear operator, the image of a line (through the origin) is a line (through the origin), the image of a plane (through the origin) is a plane (through the origin), and so on. If T is not invertible, the image of a line could be a line or a point, the image of a plane could be a plane, a line, or a point, and so on.

33.4 Definition

Let $T: \mathbb{R}^n \to \mathbb{R}^n$ be a linear operator and let W be a linear subspace of $\mathbb{R}^n$. If $T(W) \subseteq W$, W is said to be *invariant under T* or *T-invariant*.

Thus, a T-invariant subspace is one that T sends into itself.

Consider, for example, the case when T is a rotation of $\mathbb{R}^2$ through the angle $\pi/2$; in this case, no line through the origin is T-invariant. If T is the rotation of $\mathbb{R}^2$ through the angle π, then the x_1 axis is T-invariant.

33.5 Example 33.1 Revisited

Are any lines through the origin T-invariant under the transformation $T:\mathbb{R}^2 \to \mathbb{R}^2$ defined by $T(x_1, x_2) = (-2x_1 - 2x_2, -5x_1 + x_2)$? The answer is yes. The line $\mathscr{L}(1,1) = \{(x_1, x_2) \in \mathbb{R}^2: x_1 = x_2\}$ is invariant under T; the image under T of a point $\mathbf{x} = (x, x)$ on this line is $(-4x, -4x) = -4\mathbf{x}$. The line $\mathscr{L}(-\frac{2}{5}, 1) = \{(x_1, x_2) \in \mathbb{R}^2: x_1 = -\frac{2}{5}x_2\}$ is also invariant since the image under T of a point $\mathbf{x} = (-\frac{2}{5}x, x)$ on this line is $(-\frac{6}{5}x, 3x) = 3\mathbf{x}$, which is also on this line. Note how the numbers -4 and 3 arose in Exercise 33.2; you should note their reoccurrence in this geometric context. This will eventually be explained.

Let us temporarily say that an $n \times n$ matrix is *diagonalizable* if A is similar to a diagonal matrix. (This concept requires a more precise definition to be given later, but for the purpose of the present discussion, it is adequate.) In Section 32, we saw that similar matrices represent the same linear transformation with respect to different bases. The essence of the next exercise is that diagonalizability is really a property of the underlying linear operator.

33.6 Exercise

Let A and B be similar $n \times n$ matrices. Show that A is diagonalizable iff B is diagonalizable.

Because of this, the following definition makes sense.

33.7 Definition

A linear operator $T: \mathbb{R}^n \to \mathbb{R}^n$ is *diagonalizable* if any one (and hence all) of its matrix representations is a diagonalizable matrix.

This concludes the "demo." But before going on, we would like to emphasize one of the most useful aspects of diagonalizability. This is the ability to find arbitrarily large powers of a square matrix without the need to find all intermediate powers. Indeed, we can, if A is diagonalizable, obtain a formula for A^k as a function of k.

33.8 Exercise

(i) If $D = \text{diag}(d_1, d_2, \ldots, d_n)$ and k is any positive integer, then $D^k = \text{diag}(d_1^k, d_2^k, \ldots, d_n^k)$.

(ii) If $P^{-1}AP = D$, then for any positive integer k, $A^k = PD^kP^{-1}$.

33.9 Example

Let us illustrate the previous exercise by computing A^8, where A is the matrix from Example 33.1:

$$A = \begin{bmatrix} -2 & -2 \\ -5 & 1 \end{bmatrix}$$

You should verify by inverting and multiplying that

$$\begin{bmatrix} 1 & -\frac{2}{5} \\ 1 & 1 \end{bmatrix}^{-1} \begin{bmatrix} 2 & -2 \\ -5 & 1 \end{bmatrix} \begin{bmatrix} 1 & -\frac{2}{5} \\ 1 & 1 \end{bmatrix} = \begin{bmatrix} -4 & 0 \\ 0 & 3 \end{bmatrix}$$

(The exact source of the invertible P and diagonal D such that $P^{-1}AP = D$ will be given in Section 35.) Thus

$$A^8 = \begin{bmatrix} 1 & -\frac{2}{5} \\ 1 & 1 \end{bmatrix} \begin{bmatrix} (-4)^8 & 0 \\ 0 & (3)^8 \end{bmatrix} \begin{bmatrix} 1 & -\frac{2}{5} \\ 1 & 1 \end{bmatrix}^{-1}.$$

Replacing 8 by k and multiplying out will yield a formula for A^k as a function of k. Thus A^{100} can be found, without having first to find $A^2, A^3, \ldots, A^{99}$ simply by plugging 100 in for k. Specifically,

$$A^k = \begin{bmatrix} \frac{5}{7}(-4)^k + \frac{2}{7}(3)^k & \frac{2}{7}(-4)^k - \frac{2}{7}(3)^k \\ \frac{5}{7}(-4)^k - \frac{5}{7}(3)^k & \frac{2}{7}(-4)^k + \frac{5}{7}(3)^k \end{bmatrix}.$$

Problem Set 33

33.1. Go back and do any of the exercises from Section 33 that you may have skipped.

*33.2. Show that, if A is diagonalizable and if A is invertible, then A^{-1} is also diagonalizable.

34

A Real $n \times n$ Matrix A Is Diagonalizable over $\mathbb{R}$ If and Only If There Exists a Basis for $\mathbb{R}^n$ Consisting of Eigenvectors of A

For reasons that will subsequently become clear, we will use the letter $\mathbb{F}$ to denote the scalars in this section; $\mathbb{F}^n$ denotes the set of n-tuples whose coordinates are elements of $\mathbb{F}$. As you read through this section for the first time, you should continue to think that $\mathbb{F} = \mathbb{R}$ and $\mathbb{F}^n = \mathbb{R}^n$; but we are anticipating other possibilities for the "scalars" and wish to state results from the beginning in language that allows for the wider interpretation. Meanwhile, read all the proofs and do the exercises assuming $\mathbb{F} = \mathbb{R}$ and $\mathbb{F}^n = \mathbb{R}^n$.

It is customary in this part of the subject to use the Greek letter λ as a variable ranging over scalars.

34.1 Definition

Let $\mathbb{F}_{n \times n}$ denote the set of $n \times n$ matrices whose entries are elements of $\mathbb{F}$. A matrix $A \in \mathbb{F}_{n \times n}$ is *diagonalizable over* $\mathbb{F}$ iff there exist an invertible matrix $P \in \mathbb{F}_{n \times n}$ and a diagonal matrix $D \in \mathbb{F}_{n \times n}$ such that $P^{-1}AP = D$.

The usefulness of this concept was demonstrated in Section 33. Before considering techniques for deciding, given $A \in \mathbb{F}_{n \times n}$, whether such a P and D exist and for finding them when they do, we must develop a bit of theory.

To place the main ideas in better focus, we dispense with some technical details in the first theorem. Note that if P is invertible, the two equations $P^{-1}AP = D$ and $AP = PD$ are equivalent. The first theorem treats the equation $AP = PD$.

34.2 Theorem

Let A, D, and $P \in \mathbb{F}_{n\times n}$ and suppose that D is the diagonal matrix $\mathrm{diag}(d_1, d_2, \ldots, d_n)$. Let $\mathbf{p}_j$ be the jth column of P. The following are equivalent:

(i) $AP = PD$.
(ii) For each $j = 1, 2, \ldots, n$, the jth column of AP is equal to the jth column of PD.
(iii) For each $j = 1, 2, \ldots, n$, $A\mathbf{p}_j = d_j\mathbf{p}_j$.

If, in addition, P is invertible, each of these three conditions is equivalent to

(iv) $P^{-1}AP = D$.

Proof: The equivalence of (i) and (ii) is obvious; so is the equivalence of (i) and (iv) when P is invertible. The equivalence of (ii) and (iii) is immediate from the following two observations:

$$\begin{bmatrix} j^{\text{th}} \\ \text{column} \\ \text{of AP} \end{bmatrix} = A\begin{bmatrix} j^{\text{th}} \\ \text{column} \\ \text{of P} \end{bmatrix} = A\mathbf{p}_j$$

and

$$\begin{bmatrix} j^{\text{th}} \\ \text{column} \\ \text{of PD} \end{bmatrix} = P\begin{bmatrix} j^{\text{th}} \\ \text{column} \\ \text{of D} \end{bmatrix} = P\begin{bmatrix} 0 \\ \vdots \\ 0 \\ d_j \\ 0 \\ \vdots \\ 0 \end{bmatrix} = d_j\mathbf{p}_j \qquad \blacksquare$$

From this theorem, we realize that the diagonalizability of the matrix A is somehow connected with the existence of scalars and vectors that are related to A as in 34.2(iii) (i.e., such that $A\mathbf{v} = d\mathbf{v}$).

This motivates the definitions that we are about to give. Throughout all this, however, there is an important point to keep in mind.

34.3 Remark

We have seen, in Section 32, that similar matrices represent the same linear operator with respect to different bases. Thus diagonalizability is "really" an intrinsic property not of $n \times n$ matrices but of linear operators on $\mathbb{R}^n$. We must deal, at the computational level, with diagonalizability of matrices because the way to analyze a linear operator $T\colon \mathbb{F}^n \to \mathbb{F}^n$ is to choose a basis β for $\mathbb{F}^n$ and to study the matrix ${}_\beta[T]_\beta$ that represents T with respect to β. But we should not forget, in our preoccupation with matrices, that the heart of the matter is the underlying linear operator.

34.4 Definition

Let $A \in \mathbb{F}_{n \times n}$.

(i) A vector $\mathbf{v} \in \mathbb{F}^n$ is an *eigenvector of A* iff $\mathbf{v} \neq \mathbf{0}$ and, for some scalar $\lambda \in \mathbb{F}$, $A\mathbf{v} = \lambda\mathbf{v}$.
(ii) A scalar $\lambda \in \mathbb{F}$ is an *eigenvalue of A* iff, for some nonzero vector $\mathbf{v} \in \mathbb{F}^n$, $A\mathbf{v} = \lambda\mathbf{v}$.
(iii) When λ is an eigenvalue of A, the set $E_\lambda = \{\mathbf{v} \in \mathbb{F}^n\colon A\mathbf{v} = \lambda\mathbf{v}\}$ is called the *eigenspace of* λ.
(iv) The set $\sigma(A) = \{\lambda \in \mathbb{F}\colon \lambda \text{ is an eigenvalue of } A\}$ is called the *spectrum of A*.

34.5 Remark

The phrases *proper value* and *characteristic value* are frequently used instead of *eigenvalue*. The phrases *proper vector* and *characteristic vector* are common synonyms for *eigenvector*.

34.6 Exercise

Let $\mathbb{F} = \mathbb{R}$, let $A \in \mathbb{R}_{n \times n}$, let $\lambda \in \mathbb{R}$, and let $E_\lambda = \{\mathbf{v} \in \mathbb{R}^n\colon A\mathbf{v} = \lambda\mathbf{v}\}$. Show that

(i) E_λ is always a linear subspace of $\mathbb{R}^n$, whether λ is an eigenvalue of A or not.
(ii) λ is an eigenvalue of A iff $E_\lambda \neq \{\mathbf{0}\}$ (i.e., iff the subspace E_λ is nontrivial).
(iii) If λ_1 and λ_2 are distinct eigenvalues of A, then $E_{\lambda_1} \cap E_{\lambda_2} = \{\mathbf{0}\}$.

The main theorem of this section is the following characterization of diagonalizability. The special case of this theorem for $\mathbb{F} = \mathbb{R}$ is the title of Section 34.

34.7 Theorem

A matrix $A \in \mathbb{F}_{n\times n}$ is diagonalizable over $\mathbb{F}$ if and only if there exists a basis for $\mathbb{F}^n$ consisting of eigenvectors of A.

Proof: ($\leftarrow$) Suppose $\beta = \{\mathbf{v}_1, \mathbf{v}_2, \ldots, \mathbf{v}_n\}$ is a basis for $\mathbb{F}^n$ and that each $\mathbf{v}_j$ is an eigenvector of A. Suppose, to be specific, that $A\mathbf{v}_j = \lambda_j\mathbf{v}_j$. Let P be the matrix whose columns are the vectors $\mathbf{v}_1, \mathbf{v}_2, \ldots, \mathbf{v}_n$, and let $D = \operatorname{diag}(\lambda_1, \lambda_2, \ldots, \lambda_n)$. P is invertible since its columns are a basis for $\mathbb{F}^n$. It follows from 34.2 that $P^{-1}AP = D$.

($\rightarrow$) If A is diagonalizable over $\mathbb{F}$, choose an invertible $P \in \mathbb{F}_{n\times n}$ and a diagonal $D \in \mathbb{F}_{n\times n}$ such that $P^{-1}AP = D$. Let $D = \operatorname{diag}(\lambda_1, \lambda_2, \ldots, \lambda_n)$, and let $\beta = \{\mathbf{v}_1, \mathbf{v}_2, \ldots, \mathbf{v}_n\}$, where $\mathbf{v}_j$ is the jth column of P. Since P is invertible, each $\mathbf{v}_j \neq \mathbf{0}$ and β is a basis for $\mathbb{F}^n$. It follows from 34.2 that, for $j = 1, 2, \ldots, n$, $A\mathbf{v}_j = \lambda_j\mathbf{v}_j$, so the elements of β are eigenvectors of A. ■

The strength of Theorems 34.2 and 34.7 is often underestimated. The next example illustrates one of their uses.

34.8 Example

Observe that

$$\begin{bmatrix} -2 & -2 \\ -5 & 1 \end{bmatrix}\begin{bmatrix} 1 \\ 1 \end{bmatrix} = \begin{bmatrix} -4 \\ -4 \end{bmatrix} = -4\begin{bmatrix} 1 \\ 1 \end{bmatrix}$$

and that

$$\begin{bmatrix} -2 & -2 \\ -5 & 1 \end{bmatrix}\begin{bmatrix} -2 \\ 5 \end{bmatrix} = \begin{bmatrix} -6 \\ 15 \end{bmatrix} = 3\begin{bmatrix} -2 \\ 5 \end{bmatrix}$$

From this information *alone*, together with the obvious fact that $\{(1,1), (-2,5)\}$ is a basis for $\mathbb{R}^2$, you may conclude from 34.2 and 34.7 that

$$\begin{bmatrix} 1 & -2 \\ 1 & 5 \end{bmatrix}^{-1}\begin{bmatrix} -2 & -2 \\ -5 & 1 \end{bmatrix}\begin{bmatrix} 1 & -2 \\ 1 & 5 \end{bmatrix} = \begin{bmatrix} -4 & 0 \\ 0 & 3 \end{bmatrix} \qquad (*)$$

It is important to appreciate that the truth of $(*)$ need *not* be otherwise ascertained or verified by actually inverting

$$\begin{bmatrix} 1 & -2 \\ 1 & 5 \end{bmatrix}$$

and multiplying the three matrices to find their product.

It is clear from the previous theorem that obtaining linearly independent sets of eigenvectors of $A \in \mathbb{F}_{n\times n}$ will be the major objective in trying to diagonalize A. The next theorem provides the easiest, although not the only, source of such sets. In words, it asserts that a set consisting of eigenvectors that belong to distinct eigenspaces will be linearly independent.

34.9 Theorem

Let $A \in \mathbb{F}_{n\times n}$, let $\lambda_1, \lambda_2, \ldots, \lambda_p$ be distinct eigenvalues of A, and let $\mathbf{0} \neq \mathbf{v}_j \in E_{\lambda_j}$ for $j = 1, 2, \ldots, p$; then the set $\{\mathbf{v}_1, \mathbf{v}_2, \ldots, \mathbf{v}_p\}$ is linearly independent.

Proof: We view the set as an ordered set and derive a contradiction from the assumption that $(\mathbf{v}_1, \mathbf{v}_2, \ldots, \mathbf{v}_p)$ is linearly dependent. If it is linearly dependent, then either $\mathbf{v}_1 = \mathbf{0}$ or, for some $i \gneqq 1$, $\mathbf{v}_i \in \mathscr{L}(\mathbf{v}_1, \ldots, \mathbf{v}_{i-1})$. But $\mathbf{v}_1 \neq \mathbf{0}$ by definition. If q is the least such i, then $(\mathbf{v}_1, \mathbf{v}_2, \ldots, \mathbf{v}_{q-1})$ is linearly independent and there exist scalars $t_1, t_2, \ldots, t_{q-1}$ such that

$$\mathbf{v}_q = t_1\mathbf{v}_1 + t_2\mathbf{v}_2 + \cdots + t_{q-1}\mathbf{v}_{q-1} \qquad (*)$$

Note that $\mathbf{v}_q \neq \mathbf{0}$ by definition, so there exists $j \leqq q-1$ such that $t_j \neq 0$. Multiplying $(*)$ by λ_q, we get

$$\lambda_q\mathbf{v}_q = t_1\lambda_q\mathbf{v}_1 + t_2\lambda_q\mathbf{v}_2 + \cdots + t_{q-1}\lambda_q\mathbf{v}_{q-1} \qquad (\dagger)$$

Multiplying $(*)$ by A, we get

$$\begin{aligned} A\mathbf{v}_q &= t_1A\mathbf{v}_1 + t_2A\mathbf{v}_2 + \cdots + t_{q-1}A\mathbf{v}_{q-1} \\ &= t_1\lambda_1\mathbf{v}_1 + t_2\lambda_2\mathbf{v}_2 + \cdots + t_{q-1}\lambda_{q-1}\mathbf{v}_{q-1} \qquad (\ddagger) \end{aligned}$$

Because $A\mathbf{v}_q = \lambda_q\mathbf{v}_q$, we obtain upon subtracting ($\dagger$) from ($\ddagger$) that

$$\begin{aligned} \mathbf{0} = {} & t_1(\lambda_1 - \lambda_q)\mathbf{v}_1 + t_2(\lambda_2 - \lambda_q)\mathbf{v}_2 + \cdots + t_j(\lambda_j - \lambda_q)\mathbf{v}_j \\ & + \cdots + t_{q-1}(\lambda_{q-1} - \lambda_q)\mathbf{v}_{q-1}. \end{aligned}$$

Because $t_j \neq 0$ and $\lambda_j - \lambda_q \neq 0$, this is a nontrivial way to write $\mathbf{0}$ as a linear combination of $\{\mathbf{v}_1, \mathbf{v}_2, \ldots, \mathbf{v}_{q-1}\}$, contradicting the linear independence of this set. ■

As a consequence of this theorem, we find a useful *sufficient* condition for $A \in \mathbb{F}_{n\times n}$ to be diagonalizable, that is, having n-many distinct eigenvalues.

34.10 Exercise

(i) If $A \in \mathbb{F}_{n\times n}$, then A has at most n-many distinct eigenvalues.
(ii) If $A \in \mathbb{F}_{n\times n}$ and A has n-many distinct eigenvalues in $\mathbb{F}$, then A is diagonalizable over $\mathbb{F}$.

It is important to emphasize that the condition in 34.10(ii), although sufficient to guarantee diagonalizability, is *not* necessary. A condition that is both necessary and sufficient will be given in Section 37.

Problem Set 34

34.1. Go back and do any of the exercises from Section 34 that you may have skipped.

*34.2. Suppose you know that a 4×4 matrix A has distinct eigenvalues a, b, c, d and that $\beta = \{\mathbf{u}_1, \mathbf{u}_2, \mathbf{u}_3, \mathbf{u}_4\}$ is a basis for $\mathbb{R}^4$ consisting of eigenvectors of A. What additional information do you need to find a diagonal matrix D and an invertible matrix P such that $P^{-1}AP = D$?

*34.3. Find an invertible matrix P such that

$$P^{-1}\begin{bmatrix} -1 & 0 \\ 0 & 4 \end{bmatrix}P = \begin{bmatrix} 4 & 0 \\ 0 & -1 \end{bmatrix}$$

Hint: This problem requires only mental arithmetic plus an understanding of the theorems from Section 34.

34.4. Let $A \in \mathbb{R}_{n \times n}$ and $t \in \mathbb{R}$. What is the relationship between the spectrum of A and the spectrum of $A + tI_n$?

35

Computation of Eigenvalues and Eigenvectors

Let $A \in \mathbb{F}_{n \times n}$, $\mathbf{v} \in \mathbb{F}^n$, and $\lambda \in \mathbb{F}$. Note that multiplication by I_n satisfies $\lambda\mathbf{v} = \lambda I_n \mathbf{v}$. Using this fact plus results from earlier chapters about determinants and invertibility, it is easy to verify that

$$\begin{aligned}
&\lambda \text{ is an eigenvalue of A} \\
&\quad \text{iff for some } \mathbf{0} \neq \mathbf{v} \in \mathbb{F}^n,\ A\mathbf{v} = \lambda\mathbf{v} \\
&\quad \text{iff for some } \mathbf{0} \neq \mathbf{v} \in \mathbb{F}^n,\ (\lambda I_n - A)\mathbf{v} = \mathbf{0} \\
&\quad \text{iff } \det(\lambda I_n - A) = 0
\end{aligned}$$

From this, we draw two conclusions. First, the problem of finding eigenvectors is solved once the eigenvalues are known; for if λ is an eigenvalue of A, then the corresponding eigenspace, E_λ, is just the nullspace of the matrix $\lambda I_n - A$. This can be found, as in Section 12, using the Gauss–Jordan algorithm. Second, the problem of finding eigenvalues is reduced to that of computing $\det(\lambda I_n - A)$ and finding the values of λ for which this determinant is 0.

Let us return to Example 33.1 to illustrate how the eigenvalues and eigenvectors were obtained.

35.1 Example 33.1 Revisited

$$A = \begin{bmatrix} -2 & -2 \\ -5 & 1 \end{bmatrix}, \qquad \lambda I_2 = \begin{bmatrix} \lambda & 0 \\ 0 & \lambda \end{bmatrix}, \quad \text{and so}$$

$$\lambda I_2 - A = \begin{bmatrix} \lambda + 2 & 2 \\ 5 & \lambda - 1 \end{bmatrix}$$

Thus, $\det(\lambda I_2 - A) = (\lambda + 2)(\lambda - 1) - (2)(5) = \lambda^2 + \lambda - 2 - 10 = \lambda^2 + \lambda - 12 = (\lambda + 4)(\lambda - 3)$. The values of λ for which $\det(\lambda I_2 - A) = 0$ are therefore -4 and 3. Having found the eigenvalues of A, we proceed to find the eigenspaces E_{-4} and E_3 by solving the two homogeneous systems $(-4I_2 - A)\mathbf{x} = \mathbf{0}$ and $(3I_2 - A)\mathbf{x} = \mathbf{0}$.

$$-4I_2 - A = \begin{bmatrix} -2 & 2 \\ 5 & -5 \end{bmatrix}$$

Applying the Gauss–Jordan algorithm, we find

$$GJ(-4I_2 - A) = \begin{bmatrix} 1 & -1 \\ 0 & 0 \end{bmatrix}$$

Hence $E_{-4} = NS(-4I_2 - A) = \{(x_1, x_2) \in \mathbb{R}^2 \colon x_1 = x_2\}$.

Similarly,

$$GJ(3I_2 - A) = GJ\left(\begin{bmatrix} 5 & 2 \\ 5 & 2 \end{bmatrix}\right) = \begin{bmatrix} 1 & \frac{2}{5} \\ 0 & 0 \end{bmatrix}$$

and so $E_3 = NS(3I_2 - A) = \{(x_1, x_2) \in \mathbb{R}^2 \colon x_1 = -\frac{2}{5}x_2\}$.

We learned in Section 34 that, if we choose any two eigenvectors $\mathbf{v}_1 \in E_{-4}$ and $\mathbf{v}_2 \in E_3$, then the set $\{\mathbf{v}_1, \mathbf{v}_2\}$ is linearly independent. It is therefore a basis for $\mathbb{R}^2$. In Example 33.1, we chose $\mathbf{v}_1 = (1, 1)$ and $\mathbf{v}_2 = (-2/5, 1)$ and concluded, as in 34.9, that A is diagonalizable with

$$\begin{bmatrix} 1 & -\frac{2}{5} \\ 1 & 1 \end{bmatrix}^{-1} \begin{bmatrix} -2 & -2 \\ -5 & 1 \end{bmatrix} \begin{bmatrix} 1 & -\frac{2}{5} \\ 1 & 1 \end{bmatrix} = \begin{bmatrix} -4 & 0 \\ 0 & 3 \end{bmatrix}$$

Before discussing the general situation, we should see another example. We profit from this opportunity to illustrate, at the same time, that the sufficient condition for diagonalizability given in 34.10 is not necessary; that is, that a matrix $A \in \mathbb{F}_{n \times n}$ might be diagonalizable over $\mathbb{F}$ even though it does not have n-many distinct eigenvalues in $\mathbb{F}$.

35.2 Example

Consider the matrix $A \in \mathbb{R}_{4\times 4}$, where

$$A = \begin{bmatrix} -5 & 2 & 0 & 1 \\ 0 & -5 & 0 & 0 \\ -3 & 12 & 0 & 3 \\ 0 & 2 & 0 & -4 \end{bmatrix}$$

Thus,

$$\lambda I_4 - A = \begin{bmatrix} \lambda+5 & -2 & 0 & -1 \\ 0 & \lambda+5 & 0 & 0 \\ 3 & -12 & \lambda & -3 \\ 0 & -2 & 0 & \lambda+4 \end{bmatrix}$$

Evaluating the determinant by cofactor expansion along the third column, we obtain

$$|\lambda I_4 - A| = (-1)^{3+3}(\lambda)\begin{vmatrix} \lambda+5 & -2 & -1 \\ 0 & \lambda+5 & 0 \\ 0 & -2 & \lambda+4 \end{vmatrix}$$

$$= (\lambda+5)(\lambda+5)(\lambda+4)(\lambda) = (\lambda+5)^2(\lambda+4)(\lambda)$$

A has only three distinct eigenvalues, -5, -4, and 0. To find the eigenspace E_{-5}, apply the Gauss–Jordan algorithm to $-5I_4 - A$. You should verify that $E_{-5} = NS(-5I_4 - A) = \{(x_1, x_2, x_3, x_4) \in \mathbb{R}^4\colon x_1 = \frac{5}{3}x_3 - x_4,\ x_2 = -\frac{1}{2}x_4\}$. A basis for E_{-5} is therefore $\{(\frac{5}{3}, 0, 1, 0), (-1, -\frac{1}{2}, 0, 1)\}$. Similarly, by row-reducing $-4I_4 - A$ and $0I_4 - A$, you should verify that $\{(1, 0, 0, 1)\}$ is a basis for E_{-4} and $\{(0, 0, 1, 0)\}$ is a basis for E_0. You should finally verify that the set $\{(\frac{5}{3}, 0, 1, 0), (-1, -\frac{1}{2}, 0, 1), (1, 0, 0, 1), (0, 0, 1, 0)\}$ is linearly independent and is therefore a basis for $\mathbb{R}^4$ consisting of eigenvectors of A. (Remark concerning the final step: we will later prove a theorem which *guarantees* that vectors chosen this way will be linearly independent, so this final step is not really required; note that Theorem 34.9 does not apply because two of the vectors were chosen from the same eigenspace.) The conclusion is that A is diagonalizable and

$$\begin{bmatrix} \frac{5}{3} & -1 & 1 & 0 \\ 0 & -\frac{1}{2} & 0 & 0 \\ 1 & 0 & 0 & 1 \\ 0 & 1 & 1 & 0 \end{bmatrix}^{-1} \begin{bmatrix} -5 & 2 & 0 & 1 \\ 0 & -5 & 0 & 0 \\ -3 & 12 & 0 & 3 \\ 0 & 2 & 0 & -4 \end{bmatrix} \begin{bmatrix} \frac{5}{3} & -1 & 1 & 0 \\ 0 & -\frac{1}{2} & 0 & 0 \\ 1 & 0 & 0 & 1 \\ 0 & 1 & 1 & 0 \end{bmatrix}$$

$$= \begin{bmatrix} -5 & 0 & 0 & 0 \\ 0 & -5 & 0 & 0 \\ 0 & 0 & -4 & 0 \\ 0 & 0 & 0 & 0 \end{bmatrix}.$$

Returning now to the general case, with $A \in \mathbb{F}_{n\times n}$, we note that

$$\lambda I_n - A = \begin{bmatrix} \lambda - a_{11} & -a_{12} & -a_{13} & \cdots & -a_{1n} \\ -a_{21} & \lambda - a_{22} & -a_{23} & \cdots & -a_{2n} \\ \vdots & \vdots & \vdots & \cdots & \vdots \\ -a_{n1} & -a_{n2} & -a_{n3} & \cdots & \lambda - a_{nn} \end{bmatrix}$$

So

$$\begin{aligned} \det(\lambda I_n - A) &= \textstyle\sum \text{ signed elementary products from } \lambda I_n - A \\ &= (\lambda - a_{11})(\lambda - a_{22}) \cdots (\lambda - a_{nn}) \\ &\qquad + (n! - 1)\text{-many additional signed elementary products} \\ &= \lambda^n + \text{other terms involving lower powers of } \lambda \\ &= \lambda^n + c_{n-1}\lambda^{n-1} + \cdots + c_1\lambda + c_0. \end{aligned}$$

In general, where $A \in \mathbb{F}_{n\times n}$, $\det(\lambda I_n - A)$ is a monic (i.e., the leading coefficient is 1) polynomial of degree n in the variable λ. The eigenvalues of A are the roots in $\mathbb{F}$ of this polynomial.

35.3 Definition

For $A \in \mathbb{F}_{n\times n}$, the polynomial $\det(\lambda I_n - A)$ is called *the characteristic polynomial of A* and is denoted by $\chi(A)$ or, when we wish to emphasize that it is a polynomial in the variable λ, by $\chi_A(\lambda)$. See [F]35.1.

We begin our theoretical discussion of characteristic polynomials by emphasizing, once more, that diagonalizability of $A \in \mathbb{F}_{n\times n}$ is an intrinsic property not of A but of the underlying linear operator $T: \mathbb{F}^n \to \mathbb{F}^n$. We could therefore have anticipated the next theorem, which asserts that similar matrices have the same characteristic polynomial and hence the same eigenvalues with the same multiplicities.

35.4 Theorem

If $A, B, P \in \mathbb{F}_{n\times n}$, if P is invertible, and if $P^{-1}AP = B$, then $\chi(A) = \chi(B)$.

Proof:

$$\begin{aligned} \chi(B) &= \det(I_n - B) \\ &= \det(\lambda I_n - P^{-1}AP) \\ &= \det(\lambda P^{-1} I_n P - P^{-1}AP) \\ &= \det(P^{-1}(\lambda I_n - A)P) \\ &= \det P^{-1} \det(\lambda I_n - A) \det P \\ &= \frac{1}{\det P}\chi(A) \det P \\ &= \chi(A) \end{aligned}$$

■

In view of this theorem, we broaden the scope of our earlier definition.

35.5 Definition

Let $T: \mathbb{F}^n \to \mathbb{F}^n$ be a linear operator.

(i) A scalar $\lambda \in \mathbb{F}$ is called an *eigenvalue* of T if $T(\mathbf{v}) = \lambda\mathbf{v}$ for some nonzero vector $\mathbf{v} \in \mathbb{F}^n$.

(ii) A vector $\mathbf{v} \in \mathbb{F}^n$ is called an *eigenvector* of T if $\mathbf{v} \neq \mathbf{0}$ and $T(\mathbf{v}) = \lambda\mathbf{v}$ for some scalar $\lambda \in \mathbb{F}$.

(iii) $\{\lambda \in \mathbb{F}: \lambda \text{ is an eigenvalue of } T\}$ is called the *spectrum* of T.

What could prevent a matrix from being diagonalizable? We will present examples to illustrate two kinds of obstructions, and in Section 37 we will prove a theorem which asserts that these two obstructions are the only ones.

35.6 Example

Consider $B \in \mathbb{R}_{3\times 3}$, where

$$B = \begin{bmatrix} 0 & 8 & 1 \\ 1 & 0 & 0 \\ 0 & 4 & -1 \end{bmatrix}$$

So

$$\lambda I_3 - B = \begin{bmatrix} \lambda & -8 & -1 \\ -1 & \lambda & 0 \\ 0 & -4 & \lambda + 1 \end{bmatrix}$$

and

$$\begin{aligned} \chi(B) &= |(\lambda I_3 - B)| = (\lambda)(\lambda)(\lambda + 1) - 4 - 8(\lambda + 1) \\ &= \lambda^3 + \lambda^2 - 8\lambda - 12 \\ &= (\lambda - 3)(\lambda + 2)(\lambda + 2) \\ &= (\lambda - 3)(\lambda + 2)^2 \end{aligned}$$

Although $\chi(B)$ has three real roots, one of them, -2, is a double root of the characteristic polynomial, so B has only two distinct eigenvalues, -2 and 3. To find $E_{-2} = \mathrm{NS}(-2I_3 - B)$, we apply the Gauss–Jordan algorithm to

$$-2I_3 - B = \begin{bmatrix} -2 & -8 & -1 \\ -1 & -2 & 0 \\ 0 & -4 & -1 \end{bmatrix}$$

You should verify that

$$\mathrm{GJ}(-2I_3 - B) = \begin{bmatrix} 1 & 0 & -\frac{1}{2} \\ 0 & 1 & \frac{1}{4} \\ 0 & 0 & 0 \end{bmatrix}$$

so $E_{-2} = \{(x_1, x_2, x_3) \in \mathbb{R}^3\colon x_1 = \frac{1}{2}x_3, x_2 = -\frac{1}{4}x_3\}$. A basis for E_{-2} is $\{(\frac{1}{2}, -\frac{1}{4}, 1)\}$.

A similar computation shows that $E_3 = \{(x_1, x_2, x_3) \in \mathbb{R}^3\colon x_1 = 3x_3, x_2 = x_3\}$. A basis for E_3 is $\{(3,1,1)\}$. Because each of these eigenspaces is only one dimensional, it is not apparent how one could find a third eigenvector that is linearly independent from $\{(\frac{1}{2}, -\frac{1}{4}, 1), (3,1,1)\}$. In fact, this is impossible, for if $\mathrm{A}\mathbf{v} = \lambda\mathbf{v}$ then either $\lambda = -2$ or $\lambda = 3$, so either $\mathbf{v} \in E_{-2}$ or $\mathbf{v} \in E_3$. So $\mathbf{v}$ is either a multiple of $(\frac{1}{2}, -\frac{1}{4}, 1)$ or is a multiple of $(3,1,1)$. So A is not diagonalizable.

The source of the obstruction to diagonalizability lies not with E_3 but with E_{-2}. We will see in Section 37 that, for A to be diagonalizable, the dimension of E_λ must equal the multiplicity of λ as a root of $\chi(\mathrm{A})$. In this example, -2 is a double root, but $\dim E_{-2} = 1$, so A is not diagonalizable.

Note the difference between this example and the one in 35.2. In that example, the eigenvalue -5 was a double root of the characteristic polynomial, but $\dim E_{-5} = 2$. We defer to Section 37 a more general discussion of this phenomenon.

The purpose of the next two examples is to convince you that a complete treatment of diagonalizability requires the introduction of complex numbers. A brief but systematic review of complex numbers will be presented in the next section. The next two examples assume only a vague familiarity with this subject and are intended to serve as motivation.

35.7 Example

Consider the matrix $\mathrm{A} = \begin{bmatrix} 4 & 2 \\ -1 & 2 \end{bmatrix} \in \mathbb{R}_{2\times 2}$. Here, $\chi(\mathrm{A}) = |\lambda \mathrm{I}_2 - \mathrm{A}| = (\lambda - 4)(\lambda - 2) + 2 = \lambda^2 - 6\lambda + 10$. The eigenvalues of A are the roots of this polynomial, which, using the quadratic formula, are seen to be $(\frac{1}{2})[6 \pm \sqrt{36 - 40}\,] = 3 \pm i$, which are complex numbers. An example like this exposes the reason for using the letter $\mathbb{F}$ to denote the scalars, and for using the clause "over $\mathbb{F}$" in defining the concept of diagonalizability over $\mathbb{F}$. In Theorem 34.7, we saw that if $\mathrm{A} \in \mathbb{R}_{n\times n}$ is diagonalizable *over* $\mathbb{F}$, say $\mathrm{P}^{-1}\mathrm{AP} = \mathrm{D}$, where P and $\mathrm{D} \in \mathbb{F}_{n\times n}$, then the entries on the main diagonal of D must be eigenvalues of A. So the matrix $\mathrm{A} = \begin{bmatrix} 4 & 2 \\ -1 & 2 \end{bmatrix} \in \mathbb{R}_{2\times 2}$ is not diagonalizable *over* $\mathbb{R}$. The obstruction to diagonalizability over $\mathbb{R}$ is simply the lack of enough real eigenvalues. It *is*, however, diagonalizable over $\mathbb{C}$, where $\mathbb{C}$ denotes the complex numbers. In other words, although it may be impossible, for a given $\mathrm{A} \in \mathbb{R}_{n\times n}$, to find an invertible P and a diagonal D in $\mathbb{R}_{n\times n}$ with $\mathrm{P}^{-1}\mathrm{AP} = \mathrm{D}$, it may still be possible to find such a P and D in $\mathbb{C}_{n\times n}$. Viewing the matrix $\mathrm{A} = \begin{bmatrix} 4 & 2 \\ -1 & 2 \end{bmatrix}$ as an element of $\mathbb{C}_{2\times 2}$, A has two distinct eigenvalues, $3 + i$ and $3 - i$, and is therefore diagonalizable over $\mathbb{C}$ by Theorem 34.10(ii). We find the eigenspaces in the usual way, although this now involves row-reducing matrices whose entries are complex numbers. The detailed solution to Example 35.8 is presented in item 36.20. You may wish to read this item immediately if you are sufficiently familiar with complex numbers.

35.8 Example

As another simple illustration of this phenomenon, let us consider the linear operator T on $\mathbb{R}^2$, which rotates every vector in $\mathbb{R}^2$ through the angle θ, where $0 \leq \theta < 2\pi$. Thus, using 31.9,

$$A = {}_{\alpha^2}[T]_{\alpha^2} = \begin{bmatrix} \cos\theta & -\sin\theta \\ \sin\theta & \cos\theta \end{bmatrix}$$

Of course, when $\theta = 0$, $A = I_2$, T is the identity operator, $\chi(A) = (\lambda - 1)^2$, 1 is a double root of $\chi(A)$, and $E_1 = \mathbb{R}^2$.

When $\theta = \pi$, $A = \begin{bmatrix} -1 & 0 \\ 0 & -1 \end{bmatrix}$, $\chi(A) = (\lambda + 1)^2$, -1 is a double root of $\chi(A)$, and again $E_{-1} = \mathbb{R}^2$. Thus A *is* diagonalizable over $\mathbb{R}$ if $\theta = 0$ or π. (Indeed, A is itself a diagonal matrix.)

But for θ not equal to 0 or π, it is geometrically obvious that no nonzero vector in $\mathbb{R}^2$ is sent by T into a multiple of itself. This means that T has no real eigenvectors or, equivalently, that A has no real eigenvalues. We confirm this intuitive argument by calculating

$$\chi(A) = |\lambda I - A| = \begin{vmatrix} \lambda - \cos\theta & \sin\theta \\ -\sin\theta & \lambda - \cos\theta \end{vmatrix} = (\lambda - \cos\theta)^2 + \sin^2\theta$$
$$= \lambda^2 - (2\cos\theta)\lambda + 1$$

Using the quadratic formula, we find that the roots are $\cos\theta \pm i\sin\theta$; since $\sin\theta \neq 0$ for $\theta \neq 0, \pi$, these roots are indeed imaginary. Again, more detailed comments on this example are delayed to 36.21.

In Section 37, we give a necessary and sufficient condition for a matrix $A \in \mathbb{F}_{n\times n}$ to be diagonalizable over $\mathbb{F}$. Strictly speaking, Section 37 does not depend logically on Section 36 and can be understood, although not fully appreciated, by someone who does not know about complex numbers.

Problem Set 35

35.1. Go back and do any of the exercises from Section 35 that you may have skipped.

*35.2. Decide whether the matrix

$$A = \begin{bmatrix} 1 & 2 & 0 \\ -1 & 4 & 0 \\ -3 & 5 & -1 \end{bmatrix}$$

is diagonalizable and, if it is, find D and P such that $P^{-1}AP = D$.

*35.3. Decide whether the matrix

$$B = \begin{bmatrix} 2 & 0 & 3 \\ 5 & 2 & -2 \\ 0 & 0 & -1 \end{bmatrix}$$

is diagonalizable and, if it is, find D and P such that $P^{-1}BP = D$.

*35.4. Show that if $A \in \mathbb{R}_{n \times n}$ then the constant term in $\chi(A)$ is $(-1)^n \det(A)$.
Hint: Recall that the constant term in a polynomial $p(\lambda)$ is equal to $p(0)$.

*35.5. Show that if $A \in \mathbb{R}_{n \times n}$ then the coefficient of λ^{n-1} in $\chi(A)$ is the negative of the trace of A (= the sum of the diagonal entries of A; recall Problem 3.18).

35.6. Use the results from the previous two problems to find a matrix whose characteristic polynomial is $\lambda^2 - 4\lambda + 13$.

35.7. In this problem, we show that *every* monic polynomial of degree n is the characteristic polynomial of some $n \times n$ matrix. Let $p(\lambda) = \lambda^n + c_{n-1}\lambda^{n-1} + \cdots + c_1\lambda + c_0$. The matrix

$$C_p = \begin{bmatrix} 0 & 1 & 0 & \cdots & 0 \\ 0 & 0 & 1 & \cdots & 0 \\ 0 & 0 & 0 & \cdots & 0 \\ \vdots & \vdots & \vdots & \cdots & \vdots \\ 0 & 0 & 0 & \cdots & 1 \\ -c_0 & -c_1 & -c_2 & \cdots & -c_{n-1} \end{bmatrix}$$

is called the *companion matrix of* $p(\lambda)$. Show that $p(\lambda)$ is the characteristic polynomial of C_p.
Hint: The proof is by induction on n. Let $B = \lambda I_n - C_p$. Use cofactor expansion along the first column to show that $|B| = \lambda b_{11}^c + c_0$ and use the Induction Hypothesis to conclude that $b_{11}^c = \lambda^{n-1} + c_{n-1}\lambda^{n-2} + \cdots + c_2\lambda + c_1$.

35.8. A is a 2×2 matrix having 1 and -1 as eigenvalues. Given that $(2, -1) \in E_1$ and $(1, 2) \in E_{-1}$, find A.

35.9. Let $A = \begin{bmatrix} -17 & 30 \\ -10 & 18 \end{bmatrix}$.

(i) Find the characteristic polynomial of A.
(ii) Find the eigenvalues of A.
(iii) Find an invertible matrix P and a diagonal matrix D such that $D = P^{-1}AP$.
(iv) Find A^{14}.

36

An Aside on Complex Numbers

This section is *not* a comprehensive introduction to complex numbers. The most appropriate place for this is in the high school curriculum; it may alternatively constitute a component of a course in college algebra for first-year students.

Yet we must reluctantly recognize that many students of linear algebra will not have been exposed to complex numbers, especially those whose major subject is neither mathematics nor one of the physical sciences.

Because a comprehensive treatment would be out-of-place in the present text, this section is rather an outline, with very few proofs, which stresses those aspects of the theory of complex numbers that are particularly relevant for linear algebra.

Because the square of any real number is nonnegative, the equation $x^2 + 1 = 0$ has no solution in $\mathbb{R}$; equivalently, $x^2 + 1$ is a polynomial with real coefficients that has no real root.

In the eighteenth century, mathematicians artificially introduced a "number," which they denoted by i because they considered it "imaginary," which would be a root of this polynomial; thus i has the property that $i^2 = -1$. This number would be treated like any other number and could interact with reals under addition, multiplication, and so on.

Because $i^2 = -1$, $i^3 = -i$, $i^4 = 1$, $i^5 = i$, and so on, every number in this extended system can be uniquely written in the form $a + bi$, where a and b are real.

36.1 Definition

(i) The *complex number system*, denoted by $\mathbb{C}$, consists of the set of all numbers of the form $a + bi$, where $a, b \in \mathbb{R}$, together with the operations of addition and multiplication defined as follows:

$$(a + bi) + (c + di) = (a + c) + (b + d)i$$
$$(a + bi)(c + di) = (ac - bd) + (ad + bc)i$$

(ii) Let $z = a + bi$. The real number a is called the *real part of* z and is denoted by $\text{Re}\, z$. The real number b is called the *imaginary part of* z and is denoted by $\text{Im}\, z$.

(ii) Let $z = a + bi \in \mathbb{C}$; the complex number $a - bi$ is called the *complex conjugate of* z and is denoted by $\bar{z}$ (to be read: z bar).

Two complex numbers are equal iff both their real and imaginary parts are equal. It is clear that $\mathbb{R} \subsetneq \mathbb{C}$; in fact, $\mathbb{R} = \{z \in \mathbb{C}\colon \text{Im}\, z = 0\} = \{z \in \mathbb{C}\colon z = \bar{z}\}$.

Complex conjugation obeys some very simple rules, which are easy to prove and which are summarized in the next theorem.

36.2 Theorem

For all $z, z_1, z_2 \in \mathbb{C}$,

(i) $\overline{z_1 + z_2} = \bar{z}_1 + \bar{z}_2$

(ii) $\overline{z_1 z_2} = \bar{z}_1 \bar{z}_2$; in particular, $\overline{rz} = r\bar{z}$ if $r \in \mathbb{R}$

(iii) $\bar{\bar{z}} = z$. ■

Not all aspects of $\mathbb{R}$ can be generalized to $\mathbb{C}$; there is no consistent way, for example, to extend the order relation $\leq$ from $\mathbb{R}$ to $\mathbb{C}$. Declaring $i^2 = -1$ by fiat led not to a contradiction but to the very rich structure $\mathbb{C}$ extending $\mathbb{R}$; attempting, however, to declare either $i < 0$ or $i > 0$ by fiat leads to contradictions. For instance, if $0 < i$, then multiplying both sides of this inequality by the "positive" number i leads to $(0)(i) < (i)(i)$, or $0 < -1$.

We will continue to use the symbols $\leq$, $\lneq$, etc., in the context of complex numbers, but we will never write $z_1 \leq z_2$ unless z_1 and z_2 are real. In particular, the expression "$0 \leq z$" *means* that z is real and nonnegative.

The ability to express a complex number z uniquely in the form $z = a + bi$ establishes a one-to-one correspondence between the elements of $\mathbb{C}$ and those of $\mathbb{R}^2$. This correspondence provides a useful geometric picture of $\mathbb{C}$ when we identify complex numbers with vectors in the plane. This pictorial representation in Figure 36.1 is known as the *complex plane*.

The complex number $z = a + bi$ is identified with the vector whose coordinates are (a, b). In this representation, the points on the horizontal axis are identified with the real numbers (because $b = \text{Im}\, z = 0$); the points

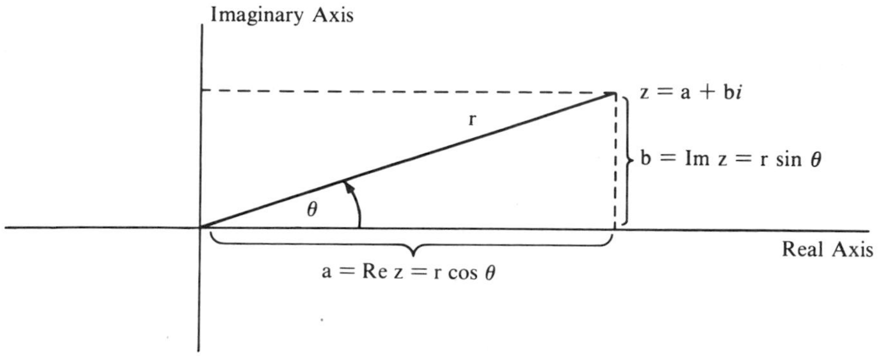

Figure 36.1

on the vertical axis are called *pure imaginary* (because $a = \operatorname{Re} z = 0$). The horizontal and vertical axes are therefore called the *real axis* and *imaginary axis*, respectively.

Figure 36.1 suggests an alternate representation of complex numbers, called the *polar representation*, which identifies a complex number by specifying the length of the vector and the angle that it makes with the real axis.

In Figure 36.1, we have

$$\cos\theta = \frac{a}{\sqrt{a^2+b^2}} \quad \text{and} \quad \sin\theta = \frac{b}{\sqrt{a^2+b^2}}$$

so we can rewrite $z = a + bi$ in the form

$$z = \sqrt{a^2+b^2}\cos\theta + \left(\sqrt{a^2+b^2}\sin\theta\right)i$$

which is conventionally rewritten as $z = |z|(\cos\theta + i\sin\theta)$, where $|z| = \sqrt{a^2+b^2}$.

36.3 Definition

Let $z = a + bi = \sqrt{a^2+b^2}\,(\cos\theta + i\sin\theta)$.

(i) The real number $\sqrt{a^2+b^2}$ is called the *modulus of z* and is denoted by $|z|$.

(ii) There are infinitely many angles θ (all differing by an integer multiple of 2π) for which the preceding equation holds; these are called the *arguments of z* and are all denoted by arg z. The *principal argument of z* is the unique argument in the interval $-\pi \lneqq \theta \leq \pi$ and is denoted by Arg z.

The reason for the notation $|z|$ is that the modulus of a complex number is the analogue of the absolute value of a real number. As the next theorem indicates, the analogy is so strong that $|z|$ is often called the *absolute value of* z. Indeed, if $z = a + bi$ happens to be real, then $z = a$ and its modulus is equal to its absolute value, so this is a consistent use of the notation and nomenclature.

36.4 Theorem

For all $z, z_1, z_2 \in \mathbb{C}$,

(i) $|z|$ is a real, nonnegative number.
(ii) $|z_1 + z_2| \leq |z_1| + |z_2|$.
(iii) $|z_1 z_2| = |z_1|\,|z_2|$. ■

The fact that $z\bar{z} = (a + bi)(a - bi) = a^2 + b^2 = |z|^2$ implies that every nonzero complex number has a multiplicative inverse, which we will denote, as for reals, by z^{-1} or by $1/z$.

$$\frac{1}{z} = z^{-1} = \frac{\bar{z}}{|z|^2}, \qquad \text{if } z \neq 0.$$

For example,

$$\frac{1}{3 - 4i} = \frac{3 + 4i}{3^2 + 4^2} = \frac{3}{25} + \frac{4}{25}i.$$

We are able, with the geometric representation (Figure 36.1) of complex numbers, to picture the effects of the various algebraic operations. For some of the operations, this is not surprising. Addition of complex numbers follows the same "parallelogram law" as addition for vectors in $\mathbb{R}^2$. Multiplication of a complex number by a *real* number has the same geometric interpretation as multiplication of a vector in $\mathbb{R}^2$ by a (real) scalar. We will not repeat the figures here; see Figures 5.1 and 5.2. Complex conjugation corresponds to reflection about the real axis; see Figure 36.2.

A distinct novelty in this context is the existence of a geometric interpretation for multiplication of complex numbers. By definition, the algebraic relationship between $z_1 = a + bi$, $z_2 = c + di$, and $z_1 z_2$ is given by Definition 36.1(i).

$$z_1 z_2 = (ac - bd) + (ad + bc)i$$

The geometric interpretation of this algebraic relationship becomes clear when we rewrite it using the polar representations for complex numbers. Thus, by definition,

$$z_1 = |z_1|(\cos\theta_1 + i\sin\theta_1)$$
$$z_2 = |z_2|(\cos\theta_2 + i\sin\theta_2)$$

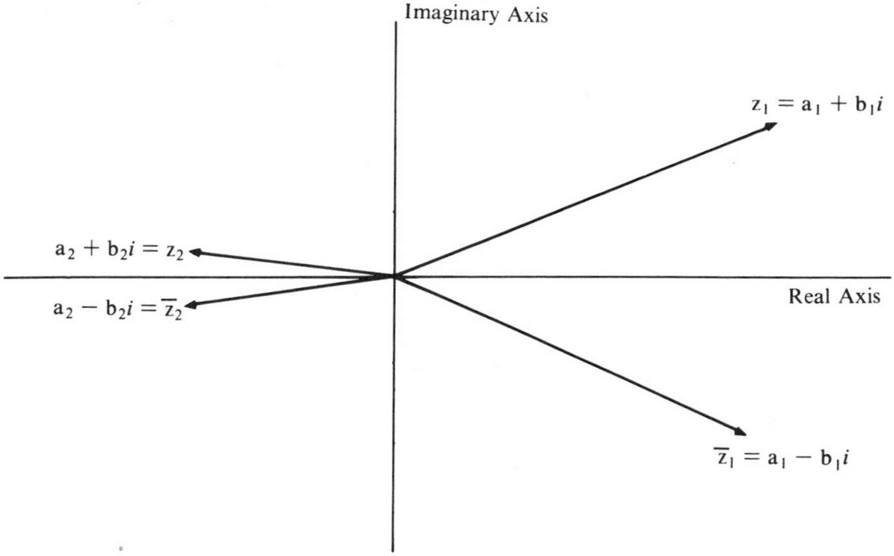

Figure 36.2

and

$$z_1 z_2 = |z_1 z_2|(\cos \theta_3 + i \sin \theta_3)$$

where θ_1, θ_2, and θ_3 are the angles that z_1, z_2, and $z_1 z_2$ make, respectively, with the real axis.

The remarkable conclusion is obtained when we recompute the polar form of $z_1 z_2$ by multiplying together the polar forms of z_1 and z_2. This computation uses the standard trigonometric identities for the sine and cosine of the sum of two angles.

$$\begin{aligned} z_1 z_2 &= |z_1|(\cos \theta_1 + i \sin \theta_1)|z_2|(\cos \theta_2 + i \sin \theta_2) \\ &= |z_1|\,|z_2|((\cos \theta_1 \cos \theta_2 - \sin \theta_1 \sin \theta_2) + i(\sin \theta_1 \cos \theta_2 + \cos \theta_1 \sin \theta_2)) \\ &= |z_1 z_2|(\cos(\theta_1 + \theta_2) + i \sin(\theta_1 + \theta_2)) \qquad \text{[Theorem 36.4(iii)]} \end{aligned}$$

Since the polar representation is unique, the conclusion is that $\theta_3 = \theta_1 + \theta_2$.

We may therefore interpret the product $z_1 z_2$ geometrically as the vector whose modulus, $|z_1 z_2|$, is the product of the moduli, $|z_1|\,|z_2|$, and whose argument is the sum of the arguments of z_1 and z_2. See Figure 36.3 for an example.

As a particular case, we should observe that multiplying a complex number z by another complex number w of modulus 1 (these are precisely the numbers of the form $w = \cos \theta + i \sin \theta$) simply rotates z through the angle θ.

The trigonometric identities used in the preceding argument can also be used to extend the definition of the exponential function e^x to complex

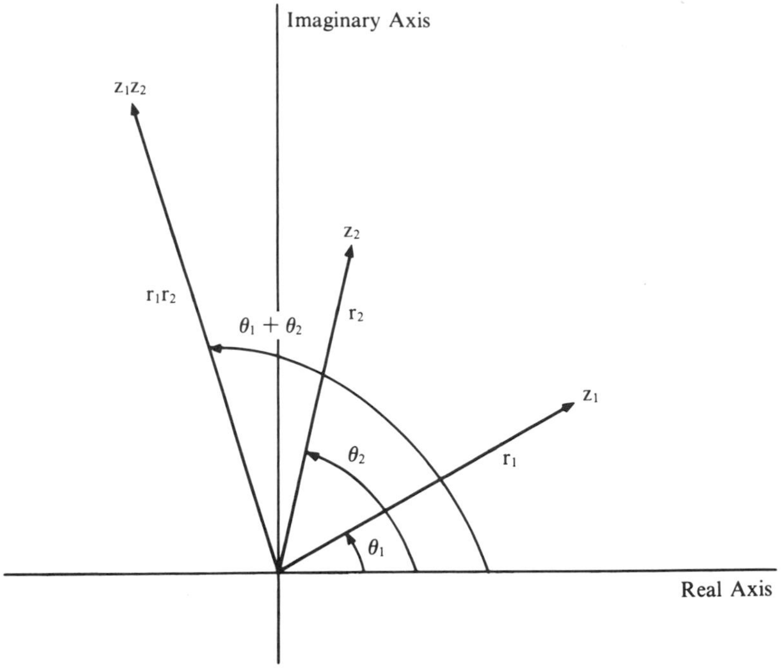

Figure 36.3

numbers. Let $z = a + bi$. The question is whether there is some useful way to define e^z. Of course, we want this definition to agree with the usual definition when z is real. We would also like the law of exponents, $e^{z_1}e^{z_2} = e^{z_1+z_2}$, to hold.

36.5 Definition

For $z = a + bi \in \mathbb{C}$, $e^z = e^a(\cos b + i \sin b)$.

Observe that, in case z is real, this definition of e^z agrees with the usual definition of e^x for real x; that is, when $b = 0$,

$$e^z = e^a(\cos 0 + i \sin 0) = e^a(1 + 0) = e^a$$

36.6 Exercise

(i) Verify that $e^{z_1}e^{z_2} = e^{z_1+z_2}$ holds when e^z is defined as in 36.5.
(ii) DeMoivre's Theorem: $(\cos\theta + i\sin\theta)^n = \cos n\theta + i \sin n\theta$ for any positive integer n.

We cannot pass up the opportunity to point out that $e^{i\pi} = -1$; this has become a popular T-shirt slogan within the mathematical fraternity. Geometric interpretation: multiplying a (complex) number by $e^{i\pi}$ (i.e., rotating it through the angle π) is the same as multiplying it by -1.

The following theorem and its corollary are the most important facts about the complex number system. $\mathbb{C}$ was created from $\mathbb{R}$ by the inclusion of the single new number i (together with all its real multiples), which was to serve as a root for the specific second-degree polynomial $x^2 + 1$; it is a truly remarkable consequence that this yields a number system in which *every* polynomial of *arbitrary* degree has a root.

Let $\mathbb{R}[x]$ ($\mathbb{C}[x]$) denote the set of all polynomials in x of arbitrary degree whose coefficients are elements of $\mathbb{R}$ (of $\mathbb{C}$).

36.7 Theorem (The Fundamental Theorem of Algebra)

If $f(x) \in \mathbb{C}[x]$ is not constant, then there exists $c \in \mathbb{C}$ such that $f(c) = 0$.

When this theorem is applied repeatedly in conjunction with the division algorithm for polynomials (i.e., if $f(x) \in \mathbb{C}[x]$, if $c \in \mathbb{C}$, and if $f(c) = 0$, then $f(x) = (x - c)g(x)$ for some $g(x) \in \mathbb{C}[x]$) the following corollary is obtained.

36.8 Corollary

If $f(x) \in \mathbb{C}[x]$ has degree $n \geq 1$, then $f(x)$ has n-many (counting multiplicities) roots in $\mathbb{C}$, and consequently factors over $\mathbb{C}$ into a product of n-many linear factors, that is,

$$f(x) = (x - c_1)^{m_1}(x - c_2)^{m_2} \cdots (x - c_p)^{m_p}$$

where $c_1, c_2, \ldots, c_p$ are the distinct roots of $f(x)$ in $\mathbb{C}$, where $m_1, m_2, \ldots, m_p$ are positive integers with m_i equal to the multiplicity of c_i as a root of $f(x)$, and where $m_1 + m_2 + \cdots + m_p = n$.

The Fundamental Theorem of Algebra has received many proofs over the years. Some depend substantially on the theory of functions of a complex variable. Others are more "elementary" and require only a minimal knowledge of complex numbers. See [BM]. ■

Since $\mathbb{R}[x] \subset \mathbb{C}[x]$, a real polynomial, which may or may not factor completely over $\mathbb{R}$, must factor completely over $\mathbb{C}$ into a product of linear factors.

36.9 Example

(i) $f(x) = x^2 + 1$.
$f(x)$ factored over $\mathbb{R}$: $x^2 + 1$.
$f(x)$ factored over $\mathbb{C}$: $(x + i)(x - i)$.
(ii) $g(x) = x^6 - 6x^5 + 27x^4 - 108x^3 + 243x^2 - 486x + 729$.
$g(x)$ factored over $\mathbb{R}$: $(x - 3)^2(x^2 + 9)^2$.
$g(x)$ factored over $\mathbb{C}$: $(x - 3)^2(x + 3i)^2(x - 3i)^2$.

We will conclude our survey of results about $\mathbb{C}$ with the fact that the complex roots of a real polynomial come in complex conjugate pairs.

36.10 Theorem

Suppose that $f(x) = a_nx^n + \cdots + a_1x + a_0 \in \mathbb{R}[x]$, that $c \in \mathbb{C}$, and that $f(c) = 0$; then $f(\bar{c}) = 0$.

Proof:

$$\begin{aligned} f(\bar{c}) &= \overline{\overline{a_n\bar{c}^n + \cdots + a_1\bar{c} + a_0}} && \text{[Theorem 36.2(iii)]} \\ &= \overline{\bar{a}_nc^n + \cdots + \bar{a}_1c + \bar{a}_0} && \text{[Theorem 36.2(i), (ii), and (iii)]} \\ &= \overline{a_nc^n + \cdots + a_1c + a_0} && [\text{since } a_i \in \mathbb{R}] \\ &= \bar{0} && [\text{since } f(c) = 0] \\ &= 0 && [\text{since } 0 \in \mathbb{R}] \end{aligned}$$ ∎

This concludes the background information on complex numbers. The remainder of the section is devoted to a sketch of the minor modifications required when the material from Chapters I to V is redone with $\mathbb{C}$ in place of $\mathbb{R}$.

Let $\mathbb{C}_{m\times n}$ denote the set of $m \times n$ matrices with entries from $\mathbb{C}$. All results from Chapter I about matrix arithmetic carry over unchanged. There is one new operation on $\mathbb{C}_{m\times n}$ arising from the presence in $\mathbb{C}$ of complex conjugates.

36.11 Definition

If $A \in \mathbb{C}_{m\times n}$, the *complex conjugate of A* is the $m \times n$ matrix, denoted by $\overline{A}$, whose $(i, j)^{\text{th}}$ entry is $\bar{a}_{ij}$.

36.12 Example

If $A = \begin{bmatrix} 3+i & 2+2i & 0 \\ 2 & i & 1-i \end{bmatrix}$, then $\overline{A} = \begin{bmatrix} 3-i & 2-2i & 0 \\ 2 & -i & 1+i \end{bmatrix}$.

The following theorem is easily proved using 36.2.

36.13 Theorem

Let $c \in \mathbb{C}$ and let A and B be complex matrices whose sizes are such that the indicated operations are defined; then

(i) $\overline{\overline{\mathrm{A}}} = \mathrm{A}$.
(ii) $\overline{\mathrm{A} + \mathrm{B}} = \overline{\mathrm{A}} + \overline{\mathrm{B}}$.
(iii) $\overline{c\mathrm{A}} = \bar{c}\overline{\mathrm{A}}$.
(iv) $\overline{\mathrm{AB}} = \overline{\mathrm{A}}\,\overline{\mathrm{B}}$. ■

The operations on $\mathbb{C}_{m \times n}$ of taking the transpose and the complex conjugate interact, and the result of this interaction will be important in later sections. It is clear that $(\overline{\mathrm{A}})^T = \overline{(\mathrm{A}^T)}$ (i.e., it makes no difference whether we conjugate first, then transpose, or transpose first, then conjugate).

36.14 Example

$$\mathrm{A} = \begin{bmatrix} 3+i & 2+2i & 0 \\ 2 & i & 1-i \end{bmatrix} \quad \overline{\mathrm{A}} = \begin{bmatrix} 3-i & 2-2i & 0 \\ 2 & -i & 1+i \end{bmatrix} \quad (\overline{\mathrm{A}})^T = \begin{bmatrix} 3-i & 2 \\ 2-2i & -i \\ 0 & 1+i \end{bmatrix}$$

$$\mathrm{A}^T = \begin{bmatrix} 3+i & 2 \\ 2+2i & i \\ 0 & 1-i \end{bmatrix} \quad \overline{(\mathrm{A}^T)} = \begin{bmatrix} 3-i & 2 \\ 2-2i & -i \\ 0 & 1+i \end{bmatrix}$$

It is convenient to have a single piece of notation for the common effect that these two operations have when performed in succession in either order.

36.15 Definition

For $\mathrm{A} \in \mathbb{C}_{m \times n}$, the matrix $(\overline{\mathrm{A}})^T = \overline{(\mathrm{A}^T)}$ is called the *Hermitian transpose* of A and is denoted by A^H.

Note that if A is a real matrix, then A^H is the same as A^T.
The proof of the following theorem is left as an easy exercise.

36.16 Theorem (Analogue of 3.15)

Let $c \in \mathbb{C}$ and let A and B be complex matrices whose sizes are such that the indicated operations are defined; then

(i) $(\mathrm{A}^H)^H = \mathrm{A}$.
(ii) $(\mathrm{A} + \mathrm{B})^H = \mathrm{A}^H + \mathrm{B}^H$.
(iii) $(c\mathrm{A})^H = \bar{c}\mathrm{A}^H$.

(iv) $(AB)^H = B^H A^H$.
(v) If A is invertible, then so is A^H and $(A^H)^{-1} = (A^{-1})^H$. ■

For square matrices, there are the possibilities that $A^H = A$ or $A^H = -A$.

36.17 Definition (Analogue of 3.17)

(i) A matrix $A \in \mathbb{C}_{n\times n}$ satisfying $A^H = A$ is called *Hermitian*.
(ii) A matrix $A \in \mathbb{C}_{n\times n}$ satisfying $A^H = -A$ is called *skew-Hermitian*.

The analogues for $\mathbb{C}^n$ of the material from Sections 5, 7, 8, and 9 on $\mathbb{R}^n$ will be covered in Chapter VIII on abstract vector spaces.

The material from Section 6 and Chapter III on solving linear systems and inverting matrices using the Gauss–Jordan algorithm carries over unchanged to the case where the coefficients are from $\mathbb{C}$. Here are a few examples.

36.18 Theorem (Analogue of 6.11)

If $C \in \mathbb{C}_{n\times n}$ is invertible, then for every $\mathbf{c} \in \mathbb{C}^n$, the linear system $C\mathbf{z} = \mathbf{c}$ has the unique solution $\mathbf{z} = C^{-1}\mathbf{c}$.

36.19 Example

Solve the linear system

$$\begin{aligned} (3+i)z_1 + (1+2i)z_2 &= 1+3i \\ -iz_1 + 2z_2 &= 2-i \end{aligned}$$

Solution:

$$\begin{bmatrix} z_1 \\ z_2 \end{bmatrix} = \begin{bmatrix} 3+i & 1+2i \\ -i & 2 \end{bmatrix}^{-1} \begin{bmatrix} 1+3i \\ 2-i \end{bmatrix}$$

$$= \frac{1}{(3+i)(2)-(1+2i)(-i)} \begin{bmatrix} 2 & -1-2i \\ i & 3+i \end{bmatrix} \begin{bmatrix} 1+3i \\ 2-i \end{bmatrix}$$

$$= \frac{1}{4+3i} \begin{bmatrix} -2+3i \\ 4 \end{bmatrix} = \frac{4-3i}{25} \begin{bmatrix} -2+3i \\ 4 \end{bmatrix} = \begin{bmatrix} \dfrac{1+18i}{25} \\ \dfrac{16-12i}{25} \end{bmatrix}.$$

If you need practice in complex arithmetic, you could substitute these values for z_1 and z_2 back into the original equations to verify that they are satisfied.

36.20 Example (Example 35.7 Revisited)

In that example, we saw that the 2×2 real matrix $A = \begin{bmatrix} 4 & 2 \\ -1 & 2 \end{bmatrix}$ has two distinct complex eigenvalues $3 + i$ and $3 - i$ and is therefore diagonalizable over $\mathbb{C}$. We find bases for the eigenspaces in the usual way, although this now requires row-reducing complex matrices.

Since $E_{3+i} = \text{NS}((3 + i)I_2 - A)$, we need to row-reduce the matrix

$$(3+i)I_2 - A = \begin{bmatrix} 3+i & 0 \\ 0 & 3+i \end{bmatrix} - \begin{bmatrix} 4 & 2 \\ -1 & 2 \end{bmatrix} = \begin{bmatrix} -1+i & -2 \\ 1 & 1+i \end{bmatrix}$$

Applying the first operation, $\dfrac{1}{-1+i}R_1$, in the Gauss–Jordan algorithm leads to the following:

$$\begin{bmatrix} 1 & \dfrac{-2}{-1+i} \\ 1 & 1+i \end{bmatrix} \quad \text{which simplifies to} \quad \begin{bmatrix} 1 & 1+i \\ 1 & 1+i \end{bmatrix}$$

The next step in the algorithm is to apply the operation $R_2 - R_1$ to this matrix; the result is $\begin{bmatrix} 1 & 1+i \\ 0 & 0 \end{bmatrix}$. Thus, $E_{3+i} = \{(z_1, z_2) \in \mathbb{C}^2 \colon z_1 = (-1 - i)z_2\}$. A basis for E_{3+i} is $\{(-1 - i, 1)\}$.

You should verify in a similar fashion that $E_{3-i} = \{(z_1, z_2) \in \mathbb{C}^2 \colon z_1 = (-1 + i)z_2\}$. A basis for E_{3-i} is $\{(-1 + i, 1)\}$.

Theorem 32.7 now guarantees that

$$\begin{bmatrix} -1-i & -1+i \\ 1 & 1 \end{bmatrix}^{-1} \begin{bmatrix} 4 & 2 \\ -1 & 2 \end{bmatrix} \begin{bmatrix} -1-i & -1+i \\ 1 & 1 \end{bmatrix} = \begin{bmatrix} 3+i & 0 \\ 0 & 3-i \end{bmatrix}$$

since $\{(-1 - i, 1), (-1 + i, 1)\}$ is a basis for $\mathbb{C}^2$ consisting of eigenvectors of A.

36.21 Example (Example 35.8 Revisited)

As another simple illustration of this phenomenon, let us consider the linear operator T on $\mathbb{R}^2$, which rotates every vector in $\mathbb{R}^2$ through the angle θ, where $0 \le \theta < 2\pi$. Thus, using 31.9,

$$A = {}_{\alpha^2}[T]_{\alpha^2} = \begin{bmatrix} \cos\theta & -\sin\theta \\ \sin\theta & \cos\theta \end{bmatrix}.$$

In Example 35.8 we found that the roots are $\cos\theta \pm i\sin\theta$, which, for $\theta \neq 0, \pi$, are not real.

As usual, let $e^{i\theta} = \cos\theta + i\sin\theta$; then $e^{-i\theta} = \cos\theta - i\sin\theta$. Since $\mathrm{A} \in \mathbb{C}_{2\times 2}$ has two distinct eigenvalues in $\mathbb{C}$, A is diagonalizable over $\mathbb{C}$ (by 34.10) and any matrix $\mathrm{P} \in \mathbb{C}_{2\times 2}$ whose first column is a nonzero vector from $E_{e^{i\theta}}$ and whose second column is a nonzero vector from $E_{e^{i\theta}}$ will satisfy

$$\mathrm{P}^{-1}\mathrm{AP} = \begin{bmatrix} e^{i\theta} & 0 \\ 0 & e^{-i\theta} \end{bmatrix}$$

To find such a P, we find bases for the eigenspaces in the usual way. Since $E_{e^{i\theta}} = \mathrm{NS}(e^{i\theta}\mathrm{I}_2 - \mathrm{A})$, we must row-reduce the matrix

$$e^{i\theta}\mathrm{I}_2 - \mathrm{A} = \begin{bmatrix} i\sin\theta & \sin\theta \\ -\sin\theta & i\sin\theta \end{bmatrix}$$

You should verify that

$$\mathrm{GJ}\left(e^{i\theta}\mathrm{I}_2 - \mathrm{A}\right) = \begin{bmatrix} 1 & -i \\ 0 & 0 \end{bmatrix}$$

Thus $E_{e^{i\theta}} = \{(z_1, z_2) \in \mathbb{C}^2\colon\ z_1 = iz_2\}$ and a basis for $E_{e^{i\theta}}$ is $\{(i, 1)\}$. Similarly, a basis for $E_{e^{-i\theta}}$ is $\{(-i, 1)\}$. The conclusion is that

$$\begin{bmatrix} i & -i \\ 1 & 1 \end{bmatrix}^{-1} \begin{bmatrix} \cos\theta & -\sin\theta \\ \sin\theta & \cos\theta \end{bmatrix} \begin{bmatrix} i & -i \\ 1 & 1 \end{bmatrix} = \begin{bmatrix} e^{i\theta} & 0 \\ 0 & e^{-i\theta} \end{bmatrix}.$$

There is an alternate way to view a complex $m \times n$ linear system $\mathbf{Cz} = \mathbf{c}$ that allows us to solve them using real arithmetic rather than complex arithmetic. Just as any $c \in \mathbb{C}$ can be uniquely written in the form $a + bi$ with $a, b \in \mathbb{R}$, it is clear that a complex vector $\mathbf{c} \in \mathbb{C}^n$ or a complex matrix $\mathrm{C} \in \mathbb{C}_{m\times n}$ can be written uniquely in the form $\mathbf{c} = \mathbf{a} + \mathbf{b}i$ and $\mathrm{C} = \mathrm{A} + \mathrm{B}i$, respectively, where $\mathbf{a}, \mathbf{b} \in \mathbb{R}^n$ and $\mathrm{A}, \mathrm{B} \in \mathbb{R}_{m\times n}$.

If we rewrite the complex $m \times n$ linear system

$$\mathbf{Cz} = \mathbf{c} \tag{†}$$

in the form $(\mathrm{A} + \mathrm{B}i)(\mathbf{x} + \mathbf{y}i) = \mathbf{a} + \mathbf{b}i$, and then multiply, we obtain

$$\mathrm{A}\mathbf{x} + \mathrm{B}\mathbf{x}i + \mathrm{A}\mathbf{y}i - \mathrm{B}\mathbf{y} = \mathbf{a} + \mathbf{b}i$$

The result of equating the real and imaginary parts of both sides is

$$\begin{aligned} \mathrm{A}\mathbf{x} - \mathrm{B}\mathbf{y} &= \mathbf{a} \\ \mathrm{B}\mathbf{x} + \mathrm{A}\mathbf{y} &= \mathbf{b} \end{aligned} \quad \text{or} \quad \left[\begin{array}{c|c} \mathrm{A} & -\mathrm{B} \\ \hline \mathrm{B} & \mathrm{A} \end{array}\right] \begin{bmatrix} [\mathbf{x}] \\ [\mathbf{y}] \end{bmatrix} = \begin{bmatrix} [\mathbf{a}] \\ [\mathbf{b}] \end{bmatrix} \tag{‡}$$

using partitioned matrices. This is an "equivalent" real linear system of $2m$ equations in the $2n$ unknowns $x_1, x_2, \ldots, x_n, y_1, y_2, \ldots, y_n$. Here, (†) and (‡) are "equivalent" not in the sense that they have the same solution space (obviously), but in the sense that knowing the solution to either one of them is tantamount to knowing the solution to the other; this is because $z_j = x_j + y_j i$ for $j = 1, 2, \ldots, n$.

To illustrate these ideas, let us resolve the system from Example 36.19 using this technique.

36.22 Example (Example 36.19 Revisited)

Solve the linear system

$$\begin{aligned} (3+i)z_1 + (1+2i)z_2 &= 1+3i \\ -iz_1 + 2z_2 &= 2-i \end{aligned}$$

Solution: This complex 2×2 system can be rewritten in the form

$$\left[\begin{bmatrix} 3 & 1 \\ 0 & 2 \end{bmatrix} + \begin{bmatrix} 1 & 2 \\ -1 & 0 \end{bmatrix} i\right]\left[\begin{bmatrix} x_1 \\ x_2 \end{bmatrix} + \begin{bmatrix} y_1 \\ y_2 \end{bmatrix} i\right] = \begin{bmatrix} 1 \\ 2 \end{bmatrix} + \begin{bmatrix} 3 \\ -1 \end{bmatrix} i$$

Equating the real and imaginary parts of both sides leads to the system

$$\begin{aligned} \begin{bmatrix} 3 & 1 \\ 0 & 2 \end{bmatrix}\begin{bmatrix} x_1 \\ x_2 \end{bmatrix} - \begin{bmatrix} 1 & 2 \\ -1 & 0 \end{bmatrix}\begin{bmatrix} y_1 \\ y_2 \end{bmatrix} &= \begin{bmatrix} 1 \\ 2 \end{bmatrix} \\ \begin{bmatrix} 1 & 2 \\ -1 & 0 \end{bmatrix}\begin{bmatrix} x_1 \\ x_2 \end{bmatrix} + \begin{bmatrix} 3 & 1 \\ 0 & 2 \end{bmatrix}\begin{bmatrix} y_1 \\ y_2 \end{bmatrix} &= \begin{bmatrix} 3 \\ -1 \end{bmatrix} \end{aligned}$$

which is "equivalent" to the real 4×4 system

$$\begin{aligned} 3x_1 + x_2 - y_1 - 2y_2 &= 1 \\ 2x_2 + y_1 \qquad &= 2 \\ x_1 + 2x_2 + 3y_1 + y_2 &= 3 \\ -x_1 \qquad + 2y_2 &= -1 \end{aligned}$$

or

$$\begin{bmatrix} 3 & 1 & -1 & -2 \\ 0 & 2 & 1 & 0 \\ 1 & 2 & 3 & 1 \\ -1 & 0 & 0 & 2 \end{bmatrix}\begin{bmatrix} x_1 \\ x_2 \\ y_1 \\ y_2 \end{bmatrix} = \begin{bmatrix} 1 \\ 2 \\ 3 \\ -1 \end{bmatrix}$$

You should verify that the solution to this system obtained using the Gauss–Jordan algorithm is $x_1 = \frac{1}{25}$, $x_2 = \frac{16}{25}$, $y_1 = \frac{18}{25}$, and $y_2 = -\frac{12}{25}$.

The conclusion is that the solution to the original complex system is

$$\begin{aligned} z_1 &= x_1 + y_1 i = \frac{1 + 18i}{25} \\ z_2 &= x_2 + y_2 i = \frac{16 - 12i}{25}. \end{aligned}$$

Problem Set 36

36.1. Go back and do any of the exercises from Section 36 that you may have skipped.

*36.2. Let $z_1 = 3 - 2i$ and $z_2 = -4 - 3i$. Simplify each of the following expressions to the form $a + bi$, where $a, b \in \mathbb{R}$.

(i) $z_1 + z_2$

(ii) $z_1 - z_2$

(iii) $z_1 z_2$

(iv) z_1/z_2

(v) $z_1\bar{z}_1$

(vi) $\overline{\bar{z}_1 z_2}$

(vii) $\dfrac{1}{z_1 z_2}$

(viii) z_1^2

(ix) z_1^3

(x) z_1^{-1}

(xi) $(2 - 3i)z_1$

(xii) $(3z_1 - iz_2)$

(xiii) $(3z_1 - iz_2)^2$

(xiv) $\dfrac{2 - 3i}{z_1}$

36.3. (Analogue of Problem 3.13.) Let $A \in \mathbb{C}_{n\times n}$.

(i) Prove that $A + A^H$ is Hermitian.

(ii) Prove that $A - A^H$ is skew-Hermitian.

(iii) Observe that $A = \frac{1}{2}(A + A^H) + \frac{1}{2}(A - A^H)$, so A can be written in at least one way as the sum of a Hermitian and a skew-Hermitian matrix.

(iv) Prove that the expression in (iii) is the unique way to write A as the sum of a Hermitian and a skew-Hermitian matrix.

36.4. Derive the following identity for real numbers, which shows that if two real numbers are each expressible as a sum of two squares, then so is their product. For all $a, b, c, d \in R$,

$$(ac - bd)^2 + (ad + bc)^2 = (a^2 + b^2)(c^2 + d^2)$$

Hint: Let $z_1 = a + bi$ and $z_2 = c + di$ and interpret the fact that $|z_1 z_2| = |z_1|\,|z_2|$. See the article by Charles W. Curtis in [A] for analogous formulas for sums of four and eight squares.

*36.5. Solve the following 1×1 complex linear system for z_1:

$$(2 + 5i)z_1 = -3 + 2i$$

*36.6. Solve the following complex linear system:

$$\begin{aligned}(2 - i)z_1 + 2iz_2 &= 5\\ z_1 - z_2 &= 3 - 2i\end{aligned}$$

36.7. (Analogue of Problem 3.20.) Let A and B be Hermitian $n \times n$ matrices. Prove that AB is Hermitian iff AB = BA.

37

A Necessary and Sufficient Condition for Diagonalizability

37.1 Definition

Let $A \in \mathbb{F}_{n\times n}$ and let λ be an eigenvalue of A.

(i) The multiplicity of λ as a root of the characteristic polynomial $\chi(A)$ is called the *algebraic multiplicity of* λ.
(ii) The dimension of the eigenspace E_λ is called the *geometric multiplicity of* λ.
(iii) If the algebraic multiplicity of λ is 1, λ is called a *simple eigenvalue.*
(iv) If the algebraic multiplicity of λ is greater than 1, λ is called a *multiple (or repeated) eigenvalue.*

37.2 Examples

(i) Let us reconsider the 4×4 matrix A from Example 35.2. Recall that $\chi(A) = (\lambda + 5)^2(\lambda + 4)(\lambda)$, so the eigenvalues of A are -5, -4, and 0. The algebraic multiplicity of the repeated eigenvalue -5 is 2. Both -4 and 0 are simple eigenvalues. The dimensions of the eigenspaces were also computed in 35.2: the geometric multiplicity of -5 is 2 and the geometric multiplicity of both -4 and 0 is 1.

(ii) Let us reconsider the 3×3 matrix B from Example 35.6. Recall that $\chi(\mathrm{B}) = (\lambda - 3)(\lambda + 2)^2$, so the eigenvalues of B are 3 and -2; their algebraic multiplicities are 1 and 2, respectively. The dimensions of E_3 and E_{-2} were calculated in 35.6: both 3 and -2 have geometric multiplicity equal to 1.

37.3 Theorem

Let $\mathrm{A} \in \mathbb{F}_{n\times n}$ and let λ be an eigenvalue of A; then the following inequalities hold:

(i) $1 \leq$ (geometric multiplicity of λ).
(ii) (geometric multiplicity of λ) $\leq$ (algebraic multiplicity of λ).
(iii) (algebraic multiplicity of λ) $\leq n$.

Proof of (i): This is immediate from the definition of eigenvalue: if λ is an eigenvalue of A, then $\mathrm{A}\mathbf{v} = \lambda\mathbf{v}$ for some $\mathbf{0} \neq \mathbf{v} \in \mathbb{R}^n$, so E_λ is at least one-dimensional.

Proof of (ii): Let r be an eigenvalue of A whose geometric multiplicity is k. We will prove that the algebraic multiplicity of r is at least k by showing that $\chi(\mathrm{A})$ has the form $(\lambda - r)^k\chi(\mathrm{C})$, where C is a matrix of size $(n-k) \times (n-k)$.

Let $(\mathbf{v}_1, \mathbf{v}_2, \ldots, \mathbf{v}_k)$ be a basis for E_r and extend this set to a basis $\beta = (\mathbf{v}_1, \mathbf{v}_2, \ldots, \mathbf{v}_k, \mathbf{w}_{k+1}, \ldots, \mathbf{w}_n)$ for $\mathbb{F}^n$. For $i = 1, 2, \ldots, k$, we have $T_\mathrm{A}(\mathbf{v}_i) = \mathrm{A}\mathbf{v}_i = r\mathbf{v}_i$ since $\mathbf{v}_i \in E_r$. When we compute the matrix that represents T_A with respect to β, we obtain, using Theorem 31.7,

$$
\begin{aligned}
{}_\beta[T_\mathrm{A}]_\beta &= \left[[T_\mathrm{A}(\mathbf{v}_1)]_\beta \;\vdots\; \cdots \;\vdots\; [T_\mathrm{A}(\mathbf{v}_k)]_\beta \;\vdots\; [T_\mathrm{A}(\mathbf{w}_{k+1})]_\beta \;\vdots\; \cdots \;\vdots\; [T_A(\mathbf{w}_n)]_\beta\right] \\
&= \left[[r\mathbf{v}_1]_\beta \;\vdots\; \cdots \;\vdots\; [r\mathbf{v}_k]_\beta \;\vdots\; [T_\mathrm{A}(\mathbf{w}_{k+1})]_\beta \;\vdots\; \cdots \;\vdots\; [T_\mathrm{A}(\mathbf{w}_n)]_\beta\right] \\
&= \left[\begin{array}{cccccc|c}
r & 0 & \cdot & \cdot & \cdot & 0 & \\
0 & r & \cdot & \cdot & \cdot & 0 & \\
\cdot & \cdot & \cdot & \cdot & \cdot & \cdot & \mathrm{B} \\
\cdot & \cdot & \cdot & \cdot & \cdot & \cdot & \\
\cdot & \cdot & \cdot & \cdot & \cdot & \cdot & \\
0 & \cdot & \cdot & \cdot & 0 & r & \\
\hline
\multicolumn{6}{c|}{0_{(n-k)\times k}} & \mathrm{C}
\end{array}\right]
\end{aligned}
$$

$$\left[\text{for } i = 1, 2, \ldots, k,\ [r\mathbf{v}_i]_\beta = r[\mathbf{v}_i]_\beta = r\mathbf{e}_i^n\right]$$

$$
= \left[\begin{array}{c|c}
r\mathrm{I}_k & \mathrm{B} \\
\hline
0_{(n-k)\times k} & \mathrm{C}
\end{array}\right].
$$

Thus

$$\begin{aligned}
\chi(A) &= \chi\left({}_\beta[T_A]_\beta\right) \quad \left[\text{by Theorem 35.4, since A and}\ {}_\beta[T_A]_\beta \text{ are similar}\right] \\
&= \left| \lambda I_n - \left[\begin{array}{c|c} rI_k & B \\ \hline 0_{(n-k)\times k} & C \end{array}\right] \right| \\
&= \left| \begin{array}{c|c} (\lambda - r)I_k & -B \\ \hline 0_{(n-k)\times k} & \lambda I_{n-k} - C \end{array} \right| \\
&= |(\lambda - r)I_k|\,|\lambda I_{n-k} - C| \qquad \text{[Problem 28.1]} \\
&= (\lambda - r)^k \chi(C)
\end{aligned}$$

Proof of (iii): This follows because the degree of the polynomial $\chi(A)$ is n if A is an $n \times n$ matrix. ■

We have seen that $A \in \mathbb{F}_{n\times n}$ can fail to be diagonalizable over $\mathbb{F}$ either because $\chi(A)$ does not have enough roots in $\mathbb{F}$ or because one of the eigenspaces does not have large enough dimension. The next theorem reveals that these are the only obstructions to diagonalizability.

37.4 Theorem

Let $A \in \mathbb{F}_{n\times n}$. A is diagonalizable over $\mathbb{F}$ if and only if both of the following conditions are met:

(i) A has n-many eigenvalues in $\mathbb{F}$, counting multiplicities. That is, $\chi(A)$ factors over $\mathbb{F}$ into a product of n-many, not necessarily distinct, linear factors.

(ii) If $r_1, r_2, \ldots, r_p \in \mathbb{F}$ are the distinct eigenvalues of A, and r_i has algebraic multiplicity m_i for $i = 1, 2, \ldots, p$, that is, if $\chi(A) = (\lambda - r_1)^{m_1}(\lambda - r_2)^{m_2} \cdots (\lambda - r_p)^{m_p}$ and $m_1 + m_2 + \cdots + m_p = n$, then, for $i = 1, 2, \ldots, p$, $\dim E_{r_i} = m_i$.

The proof will be given at the end of this section; but first, a few remarks and examples are in order.

37.5 Remarks

(i) When $\mathbb{F} = \mathbb{C}$, the first condition in 37.4 is always satisfied. This is precisely the Fundamental Theorem of Algebra, discussed in Section 36. Thus the only way a matrix $A \in \mathbb{C}_{n\times n}$ can fail to be diagonalizable over $\mathbb{C}$ is when $\chi(A)$ has a root r_i of multiplicity $m_i \gneqq 1$ and $\dim E_{r_i} \lneqq m_i$.

(ii) When $\mathbb{F} = \mathbb{R}$, a matrix $A \in \mathbb{R}_{n\times n}$ can fail to be diagonalizable over $\mathbb{R}$ if $\chi(A)$ does not have n-many real roots. In this case, A may or may not be diagonalizable over $\mathbb{C}$, depending on whether the second condition in 37.4 is met.

(iii) When $A \in \mathbb{R}_{n\times n}$ and $\chi(A)$ factors completely over $\mathbb{R}$ into n-many linear factors, A is diagonalizable over $\mathbb{R}$ provided the dimensions of the eigenspaces corresponding to any multiple eigenvalues of A are equal to their algebraic multiplicities.

37.6 Examples

(i) If $A \in \mathbb{R}_{7\times 7}$ and $\chi(A) = (\lambda - 3)(\lambda + 2)^3(\lambda - 1)(\lambda^2 + 4)$, then A is not diagonalizable over $\mathbb{R}$ since A has only five real eigenvalues, counting multiplicities: 3, -2, -2, -2, and 1. The two remaining eigenvalues are the complex roots of $\lambda^2 + 4$. Over $\mathbb{C}$, $\chi(A)$ factors as $\chi(A) = (\lambda - 3)(\lambda + 2)^3(\lambda - 1)(\lambda + 2i)(\lambda - 2i)$, so A will be diagonalizable over $\mathbb{C}$ iff $\dim E_{-2} = 3$.

(ii) If $A \in \mathbb{R}_{6\times 6}$ and $\chi(A) = (\lambda - 3)^2(\lambda^2 + 9)^2$, then A is not diagonalizable over $\mathbb{R}$ since it has only two real eigenvalues, counting multiplicities. Over $\mathbb{C}$, $\chi(A)$ factors as $\chi(A) = (\lambda - 3)^2 (\lambda + 3i)^2(\lambda - 3i)^2$, so A will be diagonalizable over $\mathbb{C}$ iff $\dim E_3 = 2$, $\dim E_{-3i} = 2$, and $\dim E_{3i} = 2$.

(iii) If $A \in \mathbb{R}_{7\times 7}$ and $\chi(A) = (\lambda + 3)^4(\lambda - 1)^2(\lambda - \frac{1}{2})$, then A has seven real eigenvalues, counting multiplicities; so A is diagonalizable over $\mathbb{R}$ iff A is diagonalizable over $\mathbb{C}$ iff $\dim E_{-3} = 4$ and $\dim E_1 = 2$.

37.7 Proof of Theorem 37.4

($\rightarrow$) Let $A \in \mathbb{F}_{n\times n}$ and suppose that $D = P^{-1}AP$, where $P, D \in \mathbb{F}_{n\times n}$ and $D = \text{diag}(r_1, r_2, \ldots, r_n)$. Then, by Theorem 35.4, since A and D are similar, we have $\chi(A) = \chi(D) = (\lambda - r_1)(\lambda - r_2) \cdots (\lambda - r_n)$; this proves that $\chi(A)$ factors over $\mathbb{F}$ into a product of n-many, not necessarily distinct, factors. It remains to show, in case some r is a repeated eigenvalue, that the geometric and algebraic multiplicities of r are equal. We already know from 37.3(ii) that the geometric multiplicity is less than or equal to the algebraic multiplicity, so we need to establish the reverse inequality. It follows from Theorem 34.3 that if $D = P^{-1}AP$, then the jth column of P belongs to E_{r_j}. Since P is invertible, the columns of P are linearly independent. Thus, if a given factor $(\lambda - r)$ is repeated k-many times in the expression $(\lambda - r_1)(\lambda - r_2) \cdots (\lambda - r_n)$, then the corresponding columns of P will constitute a linearly independent subset of cardinality k in E_r. This implies that the geometric multiplicity of r is greater than or equal to the algebraic multiplicity of r.

(←) Given that $\chi(A) = (\lambda - r_1)^{m_1}(\lambda - r_2)^{m_2} \cdots (\lambda - r_p)^{m_p}$, that $m_1 + m_2 + \cdots + m_p = n$, and that for $i = 1, 2, \ldots, p$, $\dim E_{r_i} = m_i$, let β_i be a basis for E_{r_i}. Because distinct eigenspaces have trivial intersection [recall 34.6(iii)], the β's are pairwise disjoint and $\beta = \beta_1 \cup \beta_2 \cup \cdots \cup \beta_p$ has cardinality n; thus, if β is linearly independent, it will be a basis for $\mathbb{F}^n$. Since this would be a basis consisting of eigenvectors of A, A will be diagonalizable by Theorem 34.7. It remains only to prove that, in fact, β is linearly independent. We will prove by induction on k that $\beta_1 \cup \beta_2 \cup \cdots \cup \beta_k$ is linearly independent. β_1 is linearly independent by hypothesis. We now show, assuming $\beta_1 \cup \beta_2 \cup \cdots \cup \beta_{k-1}$ is linearly independent, that $\beta_1 \cup \cdots \cup \beta_{k-1} \cup \beta_k$ is linearly independent. Let $\beta_j = (\mathbf{u}_{j1}, \mathbf{u}_{j2}, \ldots, \mathbf{u}_{jm_j})$ and assume that

$$\mathbf{0} = \sum_{i=1}^{i=k} \sum_{j=1}^{j=m_i} a_{ij}\mathbf{u}_{ij} \qquad (\dagger)$$

If we let

$$\mathbf{w}_i = \sum_{j=1}^{j=m_i} a_{ij}\mathbf{u}_{ij}$$

then $\mathbf{w}_i \in E_{r_i}$ and

$$\mathbf{w}_1 + \mathbf{w}_2 + \cdots + \mathbf{w}_{k-1} = -\mathbf{w}_k.$$

When both sides of this equation are multiplied by $A - r_k I_n$, the result is

$$(r_1 - r_k)\mathbf{w}_1 + (r_2 - r_k)\mathbf{w}_2 + \cdots + (r_{k-1} - r_k)\mathbf{w}_{k-1} = \mathbf{0}.$$

Thus

$$\sum_{i=1}^{i=k-1} \sum_{j=1}^{j=m_i} (r_i - r_k)a_{ij}\mathbf{u}_{ij} = \mathbf{0}.$$

The distinctness of the r's together with the linear independence of $\beta_1 \cup \cdots \cup \beta_{k-1}$ implies that $a_{ij} = 0$ for $i = 1, 2, \ldots, k-1$ and $j = 1, 2, \ldots, m_i$. So from the original sum (†), we are left with $\mathbf{0} = a_{k1}\mathbf{u}_{k1} + a_{k2}\mathbf{u}_{k2} + \cdots + a_{km_k}\mathbf{u}_{km_k}$ and the linear independence of β_k now ensures that these remaining coefficients are 0 as well. ■

37.8 Definition

It is convenient to have a name for matrices that have an eigenvalue whose geometric multiplicity is strictly less than its algebraic multiplicity; such a matrix is called *defective*.

Problem Set 37

37.1. Go back and do any of the exercises in Section 37 that you may have skipped.

*37.2. For each of the following matrices, decide if it is diagonalizable and, if it is, diagonalize it.

(i) $\begin{bmatrix} -6 & -10 \\ 4 & 8 \end{bmatrix}$ (ii) $\begin{bmatrix} -2 & 5 \\ -5 & 6 \end{bmatrix}$

(iii) $\begin{bmatrix} -3 & -12 & 2 \\ 2 & 7 & 0 \\ 0 & 0 & 1 \end{bmatrix}$ (iv) $\begin{bmatrix} -2 & -5 & -5 \\ 0 & 3 & 0 \\ 0 & 0 & 3 \end{bmatrix}$

(v) $\begin{bmatrix} \frac{5}{2} & 1 & -1 \\ -5 & -3 & 4 \\ -2 & -\frac{3}{2} & \frac{5}{2} \end{bmatrix}$ (vi) $\begin{bmatrix} 6 & -5 & -3 \\ 3 & -2 & -2 \\ 2 & -2 & 0 \end{bmatrix}$

*37.3. Find A^8 by diagonalizing A, where $A = \begin{bmatrix} -7 & -6 \\ 15 & 12 \end{bmatrix}$.

38

The Dot-Product in $\mathbb{R}^n$

There are two very important aspects of $\mathbb{R}^n$ that we have until now ignored: the notion of "length" of a vector and the notion of "the angle between two vectors." The reason for postponing their introduction is to emphasize how much can be done without them. This is not, however, to diminish their importance; our purpose is simply to inform the reader by keeping separate those aspects of linear algebra that depend on these notions from those that do not.

To motivate the definitions that will be given for $\mathbb{R}^n$, we will begin by reviewing some familiar facts about $\mathbb{R}^2$ and $\mathbb{R}^3$.

The Pythagorean Theorem gives us the formula for the length of a vector $\mathbf{v} = (v_1, v_2) \in \mathbb{R}^2$. Length of $\mathbf{v} = \sqrt{v_1^2 + v_2^2}$ (see Figure 38.1). Applying the Pythagorean Theorem twice gives us the formula for the length of $\mathbf{v}$ when $\mathbf{v} = (v_1, v_2, v_3) \in \mathbb{R}^3$. Length of $\mathbf{v} = \sqrt{v_1^2 + v_2^2 + v_3^2}$ (see Figure 38.2).

The notion of length can be used to define the concept of "distance between two points." Thus, if $\mathbf{u} = (u_1, u_2)$ and $\mathbf{v} = (v_1, v_2)$ are two points in the plane, the distance, $d(\mathbf{u}, \mathbf{v})$, between $\mathbf{u}$ and $\mathbf{v}$ is the length of the vector $\mathbf{v} - \mathbf{u}$ (see Figure 38.3).

$$d(\mathbf{u}, \mathbf{v}) = \sqrt{(v_1 - u_1)^2 + (v_2 - u_2)^2}$$

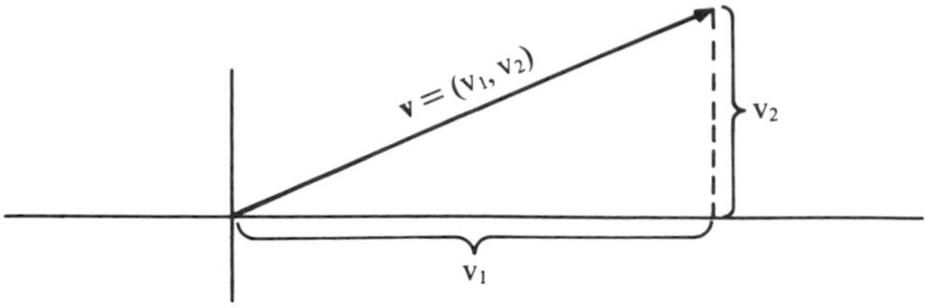

Figure 38.1

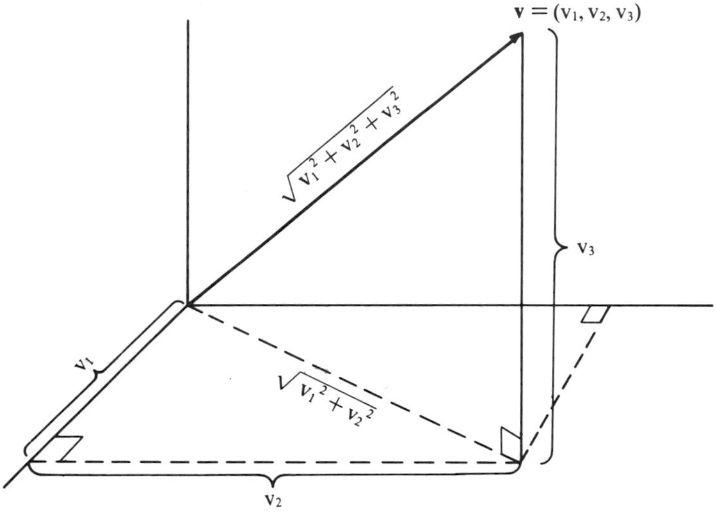

Figure 38.2

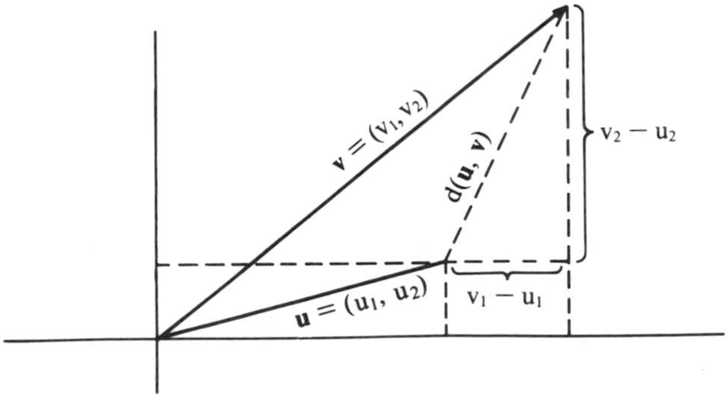

Figure 38.3

Similarly, the distance between two points $\mathbf{u} = (u_1, u_2, u_3)$ and $\mathbf{v} = (v_1, v_2, v_3)$ in $\mathbb{R}^3$ is given by the formula

$$d(\mathbf{u}, \mathbf{v}) = \sqrt{(v_1 - u_1)^2 + (v_2 - u_2)^2 + (v_3 - u_3)^2}$$

For two nonzero vectors $\mathbf{u} = (u_1, u_2)$ and $\mathbf{v} = (v_1, v_2)$ in the plane, $\mathbb{R}^2$, there is a familiar result relating the angle between $\mathbf{u}$ and $\mathbf{v}$ and the coordinates, u_1, u_2, v_1, v_2, of the endpoints of these vectors. Since this result is derived from the Law of Cosines, we begin by recalling the latter.

38.1 Theorem (Law of Cosines)

For any triangle, the square of the length of any side equals the sum of the squares of the lengths of the two remaining sides minus twice the product of the lengths of these other sides times the cosine of the angle between them (see Figure 38.4).

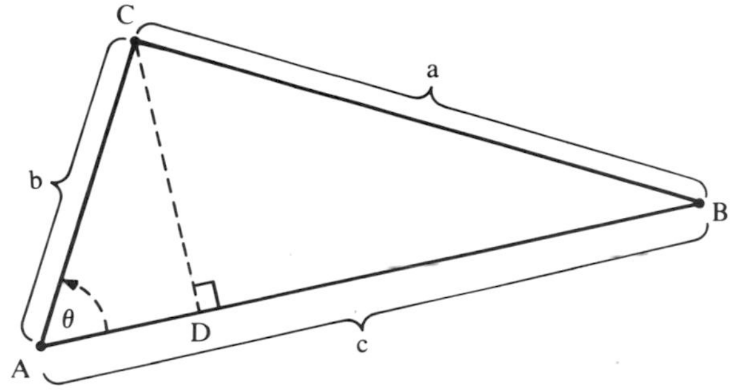

Figure 38.4

Proof: Let $\overline{\mathrm{PQ}}$ denote the length of the line segment joining points P and Q. Using the Pythagorean Theorem and the definition of cosine, the following four facts are clear:

$$a^2 = \overline{\mathrm{DC}}^2 + \overline{\mathrm{BD}}^2$$

$$\overline{\mathrm{AD}}^2 + \overline{\mathrm{DC}}^2 = b^2$$

$$\overline{\mathrm{AD}} = c - \overline{\mathrm{BD}}$$

$$\cos\theta = \frac{\overline{\mathrm{AD}}}{b}$$

Combining these four facts, we obtain

$$\begin{aligned} a^2 &= \overline{\mathrm{DC}}^2 + \overline{\mathrm{BD}}^2 \\ &= b^2 - \overline{\mathrm{AD}}^2 + \overline{\mathrm{BD}}^2 \end{aligned}$$

$$= b^2 - \left(c^2 - 2c\overline{BD} + \overline{BD}^2\right) + \overline{BD}^2$$
$$= b^2 - c^2 + 2c\overline{BD} - \overline{BD}^2 + \overline{BD}^2$$
$$= b^2 - c^2 + 2c\left(c - \overline{AD}\right)$$
$$= b^2 - c^2 + 2c(c - b\cos\theta)$$
$$= b^2 + c^2 - 2bc\cos\theta \qquad \blacksquare$$

Next we apply the Law of Cosines to obtain a formula for the cosine of the angle between two nonzero vectors $\mathbf{u} = (u_1, u_2)$ and $\mathbf{v} = (v_1, v_2)$ in $\mathbb{R}^2$ as a function of the coordinates of the vectors $\mathbf{u}$ and $\mathbf{v}$.

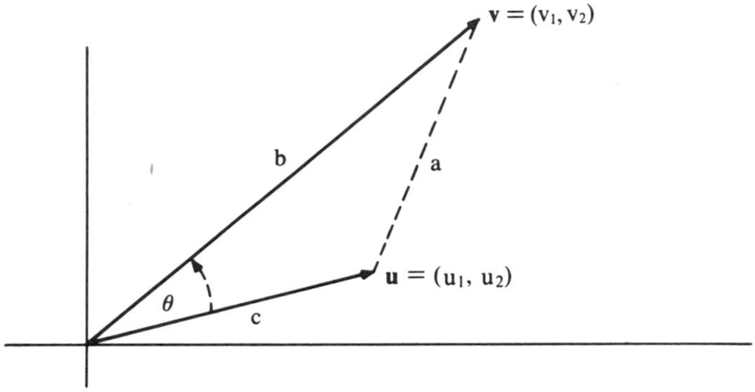

Figure 38.5

38.2 Theorem

In Figure 38.5,

$$\cos\theta = \frac{u_1v_1 + u_2v_2}{\sqrt{u_1^2 + u_2^2}\sqrt{v_1^2 + v_2^2}}$$

Proof: Let a be the distance between the two points (u_1, u_2) and (v_1, v_2), so $a = \sqrt{(v_1 - u_1)^2 + (v_2 - u_2)^2}$. Let b be the length of the vector $\mathbf{v}$, so $b = \sqrt{v_1^2 + v_2^2}$. Let c be the length of the vector $\mathbf{u}$, so $c = \sqrt{u_1^2 + u_2^2}$. The Law of Cosines yields $(v_1 - u_1)^2 + (v_2 - u_2)^2 = v_1^2 + v_2^2 + u_1^2 + u_2^2 - 2\sqrt{v_1^2 + v_2^2}\sqrt{u_1^2 + u_2^2}\cos\theta$. ■

A similar result can be proved for two nonzero vectors $\mathbf{u} = (u_1, u_2, u_3)$ and $\mathbf{v} = (v_1, v_2, v_3)$ in $\mathbb{R}^3$. If θ is the angle between these two vectors (in the plane which they determine), then

$$\cos\theta = \frac{u_1v_1 + u_2v_2 + u_3v_3}{\sqrt{u_1^2 + u_2^2 + u_3^2}\sqrt{v_1^2 + v_2^2 + v_3^2}}$$

Motivated by these phenomena in low dimensions, we are tempted to try to generalize to $\mathbb{R}^n$ the notions of length, of distance between points, and of angle between two nonzero vectors. To accomplish this, there are three functions, defined for $\mathbb{R}^n$ by analogy with the facts just presented about $\mathbb{R}^2$ and $\mathbb{R}^3$, whose properties must be analyzed:

(i) The function from $\mathbb{R}^n$ to $\mathbb{R}$ whose value at $\mathbf{u}$ is

$$\sqrt{u_1^2 + u_2^2 + \cdots + u_n^2}$$

(ii) The function from $\mathbb{R}^n \times \mathbb{R}^n$ to $\mathbb{R}$ whose value at the pair $(\mathbf{u}, \mathbf{v})$ is

$$\sqrt{(v_1 - u_1)^2 + (v_2 - u_2)^2 + \cdots + (v_n - u_n)^2}$$

(iii) The function from $\mathbb{R}^n \times \mathbb{R}^n$ to $\mathbb{R}$ whose value at the pair $(\mathbf{u}, \mathbf{v})$ is

$$\frac{u_1v_1 + u_2v_2 + \cdots + u_nv_n}{\sqrt{u_1^2 + u_2^2 + \cdots + u_n^2}\sqrt{v_1^2 + v_2^2 + \cdots + v_n^2}}$$

38.3 Remark

The three functions just mentioned have standard names and notations, but it would obscure the logical development of our undertaking to use these names prematurely. For instance, the second function is called the "distance between **u** and **v**," or more precisely, the "Euclidean distance between **u** and **v**," but the legitimacy of this name must rest on a theorem which establishes that this function has all the properties that any reasonable notion of "distance" ought to have. Similarly, to call the third function "the cosine of the angle between **u** and **v**" is clearly premature at this point; indeed, it is far from obvious that this third function only assumes values in the interval $[-1, +1]$. We do have reason to suspect, based on our experience with $\mathbb{R}^2$ and $\mathbb{R}^3$, that this will happen, but that is not the same thing as a proof.

We begin our derivation of properties of these three functions with the study of a fourth, auxiliary function that is directly involved with the third of the functions in the preceding list.

38.4 Definition

The function from $\mathbb{R}^n \times \mathbb{R}^n$ into $\mathbb{R}$ that assigns to the pair of vectors $(\mathbf{u}, \mathbf{v})$ the real number $u_1v_1 + u_2v_2 + \cdots + u_nv_n$ is called the *dot product* (or *Euclidean inner product*) of **u** and **v**. Its value at $(\mathbf{u}, \mathbf{v})$ is denoted by $\mathbf{u} \cdot \mathbf{v}$.

In the next theorem, we collect a short list of important properties of the dot-product function. The proof of the theorem is straightforward and requires only well-known facts about arithmetic in $\mathbb{R}$.

38.5 Theorem

For all $\mathbf{u}, \mathbf{v}, \mathbf{w} \in \mathbb{R}^n$ and $t \in \mathbb{R}$,

(i) $\mathbf{u} \cdot \mathbf{v} = \mathbf{v} \cdot \mathbf{u}$
(ii) $(\mathbf{u} + \mathbf{v}) \cdot \mathbf{w} = (\mathbf{u} \cdot \mathbf{w}) + (\mathbf{v} \cdot \mathbf{w})$
(iii) $(t\mathbf{u}) \cdot \mathbf{v} = t(\mathbf{u} \cdot \mathbf{v})$
(iv) $0 \leq \mathbf{v} \cdot \mathbf{v}$; moreover, $\mathbf{v} \cdot \mathbf{v} = 0$ if and only if $\mathbf{v} = \mathbf{0}$.

38.6 Exercise

Prove Theorem 38.5 and also the following easy corollaries:

(i) $\mathbf{0} \cdot \mathbf{v} = \mathbf{v} \cdot \mathbf{0} = 0$
(ii) $\mathbf{u} \cdot (\mathbf{v} + \mathbf{w}) = \mathbf{u} \cdot \mathbf{v} + \mathbf{u} \cdot \mathbf{w}$
(iii) $\mathbf{u} \cdot (t\mathbf{v}) = t(\mathbf{u} \cdot \mathbf{v})$ ■

In the next theorem, we present the principal tool needed in our analysis of the three functions mentioned earlier.

38.7 Theorem (Cauchy–Schwarz Inequality)

For all $\mathbf{u}, \mathbf{v} \in \mathbb{R}^n$,

$$(\mathbf{u} \cdot \mathbf{v})^2 \leq (\mathbf{u} \cdot \mathbf{u})(\mathbf{v} \cdot \mathbf{v}) \qquad (\dagger)$$

Proof:

Case 1: Assume $\mathbf{u} = \mathbf{0}$; then by 38.5 and 38.6, both sides of (†) are equal to 0.

Case 2: Assume $\mathbf{u} \neq \mathbf{0}$; by 38.5 and 38.6, we know that for any real number t,

$$0 \leq (t\mathbf{u} + \mathbf{v}) \cdot (t\mathbf{u} + \mathbf{v}) = (\mathbf{u} \cdot \mathbf{u})t^2 + 2(\mathbf{u} \cdot \mathbf{v})t + \mathbf{v} \cdot \mathbf{v} \qquad (\ddagger)$$

In particular, this inequality must hold for

$$t = -\frac{-\mathbf{u} \cdot \mathbf{v}}{\mathbf{u} \cdot \mathbf{u}}$$

This proves (†). ■

It is useful to point out precisely the circumstances under which equality holds in the Cauchy–Schwarz inequality.

38.8 Corollary to the Proof of 38.7

Equality holds in the Cauchy–Schwarz Inequality (†) if and only if $\mathbf{u} = \mathbf{0}$ or $\mathbf{v} = \mathbf{0}$ or $\mathbf{u}$ and $\mathbf{v}$ are scalar multiples of one another, that is, iff the set $\{\mathbf{u}, \mathbf{v}\}$ is linearly dependent.

Proof: ($\leftarrow$) This direction is trivial and is left as an exercise.

($\rightarrow$) We show that if equality holds in (†) and if $\mathbf{u}, \mathbf{v} \neq \mathbf{0}$, then $\mathbf{v} = s\mathbf{u}$ for some scalar s; indeed, our proof will show that the specific scalar s is equal to $\mathbf{v} \cdot \mathbf{u}/\mathbf{u} \cdot \mathbf{u}$.

Equality holds in (†):

iff equality holds in (‡) when $t = -\mathbf{u} \cdot \mathbf{v}/\mathbf{u} \cdot \mathbf{u}$ [as in the proof of 38.7]

iff $[(-\mathbf{u} \cdot \mathbf{v}/\mathbf{u} \cdot \mathbf{u})\mathbf{u} + \mathbf{v}] \cdot [(-\mathbf{u} \cdot \mathbf{v}/\mathbf{u} \cdot \mathbf{u})\mathbf{u} + \mathbf{v}] = 0$

iff $(-\mathbf{u} \cdot \mathbf{v}/\mathbf{u} \cdot \mathbf{u})\mathbf{u} + \mathbf{v} = \mathbf{0}$ [38.5(iv)]

iff $\mathbf{v} = (\mathbf{u} \cdot \mathbf{v}/\mathbf{u} \cdot \mathbf{u})\mathbf{u} = (\mathbf{v} \cdot \mathbf{u}/\mathbf{u} \cdot \mathbf{u})\mathbf{u}$ [38.5(i)] ■

Another immediate consequence of the Cauchy–Schwarz Inequality is that if $\mathbf{u}$ and $\mathbf{v}$ are nonzero vectors in $\mathbb{R}^n$ then

$$\frac{(\mathbf{u} \cdot \mathbf{v})^2}{(\mathbf{u} \cdot \mathbf{u})(\mathbf{v} \cdot \mathbf{v})} \leq 1$$

Taking square roots, we conclude that the real number $\mathbf{u} \cdot \mathbf{v}/\sqrt{\mathbf{u} \cdot \mathbf{u}}\sqrt{\mathbf{v} \cdot \mathbf{v}}$ belongs to the interval $[-1, +1]$. Therefore, the definition that follows makes sense.

38.9 Definition

Where $\mathbf{u}$ and $\mathbf{v}$ are nonzero vectors in $\mathbb{R}^n$, the *angle between* $\mathbf{u}$ *and* $\mathbf{v}$ is the unique angle θ between 0 and π that satisfies

$$\cos\theta = \frac{\mathbf{u} \cdot \mathbf{v}}{\sqrt{\mathbf{u} \cdot \mathbf{u}}\sqrt{\mathbf{v} \cdot \mathbf{v}}}$$

We have thus succeeded in generalizing to $\mathbb{R}^n$ the notion of "angle between two nonzero vectors" in a manner consistent with facts, such as Theorem 38.2, known for $\mathbb{R}^2$ or $\mathbb{R}^3$.

It is natural, since the notion of perpendicularity is clearly important in the familiar framework of $\mathbb{R}^2$ and $\mathbb{R}^3$, to generalize this notion as well. We are now in position to do this, since perpendicularity is defined in terms of the angle between two vectors. It is somewhat unfortunate, but historically entrenched, that a new word, orthogonality, is used for this concept in the context of $\mathbb{R}^n$. We should remind ourselves, before proceeding, that perpendicularity is a relationship that can hold between two vectors, between two lines, between a line and a plane, between two planes, between a vector and a plane, and so on. For this reason, the definition that follows is in several parts.

If we restrict our attention to θ's in the interval $[0, \pi]$, we have $\cos\theta = 0$ if and only if $\theta = \pi/2$. Thus, in view of Definition 38.9, we proceed as follows:

38.10 Definition

(i) Two vectors $\mathbf{u}, \mathbf{v} \in \mathbb{R}^n$ are called *orthogonal* if and only if $\mathbf{u} \cdot \mathbf{v} = 0$.
(ii) A subset $S = \{\mathbf{u}_1, \mathbf{u}_2, \ldots, \mathbf{u}_p\} \subset \mathbb{R}^n$ is *orthogonal* if and only if for all $i \neq j$, $\mathbf{u}_i \cdot \mathbf{u}_j = 0$.
(iii) If $\mathbf{v} \in \mathbb{R}^n$ and W is a linear subspace of $\mathbb{R}^n$, $\mathbf{v}$ is *orthogonal to* W if and only if $\mathbf{v} \cdot \mathbf{w} = 0$ for all $\mathbf{w} \in W$.
(iv) If W_1 and W_2 are linear subspaces of $\mathbb{R}^n$, W_1 and W_2 are *orthogonal* if and only if for all $\mathbf{w}_1 \in W_1$ and $\mathbf{w}_2 \in W_2$, $\mathbf{w}_1 \cdot \mathbf{w}_2 = 0$.
(v) If H_1 and H_2 are nonempty affine subspaces of $\mathbb{R}^n$, then H_1 and H_2 are *orthogonal* if and only if W_1 and W_2 are orthogonal, where W_1 and W_2 are the unique linear subspaces of which H_1 and H_2 are the translates.

We will eventually undertake further analysis of this concept, but we have still not treated the first and second of the three functions mentioned in 38.3.

We return to these issues and, specifically, to the concept of "length" of a vector. Consider the function from $\mathbb{R}^n$ to $\mathbb{R}$ whose value at the n-tuple $\mathbf{u}$ is $\sqrt{u_1^2 + u_2^2 + \cdots + u_n^2}$. It is convenient to have an abbreviation for the value of this function at $\mathbf{u}$; $\|\mathbf{u}\|$ is commonly used. In the context of $\mathbb{R}^n$, another new word, norm, is used instead of length.

38.11 Definition

For $\mathbf{u} \in \mathbb{R}^n$, $\|\mathbf{u}\|$ is called the *Euclidean norm* of $\mathbf{u}$.

Let us prove that the Euclidean norm possesses the properties that any reasonable notion of "length" ought to have. The first two properties are direct consequences of the definition: (i) the length of a vector should be a nonnegative real number; (ii) the length of a scalar multiple, $t\mathbf{u}$, of a vector $\mathbf{u}$ should be the absolute value of t times the length of $\mathbf{u}$.

38.12 Exercise

For all $\mathbf{u} \in \mathbb{R}^n$ and $t \in \mathbb{R}$,

(i) $0 \leq \|\mathbf{u}\|$; moreover, $0 = \|\mathbf{u}\|$ iff $\mathbf{u} = \mathbf{0}$.
(ii) $\|t\mathbf{u}\| = |t|\,\|\mathbf{u}\|$.

The third property gets its name from the familiar geometric interpretation that it has for vectors in $\mathbb{R}^2$ and $\mathbb{R}^3$: the sum of the lengths of any two sides of a triangle should be at least as great as the length of the third side.

38.13 Theorem (Triangle Inequality)

For all $\mathbf{u}, \mathbf{v} \in \mathbb{R}^n$,

$$\|\mathbf{u} + \mathbf{v}\| \leq \|\mathbf{u}\| + \|\mathbf{v}\|.$$

Proof: We will derive the Triangle Inequality from the Cauchy–Schwarz Inequality. (In fact, these two inequalities are equivalent, as we will show in a subsequent remark.) The following chain of equivalences shows that an arbitrary pair of vectors $\mathbf{u}, \mathbf{v} \in \mathbb{R}^n$ satisfies the Triangle Inequality iff $\mathbf{u} \cdot \mathbf{v} \leq \|\mathbf{u}\|\,\|\mathbf{v}\|$.

Recall first that for nonnegative real numbers x and y, $x \leq y$ iff $x^2 \leq y^2$. Thus,

$$\begin{aligned}
&\|\mathbf{u} + \mathbf{v}\| \leq \|\mathbf{u}\| + \|\mathbf{v}\| \\
\text{iff } &\|\mathbf{u} + \mathbf{v}\|^2 \leq (\|\mathbf{u}\| + \|\mathbf{v}\|)^2 \\
\text{iff } &(\mathbf{u} + \mathbf{v}) \cdot (\mathbf{u} + \mathbf{v}) \leq \|\mathbf{u}\|^2 + 2\|\mathbf{u}\|\,\|\mathbf{v}\| + \|\mathbf{v}\|^2 \\
\text{iff } &\mathbf{u} \cdot \mathbf{u} + \mathbf{u} \cdot \mathbf{v} + \mathbf{v} \cdot \mathbf{u} + \mathbf{v} \cdot \mathbf{v} \leq \mathbf{u} \cdot \mathbf{u} + 2\|\mathbf{u}\|\,\|\mathbf{v}\| + \mathbf{v} \cdot \mathbf{v} \\
& \qquad\qquad \text{[by 38.5(ii) and 38.6(ii)]} \\
\text{iff } &2(\mathbf{u} \cdot \mathbf{v}) \leq 2\|\mathbf{u}\|\,\|\mathbf{v}\| \qquad \text{[by 38.5(i)]} \\
\text{iff } &\mathbf{u} \cdot \mathbf{v} \leq \|\mathbf{u}\|\,\|\mathbf{v}\|.
\end{aligned}$$

Next, recall that for arbitrary real numbers x and y, $x^2 \leq y^2$ iff $|x| \leq y$. Thus it is clear that if two given vectors $\mathbf{u}, \mathbf{v} \in \mathbb{R}^n$ satisfy the Cauchy–Schwarz Inequality, then $|\mathbf{u} \cdot \mathbf{v}| \leq \|\mathbf{u}\|\,\|\mathbf{v}\|$, and so they also satisfy the Triangle Inequality, $\mathbf{u} \cdot \mathbf{v} \leq \|\mathbf{u}\|\,\|\mathbf{v}\|$. ■

38.14 Remark

Because $\mathbf{u} \cdot \mathbf{v}$ might be negative, we do not get the reverse implication quite so simply. Here is a proof that the Triangle Inequality implies the Cauchy–Schwarz Inequality. To conclude that a given pair of vectors $\mathbf{u}, \mathbf{v} \in \mathbb{R}^n$ satisfies the Cauchy–Schwarz Inequality, we need to invoke *two* instances of the Triangle Inequality: specifically, for the pair $\mathbf{u}, \mathbf{v}$ *and* for the pair $\mathbf{u}, -\mathbf{v}$. As in the proof of the previous theorem, from the fact that $\mathbf{u}$ and $\mathbf{v}$ satisfy the Triangle Inequality, we obtain $\mathbf{u} \cdot \mathbf{v} \leq \|\mathbf{u}\|\,\|\mathbf{v}\|$; similarly, from the fact that $\mathbf{u}$ and $-\mathbf{v}$ satisfy the Triangle Inequality, we obtain $\mathbf{u} \cdot (-\mathbf{v}) \leq \|\mathbf{u}\|\,\|-\mathbf{v}\|$, which is equivalent to

$$-(\mathbf{u} \cdot \mathbf{v}) \leq \|\mathbf{u}\|\,|-1|\,\|\mathbf{v}\| = \|\mathbf{u}\|\,\|\mathbf{v}\|$$

We can then combine these two consequences to obtain $|\mathbf{u} \cdot \mathbf{v}| \leq \|\mathbf{u}\|\,\|\mathbf{v}\|$, as desired.

Just as for the Cauchy–Schwarz Inequality, it is useful to point out the exact circumstances under which equality holds in the Triangle Inequality.

38.15 Theorem

Equality holds in the Triangle Inequality $\|\mathbf{u} + \mathbf{v}\| \leq \|\mathbf{u}\| + \|\mathbf{v}\|$ if and only if $\mathbf{u} = \mathbf{0}$ or $\mathbf{v} = \mathbf{0}$ or $\mathbf{u}$ and $\mathbf{v}$ are positive scalar multiples of one another.

Proof: ($\leftarrow$) Suppose that $\mathbf{u}$ and $\mathbf{v}$ are nonzero, that $\mathbf{v} = s\mathbf{u}$, and that $0 \lneqq s$; then

$$\begin{aligned} \mathbf{u} \cdot \mathbf{v} &= \mathbf{u} \cdot (s\mathbf{u}) \\ &= s(\mathbf{u} \cdot \mathbf{u}) \\ &= |s|\,\|\mathbf{u}\|^2 && \text{[since } s \text{ is positive]} \\ &= \|\mathbf{u}\|\,|s|\,\|\mathbf{u}\| \\ &= \|\mathbf{u}\|\,\|s\mathbf{u}\| && \text{[by 38.12(ii)]} \\ &= \|\mathbf{u}\|\,\|\mathbf{v}\| \end{aligned}$$

($\rightarrow$) Suppose that $\mathbf{u}$ and $\mathbf{v}$ are nonzero. The equality $\mathbf{u} \cdot \mathbf{v} = \|\mathbf{u}\|\,\|\mathbf{v}\|$ implies that $\mathbf{u} \cdot \mathbf{v}$ is positive and therefore also implies that $|\mathbf{u} \cdot \mathbf{v}| = \|\mathbf{u}\|\,\|\mathbf{v}\|$. It then follows from the proof of 38.8 that $\mathbf{v} = (\mathbf{u} \cdot \mathbf{v}/\mathbf{u} \cdot \mathbf{u})\mathbf{u}$, which concludes the argument since $\mathbf{u} \cdot \mathbf{v}$ and $\mathbf{u} \cdot \mathbf{u}$ are both positive. ■

38.16 Theorem (Generalized Pythagorean Theorem)

For all $\mathbf{u}, \mathbf{v} \in \mathbb{R}^n$, $\|\mathbf{u} + \mathbf{v}\|^2 = \|\mathbf{u}\|^2 + \|\mathbf{v}\|^2$ if and only if $\mathbf{u}$ and $\mathbf{v}$ are orthogonal.

Proof: Using properties of the dot product proved in 38.5, we have $\|\mathbf{u} + \mathbf{v}\|^2 = (\mathbf{u} + \mathbf{v}) \cdot (\mathbf{u} + \mathbf{v}) = \|\mathbf{u}\|^2 + 2(\mathbf{u} \cdot \mathbf{v}) + \|\mathbf{v}\|^2$. Hence $\|\mathbf{u} + \mathbf{v}\|^2 = \|\mathbf{u}\|^2 + \|\mathbf{v}\|^2$ if and only if $\mathbf{u} \cdot \mathbf{v} = 0$. ■

Although the result in 38.16 is usually called the Generalized Pythagorean Theorem, the following slightly stronger result really deserves the name.

38.17 Exercise

For all $\mathbf{u}_1, \mathbf{u}_2, \ldots, \mathbf{u}_p \in \mathbb{R}^n$ and $2 \leq p \leq n$, $\|\mathbf{u}_1 + \mathbf{u}_2 + \cdots + \mathbf{u}_p\|^2 = \|\mathbf{u}_1\|^2 + \|\mathbf{u}_2\|^2 + \cdots + \|\mathbf{u}_p\|^2$ if and only if $\{\mathbf{u}_1, \mathbf{u}_2, \ldots, \mathbf{u}_p\}$ is an orthogonal set of vectors.

We turn at last to the function from $\mathbb{R}^n \times \mathbb{R}^n$ into $\mathbb{R}$ whose value at the pair $(\mathbf{u}, \mathbf{v})$ is $\sqrt{(v_1 - u_1)^2 + (v_2 - u_2)^2 + \cdots + (v_n - u_n)^2}$. We will abbreviate the value of this function at $(\mathbf{u}, \mathbf{v})$ by $d(\mathbf{u}, \mathbf{v})$ and call this the *Euclidean distance between* $\mathbf{u}$ *and* $\mathbf{v}$ because, as we see in the next theorem, d possesses the properties that any reasonable notion of "distance" ought to have.

38.18 Theorem

For all $\mathbf{u}$, $\mathbf{v}$, and $\mathbf{w} \in \mathbb{R}^n$,

(i) $0 \le d(\mathbf{u}, \mathbf{v})$; moreover, $0 = d(\mathbf{u}, \mathbf{v})$ if and only if $\mathbf{u} = \mathbf{v}$.
(ii) $d(\mathbf{u}, \mathbf{v}) = d(\mathbf{v}, \mathbf{u})$.
(iii) $d(\mathbf{u}, \mathbf{v}) \le d(\mathbf{u}, \mathbf{w}) + d(\mathbf{w}, \mathbf{v})$. [the Triangle Inequality]

Proof: Because $d(\mathbf{u}, \mathbf{v}) = \|\mathbf{v} - \mathbf{u}\|$, these three properties of d are easily derived from earlier results about the Euclidean norm. ■

When working with n-tuples of real numbers it is useful to distinguish whether the vector space properties alone are under consideration or whether the additional structure provided by the dot product and the Euclidean norm and distance functions is relevant. For this reason, we will continue to use $\mathbb{R}^n$ to denote the structure consisting of n-tuples of real numbers with the operations of vector addition and scalar multiplication alone; we will use $\mathbb{E}^n$ to denote the structure consisting of n-tuples of real numbers with the operations of vector addition, scalar multiplication, and the dot product.

38.19 Definition

The structure $\mathbb{E}^n$ just defined is called *Euclidean n-space*.

It is tedious to attempt consistently to maintain the distinction between $\mathbb{R}^n$ and $\mathbb{E}^n$. This is all the more difficult because there are several abuses of language that have become common practice in this area.

Having emphasized the distinction, we will now proceed ourselves to adopt the most common of these abuses.

If $\beta = (\mathbf{v}_1, \mathbf{v}_2, \ldots, \mathbf{v}_n)$ is an ordered basis for $\mathbb{R}^n$ and $\mathbf{x} = x_1\mathbf{v}_1 + x_2\mathbf{v}_2 + \cdots + x_n\mathbf{v}_n$, we have used $[\mathbf{x}]_\beta$ to denote the $n \times 1$ matrix

$$\begin{bmatrix} x_1 \\ x_2 \\ \vdots \\ x_n \end{bmatrix}$$

(i.e., the β-coordinate matrix of $\mathbf{x}$).

The transpose, $[\mathbf{x}]_\beta^T$, of this matrix is the $1 \times n$ matrix $[x_1 x_2 \ldots x_n]$. We usually omit the subscript when the basis, β, is the standard basis, α^n, or is otherwise clearly determined by the context. Consider the following trivial but crucial observation:

38.20 Observation

The Euclidean dot product on $\mathbb{R}^n \times \mathbb{R}^n$ can be expressed as a product of matrices

$$\mathbf{x} \cdot \mathbf{y} = [\mathbf{x}]^T[\mathbf{y}]$$

provided we ignore the distinction between scalars and 1×1 matrices. (Note that, strictly speaking, $\mathbf{x} \cdot \mathbf{y} \in \mathbb{R}$, whereas $[\mathbf{x}]^T[\mathbf{y}] \in \mathbb{R}_{1 \times 1}$.) We will even occasionally omit the square brackets on the matrices and write $\mathbf{x}^T\mathbf{y}$ in place of $[\mathbf{x}]^T[\mathbf{y}]$.

This point of view has very fruitful consequences, as it allows us to bring the machinery of matrix methods to bear on the study of the dot product. Exploit this point of view to obtain the following:

38.21 Exercise

For all $\mathbf{u}, \mathbf{v} \in \mathbb{E}^n$, $(\mathrm{A}\mathbf{u}) \cdot \mathbf{v} = \mathbf{u} \cdot (\mathrm{A}^T\mathbf{v})$ and $\mathbf{u} \cdot (\mathrm{A}\mathbf{v}) = (\mathrm{A}^T\mathbf{u}) \cdot \mathbf{v}$.

38.22 Exercise

Prove for $\mathrm{A} \in \mathbb{R}_{n \times n}$ that

$$(\mathrm{A}\mathbf{u}) \cdot \mathbf{v} = \mathbf{u} \cdot (\mathrm{A}\mathbf{v}) \text{ for all } \mathbf{u}, \mathbf{v} \in \mathbb{E}^n \quad \text{iff} \quad \mathrm{A} = \mathrm{A}^T.$$

For these reasons, we will not attempt to be systematic in our use of $\mathbb{R}^n$ versus $\mathbb{E}^n$, and we will blithely write the equality symbol between any of $\mathbf{x} \cdot \mathbf{y}$, $[\mathbf{x}]^T[\mathbf{y}]$, and $\mathbf{x}^T\mathbf{y}$.

Problem Set 38

38.1. Go back and do any of the exercises from Section 38 that you may have skipped.

*38.2. Prove that an orthogonal set of nonzero vectors in $\mathbb{E}^n$ is linearly independent.

38.3. **Cross-products in $\mathbb{E}^3$**: Let $\mathbf{u} = (u_1, u_2, u_3)$ and $\mathbf{v} = (v_1, v_2, v_3) \in \mathbb{E}^3$. Define $\mathbf{w} \in \mathbb{E}^3$ by

$$\mathbf{w} = (u_2v_3 - u_3v_2, -u_1v_3 + u_3v_1, u_1v_2 - u_2v_1)$$

(i) Prove that $\mathbf{w}$ is orthogonal to both $\mathbf{u}$ and $\mathbf{v}$.
(ii) Prove that $\mathbf{w} \neq \mathbf{0}$ if and only if $\{\mathbf{u}, \mathbf{v}\}$ is linearly independent.

(iii) Prove that $\|\mathbf{w}\|^2 = \|\mathbf{u}\|^2\|\mathbf{v}\|^2 - (\mathbf{u} \cdot \mathbf{v})^2$.
(iv) Show that $\|\mathbf{w}\|$ is the area of the parallelogram determined by the vectors $\mathbf{u}$ and $\mathbf{v}$.

The preceding vector $\mathbf{w}$ is called the *cross product of* $\mathbf{u}$ *and* $\mathbf{v}$ and is denoted by $\mathbf{u} \times \mathbf{v}$. When $\{\mathbf{u}, \mathbf{v}\}$ is linearly independent, $\mathbf{u} \times \mathbf{v}$ is a vector that is perpendicular to the plane determined by $\mathbf{u}$ and $\mathbf{v}$ and whose length is the area of the parallelogram determined by $\mathbf{u}$ and $\mathbf{v}$. There are, of course, two such vectors, each the negative of the other. Of the two, $\mathbf{u} \times \mathbf{v}$ is the one determined by the right-hand screw rule.
There is a standard mnemonic device for remembering the coordinates of $\mathbf{u} \times \mathbf{v}$. This device should not be mistaken for more than it is! In particular, it is *not* the definition of $\mathbf{u} \times \mathbf{v}$. This device involves computing the "determinant" of a hybrid matrix, some of whose entries are vectors and some of whose entries are scalars. Usually, in this context, the standard basis vectors for $\mathbb{E}^3$ are denoted by $\mathbf{i}$, $\mathbf{j}$, and $\mathbf{k}$. Thus $\mathbf{i} = (1,0,0)$, $\mathbf{j} = (0,1,0)$, and $\mathbf{k} = (0,0,1)$. We then obtain $\mathbf{u} \times \mathbf{v}$ by computing the "determinant" of the matrix

$$\begin{vmatrix} \mathbf{i} & \mathbf{j} & \mathbf{k} \\ u_1 & u_2 & u_3 \\ v_1 & v_2 & v_3 \end{vmatrix}$$

by cofactor expansion along the first row. The result is the vector

$$(u_2v_3 - u_3v_2)\mathbf{i} + (-u_1v_3 + u_3v_1)\mathbf{j} + (u_1v_2 - u_2v_1)\mathbf{k}$$

which is $\mathbf{u} \times \mathbf{v}$.

*38.4. Let $\mathbf{u} = (1,0,-1)$, $\mathbf{v} = (2,1,4)$, and $\mathbf{w} = (1,-1,0)$. Compute all the following:

(i) $\mathbf{u} \times \mathbf{v}$
(ii) $\mathbf{v} \times \mathbf{u}$
(iii) $(\mathbf{u} \times \mathbf{v}) \times \mathbf{w}$
(iv) $\mathbf{u} \times (\mathbf{v} \times \mathbf{w})$
(v) $(\mathbf{u} \cdot \mathbf{w})\mathbf{v} - (\mathbf{u} \cdot \mathbf{v})\mathbf{w}$
(vi) $\mathbf{u} \times (\mathbf{v} + \mathbf{w})$
(vii) $(\mathbf{u} \times \mathbf{v}) + (\mathbf{u} \times \mathbf{w})$
(viii) $\mathbf{u} \cdot (\mathbf{v} \times \mathbf{w})$
(ix) $(\mathbf{u} \times \mathbf{v}) \cdot \mathbf{w}$
(x) $\|\mathbf{u} \times \mathbf{v}\|$

*38.5. Find the two vectors of norm 1 that are orthogonal to the plane determined by the vectors $(2,2,-2)$ and $(-3,1,3)$.

*38.6. Determine whether or not the vector $(2,-1,1)$ is orthogonal to the following subspaces of $\mathbb{E}^3$.

(i) The plane spanned by $\{(1,-2,-4)$ and $(3,5,-1)\}$.
(ii) The column space of the matrix

$$\begin{bmatrix} 4 & -1 & 1 \\ 4 & 5 & 6 \\ -4 & 7 & 5 \end{bmatrix}$$

(iii) $\{(x_1, x_2, x_3) \in \mathbb{E}^3\colon 2x_1 - x_2 + x_3 = 0\}$

38.7. Determine the angles in the triangle formed by:

(i) The following three points in $\mathbb{E}^2$: (2, 2), (4, 5), and (4, 1).

(ii) The following three points in $\mathbb{E}^3$: (4, 4, 5), (2, 2, 4), and (4, 1, 2).

*38.8. Let $\mathbf{u} = (5, 1 + x, 5 - 2x)$ and $\mathbf{v} = (2, 1 - x, 2)$.

(i) For what value(s) of x, if any, are the vectors $\mathbf{u}$ and $\mathbf{v}$ orthogonal?

(ii) For what value(s) of x, if any, is the length of the vector $\mathbf{u} - \mathbf{v}$ equal to $\sqrt{2}$?

*38.9. Let $\mathbf{u} = (1, -1, 4)$, $\mathbf{v} = (-2, -3, 0)$, and $\mathbf{w} = (5, -4, -3)$.

(i) Find the Euclidean norm of $\mathbf{w}$.

(ii) Find the dot product of $2\mathbf{u} + 3\mathbf{v}$ with $\mathbf{w}$.

(iii) Find the cosine of the angle between $\mathbf{u}$ and $\mathbf{w}$.

39

The Gram – Schmidt Algorithm

In this section, we examine how the newly introduced concepts of orthogonality and Euclidean norm and distance interact with the concepts from earlier chapters.

39.1 Definition

For any subset $S \subset \mathbb{E}^n$, let $S^\perp = \{\mathbf{v} \in \mathbb{E}^n: \mathbf{u} \cdot \mathbf{v} = 0 \text{ for all } \mathbf{u} \in S\}$. In words, $S^\perp$ consists of those vectors in $\mathbb{E}^n$ that are orthogonal to S (i.e., that are orthogonal to every vector in S). Because orthogonality generalizes perpendicularity, $S^\perp$ is called *S perp*.

Our first result is an easy exercise, reminiscent of the fact that for every $S \subset \mathbb{R}^n$, $\mathscr{L}(S)$ is a linear subspace of $\mathbb{R}^n$.

39.2 Exercise

For any subset $S \subset \mathbb{E}^n$, $S^\perp$ is a linear subspace of $\mathbb{E}^n$.

To provide a focus for this section, we immediately state the main theorem, although its proof will require several digressions and auxiliary concepts, which are important in their own right and constitute the bulk of Section 39. First, we remind you that the notion of direct sum, $W_1 \oplus W_2$, of linear subspaces W_1, W_2 of $\mathbb{R}^n$ was introduced in Problem 8.12.

39.3 Theorem (Projection Theorem)

For any linear subspace W of $\mathbb{E}^n$, $\mathbb{E}^n = W \oplus W^\perp$.

A proof of this theorem will be provided in 39.21. In the meantime, we pursue our goal of integrating the new concepts with the old ones.

The definition of matrix multiplication (recall 1.11) can obviously be expressed in terms of the dot product. If A is an $m \times n$ matrix and B is an $n \times p$ matrix, and if $\mathbf{r}_1, \mathbf{r}_2, \ldots, \mathbf{r}_m$ are the rows of A and $\mathbf{c}_1, \mathbf{c}_2, \ldots, \mathbf{c}_p$ are the columns of B viewed as vectors in $\mathbb{E}^n$, then the product, AB, is the $m \times p$ matrix whose $(i, j)^{\text{th}}$ entry is $\mathbf{r}_i \cdot \mathbf{c}_j$.

This observation gives us a way of rephrasing questions about systems of linear equations using the language of dot products and orthogonality.

39.4 Exercise

Suppose that $S = \{\mathbf{u}_1, \mathbf{u}_2, \ldots, \mathbf{u}_p\} \subset \mathbb{E}^n$ and that $W = \mathscr{L}(S)$. Then for all $\mathbf{v} \in \mathbb{E}^n$,

$$\mathbf{v} \in W^\perp \quad \text{iff} \quad \mathbf{u}_i \cdot \mathbf{v} = 0, \quad \text{for } i = 1, 2, \ldots, p.$$

In words, this exercise asserts that $\mathbf{v}$ is orthogonal to W iff $\mathbf{v}$ is orthogonal to every vector in a spanning set for W.

39.5 Exercise

(i) Let $A\mathbf{x} = \mathbf{b}$ be an $m \times n$ system of linear equations and let H be its solution space. Let $\mathbf{r}_1, \mathbf{r}_2, \ldots, \mathbf{r}_m$ denote the rows of A, viewed as vectors in $\mathbb{E}^n$. Then for all $\mathbf{v} = \mathbb{E}^n$, $\mathbf{v} \in H$ iff $\mathbf{r}_i \cdot \mathbf{v} = b_i$ for $i = 1, 2, \ldots, m$.

(ii) The special case of (i) when $\mathbf{b} = \mathbf{0}$ yields the following for homogeneous systems, $A\mathbf{x} = \mathbf{0}$: $\text{NS(A)} = \text{RS(A)}^\perp$.

Combining 39.4 with 39.5(ii), we see that a vector $\mathbf{v} \in \mathbb{E}^n$ is a solution of the $m \times n$ homogeneous system $A\mathbf{x} = \mathbf{0}$ iff $\mathbf{v}$ is orthogonal to each row of the coefficient matrix.

Next we turn our attention from systems of equations to matrices. A brief digression, however, concerning unit vectors is needed first.

39.6 Definition

A vector $\mathbf{u} \in \mathbb{E}^n$ is called a *unit vector* if $\|\mathbf{u}\| = 1$. For every nonzero vector $\mathbf{u} \in \mathbb{E}^n$, the scalar multiple $\frac{1}{\|\mathbf{u}\|}\mathbf{u}$ is a unit vector called the *unit vector in the direction of* $\mathbf{u}$. The vector $\frac{-1}{\|\mathbf{u}\|}\mathbf{u}$ is the *unit vector in the direction opposite to* $\mathbf{u}$.

39.7 Example

(i) If $\mathbf{u} = (3, 4)$, then $\|\mathbf{u}\| = \sqrt{3^2 + 4^2} = \sqrt{25} = 5$, so the unit vector in the direction $\mathbf{u}$ is $(\frac{3}{5}, \frac{4}{5})$. The unit vector in the opposite direction is $(-\frac{3}{5}, -\frac{4}{5})$.

(ii) If $\mathbf{u} = (3, -1, \sqrt{2}, \frac{1}{2})$, then $\|\mathbf{u}\| = \sqrt{9 + 1 + 2 + \frac{1}{4}} = \sqrt{\frac{49}{4}} = \frac{7}{2}$, so the unit vector in the direction of $\mathbf{u}$ is $\frac{2}{7}(3, -1, \sqrt{2}, \frac{1}{2}) = (\frac{6}{7}, -\frac{2}{7}, \frac{2\sqrt{2}}{7}, \frac{1}{7})$.

39.8 Definition

A subset of $\mathbb{E}^n$ is called *orthonormal* if it is an orthogonal set of unit vectors.

39.9 Exercise

If $S = \{\mathbf{u}_1, \mathbf{u}_2, \ldots, \mathbf{u}_p\} \subset \mathbb{E}^n$, S is orthonormal if and only if

$$\text{for all } i \text{ and } j, \quad \mathbf{u}_i \cdot \mathbf{u}_j = \begin{cases} 0, & \text{if } i \neq j \\ 1, & \text{if } i = j. \end{cases} \qquad (*)$$

The expression $\mathbf{u}_i \cdot \mathbf{u}_j = \delta_{ij}$ is often used as an abbreviation for $(*)$.

We turn next to consider bases for $\mathbb{E}^n$ that are orthonormal.

39.10 Definition

A basis $\beta = \{\mathbf{u}_1, \mathbf{u}_2, \ldots, \mathbf{u}_n\}$ for $\mathbb{E}^n$ is an *orthonormal basis* if β is an orthonormal set of vectors.

39.11 Exercise

(i) The standard basis, α^n, for $\mathbb{E}^n$ is an orthonormal basis.

(ii) The set $\beta = \{(\frac{2}{3}, \frac{1}{3}, \frac{2}{3}), (-\frac{2}{3}, \frac{2}{3}, \frac{1}{3}), (\frac{1}{3}, \frac{2}{3}, -\frac{2}{3})$ is an orthonormal basis for $\mathbb{E}^3$.

What is so special about an orthonormal basis? To answer this question, we first give an example and then state a general theorem.

39.12 Example

Let β be the orthonormal basis for $\mathbb{E}^3$ from part (ii) of the previous exercise. Suppose you want to find the β-coordinates of a typical vector, $\mathbf{v} = (3, -1, 6)$, for example. The answer to this question is usually obtained by row-reduction, as explained in Computational Procedure 15.20. For an orthonormal basis, however, we can find the coordinates very quickly simply by computing dot products.

$$(3, -1, 6) \cdot \left(\tfrac{2}{3}, \tfrac{1}{3}, \tfrac{2}{3}\right) = \tfrac{17}{3}$$

$$(3, -1, 6) \cdot \left(-\tfrac{2}{3}, \tfrac{2}{3}, \tfrac{1}{3}\right) = -\tfrac{2}{3}$$

$$(3, -1, 6) \cdot \left(\tfrac{1}{3}, \tfrac{2}{3}, -\tfrac{2}{3}\right) = -\tfrac{11}{3}$$

You should verify that $(\mathbf{v})_\beta = \left(\tfrac{17}{3}, -\tfrac{2}{3}, -\tfrac{11}{3}\right)$.

There is an easy theorem to show that this example is typical.

39.13 Theorem

If $\beta = \{\mathbf{u}_1, \mathbf{u}_2, \ldots, \mathbf{u}_n\}$ is an orthonormal basis for $\mathbb{E}^n$, then for any $\mathbf{v} \in \mathbb{E}^n$,

$$\mathbf{v} = (\mathbf{v} \cdot \mathbf{u}_1)\mathbf{u}_1 + (\mathbf{v} \cdot \mathbf{u}_2)\mathbf{u}_2 + \cdots + (\mathbf{v} \cdot \mathbf{u}_n)\mathbf{u}_n$$

Proof: Since β is a basis, there are unique scalars $t_1, t_2, \ldots, t_n$ satisfying $\mathbf{v} = t_1\mathbf{u}_1 + t_2\mathbf{u}_2 + \cdots + t_n\mathbf{u}_n$. The following calculation, which uses the orthonormality of β, shows for arbitrary $i = 1, 2, \ldots, n$ that $t_i = \mathbf{v} \cdot \mathbf{u}_i$:

$$\begin{aligned}
\mathbf{v} \cdot \mathbf{u}_i &= (t_1\mathbf{u}_1 + t_2\mathbf{u}_2 + \cdots + t_n\mathbf{u}_n) \cdot \mathbf{u}_i \\
&= t_1(\mathbf{u}_1 \cdot \mathbf{u}_i) + \cdots + t_i(\mathbf{u}_i \cdot \mathbf{u}_i) + \cdots + t_n(\mathbf{u}_n \cdot \mathbf{u}_i) \\
&= 0 + \cdots + 0 + t_i + 0 + \cdots + 0 \\
&= t_i
\end{aligned}$$

■

Because of this very desirable property of orthonormal bases, we are willing to expend considerable effort to obtain them. Fortunately, it is always possible to obtain one. This is the essence of the Gram–Schmidt algorithm.

39.14 Computational Procedure (The Gram–Schmidt Algorithm)

The Gram–Schmidt algorithm accepts as input any finite ordered set $S = (\mathbf{u}_1, \mathbf{u}_2, \ldots, \mathbf{u}_p)$ of vectors in $\mathbb{E}^n$. If S is linearly dependent, the algorithm will discover this fact and output the information either that $\mathbf{u}_1 = \mathbf{0}$ or that some specific $\mathbf{u}_i$ is a linear combination of its predecessors. If S is linearly independent, the output, which we will denote by GS(S), is an ordered set $(\mathbf{v}_1, \mathbf{v}_2, \ldots, \mathbf{v}_p)$ of vectors in $\mathbb{E}^n$ with the property that for each $j = 1, 2, \ldots, p$, $(\mathbf{v}_1, \mathbf{v}_2, \ldots, \mathbf{v}_j)$ is an orthonormal basis for the subspace spanned by $(\mathbf{u}_1, \mathbf{u}_2, \ldots, \mathbf{u}_j)$.

In particular, if the input is an arbitrary ordered basis for a p-dimensional subspace W of $\mathbb{E}^n$, then the output is an ordered orthonormal basis for W.

Before giving the details of the algorithm, we present the two very simple ideas on which it is based. The first is an observation that we have already made.

39.15 Observation

For any nonzero vector $\mathbf{u} \in \mathbb{E}^n$, the vector $\mathbf{v} = \dfrac{1}{\|\mathbf{u}\|}\mathbf{u}$ is a unit vector.

The second idea is one whose intuitive content is clear from our experience with $\mathbb{E}^2$ and $\mathbb{E}^3$ but whose precise statement for $\mathbb{E}^n$ can only be made after the Projection Theorem is proved. For nonzero vectors $\mathbf{u}, \mathbf{v} \in \mathbb{E}^2$, we know that "if you subtract from $\mathbf{u}$ the component of $\mathbf{u}$ in the direction of $\mathbf{v}$, then what is left is perpendicular to $\mathbf{v}$." See Figure 39.1. For nonzero vectors $\mathbf{u}_1, \mathbf{v}_1, \mathbf{v}_2 \in \mathbb{E}^3$, "if you subtract from $\mathbf{u}$ the component of $\mathbf{u}$ in the plane P determined by $\mathbf{v}_1$ and $\mathbf{v}_2$, then what is left is perpendicular to P." See Figure 39.2. The problem in generalizing these ideas to $\mathbb{E}^n$ is to explain and justify using the definite article in the phrase "*the* component of $\mathbf{u}$ . . . ." But this is precisely the content of the Projection Theorem, so we reserve further comment on this matter.

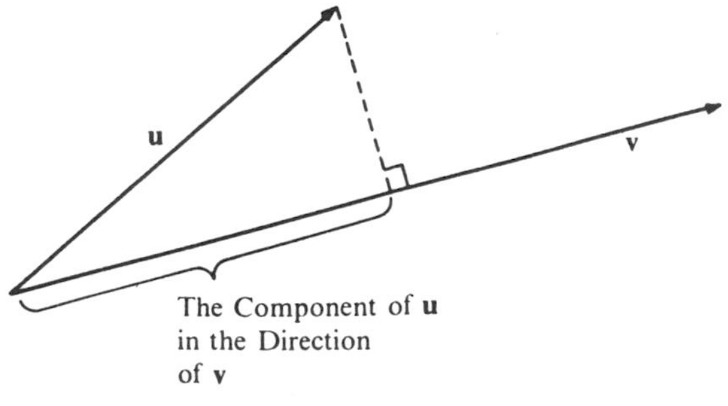

Figure 39.1

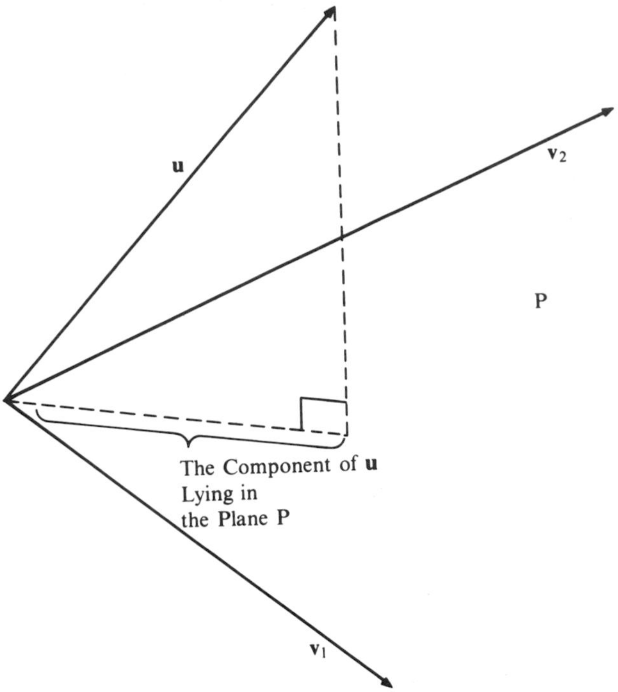

Figure 39.2

39.16 Theorem

(i) If $\mathbf{u}, \mathbf{v} \in \mathbb{E}^n$ and $\|\mathbf{v}\| = 1$, then the vector $\mathbf{w} = \mathbf{u} - (\mathbf{u} \cdot \mathbf{v})\mathbf{v}$ is orthogonal to $\mathbf{v}$.

(ii) If $\mathbf{u}, \mathbf{v}_1, \mathbf{v}_2, \ldots, \mathbf{v}_j \in \mathbb{E}^n$ and the set $S = \{\mathbf{v}_1, \mathbf{v}_2, \ldots, \mathbf{v}_j\}$ is orthonormal, then the vector

$$\mathbf{w} = \mathbf{u} - (\mathbf{u} \cdot \mathbf{v}_1)\mathbf{v}_1 - (\mathbf{u} \cdot \mathbf{v}_2)\mathbf{v}_2 - \cdots - (\mathbf{u} \cdot \mathbf{v}_j)\mathbf{v}_j$$

is orthogonal to the subspace $\mathscr{L}(S)$.

Proof of (i):

$$\begin{aligned}\mathbf{w} \cdot \mathbf{v} &= (\mathbf{u} - (\mathbf{u} \cdot \mathbf{v})\mathbf{v}) \cdot \mathbf{v} \\ &= \mathbf{u} \cdot \mathbf{v} - (\mathbf{u} \cdot \mathbf{v})(\mathbf{v} \cdot \mathbf{v}) \\ &= \mathbf{u} \cdot \mathbf{v} - (\mathbf{u} \cdot \mathbf{v})\|\mathbf{v}\|^2 \\ &= 0\end{aligned}$$

Proof of (ii): The proof for arbitrary $i = 1, 2, \ldots, j$ that $\mathbf{w} \cdot \mathbf{v}_i = 0$ is similar to the proof just given for (i). The orthonormality of S makes everything go smoothly. ■

39.17 Details of the Gram – Schmidt Algorithm

The Gram–Schmidt algorithm is a two-step loop, steps $1.k$ and $2.k$, which is traversed successively for $k = 1, 2, \ldots,$ and which will halt for some $k \leqq p$ if the input is $S = (\mathbf{u}_1, \mathbf{u}_2, \ldots, \mathbf{u}_p)$.

Step $1.k$. Set $\mathbf{w}_k = \mathbf{u}_k - (\mathbf{u}_k \cdot \mathbf{v}_1)\mathbf{v}_1 - (\mathbf{u}_k \cdot \mathbf{v}_2)\mathbf{v}_2 - \cdots - (\mathbf{u}_k \cdot \mathbf{v}_{k-1})\mathbf{v}_{k-1}$. (Note that for $k = 1$, we simply set $\mathbf{w}_1 = \mathbf{u}_1$.)

Step $2.k$. If $\mathbf{w}_k = \mathbf{0}$, halt and output the following information: in case $k = 1$, S is linearly dependent because $\mathbf{u}_1 = \mathbf{0}$; in case $k \gneqq 1$, S is linearly dependent because $\mathbf{u}_k \in \mathscr{L}(\mathbf{u}_1, \mathbf{u}_2, \ldots, \mathbf{u}_{k-1})$. If $\mathbf{w}_k \neq \mathbf{0}$, set $\mathbf{v}_k = (1/\|\mathbf{w}_k\|)\mathbf{w}_k$ and, if $k = p$, halt and output the set $(\mathbf{v}_1, \mathbf{v}_2, \ldots, \mathbf{v}_k)$; if $k \lneqq p$, go to step $1.k + 1$.

39.18 Exercise

Prove, using 39.15 and 39.16, that the Gram–Schmidt algorithm performs as claimed in 39.14.

Before attempting this exercise, you may wish to consider the following example.

39.19 Example

Here is a step-by-step description of the Gram–Schmidt algorithm applied to the set $S = ((2, 1, -2), (-1, 2, 1), (3, 3, 3)) = (\mathbf{u}_1, \mathbf{u}_2, \mathbf{u}_3)$.

Step 1.1. $\mathbf{w}_1 = \mathbf{u}_1 = (2, 1, -2)$

Step 2.1. $$\mathbf{v}_1 = \frac{1}{\|\mathbf{w}_1\|}\mathbf{w}_1 = \frac{1}{\sqrt{4+1+4}}(2, 1, -2) = (\tfrac{2}{3}, \tfrac{1}{3}, -\tfrac{2}{3})$$

Step 1.2. $$\begin{aligned}\mathbf{w}_2 &= \mathbf{u}_2 - (\mathbf{u}_2 \cdot \mathbf{v}_1)\mathbf{v}_1 \\ &= (-1, 2, 1) - ((-1)(\tfrac{2}{3}) + (2)(\tfrac{1}{3}) + (1)(-\tfrac{2}{3}))(\tfrac{2}{3}, \tfrac{1}{3}, -\tfrac{2}{3}) \\ &= (-1, 2, 1) - (-\tfrac{2}{3})(\tfrac{2}{3}, \tfrac{1}{3}, -\tfrac{2}{3}) \\ &= (-1, 2, 1) - (-\tfrac{4}{9}, -\tfrac{2}{9}, \tfrac{4}{9}) \\ &= (-\tfrac{5}{9}, \tfrac{20}{9}, \tfrac{5}{9})\end{aligned}$$

Step 2.2. $$\begin{aligned}\mathbf{v}_2 &= \frac{1}{\|\mathbf{w}_2\|}\mathbf{w}_2 = \frac{1}{\sqrt{\frac{25}{81} + \frac{400}{81} + \frac{25}{81}}}(-\tfrac{5}{9}, \tfrac{20}{9}, \tfrac{5}{9}) \\ &= \frac{9}{15\sqrt{2}}(-\tfrac{5}{9}, \tfrac{20}{9}, \tfrac{5}{9}) \\ &= \left(\frac{-\sqrt{2}}{6}, \frac{4\sqrt{2}}{6}, \frac{\sqrt{2}}{6}\right)\end{aligned}$$

Step 1.3. $\mathbf{w}_3 = \mathbf{u}_3 - (\mathbf{u}_3 \cdot \mathbf{v}_1)\mathbf{v}_1 - (\mathbf{u}_3 \cdot \mathbf{v}_2)\mathbf{v}_2$

$$= (3,3,3) - \left((3)\left(\tfrac{2}{3}\right) + (3)\left(\tfrac{1}{3}\right) + (3)\left(-\tfrac{2}{3}\right)\right)\left(\tfrac{2}{3}, \tfrac{1}{3}, -\tfrac{2}{3}\right)$$

$$- \left((3)\left(\frac{-\sqrt{2}}{6}\right) + (3)\left(\frac{4\sqrt{2}}{6}\right) + (3)\left(\frac{\sqrt{2}}{6}\right)\right)\left(\frac{-\sqrt{2}}{6}, \frac{4\sqrt{2}}{6}, \frac{\sqrt{2}}{6}\right)$$

$$= (3,3,3) - (1)\left(\tfrac{2}{3}, \tfrac{1}{3}, -\tfrac{2}{3}\right) - 2\sqrt{2}\left(\frac{-\sqrt{2}}{6}, \frac{4\sqrt{2}}{6}, \frac{\sqrt{2}}{6}\right)$$

$$= (3,3,3) - \left(\tfrac{2}{3}, \tfrac{1}{3}, -\tfrac{2}{3}\right) - \left(-\tfrac{2}{3}, \tfrac{8}{3}, \tfrac{2}{3}\right)$$

$$= (3,0,3)$$

Step 2.3. $\mathbf{v}_3 = \dfrac{1}{\|w_3\|}\mathbf{w}_3 = \dfrac{1}{\sqrt{9+0+9}}(3,0,3)$

$$= \frac{1}{3\sqrt{2}}(3,0,3) = \left(\frac{\sqrt{2}}{2}, 0, \frac{\sqrt{2}}{2}\right)$$

The algorithm halts and outputs the orthonormal set $\left(\left(\tfrac{2}{3}, \tfrac{1}{3}, -\tfrac{2}{3}\right), \left(-\sqrt{2}/6, 4\sqrt{2}/6, \sqrt{2}/6\right), \left(\sqrt{2}/2, 0, \sqrt{2}/2\right)\right)$.

The Gram–Schmidt algorithm should be viewed as a proof of the following important theorem.

39.20 Theorem

Every linear subspace W of $\mathbb{E}^n$ has an orthonormal basis.

Proof: W has a basis by Theorem 7.32. Let β be a basis for W and apply the Gram–Schmidt algorithm to β to obtain an orthonormal basis. ■

We have delayed the proof of the Projection Theorem until now because the proof we plan to give requires the previous theorem.

Theorem [The Projection Theorem (Theorem 39.3 repeated for convenience)]

For any linear subspace W of $\mathbb{E}^n$, $\mathbb{E}^n = W \oplus W^{\perp}$.

39.21 Proof of the Projection Theorem

First, we show that $\mathbb{E}^n = W + W^{\perp}$; that is, every $\mathbf{u} \in \mathbb{E}^n$ can be expressed (in at least one way) as the sum of two vectors, $\mathbf{w}_1 + \mathbf{w}_2$, where $\mathbf{w}_1 \in W$ and $\mathbf{w}_2 \in W^{\perp}$. Let $\beta = \{\mathbf{v}_1, \mathbf{v}_2, \ldots, \mathbf{v}_p\}$ be an orthonormal basis for W; such a β exists by 39.20. If $\mathbf{u} \in \mathbb{E}^n$ is given, let $\mathbf{w}_1 = (\mathbf{u} \cdot \mathbf{v}_1)\mathbf{v}_1 + (\mathbf{u} \cdot \mathbf{v}_2)\mathbf{v} + \cdots + (\mathbf{u} \cdot \mathbf{v}_p)\mathbf{v}_p$ and let $\mathbf{w}_2 = \mathbf{u} - \mathbf{w}_1$. Then the fact that $\mathbf{u} = \mathbf{w}_1 + \mathbf{w}_2$ is true by definition of $\mathbf{w}_2$; the fact that $\mathbf{w}_1 \in W$ is clear since $\mathbf{w}_1$ is a linear combination of vectors that are a basis for W; finally, since β is an orthonormal basis for W, $\mathbf{w}_2 \in W^{\perp}$ by Theorem 39.16(ii).

Second, to conclude that $\mathbb{E}^n = W \oplus W^{\perp}$, it remains to show that $W \cap W^{\perp} = \{\mathbf{0}\}$. But this fact is clear, for if $\mathbf{v} \in W \cap W^{\perp}$, then $\mathbf{v}$ is orthogonal to itself; that is, $\mathbf{v} \cdot \mathbf{v} = 0$, so $\mathbf{v} = \mathbf{0}$. ■

Combining Problem 8.12 with the Projection Theorem yields the next important corollary.

39.22 Corollary

For each linear subspace W of $\mathbb{E}^n$ and each vector $\mathbf{u} \in \mathbb{E}^n$, there is a unique pair of vectors $\mathbf{w}_1 \in W$ and $\mathbf{w}_2 \in W^{\perp}$ such that $\mathbf{u} = \mathbf{w}_1 + \mathbf{w}_2$. ■

This corollary guarantees that the following definition makes sense.

39.23 Definition

Where W is a linear subspace of $\mathbb{E}^n$ and $\mathbf{u} \in \mathbb{E}^n$, and where $\mathbf{w}_1$ and $\mathbf{w}_2$ are the unique vectors satisfying $\mathbf{u} = \mathbf{w}_1 + \mathbf{w}_2$ with $\mathbf{w}_1 \in W$ and $\mathbf{w}_2 \in W^{\perp}$, $\mathbf{w}_1$ is called the *orthogonal projection of u onto W* and is denoted by $\text{proj}_W \mathbf{u}$. It is also occasionally called the *component of u lying in W*.

This concept is the generalization to $\mathbb{E}^n$ and to subspaces of arbitrary dimension of the familiar concepts for vectors in $\mathbb{E}^2$ and $\mathbb{E}^3$ of "dropping the perpendicular to a line or to a plane." (Take another look at Figures 39.1 and 39.2.)

The next theorem shows that it makes sense to call $\|\mathbf{w}_2\| = \|\mathbf{u} - \mathbf{w}_1\| = \|\mathbf{u} - \text{proj}_W \mathbf{u}\|$ "the distance from $\mathbf{u}$ to W" and to say that $\text{proj}_W \mathbf{u}$ is "the best approximation to $\mathbf{u}$ by a vector in W." In this context, "best" means "closest" in the sense that the Euclidean distance from $\mathbf{u}$ to $\text{proj}_W \mathbf{u}$ is strictly less than the Euclidean distance from $\mathbf{u}$ to any other vector $\mathbf{w} \in W$.

39.24 Theorem

If W is a linear subspace of $\mathbb{E}^n$ and $\mathbf{u} \in \mathbb{E}^n$, then, for all $\mathbf{w} \in W$ different from $\text{proj}_W \mathbf{u}$, $\|\mathbf{u} - \text{proj}_W \mathbf{u}\| \lneqq \|\mathbf{u} - \mathbf{w}\|$.

Proof: $\mathbf{u} - \mathbf{w} = (\mathbf{u} - \text{proj}_W \mathbf{u}) + (\text{proj}_W \mathbf{u} - \mathbf{w})$. The two vectors $\mathbf{u} - \text{proj}_W \mathbf{u}$ and $\text{proj}_W \mathbf{u} - \mathbf{w}$ are orthogonal since the first belongs to $W^{\perp}$ while the second belongs to W. So by the Generalized Pythagorean Theorem, $\|\mathbf{u} - \mathbf{w}\|^2 = \|\mathbf{u} - \text{proj}_W \mathbf{u}\|^2 + \|\text{proj}_W \mathbf{u} - \mathbf{w}\|^2$. Since $\mathbf{w} \neq \text{proj}_W \mathbf{u}$,

$0 \lneq \|\text{proj}_W \mathbf{u} - \mathbf{w}\|^2$, and hence $\|\mathbf{u} - \text{proj}_W \mathbf{u}\|^2 \lneq \|\mathbf{u} - \mathbf{w}\|^2$. The conclusion now follows by taking square roots. ■

This concludes our discussion of the theoretical properties of orthogonal projections. The practical matter of computing $\text{proj}_W \mathbf{u}$ can also be settled quickly.

39.25 Computational Procedure

To find the orthogonal projection of a vector $\mathbf{u} \in \mathbb{E}^n$ onto a p-dimensional linear subspace W of $\mathbb{E}^n$, choose *any* orthonormal basis $\beta = \{\mathbf{v}_1, \mathbf{v}_2, \ldots, \mathbf{v}_p\}$ for W. Then $\text{proj}_W \mathbf{u} = (\mathbf{u} \cdot \mathbf{v}_1)\mathbf{v}_1 + (\mathbf{u} \cdot \mathbf{v}_2)\mathbf{v}_2 + \cdots + (\mathbf{u} \cdot \mathbf{v}_p)\mathbf{v}_p$.

Justification: Let $\mathbf{w}_1 = (\mathbf{u} \cdot \mathbf{v}_1)\mathbf{v}_1 + (\mathbf{u} \cdot \mathbf{v}_2)\mathbf{v}_2 + \cdots + (\mathbf{u} \cdot \mathbf{v}_p)\mathbf{v}_p$. Then $\mathbf{u} = \mathbf{w}_1 + (\mathbf{u} - \mathbf{w}_1)$, $\mathbf{w}_1 \in W$ since $\mathbf{w}_1 \in \mathscr{L}(\beta) = W$, and $\mathbf{u} - \mathbf{w}_1 \in W^\perp$ by Theorem 39.16(ii). But $\text{proj}_W \mathbf{u}$ is, by definition, the unique vector having the properties possessed by $\mathbf{w}_1$. Hence $\text{proj}_W \mathbf{u} = \mathbf{w}_1$. ■

It is important to realize the full strength of this procedure: if a *different* orthonormal basis $\beta' = \{\mathbf{v}_1', \mathbf{v}_2', \ldots, \mathbf{v}_p'\}$ for W is chosen, computing $(\mathbf{u} \cdot \mathbf{v}_1')\mathbf{v}_1' + (\mathbf{u} \cdot \mathbf{v}_2')\mathbf{v}_2' + \cdots + (\mathbf{u} \cdot \mathbf{v}_p')\mathbf{v}_p'$ will *still* yield $\text{proj}_W \mathbf{u}$.

To conclude this section, we present some examples, beginning with some from the familiar context of $\mathbb{E}^2$ and $\mathbb{E}^3$.

39.26 Example

Write the vector $(3, 8)$ as the sum of a vector on the x_1 axis plus a vector perpendicular to the x_1 axis.

Of course, we know the answer to this question: $(3, 8) = (3, 0) + (0, 8)$; we merely wish to point out how the computational procedure gives, as it must, this same answer. Let W be the x_1 axis and let $\beta = \{(1, 0)\}$; β is an orthonormal basis for W; hence $\text{proj}_W(3, 8) = ((3, 8) \cdot (1, 0))(1, 0) = 3(1, 0) = (3, 0)$. The theorems guarantee that $(3, 8) - \text{proj}_W(3, 8) = (3, 8) - (3, 0) = (0, 8)$ is perpendicular to the x_1 axis.

39.27 Example

Write $(3, 8)$ in the form $\mathbf{w}_1 + \mathbf{w}_2$, where $\mathbf{w}_1$ is a multiple of the vector $(4, 3)$ and $\mathbf{w}_2$ is perpendicular to $(4, 3)$.

The answer

$$(3, 8) = \left(\frac{144}{25}, \frac{108}{25}\right) + \left(\frac{-69}{25}, \frac{92}{25}\right)$$

is obtained using 39.25. Choose an orthonormal basis β for the line through the origin in $\mathbb{E}^2$ determined by the vector (4, 3). $\beta = \{(1/\|(4,3)\|)(4,3)\} = \{(\frac{4}{5}, \frac{3}{5})\}$ will do nicely.

The projection of (3, 8) onto this subspace is

$$\left((3,8)\cdot\left(\frac{4}{5},\frac{3}{5}\right)\right)\left(\frac{4}{5},\frac{3}{5}\right) = \frac{36}{5}\left(\frac{4}{5},\frac{3}{5}\right) = \left(\frac{144}{25},\frac{108}{25}\right)$$

The vector perpendicular to the line will be

$$(3,8) - \left(\frac{144}{25},\frac{108}{25}\right) = \left(\frac{-69}{25},\frac{92}{25}\right)$$

39.28 Example

Let W be the plane through the origin in $\mathbb{E}^3$ determined by the vectors (6, 3, 2) and (9, 4, 2). Express the vector $(8, -\frac{7}{2}, 4)$ in the form $\mathbf{w}_1 + \mathbf{w}_2$, where $\mathbf{w}_1 \in W$ and $\mathbf{w}_2 \in W^\perp$. A closely related question, whose answer requires essentially the same calculations, would be the following: What is the distance from the point $(8, -\frac{7}{2}, 4)$ to the plane determined by the three points (0, 0, 0), (6, 3, 2), and (9, 4, 2)?

The first step is to find an orthonormal basis for $W = \mathscr{L}((6,3,2),(9,4,2))$. You should check that applying the Gram–Schmidt algorithm to the set ((6, 3, 2), (9, 4, 2)) yields $((\frac{6}{7},\frac{3}{7},\frac{2}{7}),(\frac{3}{7},-\frac{2}{7},-\frac{6}{7}))$. Therefore,

$$\begin{aligned}\mathbf{w}_1 &= \operatorname{proj}_W(8,-\tfrac{7}{2},4)\\ &= ((8)(\tfrac{6}{7}) + (-\tfrac{7}{2})(\tfrac{3}{7}) + (4)(\tfrac{2}{7}))(\tfrac{6}{7},\tfrac{3}{7},\tfrac{2}{7})\\ &\quad + ((8)(\tfrac{3}{7}) + (-\tfrac{7}{2})(-\tfrac{2}{7}) + (4)(-\tfrac{6}{7}))(\tfrac{3}{7},-\tfrac{2}{7},-\tfrac{6}{7})\\ &= (\tfrac{91}{14})(\tfrac{6}{7},\tfrac{3}{7},\tfrac{2}{7}) + (1)(\tfrac{3}{7},-\tfrac{2}{7},-\tfrac{6}{7}) = (6,\tfrac{5}{2},1).\end{aligned}$$

The vector $\mathbf{w}_2 \in W^\perp$ is simply $\mathbf{w}_2 = (8, -\frac{7}{2}, 4) - (6, \frac{5}{2}, 1) = (2, -6, 3)$.

Finally, the distance from the point to the plane is $\|\mathbf{w}_2\| = \sqrt{4 + 36 + 9} = 7$.

The Projection Theorem, 39.3, and its corollary, 39.22, imply that for every $\mathbf{b} \in \mathbb{E}^n$ and every subspace $W \subset \mathbb{E}^n$ the expression

$$\mathbf{b} = \operatorname{proj}_W \mathbf{b} + (\mathbf{b} - \operatorname{proj}_W \mathbf{b})$$

is the unique way to write $\mathbf{b}$ as the sum of a vector in W plus a vector that is orthogonal to W. This fact, together with the characterization, in Theorem 39.24, of $\operatorname{proj}_W \mathbf{b}$ as the unique vector in W that is closest in Euclidean distance to $\mathbf{b}$, constitutes the foundation for one of the most important applications of linear algebra, known as *least squares approximations*. Finding a polynomial of given degree that is the "best fit" to a set of data points is a particular example of this application. Another important example from statistics is known as *regression analysis*. Here, then, are the essential ideas.

We observed in 9.4 that an $m \times n$ linear system $\mathbf{Ax} = \mathbf{b}$ is consistent iff $b \in \mathrm{CS(A)}$; each solution $\mathbf{u} = (u_1, u_2, \ldots, u_n)$ consists of coefficients that can be used to express **b** as a linear combination of the columns of A; when it exists, the solution is unique iff the columns of A are linearly independent.

Of course, **u** is a solution of $\mathbf{Ax} = \mathbf{b}$ iff $\|\mathbf{b} - \mathbf{Au}\| = 0$. This suggests something that might be done in case $\mathbf{Ax} = \mathbf{b}$ is inconsistent, that is, when $\mathbf{b} \notin \mathrm{CS(A)}$; in the absence of exact solution(s) to $\mathbf{Ax} = \mathbf{b}$, we seek, as an approximation, value(s) of **x** for which the Euclidean distance, $\|\mathbf{b} - \mathbf{Ax}\|$, between **Ax** and **b** is a minimum.

In general, problems of finding minima are solved using calculus; but the next theorem reveals that this particular minimization problem can be solved using linear algebra: the theorem shows that minimizing $\|\mathbf{b} - \mathbf{Ax}\|$ is equivalent to solving $\mathbf{Ax} = \mathrm{proj}_{\mathrm{CS(A)}}\, \mathbf{b}$. This is not surprising when we recall, from 39.24, that among the vectors in CS(A) there is a unique one, $\mathrm{proj}_{\mathrm{CS(A)}}\, \mathbf{b}$, that is closest in Euclidean distance to **b**.

39.29 Theorem

For any $m \times n$ linear system $\mathbf{Ax} = \mathbf{b}$ and for any $\mathbf{u} \in \mathbb{E}^n$, the following are equivalent:

(i) **u** is a solution of $\mathbf{Ax} = \mathrm{proj}_{\mathrm{CS(A)}}\, \mathbf{b}$
(ii) $\mathbf{Au} = \mathrm{proj}_{\mathrm{CS(A)}}\, \mathbf{b}$
(iii) $\|\mathbf{b} - \mathbf{Au}\| = \min\{\|\mathbf{b} - \mathbf{Ax}\|: \mathbf{x} \in \mathbb{E}^n\}$
(iv) $\|\mathbf{b} - \mathbf{Au}\|^2 = \min\{\|\mathbf{b} - \mathbf{Ax}\|^2: \mathbf{x} \in \mathbb{E}^n\}$

Proof: The equivalence of (i) and (ii) is trivial. The equivalence of (ii) and (iii) follows when we recall, from Observation 6.14, that for every $\mathbf{x} \in \mathbb{E}^n$, the vector $\mathbf{Ax} \in \mathrm{CS(A)}$; by 39.24, $\mathrm{proj}_{\mathrm{CS(A)}}\, \mathbf{b}$ is the unique vector in CS(A) whose Euclidean distance to **b** is a minimum. The equivalence of (iii) and (iv) follows because $\|\cdot\|^2$ is a sum of squares; thus, minimizing $\|\mathbf{b} - \mathbf{Au}\|$ is the same as minimizing $\|\mathbf{b} - \mathbf{Au}\|^2$. ■

39.30 Definition

The set $\{\mathbf{u} \in \mathbb{E}^n$: **u** satisfies any one (and hence all) of the conditions of Theorem 39.29$\}$ is the set of *least squares solutions of* $A\mathbf{x} = \mathbf{b}$.

It is from condition (iv) of Theorem 39.29 that this concept derives its name. Since $\mathrm{proj}_{\mathrm{CS(A)}}\, \mathbf{b} \in \mathrm{CS(A)}$, the linear system $\mathbf{Ax} = \mathrm{proj}_{\mathrm{CS(A)}}\, \mathbf{b}$ is consistent, so $\mathbf{Ax} = \mathbf{b}$ has at least one least squares solution. It is clear that the least squares solution is unique iff the columns of A are linearly independent; the solution in this case consists of the coefficients needed to express $\text{proj}_{CS(A)}\, \mathbf{b}$ as a linear combination of the columns of A.

When this technique is used by applied mathematicians and statisticians, the vector $\mathbf{b} - A\mathbf{x}$ is often called the *error vector* or the *residual vector*, and a least squares solution to $A\mathbf{x} = \mathbf{b}$ is one that minimizes the Euclidean norm of the error or residual. Often, but not always, A will have rank n, so there will be a unique least squares solution.

Theorem 39.29 also provides us with a way to compute the least squares solution(s): discard the inconsistent system $A\mathbf{x} = \mathbf{b}$ and instead solve the consistent system $A\mathbf{x} = \text{proj}_{CS(A)}\, \mathbf{b}$.

This approach requires a three-step process because we must begin by calculating $\text{proj}_{CS(A)}\, \mathbf{b}$.

Step 1. Apply the Gram–Schmidt algorithm to the columns of A to obtain an orthonormal basis for CS(A).
Step 2. Apply the computational procedure from 39.25 to find $\text{proj}_{CS(A)}\, \mathbf{b}$.
Step 3. Use the Gauss–Jordan algorithm to solve the system $A\mathbf{x} = \text{proj}_{CS(A)}\, \mathbf{b}$.

There is an alternate, and somewhat more familiar, approach that is based on the following theorem.

39.31 Theorem

$A\mathbf{x} = \text{proj}_{CS(A)}\, \mathbf{b}$ iff $A^T A\mathbf{x} = A^T\mathbf{b}$.

Proof: ($\rightarrow$) (Adapted from [St]) The Projection Theorem and its corollary guarantee that $\mathbf{b} - \text{proj}_{CS(A)}\, \mathbf{b}$ is orthogonal to CS(A). Thus, since for every $\mathbf{u} \in \mathbb{E}^n$, $A\mathbf{u} \in CS(A)$, we have, assuming $A\mathbf{x} = \text{proj}_{CS(A)}\, \mathbf{b}$, that

$$(A\mathbf{u}) \cdot (\mathbf{b} - A\mathbf{x}) = 0, \quad \text{for all } \mathbf{u} \in \mathbb{E}^n$$

Using Exercise 38.21, this is equivalent to

$$(A\mathbf{u})^T (\mathbf{b} - A\mathbf{x}) = 0, \quad \text{for all } \mathbf{u} \in \mathbb{E}^n$$

which, in turn, using properties of the transpose and matrix multiplication, and using 38.21 once again, is equivalent to

$$\mathbf{u} \cdot (A^T\mathbf{b} - A^T A\mathbf{x}) = 0, \quad \text{for all } \mathbf{u} \in \mathbb{E}^n$$

which is equivalent to

$$A^T\mathbf{b} - A^T A\mathbf{x} = \mathbf{0} \quad \text{or} \quad A^T A\mathbf{x} = A^T\mathbf{b}.$$

(←) Conversely, if $A^T A\mathbf{x} = A^T\mathbf{b}$, then $A^T(\mathbf{b} - A\mathbf{x}) = \mathbf{0}$, so, as in 39.5(ii), $\mathbf{b} - A\mathbf{x}$ is orthogonal to every row of A^T, and hence orthogonal to every column of A. Since $\mathbf{b} = A\mathbf{x} + (\mathbf{b} - A\mathbf{x})$ and $\mathbf{b} - A\mathbf{x} \in CS(A)^{\perp}$, it follows from Corollary 39.22 that $A\mathbf{x} = \text{proj}_{CS(A)}\, \mathbf{b}$. ■

This theorem yields a more practical way to find the least squares solution(s) since it is easier to solve $A^T A\mathbf{x} = A^T\mathbf{b}$ than to first calculate $\text{proj}_{CS(A)}\, \mathbf{b}$ and then solve $A\mathbf{x} = \text{proj}_{CS(A)}\, \mathbf{b}$. Note too that when rank(A) $= n$, $A^T A$ is invertible and the unique least squares solution is then given by $\mathbf{x} = (A^T A)^{-1} A^T\mathbf{b}$.

Statisticians usually refer to the linear system $A^T A\mathbf{x} = A^T\mathbf{b}$ as the *normal equations*. Combining Theorems 39.29 and 39.31, we see that the least squares solution(s) to $A\mathbf{x} = \mathbf{b}$ is/(are) precisely the solution(s) of the associated normal equations $A^T A\mathbf{x} = A^T\mathbf{b}$.

Many commercially available software packages for solving linear equations are programmed to give the least squares solution when the input is the augmented matrix of an inconsistent system.

We illustrate these ideas with the next two examples. In the first one, we reexamine the inconsistent system used in 15.13; in the second example, we reexamine the inconsistent system of 11.13.

39.32 Example

In 15.13, we encountered the inconsistent system $A\mathbf{x} = \mathbf{b}$, where $\mathbf{b} = (-1, 2, -3, -1)$ and

$$A = \begin{bmatrix} 1 & 2 & 1 \\ 1 & 5 & 0 \\ 0 & 3 & -1 \\ 1 & -4 & 3 \end{bmatrix}$$

From 15.13, we also notice that A only has rank 2, so we can expect to find infinitely many least squares solutions. You should verify that

$$A^T A = \begin{bmatrix} 3 & 3 & 4 \\ 3 & 54 & -13 \\ 4 & -13 & 11 \end{bmatrix} \quad \text{and} \quad A^T\mathbf{b} = \begin{bmatrix} 0 \\ 3 \\ -1 \end{bmatrix}$$

When the Gauss–Jordan algorithm is applied to the augmented matrix of the system $A^T A\mathbf{x} = A^T\mathbf{b}$, the result is

$$\begin{bmatrix} 1 & 0 & \frac{5}{3} & -\frac{1}{17} \\ 0 & 1 & -\frac{1}{3} & \frac{1}{17} \\ 0 & 0 & 0 & 0 \end{bmatrix}$$

The conclusion is that the set of least squares solutions of $A\mathbf{x} = \mathbf{b}$ is infinite:

$$\{(x_1, x_2, x_3) \in \mathbb{E}^3 : x_1 = -\tfrac{5}{3}x_3 - \tfrac{1}{17}, x_2 = \tfrac{1}{3}x_3 + \tfrac{1}{17}\}$$

We may also calculate $\text{proj}_{CS(A)}\, \mathbf{b}$ by taking a linear combination of the columns of A using the coordinates of any least squares solution as coefficients. For example, letting $x_3 = 0$, we note that $(-\frac{1}{17}, \frac{1}{17}, 0)$ is a least squares solution and so

$$-\tfrac{1}{17}(1,1,0,1) + \tfrac{1}{17}(2,5,3,-4) + 0(1,0,-1,3) = \tfrac{1}{17}(1,4,3,-5)$$
$$= \text{proj}_{\text{CS(A)}}\,\mathbf{b}$$

39.33 Example

In 11.13, we encountered the inconsistent system $\mathrm{A}\mathbf{x} = \mathbf{b}$, where $\mathbf{b} = (17,2,7)$ and $\mathrm{A} = \begin{bmatrix} 7 & 1 \\ 4 & -2 \\ 7 & 8 \end{bmatrix}$. By inspection, rank(A) = 2, so $\mathrm{A}^T\mathrm{A}$ is invertible and there is a unique least squares solution, $\mathbf{x} = (\mathrm{A}^T\mathrm{A})^{-1}\mathrm{A}^T\mathbf{b}$. You should verify that

$$(\mathrm{A}^T\mathrm{A})^{-1} = \frac{1}{4841}\begin{bmatrix} 69 & -55 \\ -55 & 114 \end{bmatrix}$$

and that the least squares solution is given by

$$x_1 = \frac{8349}{4841} \quad \text{and} \quad x_2 = \frac{-1814}{4841}$$

Consequently,

$$\text{proj}_{\text{CS(A)}}\,\mathbf{b} = \frac{8349}{4841}(7,4,7) + \frac{-1814}{4841}(1,-2,8) \approx (11.70, 7.65, 9.07)$$

39.34 Example

Find the equation of the line in $\mathbb{R}^2$ that best fits the following four points: $(-1,0)$, $(1,2)$, $(2,3)$, and $(4,7)$.

Solution: If the four points were collinear, they would all lie on the same line $y = ax + b$, so a and b would satisfy the system

$$\begin{aligned} -a + b &= 0 \\ a + b &= 2 \\ 2a + b &= 3 \\ 4a + b &= 7 \end{aligned} \qquad \text{or, equivalently,}$$

$$\begin{bmatrix} -1 & 1 \\ 1 & 1 \\ 2 & 1 \\ 4 & 1 \end{bmatrix}\begin{bmatrix} a \\ b \end{bmatrix} = \begin{bmatrix} 0 \\ 2 \\ 3 \\ 7 \end{bmatrix}.$$

Because the four points are not collinear, this system is inconsistent. In this case, there is still a unique line that best fits these points, in the sense that the sum of the squares of the errors is minimized. The values of a and b

that yield this line of best fit are

$$\begin{bmatrix} a \\ b \end{bmatrix} = (A^TA)^{-1}A^T\mathbf{b}$$

$$= \left(\begin{bmatrix} -1 & 1 & 2 & 3 \\ 1 & 1 & 1 & 1 \end{bmatrix} \begin{bmatrix} -1 & 1 \\ 1 & 1 \\ 2 & 1 \\ 4 & 1 \end{bmatrix} \right)^{-1}$$

$$\begin{bmatrix} -1 & 1 & 2 & 4 \\ 1 & 1 & 1 & 1 \end{bmatrix} \begin{bmatrix} 0 \\ 2 \\ 3 \\ 7 \end{bmatrix} = \begin{bmatrix} 18/13 \\ 12/13 \end{bmatrix}.$$

Thus the equation of the line of best fit is $y = 18/13x + 12/13$.

Problem Set 39

39.1. Go back and do any of the exercises from Section 39 that you may have skipped.

*39.2. Let $\mathbf{u} = (2, -3, 6)$ and $\mathbf{v} = (1, 1, -1)$.
 (i) Find a vector parallel to $\mathbf{u}$ with norm 1.
 (ii) Find a vector parallel to $\mathbf{v}$ with norm 1.
 (iii) Find vectors $\mathbf{w}_1$ and $\mathbf{w}_2$ such that $\mathbf{v} = \mathbf{w}_1 + \mathbf{w}_2$, where $\mathbf{w}_1$ is parallel to $\mathbf{u}$ and $\mathbf{w}_2$ is perpendicular to $\mathbf{u}$.
 (iv) Find vectors $\mathbf{w}_1$ and $\mathbf{w}_2$ such that $\mathbf{u} = \mathbf{w}_1 + \mathbf{w}_2$, where $\mathbf{w}_1$ is parallel to $\mathbf{v}$ and $\mathbf{w}_2$ is perpendicular to $\mathbf{v}$.

*39.3. Let $\mathbf{u} = \frac{1}{5}(4, 1, -2, -2)$ and $\mathbf{v} = \frac{1}{5}(1, -4, -2, 2)$, and let $W = \mathscr{L}(\mathbf{u}, \mathbf{v})$.
 (i) Show that $\{\mathbf{u}, \mathbf{v}\}$ is an orthonormal set in $\mathbb{E}^4$.
 (ii) Find the orthogonal projection of the vector $\mathbf{w} = (4, -5, -10, 3)$ onto W.
 (iii) Find the component of $\mathbf{w}$ orthogonal to W.

*39.4. Redo Examples 39.32 and 39.33 using the three-step process described just prior to Theorem 39.31 to calculate both $\text{proj}_{CS(A)}\,\mathbf{b}$ and the least squares solutions to $A\mathbf{x} = \mathbf{b}$. (You should, of course, obtain the same answers.)

*39.5. Let $W = \mathscr{L}\{(2, 0, 1, 2), (0, 0, 1, 1), (0, 0, 6, 0)\}$.
 (i) Use the Gram–Schmidt algorithm to find an orthonormal basis for W.
 (ii) Let $\mathbf{w} = (1, 1, 1, 1)$. Find vectors $\mathbf{w}_1 \in W$ and $\mathbf{w}_2 \in W^\perp$ such that $\mathbf{w} = \mathbf{w}_1 + \mathbf{w}_2$.

*39.6. (i) Is the set $S = \{(-1, 3, -1), (3, 2, 3), (1, 0, -1)\}$ orthogonal?
 (ii) What is GS(S)?

*39.7. Let W be the subspace of $\mathbb{E}^4$ spanned by the two vectors $(\frac{4}{5}, \frac{2}{5}, \frac{2}{5}, \frac{1}{5})$ and $(-\frac{1}{5}, -\frac{2}{5}, \frac{2}{5}, \frac{4}{5})$. Express the vector $\mathbf{u} = (1, 0, -2, -25)$ in the form $\mathbf{u} = \mathbf{w}_1 + \mathbf{w}_2$, where $\mathbf{w}_1 \in W$ and $\mathbf{w}_2 \in W^\perp$.

39.8. Find the least squares solution(s) of the inconsistent systems from Problems 12-3, 12-4, 12-11, 12-14, and 12-16.

40

Linear Maps That Preserve the Dot Product

Having introduced the dot product, the next order of mathematical business is, naturally, to study linear maps that preserve this additional structure.

40.1 Definition

A linear map $T:\mathbb{E}^n \to \mathbb{E}^m$ will be called *dot-product-preserving* if it satisfies

$$\text{for all } \mathbf{u}, \mathbf{v} \in \mathbb{E}^n, \quad T(\mathbf{u}) \cdot T(\mathbf{v}) = \mathbf{u} \cdot \mathbf{v} \qquad (*)$$

40.2 Remark

There is another, more common, name for these maps. This name, which will be introduced eventually, is a misnomer, which totally obscures the intuitive content of this concept; for this reason, we prefer to delay using it.

Since the notions of (i) Euclidean norm of a vector, and (ii) angle between two vectors were defined in terms of the dot product, the results in the next exercise are not surprising; they assert that a dot-product-preserving map must also preserve the Euclidean norm of a vector, as well as the angle between two vectors.

40.3 Exercise

Suppose the linear map $T:\mathbb{E}^n \to \mathbb{E}^m$ is dot-product-preserving. Then

(i) For all $\mathbf{u} \in \mathbb{E}^n$, $\|T(\mathbf{u})\| = \|\mathbf{u}\|$ [T preserves the Euclidean norm]
(ii) For all $\mathbf{u}, \mathbf{v} \in \mathbb{E}^n$, the angle between $T(\mathbf{u})$ and $T(\mathbf{v})$ is the same as the angle between $\mathbf{u}$ and $\mathbf{v}$. [T is angle-preserving]

It follows from 40.3(i) that a dot-product-preserving linear map cannot take a nonzero vector to the zero vector. Thus, from Section 13, if $T:\mathbb{E}^n \to \mathbb{E}^m$ is a dot-product-preserving linear map, then T is one-to-one, $n \leq m$, and the range of T is an n-dimensional subspace of $\mathbb{E}^m$. For this reason, there is no great loss of generality if we confine our study of dot-product-preserving linear maps to the special case when $m = n$.

To sharpen our geometric intuition about dot-product-preserving linear operators, we will immediately discuss the converse of Exercise 40.3. The converse would assert that a linear operator that preserves Euclidean norms *and* angles will preserve the dot-product. In fact, a stronger result is true: preserving Euclidean norms *alone* will guarantee that the dot product is preserved (and hence angles, as well, by 40.3).

40.4 Theorem

If the linear operator $T:\mathbb{E}^n \to \mathbb{E}^n$ satisfies $\|T(\mathbf{u})\| = \|\mathbf{u}\|$ for all $\mathbf{u} \in \mathbb{E}^n$, then T preserves the dot product.

Proof: Let $\mathbf{u}, \mathbf{v} \in \mathbb{E}^n$; then $\|\mathbf{u} + \mathbf{v}\|^2 = \|T(\mathbf{u} + \mathbf{v})\|^2$ since T is norm-preserving. So $(\mathbf{u} + \mathbf{v}) \cdot (\mathbf{u} + \mathbf{v}) = T(\mathbf{u} + \mathbf{v}) \cdot T(\mathbf{u} + \mathbf{v})$. Using the symmetry and bilinearity of the dot-product, we have

$$(\mathbf{u} + \mathbf{v}) \cdot (\mathbf{u} + \mathbf{v}) = \|\mathbf{u}\|^2 + 2\mathbf{u} \cdot \mathbf{v} + \|\mathbf{v}\|^2.$$

Again using the symmetry and bilinearity of the dot product plus, this time, the linearity of T, we have

$$T(\mathbf{u} + \mathbf{v}) \cdot T(\mathbf{u} + \mathbf{v}) = \|T(\mathbf{u})\|^2 + 2T(\mathbf{u}) \cdot T(\mathbf{v}) + \|T(\mathbf{v})\|^2.$$

But $\|T(\mathbf{u})\| = \|\mathbf{u}\|$ and $\|T(\mathbf{v})\| = \|\mathbf{v}\|$ by assumption; hence $T(\mathbf{u}) \cdot T(\mathbf{v}) = \mathbf{u} \cdot \mathbf{v}$. ■

To summarize the results obtained so far, we may say that linear operators preserve the Euclidean norm iff they preserve the dot product.

40.5 Remark

It follows that maps which preserve the Euclidean norm also preserve angles; however, angle-preserving maps need *not* preserve the Euclidean norm. Consider, for example, $T:\mathbb{E}^2 \to \mathbb{E}^2$ defined by $T(\mathbf{x}) = 2\mathbf{x}$. This map simply doubles the length of every vector, while leaving angles unchanged.

The next several items may appear to divert us from the study of linear operators that preserve the Euclidean norm, but the relevant connection will be made in Theorem 40.11.

The first item is a trivial observation relating matrix multiplication with the dot-product.

40.6 Observation

Let A be an $m \times n$ matrix, B an $n \times p$ matrix, and let $\mathbf{r}_i$ and $\mathbf{c}_j$ denote the ith row of A and j^{th} column of B, respectively, viewed as vectors in $\mathbb{E}^n$; then for all i and j, the $(i, j)^{\text{th}}$ entry of AB is $\mathbf{r}_i \cdot \mathbf{c}_j$.

40.7 Exercise

Let A be an $n \times n$ matrix.

(i) The rows of A are an orthonormal set iff $AA^T = I_n$.
(ii) The columns of A are an orthonormal set iff $A^TA = I_n$.

40.8 Corollary

The following are equivalent for an $n \times n$ matrix, A:

(i) The rows of *A* are an orthonormal set.
(ii) A is invertible and $A^{-1} = A^T$.
(iii) The columns of A are an orthonormal set.

Proof: It follows from 19.5 that $AA^T = I_n$ iff $A^TA = I_n$ iff $A^{-1} = A^T$.
■

This property of a square matrix is clearly deserving of a special name. An obvious and natural name would be "orthonormal matrix," but, unfortunately, we must live with a historically entrenched misnomer.

40.9 Definition

An $n \times n$ matrix satisfying any one (and hence all) of the three properties from 40.8 is called an *orthogonal matrix*.

40.10 Exercise

(i) If A is an orthogonal matrix, then so are A^{-1} and A^T.
(ii) The product of two orthogonal matrices is an orthogonal matrix.
(iii) A transition matrix ${}_{\alpha^n}[I]_\beta$ from a basis β to the standard basis α^n is an orthogonal matrix iff β is an orthonormal basis for $\mathbb{E}^n$.
(iv) If β and γ are both orthonormal bases for $\mathbb{E}^n$, then the transition matrix, ${}_\gamma[I]_\beta$ is orthogonal.

It is time to tie all this in with the concept of norm-preserving linear operators.

40.11 Theorem

The following are equivalent for any linear operator $T:\mathbb{E}^n \to \mathbb{E}^n$.

(i) T preserves the Euclidean norm.
(ii) The standard matrix for T, ${}_{\alpha^n}[T]_{\alpha^n}$, is an orthogonal matrix.
(iii) If β is any orthonormal basis for $\mathbb{E}^n$, then ${}_\beta[T]_\beta$ is an orthogonal matrix.

Proof: Since

$$ {}_\beta[T]_\beta = {}_\beta[I]_{\alpha^n}\,{}_{\alpha^n}[T]_{\alpha^n}\,{}_{\alpha^n}[I]_\beta = \left({}_{\alpha^n}[I]_\beta\right)^{-1}\,{}_{\alpha^n}[T]_{\alpha^n}\,{}_{\alpha^n}[I]_\beta $$

the equivalence of (ii) and (iii) follows immediately from Exercise 40.10.

(i) → (ii) By Theorem 31.7, the set of columns of ${}_{\alpha^n}[T]_{\alpha^n}$ is the set $\{T(\mathbf{e}_1^n), T(\mathbf{e}_2^n), \ldots, T(\mathbf{e}_n^n)\}$. This set is orthonormal since

$$ \begin{aligned} T(\mathbf{e}_i^n) \cdot T(\mathbf{e}_j^n) &= \mathbf{e}_i^n \cdot \mathbf{e}_j^n && \text{[Theorem 40.4]} \\ &= \delta_{ij}. \end{aligned} $$

So ${}_{\alpha^n}[T]_{\alpha^n}$ is orthonormal by 40.8.

(ii) → (i) Let $\mathbf{u} = u_1\mathbf{e}_1^n + \mathbf{u}_2\mathbf{e}_2^n + \cdots + u_n\mathbf{e}_n^n$. Then, because T is linear, $T(\mathbf{u}) = u_1T(\mathbf{e}_1^n) + u_2T(\mathbf{e}_2^n) + \cdots + u_nT(\mathbf{e}_n^n)$. Since ${}_{\alpha^n}[T]_{\alpha^n}$ is orthogonal, the set $\{T(\mathbf{e}_1^n), T(\mathbf{e}_2^n), \ldots, T(\mathbf{e}_n^n)\}$ consisting of its columns is an orthonormal set; that is,

$$ T(\mathbf{e}_i^n) \cdot T(\mathbf{e}_j^n) = \delta_{ij}. $$

Hence,

$$ \begin{aligned} \|T(\mathbf{u})\|^2 &= T(\mathbf{u}) \cdot T(\mathbf{u}) \\ &= \left(u_1T(\mathbf{e}_1^n) + u_2T(\mathbf{e}_2^n) + \cdots + u_nT(\mathbf{e}_n^n)\right) \\ &\quad \cdot\left(u_1T(\mathbf{e}_1^n) + u_2T(\mathbf{e}_2^n) + \cdots + u_nT(\mathbf{e}_n^n)\right) \\ &= \Sigma_{ij}u_iu_jT(\mathbf{e}_i^n) \cdot T(\mathbf{e}_j^n) \\ &= \Sigma_{ij}u_iu_j\delta_{ij} \\ &= u_1^2 + u_2^2 + \cdots + u_n^2 \\ &= \|\mathbf{u}\|^2 \end{aligned} $$

■

Because of this theorem, we will now introduce the more common name for such linear operators.

40.12 Definition

A linear operator $T:\mathbb{E}^n \to \mathbb{E}^n$ that preserves the Euclidean norm is called an *orthogonal operator*.

Despite this name, it is preferable to think of orthogonal operators for what they "really" are—maps that preserve the Euclidean norm (and hence also preserve dot-product and angle).

We mentioned previously that the name "orthonormal" would be an improvement; a still better name, often used in more advanced texts, is "isometry." The misleading aspect of the word orthogonal is the temptation to think that an orthogonal map is precisely one that "preserves orthogonality," that is, that satisfies $\mathbf{u} \cdot \mathbf{v} = 0$ iff $T(\mathbf{u}) \cdot T(\mathbf{v}) = 0$. As we saw, however, in 40.5, preserving right angles (indeed, preserving *all* angles) is not the same as preserving the Euclidean norm.

Problem Set 40

40.1. Go back and do any of the exercises from Section 40 that you may have skipped.

*40.2. Prove that the determinant of an orthogonal matrix is $+1$ or -1.

*40.3. Prove or give a counterexample to the following statement: if the rows of an $n \times n$ matrix form an orthogonal set of vectors, then so do the columns.

40.4. Show that for all θ the matrices

$$\begin{bmatrix} \cos\theta & -\sin\theta \\ \sin\theta & \cos\theta \end{bmatrix} \quad \text{and} \quad \begin{bmatrix} \sin\theta & \cos\theta \\ \cos\theta & -\sin\theta \end{bmatrix}$$

are orthogonal.

*40.5. In the spirit of Problem 3.6, discover all that there is to know about orthogonality for 2×2 matrices by solving the equations that arise if

$$\begin{bmatrix} a & b \\ c & d \end{bmatrix}^T = \begin{bmatrix} a & b \\ c & d \end{bmatrix}^{-1}$$

Hint 1: By Problem 40.2, this problem can be divided into two cases according as $ad - bc$ is $+1$ or -1. *Hint 2*: For all $x, y \in \mathbb{R}$, if $x^2 + y^2 = 1$, then the point (x, y) is on the unit circle; so there exists an angle θ such that $x = \cos\theta$ and $y = \sin\theta$.

41

Abstract Vector Spaces

In this section we stand back and examine carefully what we have been doing with n-tuples of real numbers to determine whether these same ideas might be useful in other contexts. Specifically, consider the concepts of linear dependence and independence, of dimension and basis, of diagonalizability and orthogonality; these have all depended on very few properties of $\mathbb{R}$ and of $\mathbb{R}^n$.

We will not attempt here to slowly retrace the steps taken by the historical development of the subject. Suffice it to say that the accumulated observations of several decades are behind the definition that we are about to give of an "abstract vector space." Given our experience with $\mathbb{R}^n$, the general idea behind this definition is very natural; some of the details, however, may be difficult to appreciate at first glance, since their purpose is merely to smooth the subsequent technical development.

Recall that a set $S \subset \mathbb{R}^n$ is linearly independent if the only way to write the zero vector, $\mathbf{0}$, as a linear combination of vectors from S is the trivial way. Entering into this definition are three ingredients:

(i) A special role played by the zero vector, $\mathbf{0}$.
(ii) The notion of "linear combination," which in turn depends only on the concepts of vector addition and of multiplication of a vector by a scalar.
(iii) A special role played by the scalar, 0, when asserting that a given linear combination is "trivial."

In Chapter II, scalars were real numbers and vectors were n-tuples of real numbers. In this section, we intend to generalize separately each of these two aspects. The scalars will now be a set of objects that behave like the elements of $\mathbb{R}$ did in Section 5, the vectors will now be a set of objects that behave like the elements of $\mathbb{R}^n$ did in Section 5, and an abstract vector space will consist of two sets, the scalars and the vectors, that interact in exactly the same way as did the elements of $\mathbb{R}$ and of $\mathbb{R}^n$ in Section 5 (as, for example, in the definition of linear combination).

Because we do not intend to develop the first of these aspects in full generality, we begin with the scalars. In analyzing the proofs of the theorems in Chapter II, the relevant feature is the fact that the real numbers under addition and multiplication form what is called a *field*. For the general definition of a field, you should consult a text in abstract algebra. In this text, we will use the word "field" in the appropriate contexts for the benefit of readers who may do further study in this area; but we will restrict our attention to two specific fields: the real numbers, $\mathbb{R}$, and the complex numbers, $\mathbb{C}$.

41.1 Remark

Roughly speaking, a field is a set consisting of things that can be "added" and "multiplied" in such a way that "addition" and "multiplication" behave in much the same way as they do in $\mathbb{R}$ and $\mathbb{C}$. The full definition is rather lengthy. Since we do not intend to present other examples than $\mathbb{R}$ and $\mathbb{C}$, no purpose would be served by further details. You should consult a text in abstract algebra, such as [BM], for the general definition.

To understand the definition of an abstract vector space that follows, it helps to keep clearly in mind the material from Section 5. The concept involves a certain amount of data that satisfies certain properties; it is instructive first to discuss the data separately from the properties that these data are required to satisfy.

41.2 Data Involved in the Definition of Vector Space

(i) A field, $\mathbb{F}$, whose elements are called *scalars*; we will continue to use lowercase letters, $a, b, c, \ldots, r, s, t, \ldots, x, y, z$ to denote scalars.

(ii) A set, V, whose elements are called *vectors*; we will continue to use boldface lowercase letters $\mathbf{a}, \mathbf{b}, \mathbf{c}, \ldots, \mathbf{u}, \mathbf{v}, \mathbf{w}, \mathbf{x}, \mathbf{y}, \mathbf{z}$ to denote vectors.

(iii) An operation on pairs of vectors called *vector addition*; the result of adding the vectors $\mathbf{u}$ and $\mathbf{v}$ is denoted by $\mathbf{u} \oplus \mathbf{v}$ and is called the *vector sum* of $\mathbf{u}$ and $\mathbf{v}$. It is required that $\mathbf{u} \oplus \mathbf{v} \in V$ (i.e., that V be closed under vector addition).

(iv) An operation that associates to each pair consisting of a scalar $t \in \mathbb{F}$ and vector $\mathbf{v} \in V$ something called the *scalar multiple of* $\mathbf{v}$ *by* t and denoted by $t \odot \mathbf{v}$; this operation is called *multiplication of a vector by a scalar* or, briefly, *scalar multiplication*. It is required that $t \odot \mathbf{v} \in V$ (i.e., that V be closed under scalar multiplication).

The reason for isolating 41.2 as a separate item is to forestall a common error; we wish to emphasize that to specify a vector space requires *four* ingredients: you must specify (i) what the scalars are, (ii) what the vectors are, (iii) how vector addition is defined, and (iv) how scalar multiplication is defined.

41.3 Definition

A *vector space* consists of data of the type specified in 41.2, satisfying the following list of axioms:

(1) $\mathbf{u} \oplus \mathbf{v} = \mathbf{v} \oplus \mathbf{u}$ for all $\mathbf{u}, \mathbf{v} \in V$. [vector addition is commutative]

(2) $\mathbf{u} \oplus (\mathbf{v} \oplus \mathbf{w}) = (\mathbf{u} \oplus \mathbf{v}) + \mathbf{w}$ for all $\mathbf{u}, \mathbf{v}, \mathbf{w} \in V$. [vector addition is associative]

(3) There exists a vector $\mathbf{u} \in V$ with the property that for all $\mathbf{v} \in V$, $\mathbf{v} \oplus \mathbf{u} = \mathbf{v}$.

At this point, we interrupt our listing because there are some consequences of these first three axioms that have a direct influence on our subsequent statements of some of the remaining axioms. In fact, our statement of axiom (4) will only make sense in the light of certain conclusions drawn from axioms (1) to (3).

41.4 Uniqueness of the Zero Vector

If $\mathbf{u}_1$ and $\mathbf{u}_2$ each satisfy the condition on $\mathbf{u}$ imposed by axiom (3), then $\mathbf{u}_1 = \mathbf{u}_2$.

Proof: $\mathbf{u}_1 \oplus \mathbf{u}_2 = \mathbf{u}_1$ by axiom (3) for $\mathbf{u}_2$ with $\mathbf{v} = \mathbf{u}_1$. Also, $\mathbf{u}_2 \oplus \mathbf{u}_1 = \mathbf{u}_2$ by axiom (3) for $\mathbf{u}_1$ with $\mathbf{v} = \mathbf{u}_2$. But $\mathbf{u}_1 \oplus \mathbf{u}_2 = \mathbf{u}_2 \oplus \mathbf{u}_1$ by axiom (1). Therefore, $\mathbf{u}_1 = \mathbf{u}_2$. ■

This shows that in any vector space there is a *unique* vector satisfying the property required of $\mathbf{u}$ in axiom (3). We are therefore justified in giving a name to this object and in introducing a special symbol to denote it. Because of the motivating example in Section 5, this object will still, in the general context, be called the *zero vector* and be denoted by $\mathbf{0}$. This usage occasionally causes confusion because, as we will see, there are examples of vector spaces in which a rather unexpected object can play the role of the zero vector.

We are now in position to continue with our listing of the axioms.

41.3 Definition (continued)

(4) For every vector $\mathbf{v} \in V$, there exists a vector $\mathbf{w} \in V$ such that $\mathbf{v} \oplus \mathbf{w} = \mathbf{0}$.

Once again, we interrupt our listing to provide an important consequence of axioms (1) to (4).

41.5 Uniqueness of the Inverse for Vector Addition

For each fixed $\mathbf{v} \in V$, if $\mathbf{w}_1$ and $\mathbf{w}_2$ both satisfy the condition on $\mathbf{w}$ imposed by axiom (4), then $\mathbf{w}_1 = \mathbf{w}_2$.

Proof: Assume $\mathbf{v} \oplus \mathbf{w}_1 = \mathbf{0}$ and $\mathbf{v} \oplus \mathbf{w}_2 = \mathbf{0}$.

Then

$$\begin{aligned}
\mathbf{w}_1 &= \mathbf{w}_1 \oplus \mathbf{0} && \text{[property of } \mathbf{0}\text{]}\\
&= \mathbf{w}_1 \oplus (\mathbf{v} \oplus \mathbf{w}_2) && \text{[by assumption]}\\
&= (\mathbf{w}_1 \oplus \mathbf{v}) \oplus \mathbf{w}_2 && \text{[by axiom (2)]}\\
&= (\mathbf{v} \oplus \mathbf{w}_1) \oplus \mathbf{w}_2 && \text{[by axiom (1)]}\\
&= \mathbf{0} \oplus \mathbf{w}_2 && \text{[by assumption]}\\
&= \mathbf{w}_2 \oplus \mathbf{0} && \text{[by axiom (1)]}\\
&= \mathbf{w}_2 && \text{[property of } \mathbf{0}\text{]} \quad \blacksquare
\end{aligned}$$

Thus, for every $\mathbf{v} \in V$, there is a *unique* vector satisfying the property required of $\mathbf{w}$ in axiom (4). Again because of the motivating example in Section 5, the unique vector with this property is called the *negative of* $\mathbf{v}$ and is denoted by $\ominus \mathbf{v}$.

We return finally to completing the list of axioms for a vector space. Note that axioms (1) to (4) concern vector addition exclusively. Axioms (5) to (8) concern scalar multiplication and its interaction with vector addition and with the arithmetic of scalars.

41.3 Definition (continued)

(5) $t \odot (\mathbf{u} \oplus \mathbf{v}) = (t \odot \mathbf{u}) \oplus (t \odot \mathbf{v})$ for all $t \in \mathbb{F}$ and all $\mathbf{u}, \mathbf{v} \in V$.
[scalar multiplication distributes from the left over vector addition]

(6) $(s + t) \odot \mathbf{v} = (s \odot \mathbf{v}) \oplus (t \odot \mathbf{v})$ for all $s, t \in \mathbb{F}$ and all $\mathbf{v} \in V$.
[scalar multiplication distributes from the right over addition of scalars]

(7) $s \odot (t \odot \mathbf{v}) = (st) \odot \mathbf{v}$ for all $s, t \in \mathbb{F}$ and all $\mathbf{v} \in V$.
[multiplication of a vector by a scalar associates with multiplication for scalars]

(8) $1 \odot \mathbf{v} = \mathbf{v}$ for all $\mathbf{v} \in V$
[the scalar 1 is an identity element for scalar multiplication]

41.6 Definition

(i) A *real vector space* is one for which $\mathbb{F} = \mathbb{R}$.
(ii) A *complex vector space* is one for which $\mathbb{F} = \mathbb{C}$.

In the remainder of the text, we will deal primarily with real vector spaces. The phrase "V is a vector space over $\mathbb{F}$" is used to indicate that $\mathbb{F}$ is the field of scalars for a given vector space V.

Before deriving further consequences of the axioms, we present some examples of vector spaces.

41.7 Example

Let n be a fixed positive integer. Let $\mathbb{F} = \mathbb{R}$, let $V = \mathbb{R}^n$, and let vector addition and scalar multiplication be defined as in 5.3 and 5.5, respectively.

This is the "original" vector space that motivated the general definition in 41.3. For any field $\mathbb{F}$, the set $\mathbb{F}^n$ of all n-tuples whose entries are elements of $\mathbb{F}$, together with vector addition and scalar multiplication defined component-wise, is a vector space over $\mathbb{F}$.

Thus, for example, the set $\mathbb{C}^3$ of all triples of complex numbers, with vector addition and scalar multiplication defined using the arithmetic of complex numbers in each coordinate, is a complex vector space.

41.8 Example

Let m and n be fixed positive integers. Let $\mathbb{F} = \mathbb{R}$, let V be the set of all $m \times n$ matrices with entries from $\mathbb{R}$, and let vector addition and scalar multiplication be defined as in 2.1 and 2.3, respectively. It follows from Theorem 3.1(i) to (viii) that this is a real vector space. We will denote this space by $\mathbb{R}_{m \times n}$.

By imitating the proof of Theorem 3.1(i) to (viii), it is easy to show, for any field $\mathbb{F}$ and positive integers m and n, that the set, $\mathbb{F}_{m \times n}$, of $m \times n$ matrices with entries from $\mathbb{F}$, with vector addition and scalar multiplication defined as in 2.1 and 2.2, is a vector space over $\mathbb{F}$.

Thus, for example, the set $\mathbb{C}_{5 \times 8}$ of all 5×8 matrices with complex entries is a complex vector space when vector addition and scalar multiplication are defined using the arithmetic of complex numbers entry by entry, as in 2.1 and 2.2.

41.9 Example

Let $\mathbb{F} = \mathbb{R}$. Let V be the set of all real-valued functions of a single real variable (i.e., the "vectors" are all functions from $\mathbb{R}$ to $\mathbb{R}$). When the function $f:\mathbb{R} \to \mathbb{R}$ is viewed as an element of V, it is useful to denote it by a different symbol; we use $\mathbf{f}$ for this purpose. If vector addition is defined for $\mathbf{f}, \mathbf{g} \in V$ by

$$(\mathbf{f} \oplus \mathbf{g})(x) = f(x) + g(x), \quad \text{for all } x \in \mathbb{R}$$

and scalar multiplication is defined for $t \in \mathbb{R}$ and $\mathbf{f} \in V$ by

$$(t \odot \mathbf{f})(x) = tf(x), \quad \text{for all } x \in \mathbb{R}$$

then V is a real vector space. We will denote it henceforth by ${}^{\mathbb{R}}\mathbb{R}$.

To see this, we must first verify the data. V is closed under vector addition and scalar multiplication since $\mathbf{f} \oplus \mathbf{g}$ and $t \odot \mathbf{f}$ are indeed functions from $\mathbb{R}$ to $\mathbb{R}$; $\mathbf{f} \oplus \mathbf{g}$ is the function whose value at x is the sum of $f(x)$ and $g(x)$; $t \odot \mathbf{f}$ is the function whose value at x is the product of t and $f(x)$.

Next we must verify the axioms. All except axioms (3) and (4) are obvious from the usual properties of arithmetic for $\mathbb{R}$. The vector satisfying axiom (3) is the constant function whose value is 0 for all x. That is, the zero vector in this vector space is the constant function with value 0. Having made this observation, we can *then* verify axiom (4) by noting that, for $\mathbf{f} \in V$, the function $-f:\mathbb{R} \to \mathbb{R}$ defined by $(-f)(x) = -f(x)$ satisfies $(\mathbf{f} \oplus -\mathbf{f})(x) = f(x) - f(x) = 0$ for all $x \in \mathbb{R}$. In other words, $\mathbf{f} \oplus -\mathbf{f} = \mathbf{0}$, so axiom (4) is satisfied by taking $\ominus \mathbf{f} = -\mathbf{f}$.

41.10 Remark

An essential feature of the argument used in 41.8 should be noted. Axiom (3) must be verified *prior to* axiom (4). The reason is that axiom (4) makes reference to the zero vector, $\mathbf{0}$, of V. You must therefore know which vector in V is the zero vector before you can verify whether, for every $\mathbf{v} \in V$, there exists a $\mathbf{w} \in V$ such that $\mathbf{v} + \mathbf{w} = \mathbf{0}$.

41.11 Example

We can generalize Example 41.7 slightly by allowing the vectors to be infinite sequences of real numbers rather than n-tuples. Let $\mathbb{F} = \mathbb{R}$, let $V = \{\mathbf{u} = (u_1, u_2, \ldots, u_i, \ldots): u_i \in \mathbb{R} \text{ for all } i\}$, define $\mathbf{u} + \mathbf{v} = (u_1 + v_1, u_2 + v_2, \ldots, u_i + v_i, \ldots)$, and define $t \cdot \mathbf{u} = (tu_1, tu_2, \ldots, tu_i, \ldots)$. It is trivial to verify the axioms. We will denote this vector space by $\mathbb{R}^\infty$.

By analogy, we have, for any field $\mathbb{F}$, the vector space over $\mathbb{F}$, denoted by $\mathbb{F}^\infty$, consisting of infinite sequences of elements of $\mathbb{F}$; vector addition and scalar multiplication are defined for $\mathbb{F}^\infty$ component-wise, using the arithmetic of $\mathbb{F}$ in each component.

To appreciate the purely formal aspect of the data and axioms for a vector space, we should see nontrivial examples in which an unexpected object plays the role of the zero vector.

41.12 Example

Let $\mathbb{F} = \mathbb{R}$ and let $V = \{x \in \mathbb{R}: 0 \lneq x\}$. The set of vectors in this example is the set of all positive real numbers. It is useful to use the symbol $\underline{x}$ to denote the positive real number x viewed as an element of V. Define vector addition for $\underline{x}, \underline{y} \in V$ by

$$\underline{x} \oplus \underline{y} = \underline{xy}$$

and define scalar multiplication for $t \in \mathbb{R}$ and $\underline{x} \in V$ by

$$t \odot \underline{x} = \underline{x^t}$$

To prove that this is a vector space, we must first verify the data. V is closed under + since the product, xy, of two positive real numbers x and y is a positive real number. V is closed under scalar multiplication since if x is a positive real number and t is any real number, then x^t is a positive real number.

Next we must verify the axioms.

(1) $\underline{u} \oplus \underline{v} = \underline{v} \oplus \underline{u}$ holds because multiplication of real numbers is commutative.
(2) $\underline{u} \oplus (\underline{v} \oplus \underline{w}) = (\underline{u} \oplus \underline{v}) \oplus \underline{w}$ holds because multiplication of real numbers is associative.
(3) There exists a vector $\underline{u} \in V$ such that $\underline{v} \oplus \underline{u} = \underline{v}$ for all $\underline{v} \in V$. This axiom is satisfied since the vector $\underline{1}$ has this property, (i.e., $\underline{x} \oplus \underline{1} = \overline{x1} = \underline{x}$. Thus, the zero vector, for this example, is $\underline{1}$ and it is correct to write $\mathbf{0} = \underline{1}$ in this context.
(4) For every $\underline{v} \in V$, there is a $\underline{w} \in V$ such that $\underline{v} \oplus \underline{w} = \mathbf{0}$. This axiom is satisfied since if $x \gneq 0$, then $1/x \gneq 0$ and so $\ominus \underline{x} = \underline{1/x}$. In other words, $\underline{x} \oplus \underline{1/x} = \overline{x1/x} = \underline{1} = \mathbf{0}$.
(5) $t \odot (\underline{u} \oplus \underline{v}) = (t \odot \underline{u}) \oplus (t \odot \underline{v})$ holds because real number arithmetic satisfies $(xy)^t = x^t y^t$.
(6) $(s + t) \odot \underline{v} = (s \odot \underline{v}) \oplus (t \odot \underline{v})$ holds because real number arithmetic satisfies $x^{s+t} = x^s x^t$.
(7) $s \odot (t \odot \underline{v}) = (st) \odot \underline{v}$ holds because real number arithmetic satisfies $(x^t)^s = x^{st}$.
(8) $1 \odot \underline{v} = \underline{v}$ holds because real number arithmetic satisfies $x^1 = x$.

We noted at the time that the proofs of certain theorems from Chapter II made use of the following fact:

for all $t \in \mathbb{R}$ and all $\mathbf{u} \in \mathbb{R}^n$, $t\mathbf{u} = \mathbf{0}$ if and only if either $t = 0$ or $\mathbf{u} = \mathbf{0}$.

Since we wish to generalize the results of Chapter II to the context of arbitrary vector spaces, we will need to know that the analogue of this fact is a consequence of the axioms for a vector space. The next item is a theorem that provides this analogue.

41.13 Theorem

Let V be a vector space over the field $\mathbb{F}$. Then for every $t \in \mathbb{F}$ and $\mathbf{v} \in V$, $t \odot \mathbf{v} = \mathbf{0}$ if and only if either $t = 0$ or $\mathbf{v} = \mathbf{0}$.

Proof: $(\leftarrow)$

Case 1: Suppose $\mathbf{v} = \mathbf{0}$ and $t \in \mathbb{F}$. By axiom (4), there is a vector $\ominus(t \odot \mathbf{0})$ such that $(t \odot \mathbf{0}) \oplus \ominus(t \odot \mathbf{0}) = \mathbf{0}$. We also have

$$\begin{aligned} t \odot \mathbf{0} &= t \odot (\mathbf{0} \oplus \mathbf{0}) && \text{[property of } \mathbf{0}\text{]} \\ &= (t \odot \mathbf{0}) \oplus (t \odot \mathbf{0}) \qquad (\dagger) && \text{[axiom (5)]} \end{aligned}$$

We may now conclude

$$\begin{aligned} \mathbf{0} &= (t \odot \mathbf{0}) \oplus \ominus(t \odot \mathbf{0}) && \text{[axiom (4)]} \\ &= [t \odot \mathbf{0} \oplus t \odot \mathbf{0}] \oplus \ominus(t \odot \mathbf{0}) && \text{[by } (\dagger)\text{]} \\ &= (t \odot \mathbf{0}) \oplus [t \odot \mathbf{0} \oplus \ominus(t \odot \mathbf{0}] && \text{[axiom (2)]} \\ &= (t \odot \mathbf{0}) \oplus \mathbf{0} && \text{[axiom (4)]} \\ &= t \odot \mathbf{0} && \text{[property of } \mathbf{0}\text{]} \end{aligned}$$

Case 2: Suppose $t = 0$ and $\mathbf{v} \in V$. By axiom (4), there is a vector $\ominus(0 \odot \mathbf{v})$ such that $(0 \odot \mathbf{v}) \oplus \ominus(0 \odot \mathbf{v}) = \mathbf{0}$. We also have

$$\begin{aligned} 0 \odot \mathbf{v} &= (0 + 0) \odot \mathbf{v} && \text{[property of 0]} \\ &= (0 \odot \mathbf{v}) \oplus (0 \odot \mathbf{v}) \qquad (\ddagger) && \text{[axiom (6)]} \end{aligned}$$

We may conclude

$$\begin{aligned} \mathbf{0} &= (0 \odot \mathbf{v}) \oplus \ominus(0 \odot \mathbf{v}) && \text{[axiom (4)]} \\ &= [(0 \odot \mathbf{v}) \oplus (0 \odot \mathbf{v})] \oplus \ominus(0 \odot \mathbf{v}) && \text{[by } (\ddagger)\text{]} \\ &= (0 \odot \mathbf{v}) \oplus [(0 \odot \mathbf{v}) \oplus \ominus(0 \odot \mathbf{v})] && \text{[axiom (2)]} \\ &= (0 \odot \mathbf{v}) \oplus \mathbf{0} && \text{[axiom (5)]} \\ &= 0 \odot \mathbf{v} && \text{[property of } \mathbf{0}\text{]} \end{aligned}$$

($\rightarrow$) Assume $t \odot \mathbf{v} = \mathbf{0}$. We will prove that if $t \neq 0$ then $\mathbf{v} = \mathbf{0}$. Assuming $t \neq 0$, it makes sense to consider $1/t$. We may conclude

$$\begin{aligned}
\mathbf{0} &= \left(\frac{1}{t}\right) \odot \mathbf{0} && \text{[this was just proved above in case 1]} \\
&= \left(\frac{1}{t}\right) \odot (t \odot \mathbf{v}) && \text{[by assumption]} \\
&= \left(\frac{1}{t} t\right) \odot \mathbf{v} && \text{[axiom (7)]} \\
&= 1 \odot \mathbf{v} && \left[\text{since } \frac{1}{t} t = 1\right] \\
&= \mathbf{v} && \text{[axiom (8)]} \quad \blacksquare
\end{aligned}$$

The final theorem in this section concerns the interaction between vector addition and scalar multiplication. The scalar field contains an element -1 satisfying $1 + (-1) = 0$. Thus for every $\mathbf{v} \in V$, the scalar multiple $(-1) \odot \mathbf{v}$ is in V; by axiom (4), there is also a vector $\ominus \mathbf{v}$ in V satisfying $\mathbf{v} \oplus \ominus \mathbf{v} = \mathbf{0}$. The theorem asserts that the two vectors $\ominus \mathbf{v}$ and $(-1) \odot \mathbf{v}$ are equal.

41.14 Theorem

Let V be a vector space over the field $\mathbb{F}$. Then for every $\mathbf{v} \in V$, $\ominus \mathbf{v} = (-1) \odot \mathbf{v}$.

Proof: Just after introducing axiom (4), we proved that $\ominus \mathbf{v}$ is the unique vector satisfying $\mathbf{v} \oplus \ominus \mathbf{v} = \mathbf{0}$. Thus we may conclude that $(-1) \odot \mathbf{v} = \ominus \mathbf{v}$ if we show that $\mathbf{v} \oplus (-1) \odot \mathbf{v} = \mathbf{0}$. But this last fact is easy to see:

$$\begin{aligned}
\mathbf{v} \oplus [(-1) \odot \mathbf{v}] &= (1 \odot \mathbf{v}) \oplus [(-1) \odot \mathbf{v}] && \text{[axiom (8)]} \\
&= (1 + (-1)) \odot \mathbf{v} && \text{[axiom (6)]} \\
&= 0 \odot \mathbf{v} && [\text{since } 1 + (-1) = 0] \\
&= \mathbf{0} && \text{[by 41.13]} \quad \blacksquare
\end{aligned}$$

The last example in this section illustrates that the same set of vectors can be used to define different vector spaces. It is, as well, another example in which the zero vector is a rather unexpected object.

41.15 Example

Let $\mathbb{F} = \mathbb{R}$ and let $V = \{(x_1, x_2)\colon x_1, x_2 \in \mathbb{R}\}$. Thus the underlying set of vectors for this example is the same as for $\mathbb{R}^2$. But vector addition and scalar multiplication are defined differently:

$$(x_1, y_1) \oplus (x_2, y_2) = (x_1 + x_2 + 1, y_1 + y_2 + 1)$$
$$t \odot (x, y) = (t + tx - 1, t + ty - 1)$$

41.16 Exercise

Prove that the data given in Example 41.15 determine a vector space.

Problem Set 41

41.1. Go back and do any of the exercises from Section 41 that you may have skipped.

*41.2. Let $\mathbb{R}$ be the scalar field and let $V = \{(x, y)\colon x \in \mathbb{R}$ and $y \in \mathbb{R}$ and $y \neq 0\}$ be the vectors. Define $\oplus$ for pairs of vectors from V by

$$(x, y) \oplus (x', y') = (xy' + yx', yy')$$

Define scalar multiplication by

$$t \odot (x, y) = \left(\frac{tx}{y}, 1\right)$$

Decide whether this structure is a vector space. If it is not, which axioms fail?

*41.3. Let $\mathbb{R}$ be the scalar field and let $V = \{(x, y)\colon x \in \mathbb{R}$ and $y \in \mathbb{R}$ and $x \neq 0\}$ be the vectors. Define $\oplus$ for pairs of vectors from V by

$$(x, y) \oplus (x', y') = (xx', xy' + yx')$$

Define scalar multiplication by

$$t \odot (x, y) = (x, ty)$$

Decide whether this structure is a vector space. If it is not, which axioms fail?

41.4. Verify that if U and V are vector spaces over the same scalar field $\mathbb{F}$, then the set, denoted by $\mathscr{L}(U, V)$, of all linear transformations from U into V can be made into a vector space over $\mathbb{F}$ by defining vector addition and scalar multiplication in the following way:

$$(S \oplus T)(\mathbf{u}) = S(\mathbf{u}) + T(\mathbf{u})$$
$$(t \odot S)(\mathbf{u}) = tS(\mathbf{u})$$

*41.5. If V is a vector space over $\mathbb{F}$, the elements of $\mathscr{L}(V, \mathbb{F})$ (defined in the previous problem) are called *linear functionals*, and the vector space $\mathscr{L}(V, \mathbb{F})$ is called the *dual space of* V and is denoted by V^*. Let $\beta = (\mathbf{v}_1, \mathbf{v}_2, \ldots, \mathbf{v}_n)$ be an ordered basis for V and, for $i = 1, 2, \ldots, n$, define $\mathbf{f}_i \in V^*$ by $\mathbf{f}_i(\mathbf{v}) = ((\mathbf{v})_\beta)_i$ (= the ith component of the β-coordinate vector of $\mathbf{v}$).
(i) Show that if $\mathbf{f} \in V^*$ then $\mathbf{f} = \Sigma_i \mathbf{f}(\mathbf{v}_i)\mathbf{f}_i$.
(ii) Show that $\beta^* = (\mathbf{f}_1, \mathbf{f}_2, \ldots, \mathbf{f}_n)$ is an ordered basis for V^*. It is called the *basis dual to* β or, simply, the *dual basis* when the basis for V is clear from the context.

*41.6. Consider the basis $\beta = ((2,1), (-1,4))$ for $\mathbb{R}^2$. Find the dual basis, β^*, for $(\mathbb{R}^2)^*$.

42

Subspaces, Linear Dependence, and Linear Independence

42.1 Definition

Let V be a vector space over $\mathbb{F}$. A nonempty subset W of V is a *subspace* of V if W is itself a vector space over $\mathbb{F}$ when vector addition and scalar multiplication are defined for W exactly as for V.

Before giving examples to illustrate this concept, it is convenient to have a theorem which shows that two very simple, and obviously necessary, conditions on a subset W of a vector space V are in fact sufficient to ensure that W is a subspace of V.

42.2 Theorem

Let V be a vector space over $\mathbb{F}$ and let $\varnothing \neq W \subset V$. Assume that W is closed under the operations of vector addition and scalar multiplication as defined for V. That is,

(i) if $\mathbf{w}_1 \in W$ and $\mathbf{w}_2 \in W$, then $\mathbf{w}_1 \oplus \mathbf{w}_2 \in W$, and
(ii) if $t \in \mathbb{F}$ and $\mathbf{w} \in W$, then $t \odot \mathbf{w} \in W$.

Then W, with these same operations restricted to W, is itself a vector space.

Proof: The hypotheses about W are necessary conditions on the data for a vector space. To see that they are sufficient, it remains only to check the axioms. Because axioms (1), (2), (5), (6), (7), and (8) are universal statements, known to hold for all elements of $\mathbb{F}$ and V, these axioms continue to hold automatically for elements of $\mathbb{F}$ and W since $W \subset V$. Axiom (3) holds for W because hypothesis (ii) implies that W must contain the zero vector of V, which will then also be the zero vector for W. Specifically, let $\mathbf{w}$ be any element of W; then $0 \odot \mathbf{w} \in W$ by (ii). But $0 \odot \mathbf{w} = \mathbf{0}$ by 41.13, so $\mathbf{0} \in W$. Similarly, axiom (4) holds for W. Specifically, for every $\mathbf{w} \in W$, there is a vector, $\ominus\mathbf{w}$, in V satisfying $\mathbf{w} \oplus \ominus\mathbf{w} = \mathbf{0}$; but $\ominus\mathbf{w} = (-1) \cdot \mathbf{w}$ by 41.14 and $(-1) \cdot \mathbf{w} \in W$ by (ii). So $\ominus\mathbf{w} \in W$. ■

42.3 Example

Because of Theorem 42.2, the subspaces of the vector space $\mathbb{R}^n$ are precisely the linear subspaces as defined in 5.15.

42.4 Example

Let $W = \{\mathbf{A} \in \mathbb{R}_{n\times n}\colon \mathbf{A}^T = \mathbf{A}\}$; that is, W is the set of all symmetric $n \times n$ matrices with real entries. Here $\mathbb{R}_{n\times n}$ is the vector space introduced in 41.8. It is trivial, using 3.15(ii) and (iii), to show that W is closed under vector addition and scalar multiplication. So by 42.2, W is a subspace of $\mathbb{R}_{n\times n}$.

42.5 Example

Let $W = \{\mathbf{f} \in {}^{\mathbb{R}}\mathbb{R}\colon f \text{ is continuous}\}$. Here, ${}^{\mathbb{R}}\mathbb{R}$ is the vector space introduced in 41.9. The fact that W is closed under vector addition is precisely the theorem of calculus that the sum of two continuous functions is a continuous function. Also, any constant multiple of a continuous function is a continuous function, so W is closed under scalar multiplication. Thus W is a subspace of ${}^{\mathbb{R}}\mathbb{R}$. We will henceforth denote this subspace by $C(\mathbb{R})$.

42.6 Example

Let $\mathscr{P} = \{\mathbf{f} \in {}^{\mathbb{R}}\mathbb{R}\colon f \text{ is a polynomial}\}$. Recall that $f\colon\mathbb{R} \to \mathbb{R}$ is a *polynomial function* if it can be expressed in the form

$$f(x) = a_n x^n + a_{n-1}x^{n-1} + \cdots + a_1 x + a_0$$

where the a_i are arbitrary real numbers. The *degree of* f is $\max\{i\colon a_i \neq 0\}$; this is the highest power of x whose coefficient is nonzero. Note that the constant function whose value is 0 for all x is a polynomial, called the *zero polynomial*; it has no degree. Since the sum of two polynomials is a polynomial and since any constant multiple of a polynomial is a polynomial, $\mathscr{P}$ is a subspace of ${}^{\mathbb{R}}\mathbb{R}$. In fact, since polynomials are continuous functions, $\mathscr{P}$ is also a subspace of $C(\mathbb{R})$.

42.7 Example

For each nonnegative integer n, let $\mathscr{P}_n = \{\mathbf{f} \in {}^{\mathbb{R}}\mathbb{R}\colon f$ is either the zero polynomial or a polynomial of degree $\leq n\}$. Each $\mathscr{P}_n$ is a subspace of $\mathscr{P}$ [and of $C(\mathbb{R})$ and of ${}^{\mathbb{R}}\mathbb{R}$] since it is closed under vector addition and scalar multiplication. Note that $\mathscr{P}_0$ is the vector space of all constant functions from $\mathbb{R}$ to $\mathbb{R}$.

42.8 Exercise

Show that $W = \{\mathbf{f} \in {}^{\mathbb{R}}\mathbb{R}\colon f(39) = 0\}$ is a subspace of ${}^{\mathbb{R}}\mathbb{R}$.

42.9 Exercise

Show that $W = \{\mathbf{A} \in \mathbb{R}_{3\times3}\colon \mathbf{A}^T = -\mathbf{A}\}$ is a subspace of $\mathbb{R}_{3\times3}$.

We turn next to the concepts of linear dependence and independence in the context of abstract vector spaces. Recall from 5.8 that the notion of "linear combination" depends only on vector addition and scalar multiplication.

It was useful, when first introducing the idea of an abstract vector space, to use $\oplus$ and $+$ to distinguish notationally between vector addition and addition for scalars; we also used $\odot$ and juxtaposition to distinguish between multiplication of a vector by a scalar and multiplication for scalars. If we were to persist with this distinction, we would have to denote a linear combination of the vectors $\mathbf{v}_1, \mathbf{v}_2, \ldots, \mathbf{v}_m$ by

$$(t_1 \odot \mathbf{v}_1) \oplus (t_2 \odot \mathbf{v}_2) \oplus \cdots \oplus (t_m \odot \mathbf{v}_m)$$

Typists are not the only people who dislike this. We will henceforth drop this distinction and use only $+$ and juxtaposition. The context will determine which operation is denoted. That is, a plus sign placed between two scalars denotes addition for scalars, while a plus sign placed between two vectors denotes vector addition. Juxtaposition is interpreted analogously. For example, when axiom (6) is restated with this convention,

$$(6) \qquad (s + t)\mathbf{v} = s\mathbf{v} + t\mathbf{v}, \quad \text{for all } s, t \in \mathbb{F} \text{ and } \mathbf{v} \in V$$

the plus sign on the left side of this equation denotes addition for scalars, while the one on the right side denotes vector addition.

Similarly, in axiom (7),

$$(7) \qquad s(t\mathbf{v}) = (st)\mathbf{v}, \quad \text{for all } s, t \in \mathbb{F} \text{ and } \mathbf{v} \in V$$

the juxtaposition, *st*, on the right side of this equation denotes multiplication for scalars, while the juxtaposition, $t\mathbf{v}$, on the left side denotes scalar multiplication.

42.10 Definition

Let V be a vector space. A *linear combination of the vectors* $\mathbf{v}_1, \mathbf{v}_2, \ldots, \mathbf{v}_m$ is a vector of the form $t_1\mathbf{v}_1 + t_2\mathbf{v}_2 + \cdots + \mathbf{t}_m\mathbf{v}_m$, where $t_1, t_2, \ldots, t_m$ are scalars. A linear combination is *nontrivial* if the vectors $\mathbf{v}_1, \mathbf{v}_2, \ldots, \mathbf{v}_m$ are distinct and $t_i \neq 0$ for at least one i.

42.11 Definition

(i) For any nonempty subset S of a vector space V, define $\mathscr{L}(S) = \{\mathbf{x} \in V$: there exist vectors $\mathbf{v}_1, \mathbf{v}_2, \ldots, \mathbf{v}_m \in S$ and scalars $t_1, t_2, \ldots, t_m$ such that $\mathbf{x} = t_1\mathbf{v}_1 + t_2\mathbf{v}_2 + \cdots + t_m\mathbf{v}_m\}$.
(ii) $\mathscr{L}(\varnothing) = \{\mathbf{0}\}$.

For any set S, $\mathscr{L}(S)$ is called the *span of S*.

42.12 Definition

A subset S of a vector space V is *linearly dependent* if the zero vector, $\mathbf{0}$, is a nontrivial linear combination of vectors from S. S is called *linearly independent* if S is not linearly dependent.

42.13 Exercise

Prove the following facts about an arbitrary vector space V. This involves checking that the proofs of the analogous results given in Chapter II for the special case $V = \mathbb{R}^n$ go through, if not verbatim then with only slight modifications, for arbitrary V. This is made easy by the facts provided in Theorems 41.13 and 41.14.

In the following, S, S_1, and S_2 are subsets of V.

(i) $S_1 \subset S_2$ implies $\mathscr{L}(S_1) \subset \mathscr{L}(S_2)$.
(ii) $S_1 \subset \mathscr{L}(S_2)$ implies $\mathscr{L}(S_1) \subset \mathscr{L}(S_2)$.
(iii) $\mathscr{L}(S)$ is a subspace of V.
(iv) The empty set, $\varnothing$, is linearly independent.
(v) If $S = \{\mathbf{u}\}$, then S is linearly dependent if and only if $\mathbf{u} = \mathbf{0}$.

(vi) S is linearly dependent if and only if there exists a vector $\mathbf{u} \in S$ such that $\mathbf{u} \in \mathscr{L}(S - \{\mathbf{u}\})$.

(vii) If S_2 is linearly independent and $S_1 \subset S_2$, then S_1 is linearly independent.

(viii) If S_1 is linearly dependent and $S_1 \subset S_2$, then S_2 is linearly dependent.

(ix) If S is linearly dependent and $\mathbf{u} \in S$ is any vector satisfying $\mathbf{u} \in \mathscr{L}(S - \{\mathbf{u}\})$, then $\mathscr{L}(S - \{\mathbf{u}\}) = \mathscr{L}(S)$.

(x) If S is linearly independent and $\mathbf{u} \notin \mathscr{L}(S)$, then $S \cup \{\mathbf{u}\}$ is linearly independent.

(xi) An ordered set $S = (\mathbf{u}_1, \mathbf{u}_2, \ldots, \mathbf{u}_m)$ is linearly dependent iff $\mathbf{u}_1 = \mathbf{0}$ or, for some $i \gneq 1$, $\mathbf{u}_i \in \mathscr{L}(\mathbf{u}_1, \mathbf{u}_2, \ldots, \mathbf{u}_{i-1})$.

Problem Set 42

42.1. Go back and do any of the exercises from Section 42 that you may have skipped.

*42.2. Which of the following subsets of $\mathbb{R}_{2\times 2}$ are subspaces?

(i) All matrices of the form $\begin{bmatrix} 0 & b \\ c & b+c \end{bmatrix}$.

(ii) All 2×2 matrices whose entries add up to zero.

(iii) All 2×2 matrices whose column sums are 0.

(iv) All 2×2 matrices with determinant equal to zero.

*42.3. Which of the following subsets of $\mathscr{P}_5$ are subspaces?

(i) $\{\mathbf{p} \in \mathscr{P}_5$: the constant term in $\mathbf{p}$ is $0\}$

(ii) $\{\mathbf{p} \in \mathscr{P}_5$: the constant term in $\mathbf{p}$ is $1\}$

(iii) $\{\mathbf{p} \in \mathscr{P}_5$: the coefficients of x^4 and of x^2 are equal$\}$

42.4. Which of the following subsets of $\mathbb{R}_{n\times n}$ are subspaces?

(i) $\{\mathrm{A} \in \mathbb{R}_{n\times n}$: A is invertible$\}$

(ii) The set of all matrices that commute with a fixed matrix B, i.e., $\{\mathrm{A} \in \mathbb{R}_{n\times n}: \mathrm{AB} = \mathrm{BA}\}$

(iii) $\{\mathrm{A} \in \mathbb{R}_{n\times n}$: A is not invertible$\}$

(iv) $\{\mathrm{A} \in \mathbb{R}_{n\times n}$: A is upper triangular$\}$

(v) $\{\mathrm{A} \in \mathbb{R}_{n\times n}$: A has either an all-zero row or an all-zero column$\}$

(vi) $\{\mathrm{A} \in \mathbb{R}_{n\times n}: \mathrm{trace(A)} = 0\}$

43

Bases and Dimension

The remaining concepts from Chapter II are also easily generalized to the context of abstract vector spaces. Again, we leave it as an exercise to check that the proofs given in Chapter II for the special case $V = \mathbb{R}^n$ apply virtually unchanged to an arbitrary vector space, V.

43.1 Definition

Let W be a subspace of a vector space V and let S be a subset of W. S is a *basis for* W if S spans W and S is linearly independent.

43.2 Exercise

Throughout the following, W is a subspace of a vector space V.

(i) If S is a linearly independent subset of W, then S is a basis for W if and only if for any S' such that $S \subsetneqq S' \subset W$, S' is linearly dependent.

(ii) If S spans W, then S is a basis for W if and only if for any S' such that $S' \subsetneqq S$, S' does not span W.

(iii) If $S = \{\mathbf{u}_1, \mathbf{u}_2, \ldots, \mathbf{u}_p\}$ spans W, then any set $S' = \{\mathbf{v}_1, \mathbf{v}_2, \ldots, \mathbf{v}_q\} \subset W$ with $p \lneqq q$ is linearly dependent.
(iv) If $S' = \{\mathbf{v}_1, \mathbf{v}_2, \ldots, \mathbf{v}_q\} \subset W$ is linearly independent, then any set $S = \{\mathbf{u}_1, \mathbf{u}_2, \ldots, \mathbf{u}_p\}$ with $p \lneqq q$ does not span W.
(v) If $S = \{\mathbf{u}_1, \mathbf{u}_2, \ldots, \mathbf{u}_p\}$ and $S' = \{\mathbf{v}_1, \mathbf{v}_2, \ldots, \mathbf{u}_q\}$ are each bases for W, then $p = q$.

The similarities with Chapter II are apparent. The definition of basis is the same; 43.2(i) asserts that a basis is a maximal linearly independent subset; 43.2(ii) asserts that a basis is a minimal spanning set; 43.2(v) asserts that, if W has a basis of cardinality n, then every basis for W has cardinality n, so it would then make sense to call this nonnegative integer the *dimension* of W.

At this point, however, there is a novel ingredient that was not present earlier. There are vector spaces than cannot be spanned by any finite subset.

43.3 Example

The vector space $\mathscr{P}$ of all polynomial functions (see 42.6) cannot be spanned by any finite set $S = \{\mathbf{p}_1, \mathbf{p}_2, \ldots, \mathbf{p}_n\}$ of polynomials. If k_i is the degree of the polynomial $\mathbf{p}_i$, and if N is the largest integer in the set $\{k_1, k_2, \ldots, k_n\}$, then $\mathscr{L}(S)$ can only contain polynomials of degree $\leq N$.

43.4 Example

The vector space $\mathbb{R}^\infty$ introduced in 41.11 contains linearly independent subsets of arbitrarily large finite cardinality. To see this, let $\mathbf{e}_n^\infty = (0, \ldots, 0, 1, 0, \ldots)$ be the vector in $\mathbb{R}^\infty$ whose n^{th} coordinate is 1 and whose other coordinates are all 0; it is clear that, for every positive integer n, the set $S_n = \{\mathbf{e}_1^\infty, \mathbf{e}_2^\infty, \ldots, \mathbf{e}_n^\infty\}$ is a linearly independent subset of size n. Thus, by 43.2(iv), $\mathbb{R}^\infty$ cannot be spanned be any finite subset.

The analogue of 8.8 is the following:

43.5 Exercise

If W is a subspace of a vector space V and if V can be spanned by a finite set of cardinality n, then W has a basis; moreover, this basis has cardinality $\leq n$.

Although we will not prove it, we will at least mention the following result:

43.6 Theorem

Every vector space has a basis.

A proof of this theorem is slightly beyond the scope of the present text. The reason is that this result belongs more to set theory than to linear algebra. Its proof makes essential use of a set-theoretic principle known as the Axiom of Choice. In fact, the theorem is equivalent to the Axiom of Choice.

It is still correct to define the dimension of a vector space V as "the" number of elements in a basis for V. But to understand what this means for infinite-dimensional vector spaces requires first explaining the concept of "the" number of elements in an infinite set. Some infinite sets have "more" elements than other infinite sets. For example, the two infinite-dimensional spaces ${}^{\mathbb{R}}\mathbb{R}$ and $\mathbb{R}^{\infty}$ discussed in 41.9 and 43.4 do not have the same dimension.

Learn more about this and related topics by reading the following:

Kaplansky, I., *Set Theory and Metric Spaces*, Allyn & Bacon, Boston, 1972.
Jech, T., *The Axiom of Choice*, Chapters 1 and 2, North-Holland, Amsterdam, 1973.
Blass, A., "Existence of bases implies the Axiom of Choice," *Contemporary Mathematics*, Volume 31, 1984, 31–33.

In the remainder of the text, we will primarily consider finite-dimensional vector spaces. It should be emphasized, however, that most of the genuine benefits derived from the study of vector spaces arise in the context of infinite-dimensional spaces. The reason will become apparent in Theorem 43.10.

43.7 Exercise

(i) If W is a vector space with $\dim W = p$ and if S is a linearly independent subset of W of cardinality p, then S is a basis for W.

(ii) If W is a vector space with $\dim W = p$ and if S is a subset of W of cardinality p such that S spans W, then S is a basis for W.

(iii) If S is any linearly independent subset of W and $\dim W = p$, then there is a subset S' of cardinality p with $S \subset S' \subset W$ such that S' is a basis for W.

(iv) The ordered set $\beta = (\mathbf{u}_1, \mathbf{u}_2, \ldots, \mathbf{u}_p)$ is a basis for the vector space W if and only if every vector $\mathbf{w} \in W$ can be expressed in a unique way as a linear combination, $\mathbf{w} = t_1\mathbf{u}_1 + t_2\mathbf{u}_2 + \cdots + t_p\mathbf{u}_p$, of the vectors in β.

Part (iv) of this result justifies introducing, as in Section 8, the *β-coordinate vector* $(\mathbf{w})_\beta = (t_1, t_2, \ldots, t_p)$ and the *β-coordinate matrix* $[\mathbf{w}]_\beta = \begin{bmatrix} t_1 \\ t_2 \\ \vdots \\ t_p \end{bmatrix}$ of $\mathbf{w}$.

43.8 Definition

Let V and W each be vector spaces over the same field $\mathbb{F}$. A function $\Psi: V \to W$ is an *isomorphism* if Ψ is both one-to-one and onto and satisfies

(i) $\Psi(\mathbf{u} + \mathbf{v}) = \Psi(\mathbf{u}) + \Psi(\mathbf{v})$, for all $\mathbf{u}, \mathbf{v} \in V$, and
(ii) $\Psi(t\mathbf{v}) = t\Psi(\mathbf{v})$, for all $t \in \mathbb{F}$, $\mathbf{v} \in V$.

V and W are called *isomorphic* if there exists an isomorphism between V and W.

43.9 Exercise

Assume that V and W are vector spaces over the field $\mathbb{F}$ and that $\Psi: V \to W$ is an isomorphism. Then

(i) For all $\mathbf{u}_1, \mathbf{u}_2, \ldots, \mathbf{u}_p, \mathbf{v} \in V$ and $t_1, t_2, \ldots, t_p \in \mathbb{F}$,

$$\mathbf{v} = t_1\mathbf{u}_1 + t_2\mathbf{u}_2 + \cdots + t_p\mathbf{u}_p$$
$$\text{iff } \Psi(\mathbf{v}) = t_1\Psi(\mathbf{u}_1) + t_2\Psi(\mathbf{u}_2) + \cdots + t_p\Psi(\mathbf{u}_p).$$

(ii) For all $\mathbf{w}_1, \mathbf{w}_2, \ldots, \mathbf{w}_p, \mathbf{v} \in W$ and $t_1, t_2, \ldots, t_p \in \mathbb{F}$,

$$\mathbf{v} = t_1\mathbf{w}_1 + t_2\mathbf{w}_2 + \cdots + t_p\mathbf{w}_p$$
$$\text{iff } \Psi^{-1}(\mathbf{v}) = t_1\Psi^{-1}(\mathbf{w}_1) + t_2\Psi^{-1}(\mathbf{w}_2) + \cdots + t_p\Psi^{-1}(\mathbf{w}_p).$$

For any subset $S \subset V$, let $\Psi(S) = \{\Psi(\mathbf{u}): \mathbf{u} \in S\}$, and for any subset $S' \subset W$, let $\Psi^{-1}(S') = \{\mathbf{u} \in S: \Psi(\mathbf{u}) \in S'\}$. Then

(iii) S is a linearly independent subset of V iff $\Psi(S)$ is a linearly independent subset of W.
(iv) $\mathscr{L}(S) = \Psi^{-1}(\mathscr{L}(\Psi(S)))$.

The next theorem, which we will state and prove for $\mathbb{F} = \mathbb{R}$ (although the same proof yields the analogous result for any field $\mathbb{F}$), asserts that any finite-dimensional real vector space V of dimension n is isomorphic to $\mathbb{R}^n$.

This theorem is simultaneously a cause for gloom and a cause for rejoicing. The gloomy aspect is the one alluded to earlier. We have gained nothing new in the finite-dimensional case by introducing abstract vector spaces; every n-dimensional real vector space is essentially identical to $\mathbb{R}^n$; the most fruitful mathematical benefits of the concept will only appear in the infinite-dimensional context. On the other hand, we have cause to rejoice because ostensibly different structures, such as the vector spaces of matrices, $\mathbb{R}_{p\times q}$, and of polynomials, $\mathscr{P}_m$, all turn out to be essentially the same as $\mathbb{R}^n$ for some n; thus we *already* know all there is to know about these spaces. Questions about these spaces can be answered using the *identical* techniques that we have been using all along for $\mathbb{R}^n$.

43.10 Theorem

Let V be a real vector space of dimension n. Then V is isomorphic to $\mathbb{R}^n$.

Proof: We must exhibit an isomorphism between V and $\mathbb{R}^n$. Actually, there are infinitely many; each ordered basis for V naturally gives rise to such an isomorphism. Let $\beta = (\mathbf{w}_1, \mathbf{w}_2, \ldots, \mathbf{w}_n)$ be an ordered basis for V. Define $\Psi: V \to \mathbb{R}^n$ by $\Psi(\mathbf{v}) = (\mathbf{v})_\beta$ for all $\mathbf{v} \in V$. Exercise 43.7(iv) guarantees that Ψ is a function from V to $\mathbb{R}^n$ that is both one-to-one and onto. Next, we must prove

$$\text{(i)} \qquad \Psi(\mathbf{u} + \mathbf{v}) = \Psi(\mathbf{u}) + \Psi(\mathbf{v}), \quad \text{for all } \mathbf{u}, \mathbf{v} \in V$$

To prove this, note that, if $\mathbf{u} = s_1\mathbf{w}_1 + s_2\mathbf{w}_2 + \cdots + s_n\mathbf{w}_n$ and $\mathbf{v} = t_1\mathbf{w}_1 + t_2\mathbf{w}_2 + \cdots + t_n\mathbf{w}_n$, then, by invoking the fact that V satisfies axioms (1), (2), and (6) for vector spaces, we may conclude that $\mathbf{u} + \mathbf{v} = (s_1 + t_1)\mathbf{w}_1 + (s_2 + t_2)\mathbf{w}_2 + \cdots + (s_n + t_n)\mathbf{w}_n$. Thus $\Psi(\mathbf{u} + \mathbf{v}) = ((s_1 + t_1), (s_2 + t_2), \ldots, (s_n + t_n)) = (s_1, s_2, \ldots, s_n) + (t_1, t_2, \ldots, t_n) = (\mathbf{u})_\beta + (\mathbf{v})_\beta = \Psi(\mathbf{u}) + \Psi(\mathbf{v})$.

Finally, to prove

$$\text{(ii)} \qquad \Psi(t\mathbf{v}) = t\Psi(\mathbf{v}), \quad \text{for all } t \in \mathbb{R}, \mathbf{v} \in V$$

proceed as for (i), this time invoking the fact that V satisfies axioms (5) and (7) for vector spaces. ■

We will now give some examples to illustrate the concepts of basis and dimension for vector spaces other than $\mathbb{R}^n$ and subspaces of $\mathbb{R}^n$.

43.11 Example

Consider the vector space $R_{2\times 2}$ of 2×2 matrices with real entries. Let

$$A_1 = \begin{bmatrix} 1 & 0 \\ 0 & 0 \end{bmatrix}, \quad A_2 = \begin{bmatrix} 0 & 1 \\ 0 & 0 \end{bmatrix}, \quad A_3 = \begin{bmatrix} 0 & 0 \\ 1 & 0 \end{bmatrix}, \quad A_4 = \begin{bmatrix} 0 & 0 \\ 0 & 1 \end{bmatrix}$$

It is clear that the set $\beta = \{\mathbf{A}_1, \mathbf{A}_2, \mathbf{A}_3, \mathbf{A}_4\}$ spans $\mathbb{R}_{2\times 2}$ since an arbitrary vector $\begin{bmatrix} a & b \\ c & d \end{bmatrix} \in \mathbb{R}_{2\times 2}$ can be written as

$$\begin{bmatrix} a & b \\ c & d \end{bmatrix} = a\mathbf{A}_1 + b\mathbf{A}_2 + c\mathbf{A}_3 + d\mathbf{A}_4$$

It is also clear that if the zero vector

$$\mathbf{0} = \begin{bmatrix} 0 & 0 \\ 0 & 0 \end{bmatrix} = t_1\mathbf{A}_1 + t_2\mathbf{A}_2 + t_3\mathbf{A}_3 + t_4\mathbf{A}_4$$

then $t_1 = t_2 = t_3 = t_4 = 0$; so β is also linearly independent. Thus $\mathbb{R}_{2\times 2}$ is four-dimensional.

43.12 Exercise

Use the ideas from the preceding example to show that $\dim(\mathbb{R}_{m\times n}) = mn$.

43.13 Example

The subspace $W = \{\mathbf{A} \in \mathbb{R}_{3\times 3}\colon\ \mathbf{A}^T = -\mathbf{A}\}$ of $\mathbb{R}_{3\times 3}$ consisting of the skew-symmetric 3×3 matrices is a three-dimensional subspace of the nine-dimensional space $\mathbb{R}_{3\times 3}$.

That W is a subspace of $\mathbb{R}_{3\times 3}$ was proved in 42.9. To see that $\dim W = 3$, observe that

$$\left\{\begin{bmatrix} 0 & 1 & 0 \\ -1 & 0 & 0 \\ 0 & 0 & 0 \end{bmatrix}, \begin{bmatrix} 0 & 0 & 1 \\ 0 & 0 & 0 \\ -1 & 0 & 0 \end{bmatrix}, \begin{bmatrix} 0 & 0 & 0 \\ 0 & 0 & 1 \\ 0 & -1 & 0 \end{bmatrix}\right\}$$

is a basis for W.

43.14 Example

The space $\mathscr{P}_n$ defined in 42.7 has dimension $n + 1$. The subset $\beta = (\mathbf{p}_0, \mathbf{p}_1, \ldots, \mathbf{p}_n)$, where $p_0(x) = 1,\ p_1(x) = x,\ p_2(x) = x^2, \ldots, p_n(x) = x^n$, is the standard basis for $\mathscr{P}_n$. An arbitrary vector $\mathbf{p} \in \mathscr{P}_n$ corresponds to a polynomial $p(x) = a_0 + a_1x + \cdots + a_nx^n$, so $\mathbf{p} = a_0\mathbf{p}_0 + a_1\mathbf{p}_1 + \cdots + a_n\mathbf{p}_n$, showing that β spans $\mathscr{P}_n$. The zero vector in $\mathscr{P}_n$ is the constant function whose value for all $x \in \mathbb{R}$ is 0. So if $\mathbf{0} = a_0\mathbf{p}_0 + a_1\mathbf{p}_1 + \cdots + a_n\mathbf{p}_n$ (i.e., if $a_0 + a_1x + \cdots + a_nx^n = 0$ for all $x \in \mathbb{R}$), then standard facts about polynomials imply $a_0 = a_1 = \cdots = a_n = 0$, showing that β is linearly independent.

43.15 Exercise

Consider the following question about the vector space $\mathscr{P}_3$. Does the polynomial $q(x) = 4x^3 - 9x^2 + x - 10$ belong to $\mathscr{L}(S)$, where $S = \{-2x^3 + 8x^2 + 4x + 4, -2x^3 - 6x^2 - 14x + 8\}$? From 43.9 and 43.10, we know that this question has the same answer as the following question about $\mathbb{R}^4$: is $(4, -9, 1, -10) \in \mathscr{L}((-2, 8, 4, 4), (-2, -6, -14, 8))$? It can therefore be answered using the computational technique presented in 15.19. What is the answer?

43.16 Exercise

Is the set

$$\left\{\begin{bmatrix} -5 & \frac{1}{2} \\ 7 & -2 \end{bmatrix}, \begin{bmatrix} 2 & 0 \\ 6 & -1 \end{bmatrix}, \begin{bmatrix} -5 & 10 \\ 14 & -\frac{5}{2} \end{bmatrix}, \begin{bmatrix} 4 & 9 \\ -3 & -2 \end{bmatrix}\right\}$$

a basis for $\mathbb{R}_{2\times 2}$? From 43.7, 43.9, 43.10 and 43.12, we know that this question has the same answer as the following question about $\mathbb{R}^4$: is the set $\{(-5, \frac{1}{2}, 7, -2), (2, 0, 6, -1), (-5, 10, 14, -\frac{5}{2}), (4, 9, -3, -2)\}$ linearly independent? It can therefore be answered by the computational procedure presented in 15.17. What is the answer?

The remainder of this section contains material that will be used only in Sections 46, 50 and 51.

It is useful to extend the concepts of spanning and independence from sets of vectors to sets of subspaces. At the same time, we extend the notion of direct sum of subspaces; the special case, for $p = 2$, of 43.20(i) to (iii) was presented in Problem 8.12.

43.17 Exercise

If $W_1, W_2, \ldots, W_p$ are subspaces of a vector space V, let $\Sigma_{i=1}^{i=p} W_i = \{\mathbf{v} \in V$: there exist vectors $\mathbf{w}_1 \in W_1, \mathbf{w}_2 \in W_2, \ldots, \mathbf{w}_p \in W_p$ such that $\mathbf{v} = \mathbf{w}_1 + \mathbf{w}_2 + \cdots + \mathbf{w}_p\}$. Prove that $\Sigma_{i=1}^{i=p} W_i$ is a subspace of V.

43.18 Definition

The subspace $\Sigma_{i=1}^{i=p} W_i$ is called the *sum* of the W_i. When the range of the index of summation is clear from the context, we will simply write $\Sigma_i W_i$.

43.19 Definition

The collection of subspaces $\{W_1, W_2, \ldots, W_p\}$ *spans* V iff $\Sigma_i W_i = V$.

Before defining independence for subspaces, we prove a theorem that establishes the equivalence of five conditions on a collection of subspaces.

43.20 Theorem

The following are equivalent for any collection $\{W_1, W_2, \ldots, W_p\}$ of subspaces of a vector space V:

(i) For every $\mathbf{v} \in \Sigma_i W_i$, there is a unique expression for $\mathbf{v}$ of the form $\mathbf{v} = \mathbf{w}_1 + \mathbf{w}_2 + \cdots + \mathbf{w}_p$ with $\mathbf{w}_i \in W_i$ for $i = 1, 2, \ldots, p$.
(ii) If $\mathbf{0} = \mathbf{w}_1 + \mathbf{w}_2 + \cdots + \mathbf{w}_p$ with $\mathbf{w}_i \in W_i$ for $i = 1, 2, \ldots, p$, then $\mathbf{w}_1 = \mathbf{w}_2 = \cdots = \mathbf{w}_p = \mathbf{0}$.
(iii) For each $i = 1, 2, \ldots, p$, $W_i \cap \Sigma_{j \neq i} W_j = \{\mathbf{0}\}$.
(iv) For every collection of bases β_1 for W_1, β_2 for $W_2, \ldots, \beta_p$ for W_p, the union, $\beta = \beta_1 \cup \beta_2 \cup \cdots \cup \beta_p$, is a basis for $\Sigma_i W_i$.
(v) $\dim(\Sigma_i W_i) = \Sigma_i \dim W_i$.

Proof: The pattern of the proof is as follows: (i) → (ii) → (iii) → (i); (ii) → (iv) → (iii); finally, (iv) → (v) → (iv).

Proof of (i) → (ii). This is trivial since (ii) is the special case of (i) for the zero vector.

Proof of (ii) → (iii). Let i be arbitrary but fixed and let $\mathbf{v} \in W_i \cap \Sigma_{j \neq i} W_j$. Then there exist vectors $\mathbf{w}_k \in W_k$ such that $\mathbf{v} = \mathbf{w}_i \in W_i$ and $\mathbf{v} = \mathbf{w}_1 + \cdots + \mathbf{w}_{i-1} + \mathbf{w}_{i+1} + \cdots + \mathbf{w}_p \in \Sigma_{j \neq i} W_j$. Hence, $\mathbf{0} = \mathbf{v} - \mathbf{v} = \mathbf{w}_1 + \cdots + \mathbf{w}_{i-1} + (-\mathbf{w}_i) + \mathbf{w}_{i+1} + \cdots + \mathbf{w}_p$. From (ii) we conclude $\mathbf{w}_1 = \mathbf{w}_2 = \cdots = \mathbf{w}_p = \mathbf{0}$; in particular, $\mathbf{v} = \mathbf{w}_i = \mathbf{0}$.

Proof of (iii) → (i). Let $\mathbf{v} = \Sigma_i \mathbf{w}_i = \Sigma_i \mathbf{w}_i'$ be two potentially different expressions for $\mathbf{v}$, with $\mathbf{w}_i, \mathbf{w}_i' \in W_i$. We will conclude that they are the same by showing, for arbitrary i, that $\mathbf{w}_i = \mathbf{w}_i'$. Note that

$$\mathbf{w}_i - \mathbf{w}_i' = \sum_{j \neq i} \mathbf{w}_j' - \sum_{j \neq i} \mathbf{w}_j = \sum_{j \neq i} (\mathbf{w}_j' - \mathbf{w}_j).$$

Because the W's are subspaces, this shows that

$$\mathbf{w}_i - \mathbf{w}_i' \in W_i \cap \sum_{j \neq i} W_j.$$

So we conclude from (iii) that $\mathbf{w}_i - \mathbf{w}_i' = \mathbf{0}$.

For the remainder of the proof, let $\beta_1, \beta_2, \ldots, \beta_p$ be arbitrary bases for $W_1, W_2, \ldots, W_p$, respectively, and let $\beta = \beta_1 \cup \beta_2 \cup \cdots \cup \beta_p$. Let $\beta_i = (\mathbf{v}_{i1}, \mathbf{v}_{i2}, \ldots, \mathbf{v}_{ik_i})$, where $k_i = \dim W_i$.

Proof of (ii) → (iv). It is obvious that β spans $\Sigma_i W_i$. It remains to prove that β is linearly independent. Since we are assuming (ii) and have proved that (ii) → (iii), we may assume (iii) as well. Using (iii), we see that the β_i are pairwise disjoint; thus, by regrouping the terms, any linear combination of vectors from β can be rewritten in a unique way as a sum of vectors from the individual subspaces, W_i. So suppose $\mathbf{0} = \Sigma_{ij} a_{ij}\mathbf{v}_{ij}$; then by letting

$$\mathbf{w}_i = \sum_{j=1}^{j=k_i} a_{ij}\mathbf{v}_{ij}$$

we obtain $\mathbf{0} = \Sigma_i \mathbf{w}_i$. Using (ii), we conclude $\mathbf{w}_i = \mathbf{0}$ for $i = 1, 2, \ldots, p$. This in turn, using the linear independence of the β_i, implies $a_{ij} = 0$ for all i and j.

Proof of (iv) → (iii). We prove the contrapositive, not(iii) → not(iv). To prove not(iv), we must find a collection of bases for the W_i whose union is not a basis for $\Sigma_i W_i$.

Case 1: Assume the given arbitrary collection of bases is pairwise disjoint; in this case, use not(iii) to choose a nonzero vector $\mathbf{v} \in W_i \cap \Sigma_{j \neq i} W_j$; then there exist $\mathbf{w}_k \in W_k$ such that $\mathbf{v} = \mathbf{w}_i$ and $\mathbf{v} = \mathbf{w}_1 + \cdots + \mathbf{w}_{i-1} + \mathbf{w}_{i+1} + \cdots + \mathbf{w}_p$. These expressions constitute two distinct ways to write $\mathbf{v}$ as a linear combination of the vectors from β (because $\mathbf{v} \neq \mathbf{0}$ and the β_i are pairwise disjoint), so β is linearly dependent.

Case 2: $\beta_i \cap \beta_j \neq \varnothing$ for some pair $i \neq j$; in this case, let $\mathbf{v} \in \beta_i \cap \beta_j$, and let β_j' be obtained from β_j by replacing the vector $\mathbf{v}$ with the vector $2\mathbf{v}$. Then $\{\beta_1, \ldots, \beta_{j-1}, \beta_j', \beta_{j+1}, \ldots, \beta_p\}$ is a collection of bases for the W_i whose union is not a basis for $\Sigma_i W_i$; the union is linearly dependent since $\mathbf{v} \in \beta_i$ and $2\mathbf{v} \in \beta_j'$.

Proof of (iv) → (v): Choose bases β_i for W_i; then by (iv), their union, β, is a basis for $\Sigma_i W_i$. By (iii) [which we may also assume since (iv) → (iii) was just proved], the β_i are pairwise disjoint; hence $|\beta| = \Sigma_i |\beta_i|$. It follows that $\dim(\Sigma_i W_i) = |\beta| = \Sigma_i |\beta_i| = \Sigma_i \dim W_i$.

Proof of (v) → (iv): We prove the contrapositive, not(iv) → not(v). Let β_i be a collection of bases for W_i whose union, β, is not a basis for $\Sigma_i W_i$. Since β spans $\Sigma_i W_i$, β must be linearly dependent. Choose $\mathbf{u} \in \beta$ such that $\mathscr{L}(\beta - \{\mathbf{u}\}) = \mathscr{L}(\beta) = \Sigma_i W_i$; then $\dim(\Sigma_i W_i) \leq |\beta - \{\mathbf{u}\}| \lneq |\beta| \leq \Sigma_i |\beta_i| = \Sigma_i \dim W_i$. ■

43.21 Definition

A collection $\{W_1, W_2, \ldots, W_p\}$ of subspaces of a vector space V is called *independent* if it satisfies any one, and hence all, of the conditions from the previous theorem.

It is clear that the dependence or independence of a collection of subspaces is not related to any particular ordering of the collection. It is possible however, as in Section 7, to get *additional* information by considering the collection as an ordered set. When we want to prove that a given collection of subspaces is independent, it is convenient to have the following condition, equivalent to the five from 43.20.

43.22 Exercise

The ordered collection of subspaces $(W_1, W_2, \ldots, W_p)$ of a vector space V is independent iff

$$\text{for all} \quad i \geq 2,\ W_i \cap \sum_{j \lneqq i} W_j = \{\mathbf{0}\}$$

Hint: Imitate the proof of Theorem 7.16.

43.23 Definition

The sum $\Sigma_i W_i$ of a collection $\{W_1, W_2, \ldots, W_p\}$ of subspaces is called their *direct sum* iff the subspaces are independent. We write $\Sigma_i W_i = W_1 \oplus W_2 \oplus \cdots \oplus W_p$ to denote this fact.

43.24 Definition

If $\{W_1, W_2, \ldots, W_p\}$ is an independent collection of subspaces that spans V, we say that $\{W_1, W_2, \ldots, W_p\}$ is a *direct-sum decomposition of* V.

In general, the union of two linearly independent sets is not linearly independent. Thus, it is significant to be able, when $V = W_1 \oplus W_2 \oplus \cdots \oplus W_p$, to construct a basis, β, for V by constructing bases, β_i, for each of the subspaces separately and taking their union.

43.25 Caution

A set of two subspaces $\{W_1, W_2\}$ is independent iff $W_1 \cap W_2 = \{\mathbf{0}\}$. When $p \geq 3$, the condition

$$W_i \cap W_j = \{\mathbf{0}\}, \quad \text{for all pairs } i \neq j$$

is *not* sufficient for the independence of $\{W_1, W_2, \ldots, W_p\}$. Consider, for example, the case of three distinct lines through the origin in $\mathbb{R}^2$.

Problem Set 43

43.1. Go back and do any of the exercises from Section 43 that you may have skipped.

*43.2. Which of the following polynomials in $\mathscr{P}_2$ is expressible as a linear combination of the two polynomials $1 + x - 2x^2$ and $2 - x + 2x^2$?
(i) 0
(ii) -1
(iii) $1 - x - x^2$
(iv) $3 - 2x + 4x^2$

*43.3. (i) Find a reduced description of the subspace of $\mathbb{R}_{2\times 2}$ spanned by the three vectors

$$\begin{bmatrix} 1 & 1 \\ 0 & 1 \end{bmatrix}, \begin{bmatrix} 2 & -1 \\ 3 & 5 \end{bmatrix}, \begin{bmatrix} 2 & 2 \\ -1 & 1 \end{bmatrix}.$$

(ii) Which of the following four matrices belong to this subspace?

$$\begin{bmatrix} 0 & 0 \\ 0 & 0 \end{bmatrix}, \begin{bmatrix} 1 & 0 \\ 0 & 1 \end{bmatrix}, \begin{bmatrix} -3 & 1 \\ 4 & 1 \end{bmatrix}, \begin{bmatrix} 2 & 4 \\ 1 & -1 \end{bmatrix}$$

*43.4. Which of the following polynomials belong to the subspace of $\mathscr{P}_2$ spanned by $\{1 - x, 2 - 3x + x^2, 4 - 9x + 5x^2\}$?
(i) $1 + x - x^2$
(ii) $5 + 6x$
(iii) $2 + 3x + 3x^2$
(iv) $4 + 5x + x^2$
(v) $1 + x^2$

*43.5. Determine whether the following sets of polynomials in $\mathscr{P}_2$ are linearly dependent or independent.
(i) $\{1 - 2x + x^2, 3 + 4x + x^2, -2 - x + x^2\}$
(ii) $\{-2 - 11x + 3x^2, 1 - 2x + 3x^2, 2 + x + 3x^2\}$
(iii) $\{-1 + 2x + 3x^2, 1 + x, 1 + x^2\}$
(iv) $\{x + x^2, 2, 3 - x + 2x^2\}$

*43.6. Return to Problems 42.2 and 42.3 and determine the dimensions of those subsets that are subspaces.

*43.7. Let $\mathbf{p}_1$, $\mathbf{p}_2$, and $\mathbf{p}_3$ be the following vectors in $\mathscr{P}_2$.

$$p_1(x) = 1 + 3x - 2x^2, \qquad p_2(x) = 1 - 2x + x^2,$$

$$p_3(x) = 2 - x^2$$

Express each of the following vectors as linear combinations of $\mathbf{p}_1$, $\mathbf{p}_2$, and $\mathbf{p}_3$.

(i) $f(x) = 6 + 2x - 2x^2$

(ii) $g(x) = 3 + x - x^2$

(iii) $h(x) = x^2$

43.8. Find a basis for $\mathscr{P}_2$ that contains the vectors $2x^2 + x - 4$ and $-x^2 - 2x - 4$.

*43.9. Let $S_1 = \{2x^2 - x + 1, x + 2, 3x^2 + 3x\} \subset \mathscr{P}_2$ and let

$$S_2 = \left\{\begin{bmatrix} 2 & 0 \\ 0 & 1 \end{bmatrix}, \begin{bmatrix} 1 & 10 \\ 2 & 0 \end{bmatrix}, \begin{bmatrix} 4 & 1 \\ 1 & 1 \end{bmatrix}\right\} \subset \mathbb{R}_{2\times 2}$$

Find reduced descriptions for $\mathscr{L}(S_1)$ and $\mathscr{L}(S_2)$.

*43.10. Is $\begin{bmatrix} -1 & 2 \\ 3 & -7 \end{bmatrix} \in \mathscr{L}\left\{\begin{bmatrix} 1 & 0 \\ -1 & 3 \end{bmatrix}, \begin{bmatrix} 2 & 5 \\ 3 & -4 \end{bmatrix}, \begin{bmatrix} 1 & 1 \\ 0 & 1 \end{bmatrix}\right\}$?

44

Linear Transformations between Abstract Vector Spaces

Once again, most of the results from Section 17 concerning linear transformations between $\mathbb{R}^n$ and $\mathbb{R}^m$ apply virtually unchanged in the general context. The proofs are left as easy exercises.

44.1 Definition

Let V and W be vector spaces over the same field $\mathbb{F}$. A function $T: V \to W$ is called a *linear transformation* if

(i) for all $\mathbf{u}, \mathbf{v} \in V$, $T(\mathbf{u} + \mathbf{v}) = T(\mathbf{u}) + T(\mathbf{v})$ (T preserves vector addition), and
(ii) for all $t \in \mathbb{F}$ and $\mathbf{v} \in V$, $T(t\mathbf{v}) = tT(\mathbf{v})$ (T preserves scalar multiplication).

44.2 Exercises

If $S: U \to V$ and $T: V \to W$ are linear transformations, then

(i) $T(\mathbf{0}) = \mathbf{0}$ (T sends the zero vector of V to the zero vector of W).
(ii) For all $t_1, t_2, \ldots, t_p \in \mathbb{F}$ and $\mathbf{u}_1, \mathbf{u}_2, \ldots, \mathbf{u}_p \in V$, if $\mathbf{v} = t_1\mathbf{u}_1 + t_2\mathbf{u}_2 + \cdots + t_p\mathbf{u}_p$, then $T(\mathbf{v}) = t_1T(\mathbf{u}_1) + t_2T(\mathbf{u}_2) + \cdots + t_pT(\mathbf{u}_p)$ (T preserves arbitrary linear combinations).
(iii) The composition $T \circ S: U \to W$ is a linear transformation.

Note that an isomorphism from V to W (see Definition 44.8) is simply a linear transformation that is both one-to-one and onto.

44.3 Definition

If $T: V \to W$ is a linear transformation, the set $\{\mathbf{v} \in V: T(\mathbf{v}) = \mathbf{0}\}$ is called the *kernel* of T and is denoted by ker T. The set $\{\mathbf{w} \in W: \mathbf{w} = T(\mathbf{v})$ for some $\mathbf{v} \in V\}$ is called the *range* of T and is denoted by ran T.

44.4 Exercise

If $T: V \to W$ is a linear transformation, then

(i) ker T is a subspace of V, and
(ii) ran T is a subspace of W.

Having established that ker T and ran T are subspaces, we may make the following definition.

44.5 Definition

If $T: V \to W$ is a linear transformation, the dimension of ker T is called the *nullity of* T; the dimension of ran T is called the *rank of* T.

Before we proceed to extend the results about linear transformations from earlier chapters to the present general context, we would like to give an example to warn you that some of these results apply only to finite-dimensional vector spaces. One of the most important results about a linear operator $T: \mathbb{R}^n \to \mathbb{R}^n$ is that such a T is one-to-one if and only if it is onto. Recall that this was proved for matrix transformations in Section 19 and extended in Section 31 to arbitrary linear transformations from $\mathbb{R}^n$ to itself after it was proved that any linear transformation $T: \mathbb{R}^n \to \mathbb{R}^n$ can be represented as a matrix transformation. We remarked at the end of Section 19 that this rare phenomenon is due, in the present instance, to linearity combined with finite-dimensionality.

The purpose of the next example is to show that this property need not hold for a linear operator $T: V \to V$ if V is infinite-dimensional.

44.6 Example

The vector space $\mathbb{R}^\infty$ was introduced in 41.11. Define $T: \mathbb{R}^\infty \to \mathbb{R}^\infty$ as follows: if $\mathbf{u} = (u_1, u_2, u_3, \ldots, u_{i-1}, u_i, u_{i+1}, \ldots)$, then $T(\mathbf{u}) = (0, u_1, u_2, \ldots, u_i, u_{i+1}, u_{i+2}, \ldots)$. It is trivial to check that T is a linear transformation and that T is one-to-one. T is not, however, onto, since, for example, $\mathbf{e}_1^\infty = (1, 0, \ldots, 0, \ldots) \notin \operatorname{ran} T$.

There are other results from earlier sections that do not generalize directly to the infinite-dimensional case. Although we will make an occasional effort to distinguish such properties, a systematic study would take us too far afield. In what follows, when we state a result for finite-dimensional spaces, we do not wish to imply one way or the other that the corresponding result for infinite-dimensional spaces is or is not true. These questions are the subject matter of another mathematics course.

44.7 Exercise (Analogue of 17.12)

If $T: V \to W$ is a linear transformation whose domain, V, is finite-dimensional, then

$$\operatorname{rank} T + \operatorname{nullity} T = \dim V.$$

When V and W are finite-dimensional, we continue to have the ability to represent linear transformations $T: V \to W$ by matrices. Here are the analogues of 31.6, 31.7, 31.8, and 31.9.

44.8 Definition

Assume $T: V \to W$ is a linear transformation, $\dim V = n$, $\dim W = m$, β is an ordered basis for V, and γ is an ordered basis for W. An $m \times n$ matrix A *represents T with respect to β and γ* if and only if

$$\text{for all } \mathbf{v} \in V, \quad [T(\mathbf{v})]_\gamma = \mathrm{A}[\mathbf{v}]_\beta \qquad (*)$$

44.9 Exercise

If $\dim V = n$ and $\dim W = m$ and $T: V \to W$ is a linear transformation, then, for every pair of ordered bases $\beta = (\mathbf{v}_1, \mathbf{v}_2, \ldots, \mathbf{v}_n)$ for V and $\gamma = (\mathbf{w}_1, \mathbf{w}_2, \ldots, \mathbf{w}_m)$ for W, there is a unique $m \times n$ matrix A satisfying

$$\text{for all } \mathbf{v} \in V, \quad [T(\mathbf{v})]_\gamma = \mathrm{A}[\mathbf{v}]_\beta \qquad (*)$$

Because of this result, it makes sense to introduce a name and notation for the unique object satisfying $(*)$.

44.10 Definition

In the context of 44.9, the unique $m \times n$ matrix satisfying $(*)$ is called the *matrix that represents T with respect to β and γ* and is denoted by ${}_{\gamma}[T]_{\beta}$.

44.11 Computational Procedure

The proof of 44.9 reveals that ${}_{\gamma}[T]_{\beta}$ is the matrix whose j^{th} column, for $j = 1, 2, \ldots, n$, is $[T(\mathbf{v}_j)]_{\gamma}$.

As in Section 31, the special case of 44.10 when $V = W$ and T is the identity transformation on V leads to the notion of the *change-of-basis matrix* (or *transition matrix*), ${}_{\gamma}[I]_{\beta}$, from the basis β to the basis γ, that satisfies $[\mathbf{v}]_{\gamma} = {}_{\gamma}[I]_{\beta}\,[\mathbf{v}]_{\beta}$ for all $\mathbf{v} \in V$.

44.12 Exercise

For any pair of ordered bases β and γ for an n-dimensional vector space V, the transition matrix ${}_{\gamma}[I]_{\beta}$ is invertible and its inverse is the transition matrix ${}_{\beta}[I]_{\gamma}$.

We conclude this section with the analogue of Theorem 32.1.

44.13 Exercise

Suppose that V and W are finite-dimensional vector spaces, that α and β are ordered bases for V, that γ and δ are ordered bases for W, and that $T: V \to W$ is a linear transformation; then

$${}_{\delta}[T]_{\alpha} = {}_{\delta}[I]_{\gamma}\,{}_{\gamma}[T]_{\beta}\,{}_{\beta}[I]_{\alpha}$$

As in Section 32, we obtain the relationship between ${}_{\alpha}[T]_{\alpha}$ and ${}_{\beta}[T]_{\beta}$, where $T: V \to V$ is a linear transformation from a finite-dimensional vector space into itself and α and β are ordered bases for V, by specializing 44.13 to the case $V = W$, $\alpha = \delta$, and $\beta = \gamma$.

44.14 Corollary

Where T, α, and β are as just described,

$${}_{\beta}[T]_{\beta} = {}_{\beta}[I]_{\alpha}\,{}_{\alpha}[T]_{\alpha}\,{}_{\alpha}[I]_{\beta}$$

Problem Set 44

44.1. Go back and do any of the exercises from Section 44 that you may have skipped.

*44.2. The linear transformation $T:\mathscr{P}_2 \to \mathscr{P}_1$ is defined by $T(a_0 + a_1x + a_2x^2) = (a_0 - a_1) + 2a_2x$.

(i) Find the matrix that represents T with respect to the standard bases.

(ii) Find the kernel and range of T.

(iii) What are the rank and nullity of T?

*44.3. U and V are both vector spaces, $\dim U = \dim V = n$, and $T:U \to V$ is a linear transformation. $S = \{\mathbf{u}_1, \mathbf{u}_2, \ldots, \mathbf{u}_n\}$ is a basis for U. Label each of the following statements as *true* or *false* and provide reasons for your answers.

(i) The set $T(S) = \{T(\mathbf{u}_1), T(\mathbf{u}_2), \ldots, T(\mathbf{u}_n)\}$ is a basis for V.

(ii) For any $\mathbf{u} \in U$, $T(\mathbf{u})$ is a linear combination of the vectors in $T(S)$.

(iii) $T(\mathbf{0}) = \mathbf{0}$.

(iv) For all $\mathbf{v} \in V$, there exists a $\mathbf{u} \in U$ such that $T(\mathbf{u}) = \mathbf{v}$.

(v) T is invertible iff $T(S)$ is linearly independent.

(vi) $T(S)$ spans the range of T.

*44.4. Let $T:\mathscr{P}_1 \to \mathscr{P}_2$ be defined by $T(a_0 + a_1x) = a_0 - 2a_1 + (2a_0 + a_1)x + (a_0 + a_1)x^2$.

(i) Find the matrix that represents T with respect to the bases $\{1, x\}$ and $\{1, x, x^2\}$.

(ii) Find the matrix that represents T with respect to the bases $\beta = \{1 - x, x\}$ and $\gamma = \{1 - x + x^2, x - x^2, 1 + x^2\}$.

(iii) Find the γ-coordinates of $T(1 - x)$.

*44.5. Define $T:\mathbb{R}_{2\times 2} \to \mathbb{R}_{2\times 2}$ by

$$T\left(\begin{bmatrix} a & b \\ c & d \end{bmatrix}\right) = \begin{bmatrix} a + 2b + 3c - d & 2a + 3b + 2c + 2d \\ 3a + c + 5d & 4a + 5b - 2c + 10d \end{bmatrix}$$

(i) Find bases for the kernel and range of T.

(ii) Does $\begin{bmatrix} 1 & 2 \\ 3 & 4 \end{bmatrix}$ belong to the range of T?

*44.6. Define $T:\mathscr{P}_2 \to \mathscr{P}_2$ by $T(a_0 + a_1x + a_2x^2) = (2a_2 - 3a_1 - 5a_0) + (3a_2 - 3a_0)x + (a_2 - a_1 - 2a_0)x^2$.

(i) Write down the standard matrix for T.

(ii) Find the kernel and range of T.

(iii) Does $-3x^2 + 3x - 3$ belong to the range of T?

45

Inner Product Spaces

Analogy: Section 41 is to Section 5 as Section 45 is to Section 38.

In other words, we are about to reexamine what was done in Section 38 for the Euclidean dot product and the Euclidean norm, to abstract the essential ideas, and to place them in a more general context. In particular, the concepts of length, distance, angle, and orthogonality will be superimposed on certain real or complex vector spaces. Indeed, the old familiar space $\mathbb{R}^n$ itself admits other notions of "distance" and "length" than the Euclidean ones.

We will first present the results for real vector spaces and leave the modifications required for complex vector spaces to the end of the section.

Begin by reviewing the proof of the Cauchy–Schwarz Inequality in 38.7; note that it depended only on four very simple properties of the Euclidean dot product listed in Theorems 38.5 and 38.6. This motivates the following definition.

45.1 Definition

Let V be a real vector space. A function $\langle \cdot , \cdot \rangle$ from $V \times V$ into $\mathbb{R}$ that associates to each ordered pair $(\mathbf{u}, \mathbf{v})$ of vectors a real number $\langle \mathbf{u}, \mathbf{v} \rangle$ is called an *inner product on* V if it satisfies the following four properties for all

$\mathbf{u}, \mathbf{v}, \mathbf{w} \in V$ and $t \in \mathbb{R}$; the common names for these properties are given at the right in brackets.

(i) $\langle \mathbf{u}, \mathbf{v} \rangle = \langle \mathbf{v}, \mathbf{u} \rangle$ [$\langle \cdot, \cdot \rangle$ is *symmetric*]
(ii) $\langle \mathbf{u} + \mathbf{v}, \mathbf{w} \rangle = \langle \mathbf{u}, \mathbf{w} \rangle + \langle \mathbf{v}, \mathbf{w} \rangle$
(iii) $\langle t\mathbf{u}, \mathbf{v} \rangle = t\langle \mathbf{u}, \mathbf{v} \rangle$ } [$\langle \cdot, \cdot \rangle$ is *linear in the first variable*]
(iv) $0 \leq \langle \mathbf{v}, \mathbf{v} \rangle$; and $\langle \mathbf{v}, \mathbf{v} \rangle = 0$ if and only if $\mathbf{v} = \mathbf{0}$. [$\langle \cdot, \cdot \rangle$ is *positive definite*]

Taken together, (ii) and (iii) assert that for every $\mathbf{w} \in V$, the function $\langle \cdot, \mathbf{w} \rangle : V \to \mathbb{R}$ is a linear map.

45.2 Exercise

Prove the following easy consequences of the definition. If $\langle \cdot, \cdot \rangle : V \times V \to \mathbb{R}$ is an inner product on a real vector space V, then

(i) $\langle \mathbf{0}, \mathbf{v} \rangle = \langle \mathbf{v}, \mathbf{0} \rangle = 0$
(ii) $\langle \mathbf{u}, \mathbf{v} + \mathbf{w} \rangle = \langle \mathbf{u}, \mathbf{v} \rangle + \langle \mathbf{u}, \mathbf{w} \rangle$
(iii) $\langle \mathbf{u}, t\mathbf{v} \rangle = t\langle \mathbf{u}, \mathbf{v} \rangle$. } [*$\langle \cdot, \cdot \rangle$ is linear in the second variable*]

Taken together, (ii) and (iii) assert that for every $\mathbf{w} \in V$, the function $\langle \mathbf{w}, \cdot \rangle : V \to \mathbb{R}$ is a linear map.

45.3 Examples

(i) Let $V = \mathbb{R}^2$. Let $\langle \mathbf{u}, \mathbf{v} \rangle = 2u_1v_1 + 5u_2v_2$.
(ii) Let $V = \mathbb{R}^2$. Let $\langle \mathbf{u}, \mathbf{v} \rangle = u_1v_2 + u_2v_1$.
(iii) Let a and b be fixed real numbers with $a < b$. Let $V = C[a, b] = \{f : f$ is a real-valued continuous function defined on the interval $[a, b]\}$. Let $\langle \mathbf{f}, \mathbf{g} \rangle = \int_a^b f(x)g(x)\,dx$.

45.4 Exercise

Prove that the three examples of 45.3 are indeed inner products. [For 45.3(iii), this requires a knowledge of calculus.]

45.5 Definition

A function from $V \times V$ into $\mathbb{R}$ that satisfies properties 45.1(ii) and (iii) and 45.2(ii) and (iii) is called *bilinear*.

Thus, a bilinear function is a function of two variables which is such that, when either one of the two variables is fixed, the resulting function of the (single) remaining variable is a linear function. This is often paraphrased by saying that a bilinear function is a function of two variables that is "linear as a function of each variable separately."

It is important to distinguish bilinearity, which is a property of functions of two variables, from linearity, which is a property of functions of one variable. Confusion may arise, for example, because $\mathbb{R} \times \mathbb{R}$ is isomorphic to $\mathbb{R}^2$, so that a function (of two variables) from $\mathbb{R} \times \mathbb{R}$ into $\mathbb{R}$ can also be viewed as a function (of one vector variable) from $\mathbb{R}^2$ into $\mathbb{R}$.

There is no general relationship between linearity and bilinearity, as the following counterexamples indicate.

45.6 Examples

(i) The function F that, to each pair of real numbers x and y, associates their product, xy, is bilinear when viewed as a function from $\mathbb{R} \times \mathbb{R}$ into $\mathbb{R}$: for all x, x_1, x_2, y, y_1, y_2, and $t \in \mathbb{R}$,

$$(x_1 + x_2)y = x_1y + x_2y \quad \text{and} \quad (tx)y = t(xy)$$
$$x(y_1 + y_2) = xy_1 + xy_2 \quad \text{and} \quad x(ty) = t(xy).$$

It is not, however, linear as a function of the single variable $\mathbf{u} = (u_1, u_2) \in \mathbb{R}^2$ because $F(\mathbf{u} + \mathbf{v}) = F(u_1 + u_2, v_1 + v_2) = (u_1 + u_2)(v_1 + v_2) = u_1v_1 + u_1v_2 + u_2v_1 + u_2v_2$, whereas $F(\mathbf{u}) + F(\mathbf{v}) = u_1u_2 + v_1v_2$.

(ii) The function G that, to each pair of real numbers x and y, associates their sum, $x + y$, is linear when viewed as a function from $\mathbb{R}^2$ into $\mathbb{R}$ because $G(\mathbf{u} + \mathbf{v}) = G(u_1 + v_1, u_2 + v_2) = u_1 + v_1 + u_2 + v_2 = u_1 + u_2 + v_1 + v_2 = G(\mathbf{u}) + G(\mathbf{v})$, and $G(t\mathbf{u}) = G((tu_1, tu_2)) = tu_1 + tu_2 = t(u_1 + u_2) = tG(\mathbf{u})$. It is not, however, bilinear as a function from $\mathbb{R} \times \mathbb{R}$ into $\mathbb{R}$. For example, for the fixed value $y = 1$, the resulting function from $\mathbb{R} \to \mathbb{R}$ is the function $G(x, 1) = x + 1$, which is not linear since $G(x_1 + x_2, 1) = x_1 + x_2 + 1$, whereas $G(x_1, 1) + G(x_2, 1) = x_1 + 1 + x_2 + 1 = x_1 + x_2 + 2$.

In the presence of symmetry, there is the following easy relationship.

45.7 Exercise

If $\langle \cdot, \cdot \rangle: V \times V \to \mathbb{R}$ is linear as a function of one of its arguments and is symmetric, then it is also linear as a function of the other argument, and thus is bilinear.

With these preliminaries aside, we can get down to the heart of the matter.

45.8 Theorem (Cauchy–Schwarz Inequality)

If $\langle \cdot , \cdot \rangle : V \times V \to \mathbb{R}$ is an inner product on a real vector space V, then for all $\mathbf{u}, \mathbf{v} \in V$,

$$\langle \mathbf{u}, \mathbf{v} \rangle^2 \leq \langle \mathbf{u}, \mathbf{u} \rangle \langle \mathbf{v}, \mathbf{v} \rangle \qquad (\dagger)$$

45.9 Exercise

Prove the Cauchy–Schwarz Inequality for arbitrary real inner products by imitating the proof of Theorem 38.7. ■

45.10 Corollary to the Proof of 45.8

Equality holds in the Cauchy–Schwarz Inequality iff $\{\mathbf{u}, \mathbf{v}\}$ is linearly dependent. ■

We may now repeat in the context of any real inner product space all that was done in Section 38 that only depended on the Cauchy–Schwarz Inequality.

In particular, if $\langle \cdot , \cdot \rangle : V \times V \to \mathbb{R}$ is an inner product on a real vector space V, then for all nonzero $\mathbf{u}, \mathbf{v} \in V$, the real number $\dfrac{\langle \mathbf{u}, \mathbf{v} \rangle}{\sqrt{\langle \mathbf{u}, \mathbf{u} \rangle}\sqrt{\langle \mathbf{v}, \mathbf{v} \rangle}}$ belongs to the interval $[-1, +1]$; so we can define the "angle between $\mathbf{u}$ and $\mathbf{v}$" to be the angle between 0 and π having this number as its cosine.

As well, each inner product gives rise to its related notions of "norm" and "distance" as in the following theorem, whose proof imitates those of 38.12 and 38.13 and is left as an exercise.

45.11 Theorem

If $\langle \cdot , \cdot \rangle : V \times V \to \mathbb{R}$ is an inner product on a real vector space V, and if $\|\mathbf{u}\| = \sqrt{\langle \mathbf{u}, \mathbf{u} \rangle}$, then for all $\mathbf{u}, \mathbf{v} \in V$ and $t \in \mathbb{R}$,

(i) $0 \leq \|\mathbf{u}\|$; and $0 = \|\mathbf{u}\|$ iff $\mathbf{u} = \mathbf{0}$ [$\|\cdot\|$ is positive definite]
(ii) $\|t\mathbf{u}\| = |t|\|\mathbf{u}\|$ [$\|\cdot\|$ is homogeneous]
(iii) $\|\mathbf{u} + \mathbf{v}\| \leq \|\mathbf{u}\| + \|\mathbf{v}\|$ [The Triangle Inequality]

■

In a similar vein, if we define $d(\mathbf{u}, \mathbf{v}) = \|\mathbf{u} - \mathbf{v}\|$, then d has all the properties of a distance function, as in Theorem 38.16.

45.12 Exercise

Consider the inner product space $C[-2, 3]$ with the inner product defined by the integral as in Example 45.3(iii).

(i) What is the angle between the functions $f(x) = x^2$ and $g(x) = x - 1$?
(ii) What is the distance between these two functions?
(iii) What is $\|\mathbf{f}\|$?

The next exercise introduces another important class of examples. It shows how each ordered basis, β, for $\mathbb{R}^n$ induces its own inner product $\langle \cdot , \cdot \rangle_\beta$ and corresponding norm $\| \cdot \|_\beta$ on $\mathbb{R}^n$.

45.13 Exercise

Let β be an ordered basis for $\mathbb{R}^n$. Define $\langle \cdot , \cdot \rangle_\beta : \mathbb{R}^n \times \mathbb{R}^n \to \mathbb{R}$ by $\langle \mathbf{u}, \mathbf{v} \rangle_\beta = (\mathbf{u})_\beta \cdot (\mathbf{v})_\beta$.

(i) Show that $\langle \cdot , \cdot \rangle_\beta$ is an inner product on $\mathbb{R}^n$.
(ii) Conclude from part (i) and Theorem 45.11 that the function $\| \cdot \|_\beta : \mathbb{R}^n \to \mathbb{R}$ defined by $\|\mathbf{u}\|_\beta = \sqrt{(\mathbf{u})_\beta \cdot (\mathbf{u})_\beta}$ is a norm on $\mathbb{R}^n$.

45.14 Definition

If β is an ordered basis for $\mathbb{R}^n$, the functions $\langle \cdot , \cdot \rangle_\beta$ and $\| \cdot \|_\beta$ are called the *β inner product* and the *β norm*, respectively. For $\mathbf{u} \in \mathbb{R}^n$, $\|\mathbf{u}\|_\beta$ is the *β norm of* $\mathbf{u}$.

When $\beta = \alpha^n$, the β inner product and the β norm are the familiar dot product and Euclidean norm; we will usually, following past practice, omit the subscript when $\beta = \alpha^n$. Occasionally, for emphasis, we will denote the Euclidean norm of $\mathbf{u}$ by $\|\mathbf{u}\|_E$.

45.15 Example

Let $\beta = ((3,6),(-1,-4))$, $\mathbf{u} = (2,-1)$, and $\mathbf{v} = (-1,2)$. When we use the methods of Section 31 to compute the β-coordinates of these vectors, we obtain

$$[\mathbf{u}]_\beta = {}_\beta[I]_{\alpha^2}\,[\mathbf{u}]_{\alpha^2} = \begin{bmatrix} 3 & -1 \\ 6 & -4 \end{bmatrix}^{-1} \begin{bmatrix} 2 \\ -1 \end{bmatrix} = \begin{bmatrix} \frac{3}{2} \\ \frac{5}{2} \end{bmatrix}$$

Similarly,

$$[\mathbf{v}]_\beta = \begin{bmatrix} 3 & -1 \\ 6 & -4 \end{bmatrix}^{-1} \begin{bmatrix} -1 \\ 2 \end{bmatrix} = \begin{bmatrix} -1 \\ -2 \end{bmatrix}$$

Now we can compare the Euclidean dot product and norm with the β inner product and β norm of these vectors.

$$\mathbf{u} \cdot \mathbf{v} = (2)(-1) + (-1)(2) = -2 - 2 = -4$$
$$\|\mathbf{u}\|_E = \|\mathbf{u}\| = \sqrt{(2)^2 + (-1)^2} = \sqrt{5}$$
$$\|\mathbf{v}\|_E = \|\mathbf{v}\| = \sqrt{(-1)^2 + (2)^2} = \sqrt{5}$$

whereas

$$\langle \mathbf{u}, \mathbf{v} \rangle_\beta = (\mathbf{u})_\beta \cdot (\mathbf{v})_\beta = (\tfrac{3}{2})(-1) + (\tfrac{5}{2})(-2) = -\tfrac{13}{2}$$
$$\|\mathbf{u}\|_\beta = \sqrt{(\tfrac{3}{2})^2 + (\tfrac{5}{2})^2} = \frac{\sqrt{34}}{2}$$
$$\|\mathbf{v}\|_\beta = \sqrt{(-1)^2 + (-2)^2} = \sqrt{5}$$

In Theorem 40.4, we showed, without further comment, that a linear operator $T:\mathbb{E}^n \to \mathbb{E}^n$ which preserves the Euclidean norm will also preserve the dot product. We may now point out the true reason for this: an inner product is completely determined by its associated norm. The next theorem is a precise formulation of this fact.

45.16 Theorem

Let $\langle \cdot , \cdot \rangle_1$ and $\langle \cdot , \cdot \rangle_2$ be two inner products on the real vector space V whose associated norms are the same, that is, which satisfy

$$\text{for all } \mathbf{w} \in V, \quad \|\mathbf{w}\|_1 = \sqrt{\langle \mathbf{w}, \mathbf{w} \rangle_1} = \sqrt{\langle \mathbf{w}, \mathbf{w} \rangle_2} = \|\mathbf{w}\|_2$$

Then for all $\mathbf{u}$ and $\mathbf{v}$ in V, $\langle \mathbf{u}, \mathbf{v} \rangle_1 = \langle \mathbf{u}, \mathbf{v} \rangle_2$.

Proof: (By analogy with the proof of Theorem 40.4.) Let $\mathbf{u}, \mathbf{v} \in V$. Then

$$\begin{aligned} \|\mathbf{u} + \mathbf{v}\|_1^2 &= \langle \mathbf{u} + \mathbf{v}, \mathbf{u} + \mathbf{v} \rangle_1 \\ &= \langle \mathbf{u}, \mathbf{u} \rangle_1 + \langle \mathbf{v}, \mathbf{u} \rangle_1 + \langle \mathbf{u}, \mathbf{v} \rangle_1 + \langle \mathbf{v}, \mathbf{v} \rangle_1 \quad &&[\text{bilinearity of } \langle \cdot , \cdot \rangle_1] \\ &= \langle \mathbf{u}, \mathbf{u} \rangle_1 + 2\langle \mathbf{u}, \mathbf{v} \rangle_1 + \langle \mathbf{v}, \mathbf{v} \rangle_1 &&[\text{symmetry of } \langle \cdot , \cdot \rangle_1] \end{aligned}$$

Similarly, using the bilinearity and symmetry of $\langle \cdot , \cdot \rangle_2$, we have

$$\|\mathbf{u} + \mathbf{v}\|_2^2 = \langle \mathbf{u}, \mathbf{u} \rangle_2 + 2\langle \mathbf{u}, \mathbf{v} \rangle_2 + \langle \mathbf{v}, \mathbf{v} \rangle_2$$

By hypothesis, $\|\mathbf{u}\|_1^2 = \|\mathbf{u}\|_2^2$, $\|\mathbf{v}\|_1^2 = \|\mathbf{v}\|_2^2$, and $\|\mathbf{u} + \mathbf{v}\|_1^2 = \|\mathbf{u} + \mathbf{v}\|_2^2$. Combining all these yields $\langle \mathbf{u}, \mathbf{v} \rangle_1 = \langle \mathbf{u}, \mathbf{v} \rangle_2$. ■

Here is an easy result about inner products that will be needed for later applications.

45.17 Theorem

Let $\langle \cdot , \cdot \rangle$ be an inner product on V and let $\mathbf{u}, \mathbf{v} \in V$.

(i) If, for every $\mathbf{w} \in V$, $\langle \mathbf{u}, \mathbf{w} \rangle = \langle \mathbf{v}, \mathbf{w} \rangle$, then $\mathbf{u} = \mathbf{v}$.
(ii) If, for every $\mathbf{w} \in V$, $\langle \mathbf{w}, \mathbf{u} \rangle = \langle \mathbf{w}, \mathbf{v} \rangle$, then $\mathbf{u} = \mathbf{v}$.

Proof of (i): Using 45.1(ii) and (iii), the hypothesis implies that $\langle \mathbf{u} - \mathbf{v}, \mathbf{w} \rangle = 0$ for every $\mathbf{w} \in V$. In particular, $\langle \mathbf{u} - \mathbf{v}, \mathbf{u} - \mathbf{v} \rangle = 0$, so $\mathbf{u} - \mathbf{v} = \mathbf{0}$ by 45.1(iv).

The second property follows from the first because $\langle \cdot , \cdot \rangle$ is symmetric. ■

45.18 Exercise

Generalize to arbitrary finite-dimensional real inner-product spaces the concepts of orthonormal basis and develop the analogue of the Gram–Schmidt algorithm in this context.

In the remainder of this section, we develop the analogues, for complex numbers, of the Euclidean dot product for $\mathbb{R}$ and outline the basic results for complex inner product spaces. The initial fundamental problem is how to define $\mathbf{u} \cdot \mathbf{v}$ for $\mathbf{u}, \mathbf{v} \in \mathbb{C}^n$ so that this function will have as many as possible of the properties of the Euclidean dot product. It should undoubtedly be a function from $\mathbb{C}^n \times \mathbb{C}^n \to \mathbb{C}$. It should also be a function whose restriction to $\mathbb{R}^n \times \mathbb{R}^n$ is the Euclidean dot product.

The natural first attempt is to imitate the definition for $\mathbb{R}^n$ *exactly*. If we do this, we would have, for example, that $(1, i) \cdot (1, i) = (1)(1) + (i)(i) = 1 - 1 = 0$. This would be a nonzero vector whose "dot-product" with itself would be zero: a violation of positive definiteness.

Here is another undesirable feature of this definition: we would like the function $\|\mathbf{u}\| = \sqrt{\mathbf{u} \cdot \mathbf{u}}$ to be a norm on $\mathbb{C}^n$, that is a notion of "length" for vectors in $\mathbb{C}^n$. The length of a vector should be a nonnegative *real* number; but with this definition we would have, if $\mathbf{v} = (0, i)$, for example, that $\|\mathbf{v}\| = \sqrt{\mathbf{v} \cdot \mathbf{v}} = \sqrt{(0)(0) + (i)(i)} = \sqrt{-1} = i$.

Experience has revealed a more fruitful path. The cost is abandoning the requirement of bilinearity in favor of a slightly more general concept, conjugate bilinearity, to be explained momentarily. The benefit is the preservation of all the desired properties of the Euclidean dot product.

As is frequently the case in the complex domain, good things come in conjugate pairs. Here, our purpose is equally served by two functions from $\mathbb{C}^n \times \mathbb{C}^n \to \mathbb{C}$, which, naturally enough, are complex conjugates of one another.

45.19 Definition

The functions $\cdot_1$ and $\cdot_2$ from $\mathbb{C}^n \times \mathbb{C}^n \to \mathbb{C}$ are defined as follows for $\mathbf{u}, \mathbf{v} \in \mathbb{C}^n$:

(i) $\mathbf{u} \cdot_1 \mathbf{v} = \bar{u}_1 v_1 + \bar{u}_2 v_2 + \cdots + \bar{u}_n v_n$
(ii) $\mathbf{u} \cdot_2 \mathbf{v} = u_1 \bar{v}_1 + u_2 \bar{v}_2 + \cdots + u_n \bar{v}_n$

For example, in $\mathbb{C}^2$, if $\mathbf{u} = (1 + i, 3i)$ and $\mathbf{v} = (-4, 2 + i)$, then

$$\mathbf{u} \cdot_1 \mathbf{v} = (1 - i)(-4) + (-3i)(2 + i) = -4 + 4i - 6i + 3 = -1 - 2i$$
$$\mathbf{u} \cdot_2 \mathbf{v} = (1 + i)(-4) + (3i)(2 - i) = -4 - 4i + 6i + 3 = -1 + 2i$$

Note that $\cdot_1$ involves conjugating the coordinates of the first argument, $\mathbf{u}$, while in $\cdot_2$ it is the coordinates of the second argument, $\mathbf{v}$, that are conjugated. If $\mathbf{u}, \mathbf{v} \in \mathbb{R}^n$, then the conjugation will have no effect, so that both these functions, restricted to $\mathbb{R}^n \times \mathbb{R}^n$, agree with the Euclidean dot product.

Before discussing the properties of these two functions, it is convenient to introduce the appropriate vocabulary.

45.20 Definition

Where V is a complex vector space, a function $T: V \to \mathbb{C}$ is *conjugate linear* if for all $\mathbf{u}, \mathbf{v} \in V$ and $c \in \mathbb{C}$,

(i) $T(\mathbf{u} + \mathbf{v}) = T(\mathbf{u}) + T(\mathbf{v})$, and
(ii) $T(c\mathbf{u}) = \bar{c}T(\mathbf{u})$.

This is a slight variant of linearity, the difference being the presence of conjugation in the interaction between T and scalar multiplication.

45.21 Definition

Let V be a complex vector space and let $\langle \cdot , \cdot \rangle : V \times V \to \mathbb{C}$.

(i) The function $\langle \cdot , \cdot \rangle$ is *Hermitian symmetric* if $\langle \mathbf{u}, \mathbf{v} \rangle = \overline{\langle \mathbf{v}, \mathbf{u} \rangle}$ for all $\mathbf{u}, \mathbf{v} \in V$.
(ii) The function $\langle \cdot , \cdot \rangle$ is *conjugate bilinear* if it is linear in one of its arguments and conjugate linear in the other.

(iii) The function $\langle \cdot , \cdot \rangle$ is *positive definite* if for all $\mathbf{v} \in V$, $0 \leq \langle \mathbf{v}, \mathbf{v} \rangle$; and, moreover, $0 = \langle \mathbf{v}, \mathbf{v} \rangle$ iff $\mathbf{v} = \mathbf{0}$.

45.22 Definition

Let V be a complex vector space.

(i) A function from $V \times V$ into $\mathbb{C}$ that is Hermitian symmetric, conjugate bilinear, and positive definite is called a *complex inner product.*
(ii) A *complex inner product space* is a complex vector space, V, together with a complex inner product on V.

45.23 Exercise

Verify that $\cdot_1$ and $\cdot_2$ are complex inner products on $\mathbb{C}^n$.

Note that $\cdot_1$ and $\cdot_2$ are complex conjugates of one another (i.e., $\mathbf{u} \cdot_2 \mathbf{v} = \overline{\mathbf{u} \cdot_1 \mathbf{v}}$). This is typical of complex inner products, which always come in conjugate pairs, as the next exercise shows.

45.24 Exercise

If $\langle \cdot , \cdot \rangle_1$ is a complex inner product on V, then so is the function $\langle \cdot , \cdot \rangle_2$ whose value at $(\mathbf{u}, \mathbf{v})$ is defined by $\langle \mathbf{u}, \mathbf{v} \rangle_2 = \overline{\langle \mathbf{u}, \mathbf{v} \rangle_1}$.

For $\mathbb{C}^n$, most authors introduce one or the other of $\cdot_1$ or $\cdot_2$ *exclusively* and never mention its mate. This suggests, misleadingly, that one or the other is the "correct" one, while in practice it makes very little difference. We prefer to deal with both since we then have the choice, in a given situation, of using the one that provides the better analogy with the corresponding situation for $\mathbb{R}^n$.

The next exercise presents some reasons why it makes very little difference, in practice, which one we use: (i) both induce the same norm on $\mathbb{C}^n$; and (ii) a subset of $\mathbb{C}^n$ is orthogonal with respect to $\cdot_1$ iff it is orthogonal with respect to $\cdot_2$.

45.25 Exercise

For all $\mathbf{u}, \mathbf{v} \in \mathbb{C}^n$

(i) $\mathbf{u} \cdot_1 \mathbf{u} = \mathbf{u} \cdot_2 \mathbf{u}$
(ii) $\mathbf{u} \cdot_1 \mathbf{v} = 0$ iff $\mathbf{u} \cdot_2 \mathbf{v} = 0$

It follows that, for $S \subset \mathbb{C}$, S is orthonormal with respect to $\cdot_1$ iff S is orthonormal with respect to $\cdot_2$.

45.26 Definition

(i) The functions $\cdot_1$ and $\cdot_2$ are called the *Euclidean inner products on* $\mathbb{C}^n$.
(ii) For $\mathbf{u} \in \mathbb{C}^n$, $\|\mathbf{u}\|_E = \sqrt{\mathbf{u} \cdot_1 \mathbf{u}} = \sqrt{\mathbf{u} \cdot_2 \mathbf{u}}$ is called the *Euclidean norm of* $\mathbf{u}$.

We will drop the subscript and simply write $\mathbf{u} \cdot \mathbf{v}$ in circumstances where $\cdot_1$ or $\cdot_2$ could be used interchangeably.

We have anticipated that the use of the word *norm* is justified in the previous definition. The next few results generalize to $\mathbb{C}^n$ the corresponding ones for the Euclidean dot product in particular and for inner products in general. Proofs will be omitted, and left as exercises, if they are easy modifications of the corresponding proofs for $\mathbb{R}^n$. Just as we used $\mathbb{E}^n$ to denote $\mathbb{R}^n$ together with the Euclidean inner product, we will now use $\mathbb{H}^n$ to denote $\mathbb{C}^n$ together with the Euclidean inner product. As in the past, we will not attempt to be systematic about this distinction, but will use $\mathbb{H}^n$ when we wish to emphasize that the Euclidean inner product is playing a role.

45.27 Theorem (Cauchy–Schwarz Inequality (Analogue of 38.7 and 45.8))

If $\langle \cdot, \cdot \rangle$ is a complex inner product on a complex vector space V, then for all $\mathbf{u}, \mathbf{v} \in V$,

$$\langle \mathbf{u}, \mathbf{v} \rangle \overline{\langle \mathbf{u}, \mathbf{v} \rangle} \leq \langle \mathbf{u}, \mathbf{u} \rangle \langle \mathbf{v}, \mathbf{v} \rangle \quad \text{or equivalently} \quad |\langle \mathbf{u}, \mathbf{v} \rangle| \leq \|\mathbf{u}\| \, \|\mathbf{v}\| \qquad (\dagger)$$

Proof: Assume that $\langle \cdot, \cdot \rangle$ is conjugate linear in the first argument and linear in the second. (The proof in case the roles are reversed is similar.)

Case 1: Assume $\mathbf{u} = \mathbf{0}$; then both sides of $(\dagger)$ are equal to 0.

Case 2: Assume $\mathbf{u} \neq \mathbf{0}$; then for any $c \in \mathbb{C}$,

$$\begin{aligned} 0 &\leq \langle c\mathbf{u} + \mathbf{v}, c\mathbf{u} + \mathbf{v} \rangle \qquad (\ddagger) && \text{[by 45.20(iii)]} \\ &= c\bar{c}\langle \mathbf{u}, \mathbf{u} \rangle + \bar{c}\langle \mathbf{u}, \mathbf{v} \rangle + c\overline{\langle \mathbf{u}, \mathbf{v} \rangle} + \langle \mathbf{v}, \mathbf{v} \rangle && \text{[by 45.20(i) and (ii)]} \end{aligned}$$

In particular, this inequality must hold for $c = -\langle \mathbf{u}, \mathbf{v} \rangle / \langle \mathbf{u}, \mathbf{u} \rangle$; this proves $(\dagger)$. ■

45.28 Theorem (Analogues of 38.12, 38.13, and 45.11 Combined)

For all $\mathbf{u}, \mathbf{v} \in V$ and $c \in \mathbb{C}$,

(i) $0 \leq \|\mathbf{u}\|$; moreover, $\|\mathbf{u}\| = 0$ iff $\mathbf{u} = \mathbf{0}$ [$\| \cdot \|$ is positive definite]
(ii) $\|c\mathbf{u}\| = |c|\,\|\mathbf{u}\|$ [$\| \cdot \|$ is homogeneous]
(iii) $\|\mathbf{u} + \mathbf{v}\| \leq \|\mathbf{u}\| + \|\mathbf{v}\|$ [the Triangle Inequality]

Proof of (iii): We will derive the Triangle Inequality from the Cauchy–Schwarz Inequality. Imitate the proof of 38.13 to conclude that an arbitrary pair of vectors $\mathbf{u}, \mathbf{v} \in V$ satisfies the Triangle Inequality iff $\operatorname{Re}\langle \mathbf{u}, \mathbf{v}\rangle \leq \|\mathbf{u}\|\,\|\mathbf{v}\|$, where Re c denotes the real part of the complex number c. From this, it is clear that if two given vectors $\mathbf{u}, \mathbf{v}$ satisfy the Cauchy–Schwarz Inequality, $|\langle \mathbf{u}, \mathbf{v}\rangle| \leq \|\mathbf{u}\|\,\|\mathbf{v}\|$, then they also satisfy the Triangle Inequality, $\langle \mathbf{u}, \mathbf{v}\rangle \leq \|\mathbf{u}\|\,\|\mathbf{v}\|$.

With these two theorems in hand, it is quite straightforward to generalize the remaining concepts to complex inner product spaces.

45.29 Exercise

Generalize to arbitrary complex inner product spaces the concepts of distance, of angle between vectors, and of orthonormal basis, and develop the analogue of the Gram–Schmidt algorithm.

Problem Set 45

45.1. Go back and do any of the exercises from Section 45 that you may have skipped.

*45.2. For all positive real numbers $t_1, t_2, \ldots, t_n$, the function $\langle \cdot, \cdot \rangle : \mathbb{R}^n \times \mathbb{R}^n \to \mathbb{R}$ defined by

$$\langle \mathbf{u}, \mathbf{v}\rangle = t_1 u_1 v_1 + t_2 u_2 v_2 + \cdots + t_n u_n v_n$$

is an inner product on $\mathbb{R}^n$.

45.3. (i) Verify that the function

$$\langle a_0 + a_1 x + a_2 x^2 + a_3 x^3, b_0 + b_1 x + b_2 x^2 + b_3 x^3 \rangle$$
$$= a_0 b_0 + a_1 b_1 + a_2 b_2 + a_3 b_3$$

is an inner product on $\mathscr{P}_3$.

(ii) Verify that the set $S = \{1 - 2x^2 + 3x^3,\ 2 + x^2,\ 3 - 6x^2 - 5x^3\}$ is orthogonal with respect to this inner product.

(iii) Find a basis for $\mathscr{P}_3$ that is orthonormal with respect to this inner product.

*45.4 (i) Show that the function $\langle \cdot, \cdot \rangle: \mathbb{R}_{n\times n} \times \mathbb{R}_{n\times n} \to \mathbb{R}$ defined by

$$\langle A, B \rangle = \text{trace}(AB^T)$$

is an inner-product on $\mathbb{R}_{n\times n}$
[Recall P3.18 for the definition and properties of the trace function.]

(ii) For the remainder of this question, fix $n = 2$.

Are the vectors $A = \begin{bmatrix} 2 & 1 \\ 0 & 1 \end{bmatrix}$ and $B = \begin{bmatrix} 1 & -2 \\ 2 & 0 \end{bmatrix}$ orthogonal?

(iii) Are the vectors $C = \begin{bmatrix} 1 & 1 \\ 2 & 0 \end{bmatrix}$ and $D = \begin{bmatrix} 2 & 1 \\ -1 & 2 \end{bmatrix}$ orthogonal?

(iv) What is $\|C\|$?

(v) What is $\|C - D\|$?

(vi) What is the cosine of the angle between C and D?

*45.5. (Analogue for $\mathbb{C}$ of Exercise 38.21) Let $\mathbf{u}, \mathbf{v} \in \mathbb{C}^n$. Observe that, ignoring the distinction between $\mathbb{C}$ and $\mathbb{C}_{1\times 1}$, we can write $\mathbf{u} \cdot_1 \mathbf{v} = [\mathbf{u}]^H[\mathbf{v}]$ and $\mathbf{u} \cdot_2 \mathbf{v} = [\mathbf{u}]^T[\bar{\mathbf{v}}]$. Use these observations plus the basic results about complex conjugation (36.2) and the transpose and Hermitian conjugate (3.15 and 36.16) to prove

(i) $A\mathbf{u} \cdot_1 \mathbf{v} = \mathbf{u} \cdot_1 A^H\mathbf{v}$ and $\mathbf{u} \cdot_1 A\mathbf{v} = A^H\mathbf{u} \cdot_1 \mathbf{v}$

(ii) $A\mathbf{u} \cdot_2 \mathbf{v} = \mathbf{u} \cdot_2 A^H\mathbf{v}$ and $\mathbf{u} \cdot_2 A\mathbf{v} = A^H\mathbf{u} \cdot_2 \mathbf{v}$

*45.6. (Analogue of Exercise 38.22 for $A \in \mathbb{C}_{n\times n}$) Prove that

$$A\mathbf{u} \cdot \mathbf{v} = \mathbf{u} \cdot A\mathbf{v} \text{ for all } \mathbf{u}, \mathbf{v} \in \mathbb{C}^n \text{ iff } A = A^H.$$

Here, $\cdot$ stands for either $\cdot_1$ or $\cdot_2$.

45.7. (Analogue of 38.8)

(i) Redo the proof of the Cauchy–Schwarz Inequality (45.26) assuming that $\langle \cdot, \cdot \rangle$ is linear in the first argument and conjugate linear in the second.

(ii) If $\langle \cdot, \cdot \rangle$ is conjugate linear in the first coordinate, then equality holds in the Cauchy–Schwarz Inequality for the vectors $\mathbf{u}$ and $\mathbf{v}$ iff $\mathbf{u} = \mathbf{0}$ or $\mathbf{v} = \mathbf{0}$ or $\mathbf{v} = c\mathbf{u}$, where $c = -\langle \mathbf{u}, \mathbf{v} \rangle / \langle \mathbf{u}, \mathbf{u} \rangle$.

(iii) If $\langle \cdot, \cdot \rangle$ is conjugate linear in the second coordinate, then equality holds in the Cauchy–Schwarz Inequality for the vectors $\mathbf{u}$ and $\mathbf{v}$ iff $\mathbf{u} = \mathbf{0}$ or $\mathbf{v} = \mathbf{0}$ or $\mathbf{v} = c\mathbf{u}$, where $c = -\overline{\langle \mathbf{u}, \mathbf{v} \rangle} / \langle \mathbf{u}, \mathbf{u} \rangle$.

*45.8. As part of Exercise 45.18, you should have obtained the following: Let S be a subset of a finite-dimensional real inner product space, V. Define $S\ perp = S^\perp = \{\mathbf{v} \in V: \langle \mathbf{v}, \mathbf{w} \rangle = 0 \text{ for all } \mathbf{w} \in S\}$.

(i) $S^\perp$ is a subspace of V.

(ii) For any subspace W of V, $V = W \oplus W^{\perp}$.
(iii) For any subspace W of V, $W = W^{\perp\perp}$.

*45.9. The facts in parts (ii) and (iii) of Problem 45.8 are not true for infinite-dimensional spaces. Show that the following are true for a subspace W of an arbitrary inner product space, V:
(i) $W \subset W^{\perp\perp}$.
(ii) $W^{\perp} = W^{\perp\perp\perp}$.

45.10. (Analogue of 38.16)
(i) Show that the Generalized Pythagorean Theorem, as stated in 38.16, is false for $\mathbb{C}^n$. Here is the correct analogue:
(ii) $\mathbf{u} \cdot \mathbf{v} = 0$ iff $\|c_1\mathbf{u} + c_2\mathbf{v}\|^2 = \|c_1\mathbf{u}\|^2 + \|c_2\mathbf{v}\|^2$ for all $c_1, c_2 \in \mathbb{C}$.

45.11. (Analogue of 45.13 and 45.14) Let β be an ordered basis for $\mathbb{C}^n$. For $i = 1$ and 2, define $\langle \cdot , \cdot \rangle_{i_\beta}: \mathbb{C}^n \times \mathbb{C}^n \to \mathbb{C}$ by $\langle \mathbf{u}, \mathbf{v} \rangle_{i_\beta} = (\mathbf{u})_\beta \cdot_i (\mathbf{v})_\beta$. Show that $\langle \cdot , \cdot \rangle_{1_\beta}$ and $\langle \cdot , \cdot \rangle_{2_\beta}$ are (a conjugate pair of) complex inner products on $\mathbb{C}^n$. These are called the *β inner products*. The common norm induced by these inner products (as in 45.24) is denoted by $\| \cdot \|_\beta$ and is called the *β norm* on $\mathbb{C}^n$.

45.12. (Analogue of 45.16) Show that if two complex inner products on V have the same associated norm, then either
(i) they are equal (which happens in the case when they are conjugate linear in the same argument), or
(ii) they are complex conjugates of one another (which happens in case they are conjugate linear in opposite arguments).

45.13. (Analogue of 38.14) Prove for arbitrary complex inner products that the Triangle Inequality implies the Cauchy–Schwarz Inequality.
Hint: You will have to use four instances of the Triangle Inequality to obtain a single instance of the Cauchy–Schwarz Inequality.

45.14. (Analogue for inner product spaces of 43.20) A collection $\{W_1, W_2, \ldots, W_p\}$ of subspaces of an inner product space, V, is *orthogonal* if $\langle \mathbf{w}_i, \mathbf{w}_j \rangle = 0$ whenever $\mathbf{w}_i \in W_i$, $\mathbf{w}_j \in W_j$, and $i \neq j$.
(i) Prove that an orthogonal collection of subspaces is independent (as defined in 43.21). If $W = W_1 \oplus W_2 \oplus \cdots \oplus W_p$ and $\{W_1, W_2, \ldots, W_p\}$ is orthogonal, we say that W is the *orthogonal direct sum* of $W_1, W_2, \ldots, W_p$.
(ii) Prove that a direct sum $W = W_1 \oplus W_2 \oplus \cdots \oplus W_p$ is an orthogonal direct sum iff whenever $\beta_1, \beta_2, \ldots, \beta_p$ are orthonormal bases for $W_1, W_2, \ldots, W_p$, respectively, then $\beta = \beta_1 \cup \beta_2 \cup \cdots \cup \beta_p$ is an orthonormal basis for W.

*45.15. A function $\| \cdot \|: V \to \mathbb{R}$ on a real vector space V is a *norm* if it satisfies the conclusion of Theorem 45.11, that is, if it is positive definite, homogeneous, and satisfies the Triangle Inequality. Theorem 45.11 asserts that every real inner product space is a normed vector space, with the norm defined from the inner product by $\|\mathbf{v}\| = \sqrt{\langle \mathbf{v}, \mathbf{v} \rangle}$. It is important for more advanced work to realize that there are normed vector spaces that are *not* inner product spaces; that is, a norm $\| \bullet \|: V \to \mathbb{R}$ on a vector space V need not arise, as above, from an inner product $\langle \bullet, \bullet \rangle: V \times V \to \mathbb{R}$. To illustrate this point, you are asked in this problem to show that a norm which arises from an inner product must satisfy an identity known as the *Parallelogram Law*. In the problem that follows this one, there is an easy example of a norm that violates the Parallelogram Law and that cannot therefore arise from an inner product.

If $\langle \bullet, \bullet \rangle$ is an inner product on a real vector space V, then its associated norm $\|\mathbf{u}\| = \sqrt{\langle \mathbf{u}, \mathbf{u} \rangle}$ satisfies the Parallelogram Law.

$$\text{For all } \mathbf{u}, \mathbf{v} \in V, \quad \|\mathbf{u} + \mathbf{v}\|^2 + \|\mathbf{u} - \mathbf{v}\|^2 = 2\left(\|\mathbf{u}\|^2 + \|\mathbf{v}\|^2\right).$$

*45.16. The *sup norm*, $\| \bullet \|_\infty$, on $\mathbb{R}^2$ is defined by

$$\|\mathbf{u}\|_\infty = \|(u_1, u_2)\|_\infty = \sup\{|u_1|, |u_2|\} = \text{the larger of } |u_1| \text{ or } |u_2|$$

(i) Prove that $\| \bullet \|_\infty$ is a norm on $\mathbb{R}^2$.
(ii) Show that $\| \bullet \|_\infty$ violates the Parallelogram Law by exhibiting a counterexample.
(iii) Conclude from the previous problem that there is *no* inner product $\langle \bullet, \bullet \rangle_\infty$ on $\mathbb{R}^2$ such that $\|\mathbf{u}\|_\infty = \sqrt{\langle \mathbf{u}, \mathbf{u} \rangle_\infty}$.

45.17. Prove that a real inner product and its associated norm are related by the following identity, known as the *Polarization Identity*.

$$\text{For all } \mathbf{u}, \mathbf{v} \in V, \quad \langle \mathbf{u}, \mathbf{v} \rangle = \tfrac{1}{4}\left(\|\mathbf{u} + \mathbf{v}\|^2 - \|\mathbf{u} - \mathbf{v}\|^2\right).$$

45.18. The previous problem suggests that, given a norm on a real vector space V, we might attempt to use the Polarization Identity to *define* an inner product from which the norm arises. We know from Problem 45.15 that for this to work it is necessary that the norm satisfy the Parallelogram Law. We will now see that this condition is also sufficient.

Prove that if $\| \bullet \|$ is a norm on a real vector space V, and if $\| \bullet \|$ satisfies the Parallelogram Law, then the function $f: V \times V \to \mathbb{R}$ defined by

$$f(\mathbf{u}, \mathbf{v}) = \tfrac{1}{4}\left(\|\mathbf{u} + \mathbf{v}\|^2 - \|\mathbf{u} - \mathbf{v}\|^2\right)$$

is an inner product on V and that $\|\mathbf{u}\| = \sqrt{f(\mathbf{u}, \mathbf{u})}$.

46

A Real $n \times n$ Matrix Is Orthogonally Diagonalizable over $\mathbb{R}$ If and Only If It Is Symmetric

We will begin this section by motivating the concept of orthogonal similarity.

Recall that two $n \times n$ matrices B and C are similar if there exists an invertible matrix P such that $C = P^{-1}BP$. Similar matrices represent the same linear operator with respect to different bases. Typically, similarity is relevant in the context where $T: \mathbb{R}^n \to \mathbb{R}^n$ is a linear operator, β and γ are bases for $\mathbb{R}^n$, $B = {}_\beta[T]_\beta$ and $C = {}_\gamma[T]_\gamma$. Thus

$$C = {}_\gamma[T]_\gamma = {}_\gamma[I]_\beta\, {}_\beta[T]_\beta\, {}_\beta[I]_\gamma = \left({}_\beta[I]_\gamma\right)^{-1} {}_\beta[T]_\beta\, {}_\beta[I]_\gamma$$

In a given application, the purpose of a change of basis is to find one, say γ, with respect to which the behavior of a given linear operator T has the simplest possible description; in ideal circumstances, ${}_\gamma[T]_\gamma$ will be a diagonal matrix.

Suppose, however, that the application is one in which it is important to be able to easily compute the Euclidean norms of vectors and angles between vectors. In such applications, it is essential to consider only those changes of basis that "preserve" these features.

When the standard coordinate system is used, these are obtained by the familiar formulas

(i) $\|\mathbf{v}\| = \sqrt{\left(v_1^2 + v_2^2 + \cdots + v_n^2\right)}$
(ii) $\cos\theta = \mathbf{u} \cdot \mathbf{v}/\|\mathbf{u}\|\,\|\mathbf{v}\|$ if θ is the angle between $\mathbf{u}$ and $\mathbf{v}$.

A change of basis may provide us with a simple description of T but at the cost of greatly complicating the computation of lengths and angles.

46.1 Example (Example 33.1 Revisited)

In Example 33.1, we saw that the linear operator $T(\mathbf{x}) = T(x_1, x_2) = (-2x_1 - 2x_2, -5x_1 + x_2)$ is represented with respect to the basis $\beta = \left((1,1),\left(-\frac{2}{5},1\right)\right)$ by the diagonal matrix $\begin{bmatrix} -4 & 0 \\ 0 & 3 \end{bmatrix}$.

To illustrate the ideas relating to similarity, we used, in that example, the vectors $\mathbf{u} = (\frac{1}{5}, 3)$ and $\mathbf{v} = (-1, -2)$. Let us reuse them in the present context.

It is easy to compute

$$\|\mathbf{u}\| = \sqrt{(1/5)^2 + (3)^2} = \frac{\sqrt{226}}{5}$$

$$\|\mathbf{v}\| = \sqrt{(-1)^2 + (-2)^2} = \sqrt{5}$$

$$\cos\theta = \frac{(\frac{1}{5})(-1) + (3)(-2)}{\sqrt{5}\,\sqrt{226}\,/5} \approx -0.922$$

where θ is the angle between $\mathbf{u}$ and $\mathbf{v}$.

It is *not* true, however, that the length of $\mathbf{v}$ is the square root of the sum of the squares of its β coordinates. For example, $(\mathbf{v})_\beta = (-\frac{9}{7}, -\frac{5}{7})$ (you should verify this) and $\sqrt{\left(-\frac{9}{7}\right)^2 + \left(-\frac{5}{7}\right)^2} = \sqrt{106}\,/7 \neq \sqrt{5}$. Nor is it true that $\cos\theta = (\mathbf{u})_\beta \cdot (\mathbf{v})_\beta / \|\mathbf{u}\|_\beta \|\mathbf{v}\|_\beta$.

We may not apply the usual formulas for length and angle to the β-coordinates of the vectors $\mathbf{u}$ and $\mathbf{v}$. Indeed, it would be a fair bit of work (which we leave as an illustrative exercise) to compute the angle between $\mathbf{u}$ and $\mathbf{v}$ given only that $\beta = \left((1,1),\left(-\frac{2}{5},1\right)\right)$, that $(\mathbf{u})_\beta = (1,2)$, and that $(\mathbf{v})_\beta = (-\frac{9}{7}, -\frac{5}{7})$.

We have seen in Section 40 that it is precisely the orthogonal operators that preserve the dot product and Euclidean norm. Thus we are motivated to make the following definition.

46.2 Definition

The $n \times n$ matrix B is *orthogonally similar* to the $n \times n$ matrix A if there exists an orthogonal matrix P such that $B = P^{-1}AP = P^TAP$.

46.3 Exercise

Verify that orthogonal similarity is reflexive (A is orthogonally similar to itself), symmetric (if B is orthogonally similar to A, then A is orthogonally similar to B), and transitive (if B is orthogonally similar to A and if C is orthogonally similar to B, then C is orthogonally similar to A).

When this concept is exploited, P is a change-of-basis matrix ${}_\gamma[I]_\beta$; so a genuine understanding of orthogonal similarity will require analyzing what it means for a change-of-basis matrix ${}_\gamma[I]_\beta$ to be orthogonal.

46.4 Theorem

A real change-of-basis matrix, ${}_\gamma[I]_\beta$, is orthogonal iff the norms induced on $\mathbb{R}^n$ by β and γ (see Definition 45.11) are the same (i.e., iff $\|\mathbf{u}\|_\gamma = \|\mathbf{u}\|_\beta$ for all $\mathbf{u} \in \mathbb{R}^n$).

Proof: $(\rightarrow)$

$$\begin{aligned}
\|\mathbf{u}\|_\gamma^2 &= (\mathbf{u})_\gamma \cdot (\mathbf{u})_\gamma && \text{[by definition]}\\
&= [\mathbf{u}]_\gamma^T\,[\mathbf{u}]_\gamma && \text{[Problem 38.4]}\\
&= \left({}_\gamma[I]_\beta\,[\mathbf{u}]_\beta\right)^T\left({}_\gamma[I]_\beta\,[\mathbf{u}]_\beta\right) && \text{[31.13]}\\
&= [\mathbf{u}]_\beta^T\,{}_\gamma[I]_\beta^T\,{}_\gamma[I]_\beta\,[\mathbf{u}]_\beta && \text{[3.15(iv)]}\\
&= [\mathbf{u}]_\beta^T\,{}_\gamma[I]_\beta^{-1}\,{}_\gamma[I]_\beta\,[\mathbf{u}]_\beta && \left[\text{orthogonality of } {}_\gamma[I]_\beta\right]\\
&= [\mathbf{u}]_\beta^T\,[\mathbf{u}]_\beta &&\\
&= (\mathbf{u})_\beta \cdot (\mathbf{u})_\beta && \text{[Problem 38.4]}\\
&= \|\mathbf{u}\|_\beta^2 && \text{[by definition]}
\end{aligned}$$

$(\leftarrow)$ Recall from Section 31 that ${}_\gamma[I]_\beta = [[\mathbf{u}_1]_\gamma \,\vdots\, [\mathbf{u}_2]_\gamma \,\vdots\, \cdots \,\vdots\, [\mathbf{u}_n]_\gamma]$, where $\beta = (\mathbf{u}_1, \mathbf{u}_2, \ldots, \mathbf{u}_n)$. To prove that ${}_\gamma[I]_\beta$ is orthogonal, it suffices by 40.7 to show that its columns form an orthonormal set. The hypothesis $\|\cdot\|_\beta = \|\cdot\|_\gamma$ implies, by Theorem 45.16, that for all $\mathbf{u}, \mathbf{v} \in \mathbb{R}^n$, $(\mathbf{u})_\gamma \cdot (\mathbf{v})_\gamma = (\mathbf{u})_\beta \cdot (\mathbf{v})_\beta$. Using this, we have for all i and j that $(\mathbf{u}_i)_\gamma \cdot (\mathbf{u}_j)_\gamma = (\mathbf{u}_i)_\beta \cdot (\mathbf{u}_j)_\beta = (\mathbf{e}_i^n) \cdot (\mathbf{e}_j^n) = \delta_{ij}$. ■

The special case of this result in which one of the two bases is the standard basis is worth mentioning explicitly.

46.5 Corollary

The following are equivalent:

(i) β is an orthonormal basis for $\mathbb{R}^n$.
(ii) The transition matrices ${}_{\alpha^n}[I]_\beta$ and ${}_\beta[I]_{\alpha^n}$ are orthogonal.
(iii) $\|\cdot\|_\beta = \|\cdot\|_E$ (i.e., the β norm is equal to the Euclidean norm).

Proof: The equivalence of (i) and (ii) was established in 40.10. The equivalence of (ii) and (iii) follows from 46.4. ■

This corollary pinpoints one of the essential features of orthonormal bases: when β is orthonormal, it is still possible to compute the Euclidean norm of a vector and the angle between vectors *using the old familiar formulas* applied to the β-coordinates of the vectors rather than their standard coordinates.

Thus it is a very desirable situation when the standard matrix of a linear operator is similar to a diagonal matrix via an orthogonal change of basis.

46.6 Definition

A matrix $A \in \mathbb{R}_{n\times n}$ is *orthogonally diagonalizable over* $\mathbb{R}$ iff there exist in $\mathbb{R}_{n\times n}$ a diagonal matrix D and an orthogonal matrix P such that $D = P^{-1}AP = P^{T}AP$.

Here is an easy theorem that is the analogue for orthogonal diagonalizability of Theorem 34.7.

46.7 Theorem

A matrix $A \in \mathbb{R}_{n\times n}$ is orthogonally diagonalizable over $\mathbb{R}$ iff there exists an orthonormal basis for $\mathbb{R}^n$ consisting of eigenvectors of A.

Proof: Combine 34.7 with 46.6. ■

The title of this section is a statement of the main theorem. This theorem precisely characterizes the class of real matrices that are orthogonally diagonalizable over $\mathbb{R}$.

46.8 Theorem

The following are equivalent for $A \in \mathbb{R}_{n\times n}$:

(i) A is orthogonally diagonalizable over $\mathbb{R}$.
(ii) A is symmetric.

We will devote the rest of this section to a discussion and proof of this theorem. Let us dispense immediately with the trivial direction.

46.9 Proof of (i) → (ii) from 46.8

Suppose that $A \in \mathbb{R}_{n\times n}$, that $D \in \mathbb{R}_{n\times n}$ is diagonal, that $P \in \mathbb{R}_{n\times n}$ is orthogonal, and that $D = P^{-1}AP = P^{T}AP$. Then $A = PDP^{-1}$, so

$$\begin{aligned} A^T &= (PDP^{-1})^T \\ &= (P^{-1})^T D^T P^T && [3.15(iv)] \\ &= (P^T)^T DP^{-1} \text{ [since } P^{-1} = P^T \text{ and } D^T = D] \\ &= PDP^{-1} && [3.15(i)] \\ &= A \end{aligned}$$

which shows that A is symmetric. ■

Because the proof of the reverse implication is lengthy, we will provide an outline. Each step in the proof will then be discussed separately, as appropriate. Before beginning, however, we should stress that, from the hypothesis that A is symmetric, we are getting a two-part conclusion: that A is diagonalizable over $\mathbb{R}$ and, moreover, that A is orthogonally diagonalizable. It is the first part of this conclusion, the diagonalizability over $\mathbb{R}$ of A, that is the more significant one; the additional fact that there is an *orthogonal* P that diagonalizes A comes, almost, as a bonus.

46.10 Outline of the Proof of (ii) → (i) from 46.8

Recall, from Theorem 37.4, that there are two possible obstructions to diagonalizability over $\mathbb{R}$.

Step 1. Although $A \in \mathbb{R}_{n\times n}$, we will view A as a complex matrix. Viewing $A \in \mathbb{C}_{n\times n}$ allows us to apply the Fundamental Theorem of Algebra (36.7) to conclude that $\chi(A)$ has n-many roots in $\mathbb{C}$ (i.e., A has n-many complex eigenvalues). What we need to know, however, is that A has n-many real eigenvalues. This is the purpose of the next step, which uses the symmetry of A.

Step 2. Prove that all the eigenvalues of a real symmetric matrix are real.
These first two steps remove the first obstruction to diagonalizability over $\mathbb{R}$.

Step 3. (This is the hard one!) Prove that if r is an eigenvalue of a symmetric matrix then its geometric multiplicity is equal to its algebraic multiplicity.
This step removes the second obstruction and is exactly what is needed to conclude, from Theorem 37.4, that A is diagonalizable over $\mathbb{R}$.

Step 4. At this stage, we have, according to Theorem 34.7, a basis for $\mathbb{R}^n$ consisting of eigenvectors of A; but we still have to produce an orthonormal basis of eigenvectors of A. The naive approach—just apply the Gram–Schmidt algorithm—fails; but it turns out that we can use the naive approach *within each eigenspace separately* to obtain orthonormal bases for each of the eigenspaces. The fact that these bases can then be combined to form an orthonormal basis for the whole of $\mathbb{R}^n$ requires still another argument that relies on the symmetry of A.

Now we will see the details. Step 1 requires no further explanation. The following theorem completes step 2.

46.11 Theorem

If $A \in \mathbb{C}_{n\times n}$ is Hermitian (Definition 36.17) (in particular, if $A \in \mathbb{R}_{n\times n}$ and is symmetric), then the eigenvalues of A are real.

Proof: Suppose that λ is an eigenvalue of the Hermitian matrix A. Let $\mathbf{u}$ be any eigenvector from the eigenspace of λ; then

$$
\begin{aligned}
\lambda(\mathbf{u} \cdot_2 \mathbf{u}) &= \lambda\mathbf{u} \cdot_2 \mathbf{u} && [\cdot_2 \text{ is linear in the first variable}] \\
&= A\mathbf{u} \cdot_2 \mathbf{u} \\
&= \mathbf{u} \cdot_2 A^H\mathbf{u} && [\text{by Problem 45.5}] \\
&= \mathbf{u} \cdot_2 A\mathbf{u} && [\text{since A is Hermitian}] \\
&= \mathbf{u} \cdot_2 \lambda\mathbf{u} \\
&= \bar{\lambda}(\mathbf{u} \cdot_2 \mathbf{u}). && [\cdot_2 \text{ is conjugate linear in the second variable}]
\end{aligned}
$$

Since $\mathbf{u} \cdot_2 \mathbf{u} \neq 0$, we may conclude that $\lambda = \bar{\lambda}$. ■

For step 3, we first recall Theorem 37.2(ii): if $A \in \mathbb{R}_{n\times n}$ and r is an eigenvalue of A, then the geometric multiplicity of r is less than or equal to the algebraic multiplicity of r. It is essential to reread the proof of that theorem because we are about to modify it slightly, strengthening the hypothesis to include the symmetry of A, while strengthening the conclusion to obtain $=$ in place of $\leq$.

46.12 Theorem

Suppose that $A \in \mathbb{R}_{n\times n}$ is symmetric and r is an eigenvalue of A; then

$$(\text{geometric multiplicity of } r) = (\text{algebraic multiplicity of } r).$$

Proof: Suppose dim $E_r = k$. We already know, from 37.2(ii), that the multiplicity of r as a root of $\chi(\mathrm{A})$ is at least k. The problem is to show that it is exactly k. We proceed as in the proof of 37.2(ii), but we insist on working this time with orthonormal bases. Thus let $(\mathbf{v}_1, \mathbf{v}_2, \ldots, \mathbf{v}_k)$ be an orthonormal basis for E_r and extend this to an orthonormal basis $\beta = (\mathbf{v}_1, \ldots, \mathbf{v}_k, \mathbf{v}_{k+1}, \ldots, \mathbf{v}_n)$ for $\mathbb{R}^n$. As before, we conclude that

$$_\beta[T_\mathrm{A}]_\beta = \left[\begin{array}{c|c} r\mathrm{I}_k & \mathrm{B} \\ \hline 0_{(n-k)\times k} & \mathrm{C} \end{array}\right]$$

and that $\chi(\mathrm{A}) = (\lambda - r)^k \chi(\mathrm{C})$. To show that the algebraic multiplicity of r is exactly equal to k, we need to prove that r *cannot* be a root of $\chi(\mathrm{C})$, that is, that r is *not* an eigenvalue of the $(n-k) \times (n-k)$ matrix C. This is equivalent to showing that the matrix $r\mathrm{I}_{n-k} - \mathrm{C}$ has full rank, that is, $n - k$.

Because β is an orthonormal basis, the matrix $\mathrm{P} = _{\alpha^n}[I]_\beta$ is orthogonal. Thus $_\beta[T_\mathrm{A}]_\beta = \mathrm{P}^{-1}\mathrm{AP} = \mathrm{P}^T\mathrm{AP}$ is symmetric since it is orthogonally similar to the symmetric matrix, A. Hence $\mathrm{B} = 0_{k\times(n-k)}$.

By hypothesis, $k = \dim E_r = \text{nullity}\,(r\mathrm{I}_n - \mathrm{A})$. Hence, $\text{rank}(r\mathrm{I}_n - \mathrm{A}) = n - k$.

Since similar matrices have the same rank, we obtain $\text{rank}(\mathrm{N}) = n - k$, where $\mathrm{N} = \mathrm{P}^{-1}(r\mathrm{I}_n - \mathrm{A})\mathrm{P}$. But

$$\begin{aligned} \mathrm{N} &= \mathrm{P}^{-1}(r\mathrm{I}_n - \mathrm{A})\mathrm{P} \\ &= r\mathrm{P}^{-1}\mathrm{I}_n\mathrm{P} - \mathrm{P}^{-1}\mathrm{AP} \\ &= r\mathrm{I}_n - {}_\beta[T_\mathrm{A}]_\beta \\ &= r\mathrm{I}_n - \left[\begin{array}{c|c} r\mathrm{I}_k & 0_{k\times(n-k)} \\ \hline 0_{(n-k)\times k} & \mathrm{C} \end{array}\right] = \left[\begin{array}{c|c} 0_{k\times k} & 0_{k\times(n-k)} \\ \hline 0_{(n-k)\times k} & r\mathrm{I}_{n-k} - \mathrm{C} \end{array}\right] \end{aligned}$$

Thus $n - k = \text{rank}(\mathrm{N}) = \text{rank}(r\mathrm{I}_{n-k} - \mathrm{C})$, as desired. ■

We turn, finally, to step 4. The first three steps allow us to conclude, from Theorem 37.4, that A is diagonalizable over $\mathbb{R}$ and that there consequently exists a basis for $\mathbb{R}^n$ consisting of eigenvectors of A. In the initial outline of the proof, we mentioned that simply applying the Gram–Schmidt algorithm to this basis would not suffice to yield an orthonormal basis of eigenvectors of A. The problem is that the output of the algorithm, although it will be an orthonormal basis for $\mathbb{R}^n$, may no longer consist of eigenvectors of A.

46.13 Exercise

Give an example of a diagonalizable $n \times n$ matrix B and a basis β for $\mathbb{R}^n$ consisting of eigenvectors of B for which one or more of the vectors in the orthonormal basis $\mathrm{GS}(\beta)$ is not an eigenvector of B.
Hint: Keep the geometric interpretations of these concepts in mind.

46.14 Exercise

Prove that if r is a fixed eigenvalue of B and if $\gamma = (\mathbf{v}_1, \mathbf{v}_2, \ldots, \mathbf{v}_k)$ is a basis for the subspace E_r, then $\mathrm{GS}(\gamma)$ is an orthonormal basis for E_r.

As a result of the previous exercise, we may apply the Gram–Schmidt algorithm separately, within each eigenspace, to obtain, where $r_1, r_2, \ldots, r_p$ are the distinct eigenvalues of A, orthonormal bases $\beta_1, \beta_2, \ldots, \beta_p$ for each of the corresponding eigenspaces. Since $\mathbb{R}^n$ is the direct sum of the eigenspaces, $\gamma = \beta_1 \cup \beta_2 \cup \cdots \cup \beta_p$ is a basis for $\mathbb{R}^n$ by 43.20. Moreover, all the vectors in γ are eigenvectors of A. In general, the basis, γ, produced by this construction will not be orthonormal: all we are guaranteed by the construction is that the vectors in γ are unit vectors, and that $\mathbf{u} \cdot \mathbf{v} = 0$ if $\mathbf{u}$ and $\mathbf{v}$ come from the *same* β_i; fortunately, when A is symmetric, the finishing touch is provided by the next theorem, which guarantees that eigenvectors from distinct eigenspaces of a symmetric matrix are orthogonal.

46.15 Theorem

Assume that $A \in \mathbb{R}_{n \times n}$ is symmetric; we already know from 46.12 that the eigenvalues of A are real, so assume that $r_1 \neq r_2 \in \mathbb{R}$, that $\mathbf{u}, \mathbf{v} \in \mathbb{R}^n$, that $A\mathbf{u} = r_1\mathbf{u}$, and that $A\mathbf{v} = r_2\mathbf{v}$; then $\mathbf{u} \cdot \mathbf{v} = 0$.

Proof:

$$
\begin{aligned}
r_1(\mathbf{u} \cdot \mathbf{v}) &= (r_1\mathbf{u}) \cdot \mathbf{v} \\
&= (A\mathbf{u}) \cdot \mathbf{v} \\
&= (\mathbf{u}) \cdot (A^T\mathbf{v}) && \text{[Exercise 38.21]} \\
&= (\mathbf{u}) \cdot (A\mathbf{v}) && \text{[A is symmetric]} \\
&= \mathbf{u} \cdot (r_2\mathbf{v}) \\
&= r_2(\mathbf{u} \cdot \mathbf{v})
\end{aligned}
$$

Thus, $(r_1 - r_2)(\mathbf{u} \cdot \mathbf{v}) = 0$, and since $r_1 \neq r_2$, $\mathbf{u} \cdot \mathbf{v} = 0$. ∎

The analogues of all the preceding results for a complex matrix $A \in \mathbb{C}_{n \times n}$ are too important and a bit too difficult to leave for the problem sets. We will therefore close this section with a brief treatment. If a proof differs substantially from that of the corresponding result for $\mathbb{R}_{n \times n}$, it will be given in full. Omitted proofs are intended as exercises; usually, these are straightforward modifications of the corresponding proof for $\mathbb{R}_{n \times n}$.

As the next few exercises show, the natural analogue for $A \in \mathbb{C}_{n \times n}$ of orthogonality for real matrices is the property $A^{-1} = A^H$.

46.16 Exercise (Analogue of 40.7)

Let $A \in \mathbb{C}_{n\times n}$. Prove that

(i) The rows of A are an orthonormal set with respect to $\bullet_2$ iff $AA^H = I_n$.
(ii) The columns of A are an orthonormal set with respect to $\bullet_1$ iff $A^HA = I_n$.

46.17 Corollary (Analogue of 40.8)

Prove that the following are equivalent for $A \in \mathbb{C}_{n\times n}$:

(i) The rows of A are orthonormal with respect to both of the Euclidean inner products $\bullet_1$ and $\bullet_2$ on $\mathbb{C}^n$.
(ii) A is invertible and $A^{-1} = A^H$.
(iii) The columns of A are orthonormal with respect to both of the Euclidean inner products $\bullet_1$ and $\bullet_2$ on $\mathbb{C}^n$.

The next definition gives a name to this property.

46.18 Definition (Analogue of 40.9)

A matrix $A \in \mathbb{C}_{n\times n}$ satisfying any one (and hence all) of the properties from 46.17 is called a *unitary matrix.*

46.19 Exercise (Analogue of 40.10)

(i) If A is a unitary matrix, then so are A^{-1} and A^H.
(ii) The product of two unitary matrices is a unitary matrix.
(iii) A transition matrix ${}_{\alpha^n}[I]_\beta$ from a basis β to the standard basis α^n for $\mathbb{C}^n$ is unitary iff β is an orthonormal basis for $\mathbb{C}^n$.
(iv) If β and γ are both orthonormal bases for $\mathbb{C}^n$, then the transition matrix ${}_\gamma[I]_\beta$ is unitary.

46.20 Theorem (Analogue of 40.11)

The following are equivalent for any linear operator $T:\mathbb{C}^n \to \mathbb{C}^n$.

(i) T preserves the Euclidean norm; that is, $\|T(\mathbf{u})\|_E = \|\mathbf{u}\|_E$ for all $\mathbf{u} \in \mathbb{C}^n$.
(ii) The standard matrix for T is a unitary matrix.
(iii) If β is any orthonormal basis for $\mathbb{C}^n$, then ${}_\beta[T]_\beta$ is a unitary matrix.

46.21 Definition (Analogue of 40.12)

A linear operator $T:\mathbb{C}^n \to \mathbb{C}$ that preserves the Euclidean norm is called a *unitary operator.*

The ideas that motivated the concept of orthogonal similarity for real matrices are equally valid in the complex domain and give rise to the following definition.

46.22 Definition (Analogue of 46.2)

The matrix $B \in \mathbb{C}_{n\times n}$ is *unitarily similar* to the matrix $A \in \mathbb{C}_{n\times n}$ if there exists a unitary matrix P such that $B = P^{-1}AP = P^{H}AP$.

46.23 Exercise (Analogue of 46.3)

Verify that unitary similarity is reflexive, symmetric, and transitive.

46.24 Theorem (Analogue of 46.4)

A complex change-of-basis matrix, ${}_{\gamma}[I]_{\beta}$, is unitary iff the norms induced on $\mathbb{C}^n$ by β and γ (see Problem 45.8) are the same.

46.25 Corollary (Analogue of 46.5)

The following are equivalent:

(i) β is an orthonormal basis for $\mathbb{C}^n$.
(ii) The transition matrices ${}_{\alpha^n}[I]_{\beta}$ and ${}_{\beta}[I]_{\alpha^n}$ are unitary.
(iii) $\|\cdot\|_{\beta} = \|\cdot\|_{E}$ (i.e., the β norm is equal to the Euclidean norm on $\mathbb{C}^n$).

46.26 Definition (Analogue of 46.6)

A matrix $A \in \mathbb{C}_{n\times n}$ is *unitarily diagonalizable over* $\mathbb{C}$ if there exist in $\mathbb{C}_{n\times n}$ a diagonal matrix D and a unitary matrix P such that $D = P^{-1}AP = P^{H}AP$.

46.27 Theorem (Analogue of 46.7)

A matrix $A \in \mathbb{C}_{n\times n}$ is unitarily diagonalizable over $\mathbb{C}$ iff there exists an orthonormal basis for $\mathbb{C}^n$ consisting of eigenvectors of A.

At last, we come to the main problem: characterize the class of complex $n \times n$ matrices that are unitarily diagonalizable over $\mathbb{C}$. It is very tempting to conjecture, by analogy with 46.8, that A is unitarily diagonalizable over $\mathbb{C}$ iff A is Hermitian. (This conjecture is so tempting that it can be found in print stated as a theorem.) This statement is, however, false.

Here is a simple counterexample. The matrix $\begin{bmatrix} i & 0 \\ 0 & i \end{bmatrix}$ is unitarily diagonalizable over $\mathbb{C}$; indeed, it is already itself a diagonal matrix. It is not Hermitian since the entries on the main diagonal of a Hermitian matrix must be real.

Let us reexamine the proof that a real matrix which is orthogonally diagonalizable over $\mathbb{R}$ must be symmetric (46.9) to see where it breaks down when we attempt to use it to prove the false conjecture. Suppose that $A \in \mathbb{C}_{n \times n}$, that $D \in \mathbb{C}_{n \times n}$ is diagonal, that $P \in \mathbb{C}_{n \times n}$ is unitary, and that $D = P^{-1}AP = P^{H}AP$. An attempt to show that $A^{H} = A$ by imitating the argument from 46.9 that $A^{T} = A$ fails because we do not have $D^{H} = D$. For a diagonal matrix $D \in \mathbb{R}_{n \times n}$, we have $D^{T} = D$. But for a diagonal matrix $D \in \mathbb{C}_{n \times n}$, we do not, in general, have $D^{H} = D$; this happens iff all the entries on the main diagonal of D are real.

The correct analogue of Theorem 46.8 requires a concept that is slightly more general than that of being Hermitian.

46.28 Definition

A matrix $A \in \mathbb{C}_{n \times n}$ is *normal* iff $A^{H}A = AA^{H}$ (i.e., iff A commutes with its Hermitian conjugate).

It is clear that Hermitian, skew-Hermitian, and unitary matrices are normal. Here, then, is the correct analogue of Theorem 46.8.

46.29 Theorem (Analogue of 46.8)

The following are equivalent for $A \in \mathbb{C}_{n \times n}$:

(i) A is unitarily diagonalizable over $\mathbb{C}$.
(ii) A is normal.

As before, let us first dispense with the proof of this in the easy direction.

46.30 Proof of (i) → (ii) from 46.29

Suppose that A, D, and P are complex $n \times n$ matrices, that D is diagonal, that P is unitary, and that $D = P^{-1}AP = P^{H}AP$. Then $A = PDP^{-1} = PDP^{H}$

and $A^H = (PDP^H)^H = PD^HP^H$, so

$$\begin{aligned} A^HA &= (PD^HP^H)(PDP^H) \\ &= PD^HDP^H && [\text{associative law plus fact that } P^{-1} = P^H] \\ &= PDD^HP^H && [\text{any two diagonal matrices commute}] \\ &= (PDP^{-1})(PD^HP^H) && [\text{associative law plus fact that } P^{-1}P = I_n] \\ &= AA^H \end{aligned}$$

■

The proof of 46.29 in the difficult direction follows the same outline as the proof that real symmetric matrices are orthogonally diagonalizable over ℝ (see 46.10). We need to generalize some of the steps, replacing the assumption that A is a real, symmetric matrix with the assumption that A is a complex, normal matrix.

Step 1 is the same. Since we are working over ℂ, the first obstruction to diagonalizability does not arise. The characteristic polynomial $\chi(A)$ has n-many roots in ℂ by the Fundamental Theorem of Algebra.

Step 2 is no longer required, nor even possible; we are not insisting, nor may we insist, that the eigenvalues of A be real.

Step 3 is, once again, the difficult one.

To complete step 3, we will need the following result, which generalizes the fact for real matrices that if B is orthogonally similar to a symmetric matrix A, then B is symmetric.

46.31 Exercise

Suppose that $A \in \mathbb{C}_{n\times n}$ is normal, that $P \in \mathbb{C}_{n\times n}$ is unitary, and that $B = P^{-1}AP = P^HAP$; show that B is normal.

46.32 Theorem (Analogue of 46.12)

Suppose that $A \in \mathbb{C}_{n\times n}$ is normal and that $c \in \mathbb{C}$ is an eigenvalue of A; then

$$(\text{geometric multiplicity of } c) = (\text{algebraic multiplicity of } c).$$

Proof: Begin by rereading the proof of 46.12, for we are about to modify it slightly. Suppose $\dim E_c = k$, let $(\mathbf{v}_1, \mathbf{v}_2, \ldots, \mathbf{v}_k)$ be an orthonormal basis for E_c, and extend this to an orthonormal basis $(\mathbf{v}_1, \ldots, \mathbf{v}_k, \mathbf{v}_{k+1}, \ldots, \mathbf{v}_n)$ for $\mathbb{C}^n$. Conclude, as before, that

$$M = {}_\beta[T_A]_\beta = \left[\begin{array}{c|c} cI_k & B \\ \hline 0_{(n-k)\times k} & C \end{array}\right]$$

and that $\chi(A) = \chi(M) = (\lambda - c)^k\chi(C)$. If we can show that $B = 0_{k\times(n-k)}$, then we obtain the desired conclusion, that c is *not* a root of $\chi(C)$, exactly as in the proof of 46.12. So it remains to show that $B = 0_{k\times(n-k)}$.

In the earlier proof, we showed this using the symmetry of A and the orthogonality of P. This time we have that A is normal, by assumption, and that $P = {}_{\alpha^n}[I]_\beta$ is unitary, by 46.25. Now

$$M = {}_\beta[T_A]_\beta = \left[\begin{array}{c|c} cI_k & B \\ \hline 0_{(n-k)\times k} & C \end{array}\right] \quad \text{so} \quad M^H = \left[\begin{array}{c|c} \bar{c}I_k & 0_{k\times(n-k)} \\ \hline B^H & C^H \end{array}\right]$$

Also, since $M = P^{-1}AP = P^HAP$, M is normal by Exercise 46.31. Thus $M^HM = MM^H$.

The desired conclusion, $B = 0_{k\times(n-k)}$, is equivalent to the statement that $m_{ij} = 0$ for $1 \le i \le k \lneq j \le n$.

If i satisfies $1 \le i \le k$, then the $(i, i)^{\text{th}}$ entry in MM^H is $c\bar{c} + m_{ik+1}\overline{m}_{ik+1} + m_{ik+2}\overline{m}_{ik+2} + \cdots + m_{in}\overline{m}_{in}$, while the $(i, i)^{\text{th}}$ entry in M^HM is $\bar{c}c$.

Because $MM^H = M^HM$, these two expressions must be equal; so we conclude that

$$0 = \sum_{j=k+1}^{j=n} m_{ij}\overline{m}_{ij} = \sum_{j=k+1}^{j=n} |m_{ij}|^2$$

and so $m_{ik+1} = m_{ik+2} = \cdots = m_{in} = 0$, as desired. ■

Step 4 requires that we overcome the same problem we had before. We can apply the Gram–Schmidt algorithm to obtain orthonormal bases for each of the eigenspaces separately. The union of these bases will not be an orthonormal basis for $\mathbb{C}^n$ unless the eigenvectors from distinct eigenspaces are orthogonal. This is not true in general, but it is true when A is normal, as we see in part (iii) of the next theorem.

46.33 Theorem (Generalization of 46.11 and 46.15)

Let $A \in \mathbb{C}_{n\times n}$ and $\lambda \in \mathbb{C}$.

(i) If A is normal, then $\|A\mathbf{u}\| = \|A^H\mathbf{u}\|$ for all $\mathbf{u} \in \mathbb{C}^n$.
(ii) If A is normal, then for all $\mathbf{u} \in \mathbb{C}^n$, $A\mathbf{u} = \lambda\mathbf{u}$ iff $A^H\mathbf{u} = \bar{\lambda}\mathbf{u}$.
(iii) For all $\mathbf{u}, \mathbf{v} \in \mathbb{C}^n$, if A is normal, if $c_1 \neq c_2 \in \mathbb{C}$, if $A\mathbf{u} = c_1\mathbf{u}$, and if $A\mathbf{v} = c_2\mathbf{v}$, then $\mathbf{u} \cdot \mathbf{v} = 0$.

Proof of (i):
$$\begin{aligned} \|A\mathbf{u}\|^2 &= (A\mathbf{u}) \cdot (A\mathbf{u}) \\ &= (A^HA\mathbf{u}) \cdot \mathbf{u} && \text{[Problem 45.5]} \\ &= (AA^H\mathbf{u}) \cdot \mathbf{u} && \text{[A is normal]} \\ &= (A^H\mathbf{u}) \cdot (A^H\mathbf{u}) && \text{[Problem 45.5]} \\ &= \|A^H\mathbf{u}\|^2 \end{aligned}$$

We should mention in passing that the converse of 46.33(i) is also true; but we do not need it at this point. Moreover, it will follow easily from a later result so we postpone it.

Proof of (ii): It is easy to check that if A is normal then so is $\lambda I_n - A$. So we may apply the result from part (i) to $\lambda I_n - A$ to conclude that

$$\|(\lambda I_n - A)\mathbf{u}\| = \|(\bar{\lambda} I_n - A^H)\mathbf{u}\| \quad \text{for all } \mathbf{u} \in \mathbb{C}^n$$

From this, (ii) follows easily.

Proof of (iii): It makes no difference which of $\cdot_1$ or $\cdot_2$ we use.

$$\begin{aligned} c_1(\mathbf{u} \cdot_2 \mathbf{v}) &= (c_1\mathbf{u}) \cdot_2 \mathbf{v} \\ &= (A\mathbf{u}) \cdot_2 \mathbf{v} \\ &= \mathbf{u} \cdot_2 (A^H\mathbf{v}) && [\text{Problem 45.5}] \\ &= \mathbf{u} \cdot (\bar{c}_2\mathbf{v}) \\ &= c_2(\mathbf{u} \cdot_2 \mathbf{v}) && [\text{by part (ii) since } A\mathbf{v} = c_2\mathbf{v}] \end{aligned}$$

Thus $(c_1 - c_2)(\mathbf{u} \cdot_2 \mathbf{v}) = 0$, and since $c_1 \neq c_2$, $\mathbf{u} \cdot_2 \mathbf{v} = 0$. ■

Problem Set 46

46.1. Go back and do any of the exercises from Section 46 that you may have skipped.

*46.2. Explain precisely how Theorem 46.11 is a special case of Theorem 46.33(ii).

*46.3. Let $A \in \mathbb{C}_{n\times n}$. Prove that A is Hermitian iff A is normal and all its eigenvalues are real.

*46.4. *True* or *false*: If $A \in \mathbb{R}_{n\times n}$ and if $AA^T = A^TA$, then $A = A^T$?

*46.5. A matrix $A \in \mathbb{R}_{n\times n}$ is *idempotent* iff $A^2 = A$. Show that if A is symmetric and idempotent, then rank A = trace A.

46.6. (Analogue of 38.14) Let $\langle \cdot, \cdot \rangle$ be a complex inner product and let $\| \cdot \|$ be its associated norm. Prove that if $\| \cdot \|$ satisfies the Triangle Inequality, then $\langle \cdot, \cdot \rangle$ satisfies the Cauchy-Schwarz Inequality.

*46.7. Explain, on theoretical grounds alone, why there must be at least one real value of λ for which the matrix $\begin{bmatrix} \lambda - 3 & 4 \\ 4 & \lambda - 5 \end{bmatrix}$ is not invertible. [No numerical calculations are permitted.]

46.8. Find an orthonormal basis for $\mathbb{R}^3$ consisting of eigenvectors of the matrix

$$\begin{bmatrix} -7/3 & -2/3 & 4/3 \\ -2/3 & -7/3 & -4/3 \\ 4/3 & -4/3 & -1/3 \end{bmatrix}.$$

47

Bilinear Forms

To put yourself in the proper frame of mind for the present section, you should begin by reviewing Sections 31 and 32, especially items 31.6 to 31.21, 32.1, and 32.5, as well as item 45.1.

We have seen the importance of the concepts of similarity and diagonalizability for square matrices. So far, these have been used in the study of linear operators on a vector space, V, with emphasis on the special case $V = \mathbb{R}^n$.

We will now exploit these same ideas to study a different kind of function related to inner products. Once again, the emphasis will be on square matrices and on the vector space $\mathbb{R}^n$. If you have studied the text this far, however, you should be convinced that many facts about square matrices and linear operators are better understood as special cases, when $m = n$, of some more general facts about arbitrary $m \times n$ matrices and arbitrary linear maps (e.g., that Corollary 32.5 is clearer when viewed as a special case of Theorem 32.1).

You are also by now aware that results proved about $\mathbb{R}^n$ are quite readily transferred to arbitrary real vector spaces because all real n-dimensional vector spaces are isomorphic to $\mathbb{R}^n$. The transfer to arbitrary vector spaces over an arbitrary field of scalars is not usually problematic.

For these reasons, we will continue in this and the next two sections to state results for $\mathbb{R}^m$ and $\mathbb{R}^n$. At the ends of the sections and in the problem sets, we will outline the minor modifications required when the scalars are $\mathbb{C}$ or some other field.

47.1 Definition

A function $p:\mathbb{R}^m \times \mathbb{R}^n \to \mathbb{R}$ that, to each pair $(\mathbf{x}, \mathbf{y})$ of vectors satisfying $\mathbf{x} \in \mathbb{R}^m$ and $\mathbf{y} \in \mathbb{R}^n$, associates a scalar, $p(\mathbf{x}, \mathbf{y}) \in \mathbb{R}$, will be called *a scalar product on* $\mathbb{R}^m \times \mathbb{R}^n$.

The prototypical example is, of course, the Euclidean dot product on $\mathbb{R}^n \times \mathbb{R}^n$ that associates the scalar $x_1y_1 + x_2y_2 + \cdots + x_ny_n$ to the pair $(\mathbf{x}, \mathbf{y})$. An important class of examples consists of real inner products defined on $\mathbb{R}^n \times \mathbb{R}^n$; these are the scalar products that satisfy the three conditions from Definition 45.1 (i.e., they are bilinear, symmetric, and positive definite).

Among the questions we wish to treat in this and the next two sections are the following:

(i) Can anything interesting or useful be said about scalar products that are not bilinear or not symmetric?
(ii) What is the role of positive definiteness in the theory of inner-product spaces?
(iii) More specifically, what can be said about scalar products that are bilinear and symmetric but not necessarily positive definite?
(iv) Under what conditions will a bilinear, symmetric scalar product be an inner product?

In very brief outline, the points to be made are the following:

(i) Recall that the functions from $\mathbb{R}^n \to \mathbb{R}^m$ that are *linear* are precisely those that can be represented by matrices (Theorems 11.6 and 31.7). In exact analogy with this fact, the scalar products on $\mathbb{R}^m \times \mathbb{R}^n$ that are *bilinear* are precisely those that can be "represented" by matrices, in a sense to be explained subsequently. The analogy with 31.7 is truly exact in that a bilinear scalar product can be "represented" by infinitely many "different" matrices: there will be a different matrix corresponding to each pair of bases β for $\mathbb{R}^m$ and γ for $\mathbb{R}^n$.
(ii) The bilinear scalar products of $\mathbb{R}^n \times \mathbb{R}^n$ that are also symmetric [i.e., $p(\mathbf{x}, \mathbf{y}) = p(\mathbf{y}, \mathbf{x})$ for all $\mathbf{x}, \mathbf{y} \in \mathbb{R}^n$] will turn out to be those whose matrix representations can be chosen to be symmetric matrices.
(iii) In Section 49, we will isolate a property of matrices such that a bilinear, symmetric scalar product is an inner product iff one of its representing matrices has this property. The name we will eventually give to this property of matrices is positive definiteness.

(iv) Section 48 contains material on quadratic forms and an important application known as the Principal Axes Theorem (see Theorem 48.19).

47.2 Definition

A scalar product $p:\mathbb{R}^m \times \mathbb{R}^n \to \mathbb{R}$ is *bilinear* if for all $\mathbf{x}, \mathbf{x}_1, \mathbf{x}_2 \in \mathbb{R}^m$, $\mathbf{y}, \mathbf{y}_1, \mathbf{y}_2 \in \mathbb{R}^n$ and $s, t \in \mathbb{R}$,

(i) $p(\mathbf{x}_1 + \mathbf{x}_2, \mathbf{y}) = p(\mathbf{x}_1, \mathbf{y}) + p(\mathbf{x}_2, \mathbf{y})$

(ii) $p(s\mathbf{x}, \mathbf{y}) = sp(\mathbf{x}, \mathbf{y})$

(ii) $p(\mathbf{x}, \mathbf{y}_1 + \mathbf{y}_2) = p(\mathbf{x}_1, \mathbf{y}_1) + p(\mathbf{x}, \mathbf{y}_2)$

(iv) $p(\mathbf{x}, t\mathbf{y}) = tp(\mathbf{x}, \mathbf{y})$

Conditions (i) and (ii) assert that p is linear as a function of the first variable alone; conditions (iii) and (iv) assert that p is linear as a function of the second variable alone.

It is important to keep in mind the distinction between linearity and bilinearity. Recall that the function $p:\mathbb{R} \times \mathbb{R} \to \mathbb{R}$ defined by $p(x_1, x_2) = x_1 x_2$ is bilinear but is not linear (as a function from $\mathbb{R}^2 \to \mathbb{R}$).

Scalar products that are bilinear are also called *bilinear forms*. We will henceforth adopt this term. It should be emphasized that bilinear forms are *scalar-valued* functions of *two vector* variables.

47.3 Exercise (Analogue of Theorem 17.6)

For every matrix $\mathbf{A} \in \mathbb{R}_{m \times n}$, the function $p_{\mathbf{A}}:\mathbb{R}^m \times \mathbb{R}^n \to \mathbb{R}$ defined by $p_{\mathbf{A}}(\mathbf{x}, \mathbf{y}) = [\mathbf{x}]^T \mathbf{A} [\mathbf{y}]$ is a bilinear form.

47.4 Exercise

(i) Let $\mathbf{A} = \begin{bmatrix} 0 & 1 & 2 \\ 3 & 4 & 5 \end{bmatrix}$ and let $\mathbf{B} = \begin{bmatrix} 1 & 2 \\ 0 & 1 \end{bmatrix}$. Write out the general expressions for $p_{\mathbf{A}}(\mathbf{x}, \mathbf{y})$ and $p_{\mathbf{B}}(\mathbf{x}, \mathbf{y})$.

(ii) Let

$$\mathbf{C} = \begin{bmatrix} 2 & 1 & 7 \\ -3 & 9 & 10 \\ -2 & -4 & 6 \\ 8 & 8 & 1 \end{bmatrix}$$

Write out the general expression for $p_{\mathbf{C}}(\mathbf{x}, \mathbf{y})$. Let $\mathbf{u} = (0, -2, 3, -3)$ and $\mathbf{v} = (2, 0, 1)$. Evaluate $p_{\mathbf{C}}(\mathbf{u}, \mathbf{v})$.

47.5 Exercise (Analogue of 17.2)

Let $p:\mathbb{R}^m \times \mathbb{R}^n \to \mathbb{R}$ be a bilinear form; let $\beta = (\mathbf{u}_1, \mathbf{u}_2, \ldots, \mathbf{u}_m)$ and $\gamma = (\mathbf{v}_1, \mathbf{v}_2, \ldots, \mathbf{v}_n)$ be bases for $\mathbb{R}^m$ and $\mathbb{R}^n$, respectively. Show that the values of $p(\mathbf{x}, \mathbf{y})$ are determined by the values $p(\mathbf{u}_i, \mathbf{v}_j)$ of p on pairs of basis vectors from the bases β and γ.

Exercise 47.5 suggests that bilinear forms can be represented by matrices; the following theorem confirms this.

47.6 Theorem (Analogue of Theorem 31.7)

For every bilinear form $p:\mathbb{R}^m \times \mathbb{R}^n \to \mathbb{R}$ and for every pair of ordered bases $\beta = (\mathbf{u}_1, \mathbf{u}_2, \ldots, \mathbf{u}_m)$ for $\mathbb{R}^m$ and $\gamma = (\mathbf{v}_1, \mathbf{v}_2, \ldots, \mathbf{v}_n)$ for $\mathbb{R}^n$, there is a unique $m \times n$ matrix A satisfying

$$\text{for all } \mathbf{x} \in \mathbb{R}^m \text{ and } \mathbf{y} \in \mathbb{R}^n, \quad p(\mathbf{x}, \mathbf{y}) = [\mathbf{x}]_\beta^T \, \mathrm{A} \, [\mathbf{y}]_\gamma \qquad (*)$$

47.7 Exercise

Prove the preceding theorem by imitating the argument given for 31.7. *Hint*: The required matrix A is the one whose $(i,\ j)^{\text{th}}$ entry is $p(\mathbf{u}_i, \mathbf{v}_j)$. ■

Theorem 47.6 guarantees that the following definition makes sense.

47.8 Definition (Analogue of 31.8)

For a bilinear form $p:\mathbb{R}^m \times \mathbb{R}^n \to \mathbb{R}$ and ordered bases β for $\mathbb{R}^m$ and γ for $\mathbb{R}^n$, the unique $m \times n$ matrix satisfying $(*)$ from Theorem 47.6 is called the *matrix that represents p with respect to β and γ* and is denoted by ${}_\beta[p]_\gamma$ (this symbol is to be read p-beta-gamma). The matrix ${}_{\alpha^m}[p]_{\alpha^n}$ is called the *standard matrix for p*.

47.9 Exercise

Show that the standard matrix for the Euclidean dot product on $\mathbb{R}^n \times \mathbb{R}^n$ is simply the identity matrix, I_n.

If $p:\mathbb{R}^m \times \mathbb{R}^n \to \mathbb{R}$ is a bilinear form, if α and β are bases for $\mathbb{R}^m$, and if γ and δ are bases for $\mathbb{R}^n$, the relationship between the matrices ${}_\beta[p]_\gamma$ and ${}_\alpha[p]_\delta$ can be discovered using the following argument:

By Theorem 47.6, ${}_\beta[p]_\gamma$ is the unique matrix satisfying $p(\mathbf{x}, \mathbf{y}) = [\mathbf{x}]_\beta^T \, {}_\beta[p]_\gamma \, [\mathbf{y}]_\gamma$ for all $\mathbf{x} \in \mathbb{R}^m$ and $\mathbf{y} \in \mathbb{R}^n$. Using the appropriate change-of-basis matrices, we may reexpress this as follows: for all $\mathbf{x} \in \mathbb{R}^m$ and $\mathbf{y} \in \mathbb{R}^n$,

$$p(\mathbf{x},\mathbf{y}) = \left({}_\beta[I]_\alpha\,[\mathbf{x}]_\alpha\right)^T {}_\beta[p]_\gamma\left({}_\gamma[I]_\delta\,[\mathbf{y}]_\delta\right) \qquad \text{[Exercise 31.13]}$$

$$= [\mathbf{x}]_\alpha^T\left({}_\beta[I]_\alpha^T\,{}_\beta[p]_\gamma\,{}_\gamma[I]_\delta\right)[\mathbf{y}]_\delta. \quad \text{[Associative Law + 3.15(iv)]}$$

But also by Theorem 47.7, ${}_\alpha[p]_\delta$ is the unique matrix satisfying

$$p(\mathbf{x},\mathbf{y}) = [\mathbf{x}]_\alpha^T\,{}_\alpha[p]_\delta\,[\mathbf{y}]_\delta, \quad \text{for all } \mathbf{x} \in \mathbb{R}^m \text{ and } \mathbf{y} \in \mathbb{R}^n.$$

Since we just showed that the matrix ${}_\beta[I]_\alpha^T\;{}_\beta[p]_\gamma\;{}_\gamma[I]_\delta$ satisfies the condition that uniquely defines ${}_\alpha[p]_\delta$, we may conclude that these two matrices are equal.

The preceding remarks constitute a proof of the following theorem.

47.10 Theorem (Analogue of Theorem 32.1)

If $p\colon \mathbb{R}^m \times \mathbb{R}^n \to \mathbb{R}$ is a bilinear form, if α and β are bases for $\mathbb{R}^m$, and if γ and δ are bases for $\mathbb{R}^n$, then

$${}_\alpha[p]_\delta = {}_\beta[I]_\alpha^T\,{}_\beta[p]_\gamma\,{}_\gamma[I]_\delta \qquad \blacksquare$$

Here is an example whose purpose is to illustrate Theorems 47.6 and 47.10.

47.11 Example

Let $p\colon\mathbb{R}^2 \times \mathbb{R}^3 \to \mathbb{R}$ be defined by $p(\mathbf{x},\mathbf{y}) = 3x_2y_1 + x_1y_2 + 4x_2y_2 + 2x_1y_3 + 5x_2y_3$. This is the bilinear form whose standard matrix is the matrix A from Exercise 47.5(i). Thus

$$\mathrm{A} = \begin{bmatrix} 0 & 1 & 2 \\ 3 & 4 & 5 \end{bmatrix} = {}_{\alpha^2}[p]_{\alpha^3}$$

Consider different bases for $\mathbb{R}^2$ and $\mathbb{R}^3$, say, $\alpha = ((3,6),(-1,-4))$ and $\delta = ((1,0,0),(1,2,0),(1,2,3))$. Then

$${}_\alpha[p]_\delta = {}_{\alpha^2}[I]_\alpha^T\,{}_{\alpha^2}[p]_{\alpha^3}\,{}_{\alpha^3}[I]_\delta \qquad \text{[Theorem 47.10]}$$

$$= \begin{bmatrix} 3 & -1 \\ 6 & -4 \end{bmatrix}^T \begin{bmatrix} 0 & 1 & 2 \\ 3 & 4 & 5 \end{bmatrix} \begin{bmatrix} 1 & 1 & 1 \\ 0 & 2 & 2 \\ 0 & 0 & 3 \end{bmatrix} \qquad \text{[31.13]}$$

$$= \begin{bmatrix} 3 & 6 \\ -1 & -4 \end{bmatrix} \begin{bmatrix} 0 & 2 & 8 \\ 3 & 11 & 26 \end{bmatrix}$$

$$= \begin{bmatrix} 18 & 72 & 180 \\ -12 & -46 & -112 \end{bmatrix}$$

To illustrate the theorems, you should verify that the same answer is obtained when ${}_{\alpha}[p]_{\delta}$ is computed directly from the definition; for example, the (2, 3)-entry in the matrix is $p((-1,-4),(1,2,3)) = (3)(-4)(1) + (-1)(2) + (4)(-4)(2) + (2)(-1)(3) + (5)(-4)(3) = -12 - 2 - 32 - 6 - 60 = -112$.

Let us further illustrate the theorems using the specific pair of vectors $\mathbf{u} = (2, -1)$ and $\mathbf{v} = (3, 0, -2)$.

$$\begin{aligned} p(\mathbf{u},\mathbf{v}) &= (3)(-1)(3) + (2)(0) + (4)(-1)(0) + (2)(2)(-2) + (5)(-1)(-2) \\ &= -9 - 8 + 10 = -7 \end{aligned}$$

Theorem 47.10 guarantees that $[\mathbf{u}]_{\alpha}^{T}\, {}_{\alpha}[p]_{\delta}\, [\mathbf{v}]_{\delta} = -7$. To see that this is indeed the case, you should verify that

$$[\mathbf{u}]_{\alpha} = \begin{bmatrix} \frac{3}{2} \\ \frac{5}{2} \end{bmatrix} \quad \text{and} \quad [\mathbf{v}]_{\delta} = \begin{bmatrix} 3 \\ \frac{2}{3} \\ -\frac{2}{3} \end{bmatrix}$$

and that

$$\begin{bmatrix} \frac{3}{2} & \frac{5}{2} \end{bmatrix} \begin{bmatrix} 18 & 72 & 180 \\ -12 & -46 & -112 \end{bmatrix} \begin{bmatrix} 3 \\ \frac{2}{3} \\ -\frac{2}{3} \end{bmatrix} = -7$$

All along, our principal interest has been the special case of the previous theorem in which $m = n$, $\alpha = \delta$, and $\beta = \gamma$.

47.12 Corollary

If $p:\mathbb{R}^n \times \mathbb{R}^n \to \mathbb{R}$ is a bilinear form, if α and β are bases for $\mathbb{R}^n$, then

$${}_{\beta}[p]_{\beta} = {}_{\alpha}[I]_{\beta}^{T}\, {}_{\alpha}[p]_{\alpha}\, {}_{\alpha}[I]_{\beta} \qquad \blacksquare$$

Because transition matrices are invertible, we conclude from Corollary 47.12 that a pair of matrices A and B representing the same bilinear form with respect to different bases will satisfy

there exists an invertible matrix P such that $B = P^{T}AP$ (#)

Conversely, because any invertible matrix can be interpreted as a change-of-basis matrix, we know that, if a pair of matrices A and B satisfies condition (#), then A and B represent the same bilinear form with respect to different bases. More precisely, if P is invertible and if $B = P^TAP$, then if A represents the bilinear form p with respect to the basis α, then B represents p with respect to β, where β is the unique basis such that $P = {}_\alpha[I]_\beta$ (recall Theorem 31.21).

This concept is obviously important, so we give it a name.

47.13 Definition (Analogue of 32.6)

The $n \times n$ matrix B is *conjugate* to the $n \times n$ matrix A iff there exists an invertible matrix P such that $B = P^TAP$.

47.14 Exercise (Analogue of Exercise 32.7)

Prove that conjugacy is reflexive (A is conjugate to itself), symmetric (if B is conjugate to A, then A is conjugate to B), and transitive (if B is conjugate to A and C is conjugate to B, then C is conjugate to A).

This concludes the fairly lengthy analogy that we have developed between conjugacy and similarity and that we may summarize as follows:

(i) Just as two real $n \times n$ matrices are similar iff they represent the same linear operator on $\mathbb{R}^n$ with respect to different bases,
(ii) two real $n \times n$ matrices are conjugate iff they represent the same bilinear form on $\mathbb{R}^n$ with respect to different bases.

We turn next to considerations of symmetry, which are important (and make sense only) when $m = n$.

47.15 Definition

A bilinear form $p:\mathbb{R}^n \times \mathbb{R}^n \to \mathbb{R}$ is:

(i) *Symmetric* if $p(\mathbf{x}, \mathbf{y}) = p(\mathbf{y}, \mathbf{x})$ for all $\mathbf{x}$ and $\mathbf{y} \in \mathbb{R}^n$.
(ii) *Skew-symmetric* if $p(\mathbf{x}, \mathbf{y}) = -p(\mathbf{y}, \mathbf{x})$ for all $\mathbf{x}$ and $\mathbf{y} \in \mathbb{R}^n$.

It is not a great loss of generality to restrict our attention to symmetric bilinear forms. This is because, as we will shortly observe, an arbitrary bilinear form can be written, in a unique way, as the sum of a symmetric bilinear form plus a skew-symmetric bilinear form. Most results about symmetric bilinear forms have fairly straightforward variants that apply to skew-symmetric bilinear forms; thus a thorough understanding of the symmetric case is, for most purposes, adequate. For one of the main applications, to the study of quadratic forms in Section 48, it is fully adequate.

47.16 Exercise

Let $p:\mathbb{R}^n \times \mathbb{R}^n \to \mathbb{R}$ be a bilinear form. Show that the following are equivalent:

(i) p is a symmetric bilinear form.
(ii) The standard matrix for p is a symmetric matrix.
(iii) If β is any orthonormal basis for $\mathbb{R}^n$, then ${}_\beta[p]_\beta$ is a symmetric matrix.

The corresponding results with "skew-symmetric" in place of "symmetric" are also true.

47.17 Exercise

Use the results from Problem 3.13 to show that an arbitrary bilinear form p is expressible, in a unique way, as the sum of a symmetric bilinear form, p_1, plus a skew-symmetric bilinear form, p_2.

This exercise guarantees that it makes sense to talk about "the" symmetric and skew-symmetric parts of a bilinear form.

47.18 Definition

If $p:\mathbb{R}^n \times \mathbb{R}^n \to \mathbb{R}$ is a bilinear form and if p_1 and p_2 are respectively symmetric and skew-symmetric bilinear forms with $p = p_1 + p_2$, then p_1 is called the *symmetric part of p* and p_2 is called the *skew-symmetric part of p*.

47.19 Exercise

Use the results from Problem 3.13 to find the symmetric and skew-symmetric parts of p_A, where $A = \begin{bmatrix} 3 & -2 \\ 1 & 5 \end{bmatrix}$.

Problem Set 47

47.1. Go back and do any of the exercises from Section 47 that you may have skipped.

*47.2. We have seen in Theorem 47.10 that

$${}_\alpha[p]_\delta = {}_\beta[I]_\alpha^T \, {}_\beta[p]_\gamma \, {}_\gamma[I]_\delta$$

is the correct analogue, for a bilinear form p, of Theorem 32.1,

$${}_\delta[T]_\alpha = {}_\delta[I]_\gamma \, {}_\gamma[T]_\beta \, {}_\beta[I]_\alpha$$

for a linear transformation T. If we had not seen 47.10, we might have been tempted to try

$$_{\alpha}[p]_{\delta} = {}_{\alpha}[I]_{\beta}\, {}_{\beta}[p]_{\gamma}\, {}_{\gamma}[I]_{\delta}$$

which is false in general. Attempt to prove this by imitating the proof of 32.1 and pinpoint the step where the argument breaks down.

*47.3. Show that $[\mathbf{u}]_{\alpha}^{T}\, {}_{\alpha}[I]_{\beta} = [\mathbf{u}]_{\beta}^{T}$ for all $\mathbf{u} \in \mathbb{R}^n$ iff ${}_{\alpha}[I]_{\beta}$ is orthogonal.

*47.4. Define $p:\mathbb{R}^3 \times \mathbb{R}^3 \to \mathbb{R}$ by

$$p(\mathbf{u}, \mathbf{v}) = 3\mathbf{u}_1\mathbf{v}_2 + 2\mathbf{u}_2\mathbf{v}_1 - \mathbf{u}_3\mathbf{v}_3.$$

(i) Find the standard matrix for p.

(ii) Let $\delta = ((0,0,1),(1,0,1),(0,2,1))$. Find ${}_{\delta}[p]_{\delta}$.

48

Quadratic Forms

In this section, we will apply the preceding abstract ideas to the mathematically down-to-earth subject of polynomials of degree ≤ 2.

First, we recall the familiar polynomial functions of one variable, $f:\mathbb{R} \to \mathbb{R}$. The simplest polynomials are those of degree 0, the constants: $f(x) = c$. The next simplest are those of degree 1, the affine functions: $f(x) = bx + c$, where $b \neq 0$. These consist of a linear part, bx, plus a constant. Next come those of degree 2, the quadratics: $f(x) = ax^2 + bx + c$, where $a \neq 0$. These consist of a pure quadratic part, ax^2, plus terms of lower degree. More generally, if $a_p \neq 0$, $f(x) = a_p x^p + a_{p-1}x^{p-1} + \cdots + a_1 x_1 + a_0$ is a polynomial function of degree p.

This being a text on $\mathbb{R}^n$ rather than $\mathbb{R}$, we wish to study polynomial functions of more than one variable.

48.1 Examples

(i) $-2x_1x_4^2 + 3x_1x_3 + x_2x_3 - 4x_3^3 + 3x_2 - 5$
(ii) $x_1x_2x_3 - x_1^2 - x_2^2 - x_3^2 + 1$
(iii) $2x_1x_2 + x_3 - 4$
(iv) $3x_2^2x_5^3 + 9x_1x_4^2 - x_5 + \frac{1}{2}$
(v) $x_5^2 - 1$

Roughly speaking, a polynomial in the variables $x_1, x_2, \ldots, x_n$ is a sum of terms, each term consisting of a scalar coefficient times a product of some of the variables to integer powers. There is a fairly obvious way to classify such functions by degree. Here is the exact definition.

48.2 Definition

(i) A function $f:\mathbb{R}^n \to \mathbb{R}$ is a *polynomial in the variables* $x_1, x_2, \ldots, x_n$ if it is a sum of terms of the form

$$s_{i_1 i_2 \cdots i_p} x_{i_1}^{k_1} x_{i_2}^{k_2} \cdots x_{i_p}^{k_p}$$

where $1 \le p \le n$, $1 \le i_1 \le i_2 \le \cdots \le i_p \le n$, the coefficients, $s_{i_1 i_2 \cdots i_p}$ are real numbers, and the k's are nonnegative integers.

(ii) The *degree of such a term* is the sum $k_1 + k_2 + \cdots + k_p$ of the exponents.

(iii) The *degree of* f is the maximum degree of a term of f whose coefficient is nonzero.

48.3 Examples

Consider the polynomials from Example 48.1

(i) is a third-degree polynomial in the variables x_1, x_2, x_3, x_4.
(ii) is a third-degree polynomial in x_1, x_2, x_3.
(iii) is a second-degree polynomial in x_1, x_2, x_3.
(iv) is a fifth-degree polynomial in x_1, x_2, x_3, x_4, x_5.
(v) is a second-degree polynomial in x_1, x_2, x_3, x_4, x_5.

We will be restricting our attention to polynomials of degree ≤ 2; these are commonly called quadratic.

48.4 Definition

(i) A function $f:\mathbb{R}^n \to \mathbb{R}$ is a *quadratic function of* $x_1, x_2, \ldots, x_n$ if it is a sum of

terms of the form $s_{ij}x_i x_j$, where $1 \le i \le j \le n$
plus terms of the form $t_i x_i$
plus a constant term, c

(ii) Such a function is called a *quadratic form in* $x_1, x_2, \ldots, x_n$ if all the terms of degree $\lneq 2$ are zero (i.e., if $c = t_1 = t_2 = \cdots = t_n = 0$).

(iii) The terms $s_{ij}x_i x_j$ of a quadratic form are called *squares* when $i = j$ and are called *cross-products* when $i \neq j$.

(iv) If f is a quadratic function, the quadratic form consisting of the second degree terms of f is called the *quadratic form associated with f*.

48.5 Examples

(i) $x_1^2 + x_2^2$ is a quadratic form in x_1, x_2.
(ii) $x_1^2 + 2x_1x_2 + x_2^2$ is a quadratic form in x_1, x_2.
(iii) $x_1^2 + 2x_1x_2 + x_2^2 - x_1 - x_2 + 1$ is a quadratic function of x_1, x_2 whose associated quadratic form is $x_1^2 + 2x_1x_2 + x_2^2$.
(iv) $x_1^2 + 2x_2^2 + x_3^2 + 4x_1x_2 + 5x_1x_3 + 6x_2x_3 + 7x_1 + 8x_2 + 9x_3 + 10$ is a quadratic function of x_1, x_2, x_3 whose associated quadratic form consists of the first six terms.
(v) $2x_1^2 - x_3^2 + 4x_1x_2x_3 - 2x_3 + 7$ is not a quadratic function since it contains a term, $4x_1x_2x_3$, of degree 3.

The key observation, which allows us to apply the earlier results from this section to the study of quadratic functions, is that the language of vectors and matrices provides a different way to express quadratic functions. To motivate this observation, we can suggest the principle that, in seeking to generalize a fact about $\mathbb{R}$ to $\mathbb{R}^n$, we should think about real numbers as 1×1 matrices and n-tuples as $1 \times n$ or $n \times 1$ matrices.

48.6 Observation

If $f:\mathbb{R}^n \to \mathbb{R}$ is a quadratic function, say $f(x_1, x_2, \ldots, x_n) = \Sigma s_{ij}x_ix_j + \Sigma t_ix_i + c$, then letting

$$A = \begin{bmatrix} s_{11} & s_{12}/2 & \cdot & \cdot & \cdot & s_{1n}/2 \\ s_{12}/2 & s_{22} & \cdot & \cdot & \cdot & s_{2n}/2 \\ \cdot & \cdot & \cdot & \cdot & \cdot & \cdot \\ \cdot & \cdot & \cdot & \cdot & s_{ij}/2 & \cdot \\ & & \cdot & \cdot & \cdot & \cdot \\ \cdot & \cdot & s_{ij}/2 & \cdot & \cdot & \cdot \\ \cdot & \cdot & \cdot & \cdot & \cdot & \cdot \\ s_{1n}/2 & \cdot & \cdot & \cdot & \cdot & s_{nn} \end{bmatrix}$$

and $B = [t_1t_2 \ldots t_n]$ and $[\mathbf{x}]$ = the standard α^n-coordinate matrix of $\mathbf{x}$,

we can write (again, provided we ignore the distinction between 1×1 matrices and scalars)

$$f(x_1, x_2, \ldots, x_n) = [\mathbf{x}]^T A[\mathbf{x}] + B[\mathbf{x}] + c$$

Observe also that the matrix A is symmetric.

48.7 Examples

Let us rewrite, in matrix form, the quadratic functions from Example 48.5.

(i) $\begin{bmatrix} x_1 & x_2 \end{bmatrix}\begin{bmatrix} 1 & 0 \\ 0 & 1 \end{bmatrix}\begin{bmatrix} x_1 \\ x_2 \end{bmatrix}$

(ii) $\begin{bmatrix} x_1 & x_2 \end{bmatrix}\begin{bmatrix} 1 & 1 \\ 1 & 1 \end{bmatrix}\begin{bmatrix} x_1 \\ x_2 \end{bmatrix}$

(iii) $\begin{bmatrix} x_1 & x_2 \end{bmatrix}\begin{bmatrix} 1 & 0 \\ 0 & 1 \end{bmatrix}\begin{bmatrix} x_1 \\ x_2 \end{bmatrix} + \begin{bmatrix} -1 & -1 \end{bmatrix}\begin{bmatrix} x_1 \\ x_2 \end{bmatrix} + 1$

(iv) $\begin{bmatrix} x_1 & x_2 & x_3 \end{bmatrix}\begin{bmatrix} 1 & 2 & \frac{5}{2} \\ 2 & 2 & 3 \\ \frac{5}{2} & 3 & 1 \end{bmatrix}\begin{bmatrix} x_1 \\ x_2 \\ x_2 \end{bmatrix} + \begin{bmatrix} 7 & 8 & 9 \end{bmatrix}\begin{bmatrix} x_1 \\ x_2 \\ x_2 \end{bmatrix} + 10$

Since every quadratic function $f:\mathbb{R}^n \to \mathbb{R}$ is the sum of a quadratic form and an affine function, and since affine functions have been thoroughly studied in the earlier chapters, we will concentrate, for the remainder of this section, on quadratic forms.

From Observation 48.6, we know that if $q:\mathbb{R}^n \to \mathbb{R}$ is a quadratic form on $\mathbb{R}^n$ then there is a symmetric matrix A such that

$$\text{for all } \mathbf{x} \in \mathbb{R}^n, \quad q(\mathbf{x}) = [\mathbf{x}]^T \mathrm{A}[\mathbf{x}]$$

48.8 Definition

If β is a basis for $\mathbb{R}^n$, let us say that an $n \times n$ matrix B *represents the quadratic form* $q:\mathbb{R}^n \to \mathbb{R}$ *with respect to* β if

$$\text{for all } \mathbf{x} \in \mathbb{R}^n, \quad q(\mathbf{x}) = [\mathbf{x}]_\beta^T \mathrm{B}[\mathbf{x}]_\beta \qquad (*)$$

Thus, the matrix A, as defined in Observation 48.6, represents q with respect to the standard basis α^n.

It would be premature, at this point, to begin using the symbol ${}_\beta[q]_\beta$ to stand for "the" matrix that represents q with respect to β. We must first ask whether, for every basis β, there is a unique matrix satisfying $(*)$. The answer to this question is *no*. This is because the values of a quadratic form on $\mathbb{R}^n$ are not determined by its values on a basis; this is, indeed, the essential difference between quadratic forms and linear transformations.

48.9 Exercise

(i) Show that the two different quadratic forms $x_1^2 + x_2^2$ and $x_1^2 + 2x_1x_2 + x_2^2$ agree on the standard basis vectors $(1, 0)$ and $(0, 1)$.

(ii) Show that the three matrices

$$\begin{bmatrix} 1 & 1 \\ 1 & 1 \end{bmatrix}, \quad \begin{bmatrix} 1 & 3 \\ -1 & 1 \end{bmatrix}, \quad \text{and} \quad \begin{bmatrix} 1 & 2 \\ 0 & 1 \end{bmatrix}$$

all represent $x_1^2 + 2x_1x_2 + x_2^2$ with respect to the standard basis.

The next theorem, however, shows that for every quadratic form $q:\mathbb{R}^n \to \mathbb{R}$ and for every basis β there is a unique *symmetric* matrix representing q with respect to β.

48.10 Lemma

Let $\beta = (\mathbf{u}_1, \mathbf{u}_2, \ldots, \mathbf{u}_n)$ be a basis for $\mathbb{R}^n$, and let B and C be $n \times n$ matrices that satisfy

$$\text{for all } \mathbf{x} \in \mathbb{R}^n, \quad [\mathbf{x}]_\beta^T \mathrm{B}[\mathbf{x}]_\beta = [\mathbf{x}]_\beta^T \mathrm{C}[\mathbf{x}]_\beta \tag{$*$}$$

Then $\mathrm{B} + \mathrm{B}^T = \mathrm{C} + \mathrm{C}^T$.

Proof: First we show that the main diagonal entries of B and C must actually be equal. Because $(*)$ holds for all $\mathbf{x}$, it holds in particular for $\mathbf{u}_i$, and since $(\mathbf{u}_i)_\beta = \mathbf{e}_i^n$, we have

$$b_{ii} = [\mathbf{e}_1^n]^T \mathrm{B}[\mathbf{e}_1^n] = [\mathbf{u}_i]_\beta^T \mathrm{B}[\mathbf{u}_i]_\beta = [\mathbf{u}_i]_\beta^T \mathrm{C}[\mathbf{u}_i]_\beta = [\mathbf{e}_i^n]^T \mathrm{C}[\mathbf{e}_i^n] = c_{ii}$$

Next, because $(*)$ must also hold for the vector $\mathbf{u}_i + \mathbf{u}_j$, and because

$$(\mathbf{u}_i + \mathbf{u}_j)_\beta = (0, \ldots, 0, 1, 0, \ldots, 0, 1, 0, \ldots, 0)$$

we have

$$\begin{aligned} b_{ii} + b_{ij} + b_{ji} + b_{jj} &= [\mathbf{u}_i + \mathbf{u}_j]_\beta^T \mathrm{B}[\mathbf{u}_i + \mathbf{u}_j]_\beta = [\mathbf{u}_i + \mathbf{u}_j]_\beta^T \mathrm{C}[\mathbf{u}_i + \mathbf{u}_j]_\beta \\ &= c_{ii} + c_{ij} + c_{ji} + c_{jj} \end{aligned}$$

Since we already know that $b_{ii} = c_{ii}$ and $b_{jj} = c_{jj}$, we conclude $b_{ij} + b_{ji} = c_{ij} + c_{ji}$. ■

48.11 Theorem

If $q:\mathbb{R}^n \to \mathbb{R}$ is a quadratic form, if β is a basis for $\mathbb{R}^n$, and if B and C are symmetric matrices that each represent q with respect to β, then $\mathrm{B} = \mathrm{C}$.

Proof: From the lemma, we conclude $\mathrm{B} + \mathrm{B}^T = \mathrm{C} + \mathrm{C}^T$. But since $\mathrm{B} = \mathrm{B}^T$ and $\mathrm{C} = \mathrm{C}^T$, this amounts to $2\mathrm{B} = 2\mathrm{C}$, so $\mathrm{B} = \mathrm{C}$. ■

We may now safely introduce the symbol ${}_\beta[q]_\beta$ with a precise meaning.

48.12 Definition

If $q:\mathbb{R}^n \to \mathbb{R}$ is a quadratic form and β is a basis for $\mathbb{R}^n$ the *matrix that represents q with respect to β* will be denoted by ${}_\beta[q]_\beta$ and is the unique symmetric matrix satisfying

$$\text{for all } \mathbf{x} \in \mathbb{R}^n \quad [\mathbf{x}]_\beta^T \, {}_\beta[q]_\beta \, [\mathbf{x}]_\beta \qquad (*)$$

We have yet to discuss the natural relationship between bilinear forms and quadratic forms.

48.13 Definition

The *diagonal*, Δ_S, of a set of pairs $S \times S = \{(s_1, s_2): s_1, s_2 \in S\}$ is the subset consisting of those pairs whose first and second coordinates are equal. Thus $\Delta_S = \{(s_1, s_2): s_1 = s_2\}$.

When it is clear from the context which set S is involved, we will simply refer to the diagonal, Δ, and omit the subscript.

48.14 Definition

If $p:\mathbb{R}^n \times \mathbb{R}^n \to \mathbb{R}$ is a bilinear form on $\mathbb{R}^n \times \mathbb{R}^n$, it is clear that the restriction of p to the diagonal, $p \restriction \Delta$, is a quadratic form on $\mathbb{R}^n$. Let us denote this by q_p and call it the *quadratic form derived from p*. Thus $q_p(\mathbf{x}) = p(\mathbf{x}, \mathbf{x})$. If $\mathrm{A} \in \mathbb{R}_{n \times n}$, the quadratic form derived from the bilinear form p_A will be denoted by q_A.

48.15 Example

If the bilinear form p is the Euclidean dot product, $p(\mathbf{x}, \mathbf{y}) = \mathbf{x} \cdot \mathbf{y}$, then the derived quadratic form is just the square of the Euclidean norm, $q_p(\mathbf{x}) = p(\mathbf{x}, \mathbf{x}) = \mathbf{x} \cdot \mathbf{x} = \|\mathbf{x}\|^2$.

Recall from 47.17 that an arbitrary bilinear form p can be expressed in a unique way as the sum of a symmetric bilinear form, p_1, plus a skew-symmetric bilinear form p_2. Since the skew-symmetric part of a bilinear form must vanish on the diagonal, [i.e., by skew-symmetry, $p_2(\mathbf{x}, \mathbf{x}) = -p_2(\mathbf{x}, \mathbf{x})$, and therefore $p_2(\mathbf{x}, \mathbf{x}) = 0$], the quadratic form q_p derived from a bilinear form p depends exclusively on the symmetric part of p.

One possible way, therefore, in which the same quadratic form q might be derived from two different bilinear forms p and p' is for p and p' to have the same symmetric part but different skew-symmetric parts. The essence of Theorem 48.11 is that this is the only way this phenomenon can happen.

48.16 Theorem (Reinterpretation of Theorem 48.11)

If p and p' are two symmetric bilinear forms on $\mathbb{R}^n \times \mathbb{R}^n$ that agree on the diagonal Δ (i.e., that have the same derived quadratic form), then $p = p'$.

Proof: No new proof is needed. ■

The following special case of this theorem was previously encountered in Section 45, where it was given a different proof.

48.17 Corollary (Theorem 45.6 Revisited)

The values of a real inner product are completely determined by the values of its associated norm. ■

We conclude this section with the Principal Axes Theorem for $\mathbb{R}^n$. The description with respect to the standard basis of a quadratic form $q:\mathbb{R}^n \to \mathbb{R}$ will, in general, involve cross-product terms. From the matrix point of view, this corresponds to the fact that the standard matrix for q, ${}_{\alpha^n}[q]_{\alpha^n}$, is not, in general, a diagonal matrix; the nonzero off-diagonal entries contribute the coefficients of the cross-product terms.

If, however, β is a basis for which ${}_{\beta}[q]_{\beta} = \text{diag}(\lambda_1, \lambda_2, \ldots, \lambda_n)$, the description of q with respect to β is greatly simplified: $q(\mathbf{x})$ is a weighted sum of the squares of the β-coordinates of $\mathbf{x}$. Specifically, if $((\mathbf{x})_\beta)_i$ denotes the ith coordinate of $(\mathbf{x})_\beta$, then

$$q(\mathbf{x}) = [\mathbf{x}]_\beta^T \, {}_{\beta}[q]_{\beta} \, [\mathbf{x}]_\beta = \sum_{i=1}^{i=n} \lambda_i ((\mathbf{x})_\beta)_i^2.$$

We know from Observation 48.6 that there is always a symmetric matrix that represents q with respect to α^n. The main result from Section 46 was that any real symmetric matrix is orthogonally diagonalizable over $\mathbb{R}$. Once we check that orthogonally similar matrices represent the same quadratic form, we will have proved that every real quadratic form can be represented as a sum of squares.

48.18 Theorem

If $q:\mathbb{R}^n \to \mathbb{R}$ is a quadratic form whose standard matrix is A and if P is an orthogonal matrix, then $\text{B} = \text{P}^{-1}\text{AP} = \text{P}^T\text{AP}$ represents q with respect to the basis β consisting of the columns of P. That is,

$$\text{for all } \mathbf{x} \in \mathbb{R}^n, \quad [\mathbf{x}]^T A[\mathbf{x}] = [\mathbf{x}]_\beta^T B[\mathbf{x}]_\beta.$$

Proof: This is left as an exercise. Use the fact that $P = {}_{\alpha^n}[I]_\beta$ and imitate the proofs of 32.1 and 47.10. ■

48.19 Theorem (Principal Axes Theorem for $\mathbb{R}^n$)

For any quadratic form $q:\mathbb{R}^n \to \mathbb{R}$ there is a norm-and-angle-preserving change of basis from α^n to an orthonormal basis β with respect to which ${}_\beta[q]_\beta$ is a diagonal matrix.

Proof: As outlined previously, the columns of any orthogonal matrix P that orthodiagonalizes the standard symmetric matrix for q will be such a basis. That such a P exists follows from Theorem 46.8. ■

48.20 Example

Let $q:\mathbb{R}^2 \to \mathbb{R}$ be defined by $q(\mathbf{x}) = q(x_1, x_2) = 2x_1^2 + 8x_1x_2 + 2x_2^2$. Then

$$q(\mathbf{x}) = [x_1 \quad x_2]\begin{bmatrix} 2 & 4 \\ 4 & 2 \end{bmatrix}\begin{bmatrix} x_1 \\ x_2 \end{bmatrix} = [\mathbf{x}]^T A[\mathbf{x}]$$

To find a basis with respect to which q can be expressed as a sum of squares, we orthodiagonalize A using the techniques of Sections 35 and 46. You should verify independently that these techniques yield the equation

$$\begin{bmatrix} 1/\sqrt{2} & 1/\sqrt{2} \\ -1/\sqrt{2} & 1/\sqrt{2} \end{bmatrix}\begin{bmatrix} 2 & 4 \\ 4 & 2 \end{bmatrix}\begin{bmatrix} 1/\sqrt{2} & -1/\sqrt{2} \\ 1/\sqrt{2} & 1/\sqrt{2} \end{bmatrix} = \begin{bmatrix} 6 & 0 \\ 0 & -2 \end{bmatrix}$$

Thus $\beta = ((1/\sqrt{2}, 1/\sqrt{2}), (-1/\sqrt{2}, 1/\sqrt{2}))$ is an orthonormal basis for $\mathbb{R}^2$ with the property that, for all $\mathbf{x} \in \mathbb{R}^2$, if $(\mathbf{x})_\beta = (\bar{x}_1, \bar{x}_2)$, then $q(\mathbf{x}) = 6\bar{x}_1^2 - 2\bar{x}_2^2$.

Let us illustrate this conclusion using the vector $\mathbf{x} = (-1, 3)$. By direct computation, $q(\mathbf{x}) = q(-1, 3) = -4$. Now

$$[\mathbf{x}]_\beta = {}_\beta[I]_{\alpha^2}[\mathbf{x}]_{\alpha^2} = \begin{bmatrix} 1/\sqrt{2} & 1/\sqrt{2} \\ -1/\sqrt{2} & 1/\sqrt{2} \end{bmatrix}\begin{bmatrix} -1 \\ 3 \end{bmatrix} = \begin{bmatrix} -2/\sqrt{2} \\ 4/\sqrt{2} \end{bmatrix}$$

The theorem guarantees that $q(\mathbf{x}) = 6(-2/\sqrt{2})^2 - 2(4/\sqrt{2})^2$, which does indeed, as you may easily check, equal -4.

48.21 Caution

It is essential to *ortho*diagonalize A. It is not sufficient merely to diagonalize A using the techniques of Section 35. Returning to the previous example, if we merely diagonalize A, we find $P^{-1}AP = \text{diag}(6, -2)$, where $P = \begin{bmatrix} 1 & -1 \\ 1 & 1 \end{bmatrix}$. But it is false, for the basis $\gamma = ((1,1),(-1,1))$, that

$$[\mathbf{x}]_\gamma^T \begin{bmatrix} 6 & 0 \\ 0 & -2 \end{bmatrix} [\mathbf{x}]_\gamma = q(\mathbf{x}), \quad \text{for all } \mathbf{x} \in \mathbb{R}^2$$

For example, if $\mathbf{x} = (-1, 3)$, then $(\mathbf{x})_\gamma = (1, 2)$ (you should check this), but

$$\begin{bmatrix} 1 & 2 \end{bmatrix} \begin{bmatrix} 6 & 0 \\ 0 & -2 \end{bmatrix} \begin{bmatrix} 1 \\ 2 \end{bmatrix} = -2 \neq -4 = q(-1, 3)$$

Problem Set 48

48.1. Go back and do any of the exercises from Section 48 that you may have skipped.

*48.2. For each of the following quadratic functions, find the associated quadratic form and the symmetric matrix that represents this quadratic form with respect to the standard basis.

(i) $3x_1^3 + 4x_1x_2 + 5x_2^2 + 3x_1 - 2x_2 + 7$

(ii) $2x_1^2 - 5x_2^2 + 15$

(iii) $x_1^2 - 3x_2^2 + 7x_3^2 - 3x_1x_2 + 3x_2x_3 - 5x_1x_3 + x_1 - 2x_3 - 3$

(iv) $2x_4^2 - 3x_2^2 - x_1x_3 + 3x_2x_3 - 8x_2x_4$

*48.3. By diagonalizing the appropriate matrix, find a basis for $\mathbb{R}^2$ with respect to which the quadratic form $4x_1^2 - 3x_2^2 + 7x_1x_2$ is represented as a sum of squares.

49

Positive Definite Forms

Having studied bilinearity and symmetry, we turn our attention to the third of the properties defining inner products: positive definiteness.

Recall [45.1(iv)] that an inner product $\langle \cdot , \cdot \rangle$ on $\mathbb{R}^n$ is positive definite iff

$$\text{for all } \mathbf{v} \in \mathbb{R}^n, 0 \leq \langle \mathbf{v}, \mathbf{v} \rangle \quad \text{and} \quad 0 = \langle \mathbf{v}, \mathbf{v} \rangle \text{ iff } v = \mathbf{0}.$$

Having seen that inner products are bilinear forms with certain additional properties, and that bilinear forms can be represented by matrices, we will only consider positive definiteness for functions of the form p_{A}, where $\mathrm{A} \in \mathbb{R}_{n \times n}$. For such a function, positive definiteness is equivalent to the property

$$p_{\mathrm{A}}(\mathbf{x}) = \mathbf{x}^T \mathrm{A} \mathbf{x} \gneq 0, \quad \text{for all } \mathbf{0} \neq \mathbf{x} \in \mathbb{R}^n \qquad (\dagger)$$

In this section, the standard basis is the one most frequently used, so we have simplified the notation by writing $\mathbf{x}^T\mathrm{A}\mathbf{x}$ in place of $[\mathbf{x}]^T\mathrm{A}[\mathbf{x}]$.

Our goal is to study (†) as a property of matrices. Because (†) refers only to values assumed by p_{A} on the diagonal, it is clear that (†) is really a property of the derived quadratic form $q_{\mathrm{A}} = p_{\mathrm{A}} \upharpoonright \Delta$. Since quadratic forms can always be represented by symmetric matrices, there is no loss of generality in restricting our study of (†) to symmetric matrices. Thus we will make symmetry part of the definition of positive definiteness for real matrices.

49.1 Definition

A matrix $A \in \mathbb{R}_{n\times n}$ is *positive definite* if A is symmetric and $x^TAx \gneqq 0$ for all $0 \neq x \in \mathbb{R}^n$.

49.2 Remark

There is another, deeper reason why it is natural to incorporate symmetry into the definition of positive definiteness for real matrices. For a complex matrix $A \in \mathbb{C}_{n\times n}$, the analogue of positive definiteness

$$x^HAx \gneqq 0, \quad \text{for all } 0 \neq x \in \mathbb{C}^n$$

implies, *by itself*, that A is Hermitian. See Problem 49.4.

Since positive definiteness is really a property of the underlying quadratic form, we expect that if a matrix A is positive definite, and if B represents the same quadratic form as A with respect to a different basis (i.e., if B is conjugate to A), then B should be positive definite. The next theorem confirms this.

49.3 Theorem

If A is positive definite and if $B = P^TAP$ for some invertible matrix P, then B is positive definite.

Proof: B is symmetric by Problem 3.23(i). If $x \neq 0$, then $x^TBx = x^TP^TAPx = (Px)^TA(Px) \gneqq 0$ since A is positive definite and $Px \neq 0$. ■

This definition of positive definiteness for matrices is the most natural one; it is not useful for computational purposes, however, since there are infinitely many things to check to decide whether A satisfies (†). We would like a theorem that connects (†), which is a property of the function q_A, with some property of A, as a matrix.

Let us begin with a collection of necessary conditions.

49.4 Theorem

Assume $A \in \mathbb{R}_{n\times n}$ is positive definite; then

(i) $a_{ii} \gneqq 0$.
(ii) $a_{ii}a_{jj} \gneqq a_{ij}^2$ whenever $i \neq j$.

(iii) $|a_{ij}| \leq \max\{a_{11}, a_{22}, \ldots, a_{nn}\}$ whenever $i \neq j$.
(iv) All the eigenvalues of A are positive.
(v) $\det A \gneq 0$.

Proof:
(i) $a_{ii} = \mathbf{e}_i^{nT} A \mathbf{e}_i^n$.
(ii) Let $\mathbf{u} = (-a_{ij}/a_{ii})\mathbf{e}_i^n + \mathbf{e}_j^n$. Using part (i), show that (ii) is equivalent to $\mathbf{u}^T A \mathbf{u} \gneq 0$.
(iii) $a_{ij}^2 \lneq a_{ii}a_{jj}$ [from part (ii)]
$\leq (\max\{a_{11}, a_{22}, \ldots, a_{nn}\})^2$
Hence $|a_{ij}| \leq |\max\{a_{11}, a_{22}, \ldots, a_{nn}\}|$
$= \max\{a_{11}, a_{22}, \ldots, a_{nn}\}$ [from part (i)]
(iv) Assume r is an eigenvalue of A and let $\mathbf{0} \neq \mathbf{x} \in E_r$; then $0 \lneq \mathbf{x}^T A \mathbf{x} = \mathbf{x}^T(r\mathbf{x}) = r\mathbf{x}^T\mathbf{x} = r\|\mathbf{x}\|^2$, which holds iff $r \gneq 0$.
(v) This follows from (iv), using the fact that the determinant of A is the product of the eigenvalues of A. ■

These necessary conditions are useful for determining, by inspection, that certain symmetric matrices are not positive definite.

49.5 Examples

$$A = \begin{bmatrix} 3 & -\frac{5}{2} \\ -\frac{5}{2} & 2 \end{bmatrix}$$

is not positive definite since its determinant is negative.

$$B = \begin{bmatrix} -5 & 2 \\ 2 & -1 \end{bmatrix}$$

is not positive definite; it has a negative entry on the main diagonal, violating (i).

$$C = \begin{bmatrix} 2 & -2 & 1 \\ -2 & 1 & -3 \\ 1 & -3 & \frac{1}{2} \end{bmatrix}$$

is not positive definite since it violates (iii); one of its off-diagonal entries, -3, is greater in absolute value than the largest of its main diagonal entries, 2.

Concerning the previous examples, observe that A satisfies 48.4(i) and (iii) and that B satisfies (ii) and (v). Thus none of these conditions is sufficient for positive definiteness. On the other hand, condition (iv) *is* sufficient for a real symmetric matrix to be positive definite.

49.6 Theorem

Assume $A \in \mathbb{R}_{n\times n}$ is symmetric; then the following are equivalent:

(i) A is positive definite.
(ii) All the eigenvalues of A are positive.

Proof: (i) → (ii) was proved in 49.4, so we need only prove (ii) → (i). Since by 46.8, a real symmetric matrix is orthogonally diagonalizable over $\mathbb{R}$, there exist in $\mathbb{R}_{n\times n}$ a diagonal matrix $D = \mathrm{diag}(\lambda_1, \lambda_2, \ldots, \lambda_n)$, whose diagonal entries are the eigenvalues of A, and an orthogonal matrix P such that $A = P^{-1}AP = P^TAP$. Let β be the basis for $\mathbb{R}^n$ consisting of the columns of P. Then for any $\mathbf{x} \neq \mathbf{0}$, if we let $((\mathbf{x})_\beta)_i$ denote the ith coordinate of $(\mathbf{x})_\beta$, we have

$$\begin{aligned} \mathbf{x}^TA\mathbf{x} &= [\mathbf{x}]_\beta^T D[\mathbf{x}]_\beta && \text{[Theorem 48.18]} \\ &= \Sigma\lambda_i((\mathbf{x})_\beta)_i^2 \\ &\gneqq 0 && [\text{since each } \lambda_i \gneqq 0] \end{aligned}$$

■

There is another useful characterization of positive definiteness whose statement requires an initial definition.

49.7 Definition

Let $A \in \mathbb{R}_{n\times n}$. For $k = 1, 2, \ldots, n$, let A_k denote the submatrix consisting of the first k-many rows and columns of A (i.e., A_k is the upper-left block from A of size $k \times k$). The *principal diagonal minors of A* are the determinants of these submatrices,

$$|A_1| = a_{11}, \qquad |A_2| = \begin{vmatrix} a_{11} & a_{12} \\ a_{21} & a_{22} \end{vmatrix},$$

$$|A_3| = \begin{vmatrix} a_{11} & a_{12} & a_{13} \\ a_{21} & a_{22} & a_{23} \\ a_{31} & a_{32} & a_{33} \end{vmatrix}, \ldots, |A_n| = |A|$$

49.8 Theorem

Let $A \in \mathbb{R}_{n\times n}$ be symmetric; then the following are equivalent:

(i) A is positive definite.
(ii) The principal diagonal minors of A are all positive.

Proof: ($\rightarrow$) Let k be arbitrary but fixed. Observe that if $\mathbf{x} \in \mathbb{R}^k$, then

$$\mathbf{x}^T \mathrm{A}_k \mathbf{x} = \begin{bmatrix} x_1 & \dots & x_k & 0 & \dots & 0 \end{bmatrix} \mathrm{A} \begin{bmatrix} x_1 \\ \vdots \\ x_k \\ 0 \\ \vdots \\ 0 \end{bmatrix}$$

Thus, if A is positive definite, so is A_k; and so $|\mathrm{A}_k| \gneq 0$ by Theorem 49.4.

($\leftarrow$) The proof is by induction on n. For $n = 1$, we have $a_{11} \gneq 0$ and the result is immediate. We will now prove the result for $n \times n$ matrices assuming, inductively, that it is true for matrices of size $(n - 1) \times (n - 1)$. Note that the principal diagonal minors of A_{n-1} are among those of A, so the induction assumption implies that A_{n-1} is positive definite. If we denote the column vector

$$\begin{bmatrix} a_{1n} \\ a_{2n} \\ \vdots \\ a_{n-1\,n} \end{bmatrix}$$

by $\mathbf{a}$, we may observe, since A is symmetric, that A can be written as the partitioned matrix

$$\mathrm{A} = \left[\begin{array}{c|c} \mathrm{A}_{n-1} & \mathbf{a} \\ \hline \mathbf{a}^T & a_{nn} \end{array}\right]$$

By assumption, $|\mathrm{A}_{n-1}| \gneq 0$, so A_{n-1} is invertible. If we let $d = a_{nn} - \mathbf{a}^T \mathrm{A}_{n-1}^{-1} \mathbf{a}$, it is routine to check, using the symmetry of A_{n-1} and the special case of the Block Multiplication Theorem [Problem 3.24(ii)], that

$$\mathrm{A} = \left[\begin{array}{c|c} \mathrm{I}_{n-1} & 0_{(n-1)\times 1} \\ \hline \mathbf{a}^T (\mathrm{A}_{n-1}^{-1})^T & 1 \end{array}\right] \left[\begin{array}{c|c} \mathrm{A}_{n-1} & 0_{(n-1)\times 1} \\ \hline 0_{1\times(n-1)} & d \end{array}\right] \left[\begin{array}{c|c} \mathrm{I}_{n-1} & \mathrm{A}_{n-1}^{-1}\mathbf{a} \\ \hline 0_{1\times(n-1)} & 1 \end{array}\right]$$

$$= \mathrm{P}^T \mathrm{B} \mathrm{P}.$$

Note that $|\mathrm{P}^T| = |\mathrm{P}| = 1$, so $|\mathrm{A}| = |\mathrm{B}| = d|\mathrm{A}_{n-1}|$. Since $|\mathrm{A}|$ and $|\mathrm{A}_{n-1}|$ are positive by assumption, d is positive; thus, since

$$\mathbf{x}^T\mathrm{B}\mathbf{x} = [x_1 \ldots x_{n-1}]A_{n-1}\begin{bmatrix} x_1 \\ \vdots \\ x_{n-1} \end{bmatrix} + dx_n^2$$

and since A_{n-1} is positive definite, B is positive definite. Finally, since A is conjugate to B, A is positive definite as well by Theorem 49.3. ■

There is a whole assortment of closely related concepts that should be mentioned.

49.9 Definition

Let $\mathrm{A} \in \mathbb{R}_{n\times n}$ be symmetric.

(i) A is *positive semidefinite* iff $\mathbf{x}^T\mathrm{A}\mathbf{x} \geq 0$ for all $\mathbf{x} \in \mathbb{R}^n$.
(ii) A is *negative definite* iff $\mathbf{x}^T\mathrm{A}\mathbf{x} \lneq 0$ for all $\mathbf{0} \neq \mathbf{x} \in \mathbb{R}^n$.
(iii) A is *negative semidefinite* iff $\mathbf{x}^T\mathrm{A}\mathbf{x} \leq 0$ for all $\mathbf{x} \in \mathbb{R}^n$.
(iv) A is *indefinite* if q_A assumes both positive and negative values.

49.10 Exercise

Let $\mathrm{A} \in \mathbb{R}_{n\times n}$ be symmetric. Show that properties (i) to (iv) from Definition 49.9 are equivalent, respectively, to the following:

(i) All the eigenvalues of A are ≥ 0.
(ii) All the eigenvalues of A are $\lneq 0$.
(iii) All the eigenvalues of A are ≤ 0.
(iv) A has both positive and negative eigenvalues.

We close this section with a characterization of positive definiteness that is important for statistical applications.

49.11 Theorem

$\mathrm{A} \in \mathbb{R}_{n\times n}$ is (positive definite) positive semidefinite iff there exists an (invertible) $\mathrm{B} \in \mathbb{R}_{n\times n}$ such that $\mathrm{A} = \mathrm{B}^T\mathrm{B}$.

Proof: ($\leftarrow$) Any matrix of the form $\mathrm{B}^T\mathrm{B}$ is symmetric. For any $\mathbf{0} \neq \mathbf{x} \in \mathbb{R}^n$, $\mathbf{x}^T\mathrm{A}\mathbf{x} = \mathbf{x}^T\mathrm{B}^T\mathrm{B}\mathbf{x} = (\mathrm{B}\mathbf{x}) \cdot (\mathrm{B}\mathbf{x}) = \|\mathrm{B}\mathbf{x}\|^2 \geq 0$ ($\gneq 0$ if B is invertible).

($\rightarrow$) Using Theorem 46.8, choose a $\mathrm{P} \in \mathbb{R}_{n\times n}$ that orthodiagonalizes A over $\mathbb{R}$, and let $\mathrm{D} = \mathrm{diag}(\lambda_1, \lambda_2, \ldots, \lambda_n) = \mathrm{P}^{-1}\mathrm{AP} = \mathrm{P}^T\mathrm{AP}$. Then $\lambda_i \geq 0$ ($\lambda_i \gneq 0$ if A is positive definite) by Exercise 49.10 (by Theorem 49.4). Let $\sqrt{\lambda_i}$

be the positive square root of λ_i, let $\sqrt{D}$ denote the matrix $\mathrm{diag}(\sqrt{\lambda_1}, \sqrt{\lambda_2}, \ldots, \sqrt{\lambda_n})$, and let $B = P\sqrt{D}\,P^T = P\sqrt{D}\,P^{-1}$. Then

$$\begin{aligned} B^TB &= (P\sqrt{D}\,P^T)^T P\sqrt{D}\,P^{-1} \\ &= P(\sqrt{D}\,)^T P^T P\sqrt{D}\,P^{-1} \\ &= P\sqrt{D}\,\sqrt{D}\,P^{-1} && \text{[symmetry of } \sqrt{D} \text{ and orthogonality of P]} \\ &= PDP^{-1} \\ &= A \end{aligned}$$

Moreover, if A is positive definite, each λ_i is greater than zero, so B is invertible in this case. ■

49.12 Remarks on Theorem 49.11

(i) The factorization of a positive definite matrix A into a product of the form B^TB is not unique. For example, instead of the matrix B, we might have taken $B' = \sqrt{D}\,P^T$. You should check that $(B')^TB' = A$. B′ is the matrix used for the statistical procedure known as *principal component analysis*.

(ii) Recall that it does not make sense to talk about *the* square root of an arbitrary matrix. Thus the following features of the matrix B are notable: $B^2 = B$, B is symmetric, and B, being orthogonally similar to $\sqrt{D}$, is positive definite if A is positive definite, by Theorem 49.6. It can be shown that B is the unique positive definite matrix with these properties. For these reasons, B could legitimately be called *the* square root of the positive definite matrix A.

Problem Set 49

49.1. Go back and do any of the exercises from Section 49 that you may have skipped.

*49.2. Find necessary and sufficient conditions on a, b, c, and d for the matrix $\begin{bmatrix} a & b \\ c & d \end{bmatrix}$ to be positive definite.

*49.3. (Generalization of the right-to-left implication of 49.11 to arbitrary rectangular matrices) If $B \in \mathbb{R}_{m\times n}$, then:
(i) Both B^TB and BB^T are positive semidefinite.
(ii) B^TB is positive definite iff $\mathrm{rank}(B) = n$.
(iii) BB^T is positive definite iff $\mathrm{rank}(B) = m$.

*49.4. Let $A \in \mathbb{C}_{n\times n}$. Prove that the following are equivalent:
(i) $\mathbf{x}^H A\mathbf{x}$ is real for all $\mathbf{x} \in \mathbb{C}^n$.
(ii) A is Hermitian.

49.5. Supply the details of the step from the proof of 49.8 that invokes the Block Multiplication Theorem.

49.6. Let A be a real symmetric matrix. Prove that there exists an $s \in \mathbb{R}$ such that whenever $t > s$, the matrix $t\mathrm{I}_n + \mathrm{A}$ is positive definite.

Hint: Problem 34.4 is relevant.

49.7. For each of the matrices below, decide whether it is positive definite; if it is, express it in the form $\mathrm{B}^T\mathrm{B}$.

(i) $\begin{bmatrix} 3 & -2 \\ -2 & 1 \end{bmatrix}$ (ii) $\begin{bmatrix} -2 & 3/2 \\ 3/2 & -2 \end{bmatrix}$

(iii) $\begin{bmatrix} 1 & -3/2 \\ -3/2 & 4 \end{bmatrix}$ (iv) $\begin{bmatrix} 7 & 2 & -4 \\ 2 & 7 & 4 \\ -4 & 4 & 1 \end{bmatrix}$

(v) $\begin{bmatrix} 5/2 & 2 & 1 \\ 2 & 5/2 & 1 \\ 1 & 1 & 1 \end{bmatrix}$ (vi) $\begin{bmatrix} 8 & -3 & 1 \\ -3 & 2 & 1 \\ 1 & 1 & 3 \end{bmatrix}$.

50

The Spectral Theorems

We return in this section to the fundamentals. Matrices are, after all, merely representations of linear transformations. The results from Section 46 on the orthogonal diagonalizability over ℝ of real, symmetric matrices (and the unitary diagonalizability over ℂ of complex, normal matrices) can and should be recast as results about linear operators. We chose to initially present the versions for matrices because we felt that this approach is more tangible for beginning students. The essential nature of these results, however, concerns the ability to decompose the domain, V, of a linear operator, T, into a direct sum, $V = W_1 \oplus W_2 \oplus \cdots \oplus W_p$, of subspaces in such a way that the restrictions of T to these subspaces are operators of a particularly simple kind. In more advanced courses, where infinite-dimensional vector spaces are studied, this "operator-theoretic" point of view, which we are about to present, takes precedence over the "matrix-theoretic" one, because matrices of infinite size are not significantly easier to deal with than the operators themselves.

Most of the proofs in this section are either omitted or only sketched; this is because the results are, by and large, routine translations from the language of matrices to the language of linear operators, of earlier results about matrices that were proved in detail. There is a bit of work to do at the beginning; we must introduce the operator-theoretic concepts and language and establish the translation table; the translation itself is then quite automatic.

Let $T: V \to V$ be a linear operator on a vector space V, and let W be a subspace of V. Of course, $\mathrm{dom}(T \upharpoonright W) = W$. It may or may not happen that $\mathrm{ran}(T \upharpoonright W) \subset W$ (i.e., $T \upharpoonright W$ may or may not be a linear operator *on* W).

50.1 Definition

Let $T: V \to V$ be a linear operator on a vector space V, and let W be a subspace of V. W is called a *T-invariant subspace* if $T \upharpoonright W: W \to W$ (i.e., if $T \upharpoonright W$ is a linear operator on W).

We want to study restrictions of T to T-invariant subspaces; but it is more convenient, notationally, to work with operators whose domain is V. So we artificially extend $T \upharpoonright W$ to V by defining it to be zero off W.

50.2 Definition

Where $T: V \to V$ is a linear operator and W is a subspace of V, define $T_W: V \to V$ by

$$T_W(\mathbf{v}) = \begin{cases} T(\mathbf{v}), & \text{if } \mathbf{v} \in W \\ \mathbf{0}, & \text{if } \mathbf{v} \notin W \end{cases}$$

A special name is given to the preceding when T is the identity operator I.

50.3 Definition

I_W is called the *projection of V onto W*.

Thus I_W sends $\mathbf{v}$ into itself if $\mathbf{v} \in W$ and into $\mathbf{0}$ otherwise.

50.4 Exercise

Prove that

(i) For all $\mathbf{v} \in V$, $I_W^2 = I_W \circ I_W = I_W$.
(ii) If $V = W_1 \oplus W_2$, then $I_{W_1} \circ I_{W_2} = I_{W_2} \circ I_{W_1} = 0$.

50.5 Definition

Let $T: V \to V$ be a linear operator on a vector space V and let W be a subspace of V. A direct-sum decomposition of V, $V = W_1 \oplus W_2 \oplus \cdots \oplus W_p$, is *T-invariant* (abbreviation: T-inv. d.s.d.) if each W_i is a T-invariant subspace.

Having a T-inv. d.s.d. of V reduces the study of T to the independent study of the restrictions, T_{W_i}. This is because each $\mathbf{v} \in V$ is uniquely expressible (recall 43.20) in the form $\mathbf{v} = \mathbf{w}_1 + \mathbf{w}_2 + \cdots + \mathbf{w}_p$, where $\mathbf{w}_i \in W_i$. So

$$\begin{aligned} T(\mathbf{v}) &= T(\mathbf{w}_1 + \mathbf{w}_2 + \cdots + \mathbf{w}_p) \\ &= T(\mathbf{w}_1) + T(\mathbf{w}_2) + \cdots + T(\mathbf{w}_p) && [\text{linearity of } T] \\ &= T_{W_1}(\mathbf{w}_1) + T_{W_2}(\mathbf{w}_2) + \cdots + T_{W_p}(\mathbf{w}_p) && [\text{definition of } T_{W_i}] \end{aligned}$$

Thus, when $V = W_1 \oplus W_2 \oplus \cdots \oplus W_p$ is a T-inv. d.s.d., we are justified in writing

$$T = \sum_{i=1}^{i=p} T_{W_i}.$$

It is useful to broaden the use of this notation in the following way.

50.6 Definition

The expression $T = \Sigma_{i=1}^{i=p} T_i$ means that T and each T_i are linear operators on a vector space V and that there exist T-invariant subspaces W_i, $1 \leq i \leq p$, such that $V = W_1 \oplus W_2 \oplus \cdots \oplus W_p$, ran $T_i \subseteq W_i$ for all i, and $T_i \upharpoonright W_j = 0$ if $i \neq j$.

Before turning to specific examples of T-invariant direct-sum decompositions, we present a theorem that characterizes when a d.s.d. is T-invariant.

50.7 Theorem

Let $T: V \to V$ be a linear operator and let $V = W_1 \oplus W_2 \oplus \cdots \oplus W_p$ be a direct-sum decomposition of V. This d.s.d. is T-invariant iff $T \circ I_{W_i} = I_{W_i} \circ T$ for $i = 1, 2, \ldots, p$; that is, iff T commutes with each of the projections onto the subspaces in question.

Proof: ($\leftarrow$) For each i and for each $\mathbf{v} \in W_i$,

$$\begin{aligned} T(\mathbf{v}) &= T\big(I_{W_i}(\mathbf{v})\big) && [\text{because } \mathbf{v} \in W_i] \\ &= I_{W_i}(T(\mathbf{v})) && [T \text{ commutes with } I_{W_i}] \\ &\in W_i && [\text{by definition of } I_{W_i}] \end{aligned}$$

($\rightarrow$) Fix $\mathbf{v} \in V$ and let $\mathbf{v} = \mathbf{w}_1 + \mathbf{w}_2 + \cdots + \mathbf{w}_p$ be the unique expression for $\mathbf{v}$ such that $\mathbf{w}_i \in W_i$ for $i = 1, 2, \ldots, p$. Then

$$\begin{aligned} T(\mathbf{v}) &= \Sigma T(\mathbf{w}_j) && [\text{linearity of } T] \\ &= \Sigma I_{W_j}\big(T(\mathbf{w}_j)\big) && [\text{each } W_j \text{ is } T\text{-invariant}] \end{aligned}$$

Using this, we obtain

$$\begin{aligned}(I_{W_i}\circ T)(\mathbf{v}) &= I_{W_i}(T(\mathbf{v}))\\ &= I_{W_i}\big(\Sigma I_{W_j}(T(\mathbf{w}_j))\big)\\ &= \Sigma\big(I_{W_i}\circ I_{W_j}\big)\big(T(\mathbf{w}_j)\big)\\ &= I_{W_i}(T(\mathbf{w}_i)) && \text{[Exercise 50.4(ii)]}\\ &= T(\mathbf{w}_i) && [W_i \text{ is } T\text{-invariant}]\\ &= T\big(I_{W_i}(\mathbf{v})\big)\\ &= \big(T\circ I_{W_i}\big)(\mathbf{v}) && \blacksquare\end{aligned}$$

Diagonalizable operators provide the best examples of T-invariant direct-sum decompositions; here, the restrictions are as simple as possible: they are scalar multiples of the identity.

50.8 Theorem

A linear operator $T:V \to V$ is diagonalizable iff $T = \Sigma\lambda_i I_{E_i}$, where $\lambda_1, \ldots, \lambda_p$ are the distinct eigenvalues of T and $E_1, \ldots, E_p$ are the corresponding eigenspaces.

Remark: We are saving one level of subscripting by writing E_i in place of E_{λ_i}. In the theorem, we should really have written $T = \Sigma\lambda_i I_{E_{\lambda_i}}$.

Proof outline: Use the results about direct sums (43.20) and about the independence of distinct eigenspaces (34.9) to show that this theorem is just a rephrasing, in operator-theoretic language, of 34.7 and 37.4 combined. ■

The Spectral Theorem for $\mathbb{E}^n$ (for $\mathbb{H}^n$) is a translation into operator-theoretic language of the fundamental result from Section 46 on the orthogonal diagonalizability over $\mathbb{R}$ (the unitary diagonalizability over $\mathbb{C}$) of a symmetric (normal) matrix.

Symmetry (normality) is a property of a matrix. Before we can make the translation, we need to know what it means in operator-theoretic terms to say that the standard matrix for $T:\mathbb{E}^n \to \mathbb{E}^n$ (for $T:\mathbb{H}^n \to \mathbb{H}^n$) is symmetric (normal). This is the purpose of the next few pages.

The definitions that we are about to give make sense for operators on arbitrary inner product spaces, and we prefer to state them in their general form. Later, we will specialize to $\mathbb{E}^n$ and $\mathbb{H}^n$ (i.e., to the Euclidean inner products on $\mathbb{R}^n$ and $\mathbb{C}^n$).

We deal first with real inner product spaces; complex inner product spaces are considered at the end of the section.

If V is a real inner product space, then among the linear functionals on V (recall the definition of the dual space, V^*, in Problem 41.5) are those that are defined "by taking the inner product with a fixed vector, **w**." To understand what this means, recall that an inner product is linear in each variable separately; thus, for each $\mathbf{w} \in V$, the function $f_{\mathbf{w}}$ defined by $f_{\mathbf{w}}(\mathbf{v}) = \langle \mathbf{v}, \mathbf{w} \rangle$ is a linear functional on V.

The point of the next theorem is that when V is a finite-dimensional inner product space, these are the *only* linear functionals on V (i.e., every element of V^* is of the form $f_{\mathbf{w}}$ for some $\mathbf{w} \in V$). We have explicitly mentioned the finite-dimensionality of V in this context because the theorem is false in general for infinite-dimensional spaces.

50.9 Theorem

Let V be a finite-dimensional real inner product space and let $f \in V^*$; then there is a unique $\mathbf{w} \in V$ such that $f = f_{\mathbf{w}}$.

Proof. Existence: Let $\beta = (\mathbf{u}_1, \mathbf{u}_2, \ldots, \mathbf{u}_n)$ be a basis for V that is orthonormal with respect to the given inner product, and let $\mathbf{w} = \Sigma f(\mathbf{u}_i)\mathbf{u}_i$. We will show that f and $f_{\mathbf{w}}$ agree on each of the basis vectors in β, and hence that $f = f_{\mathbf{w}}$. For each fixed j,

$$\begin{aligned} f_{\mathbf{w}}(\mathbf{u}_j) &= \langle \Sigma f(\mathbf{u}_i)\mathbf{u}_i, \mathbf{u}_j \rangle \\ &= \Sigma f(\mathbf{u}_i)\langle \mathbf{u}_i, u_j \rangle && [\text{bilinearity of } \langle \cdot , \cdot \rangle] \\ &= \Sigma f(\mathbf{u}_i)\delta_{ij} && [\text{orthonormality of } \beta] \\ &= f(\mathbf{u}_j) \end{aligned}$$

Uniqueness: We have to show that $f_{\mathbf{w}_1} = f_{\mathbf{w}_2}$ implies $\mathbf{w}_1 = \mathbf{w}_2$. But this is a general fact about inner products (recall 45.17). ■

The next theorem provides the foundation for the definition (to follow in 50.11) of the adjoint of a linear operator on an inner product space.

50.10 Theorem

Let V be a finite-dimensional real inner product space. For every linear operator T on V, there is a unique linear operator on V, denoted by T^*, that satisfies

$$\langle T(\mathbf{u}), \mathbf{v} \rangle = \langle \mathbf{u}, T^*(\mathbf{v}) \rangle \qquad \text{for all } \mathbf{u}, \mathbf{v} \in V.$$

Proof. Existence: Here is the process for defining $T^*(\mathbf{v})$: first, observe that the function $f: V \to \mathbb{R}$, which depends on T and $\mathbf{v}$ and is defined by $f(\mathbf{u}) = \langle T(\mathbf{u}), \mathbf{v} \rangle$, is a linear functional on V; second, use this observation together with Theorem 50.9 to conclude that there is a unique $\mathbf{w} \in V$ such that $f(\mathbf{u}) = \langle \mathbf{u}, \mathbf{w} \rangle$ for all $\mathbf{u} \in V$. If we define $T^*(\mathbf{v})$ to be this unique $\mathbf{w}$, then $\langle T(\mathbf{u}), \mathbf{v} \rangle = \langle \mathbf{u}, T^*(\mathbf{v}) \rangle$ is satisfied.

Uniqueness: Suppose T_1 and T_2 satisfy

$$\langle T(\mathbf{u}), \mathbf{v} \rangle = \langle \mathbf{u}, T_1(\mathbf{v}) \rangle = \langle \mathbf{u}, T_2(\mathbf{v}) \rangle, \quad \text{for all } \mathbf{u}, \mathbf{v} \in V.$$

Then, in particular, for each fixed $\mathbf{v} \in V$ we have that

$$\langle \mathbf{u}, T_1(\mathbf{v}) \rangle = \langle \mathbf{u}, T_2(\mathbf{v}) \rangle, \quad \text{for all } \mathbf{u} \in V.$$

This implies, by 45.17, that $T_1(\mathbf{v}) = T_2(\mathbf{v})$. Since this is true for each $\mathbf{v} \in V$, $T_1 = T_2$. ■

Note that the proof of uniqueness does not depend on the finite dimensionality of V, so we may broaden the scope of the definition that follows.

50.11 Definition

Let T be a linear operator on an inner product space V. When it exists, the unique linear operator T^* on V satisfying

$$\langle T(\mathbf{u}), \mathbf{v} \rangle = \langle \mathbf{u}, T^*(\mathbf{v}) \rangle, \quad \text{for all } \mathbf{u}, \mathbf{v} \in V$$

is called the *adjoint of T.*

If V is infinite dimensional, the adjoint of a given linear operator T may or may not exist; when V is finite dimensional, the existence of T^* is guaranteed by Theorem 50.10.

50.12 Definition

A linear operator T on an inner-product space V is called *self-adjoint* if $T = T^*$.

With this important definition in place, we can now, by specializing it to the case of a linear operator T on $\mathbb{E}^n$, provide an operator-theoretic equivalent of the statement that the standard matrix for T is symmetric.

50.13 Theorem

The following are equivalent for any linear operator $T:\mathbb{E}^n \to \mathbb{E}^n$.

(i) T is self-adjoint.
(ii) ${}_{\alpha^n}[T]_{\alpha^n}$ is symmetric.
(iii) If β is any orthonormal basis for $\mathbb{E}^n$, then ${}_{\beta}[T]_{\beta}$ is symmetric.

Proof: The equivalence of (ii) and (iii) is immediate since ${}_{\alpha^n}[T]_{\alpha^n}$ and ${}_{\beta}[T]_{\beta}$ are orthogonally similar.

Proof of (i) ↔ (ii): Let A denote the standard matrix for T. Recall that the inner product involved here is the Euclidean dot product. We have

$$\begin{aligned} &T = T^* \\ &\quad \text{iff } T(\mathbf{u}) \cdot \mathbf{v} = \mathbf{u} \cdot T(\mathbf{v}), \quad \text{for all } \mathbf{u}, \mathbf{v} \in \mathbb{E}^n \quad [\text{definition of } T^*] \\ &\quad \text{iff } \mathrm{A}\mathbf{u} \cdot \mathbf{v} = \mathbf{u} \cdot \mathrm{A}\mathbf{v}, \quad \text{for all } \mathbf{u}, \mathbf{v} \in \mathbb{E}^n \quad [\text{definition of A}] \\ &\quad \text{iff } \mathrm{A} = \mathrm{A}^T \quad [\text{Problem 45.4}] \ \blacksquare \end{aligned}$$

50.14 Theorem (Spectral Theorem for $\mathbb{E}^n$)

If T is a self-adjoint linear operator on $\mathbb{E}^n$, then $T = \Sigma_i \lambda_i I_{E_i}$, where $\lambda_1, \lambda_2, \ldots, \lambda_p$ are the distinct eigenvalues of T and $E_1, E_2, \ldots, E_p$ are the corresponding eigenspaces.

Proof outline: Use Theorems 50.8 and 50.13 to show that the Spectral Theorem for $\mathbb{E}^n$ is just a translation of the fact that a real symmetric matrix is orthogonally diagonalizable over $\mathbb{R}$. ■

The expression $T = \Sigma_i \lambda_i I_{E_i}$ is known as the *spectral resolution of T*. Note that the T-inv. d.s.d. $\mathbb{E}^n = E_1 \oplus E_2 \oplus \cdots \oplus E_p$ is actually, in this case, a T-invariant *orthogonal* direct-sum decomposition.

Very few modifications are needed to obtain the analogues of all these results for complex inner product spaces. We will sketch the details, assuming that the complex inner product $\langle \cdot , \cdot \rangle$ is linear in the first argument and conjugate linear in the second. In case the reverse is true, some trivial changes suffice; these are left to the reader.

If V is a complex inner product space, then for each $\mathbf{w} \in V$, the function $f_{\mathbf{w}}: V \to \mathbb{C}$ defined by $f_{\mathbf{w}}(\mathbf{v}) = \langle \mathbf{v}, \mathbf{w} \rangle$ is a linear functional on V.

50.15 Theorem (Analogue of 50.9)

Let V be a finite-dimensional complex inner product space, and let $f \in V^*$ be any linear functional on V. Then there is a unique $\mathbf{w} \in V$ such that $f = f_{\mathbf{w}}$.

Proof: Except for the presence of complex conjugation, the proof is identical to that of 50.9. Here we must take $\mathbf{w} = \Sigma_i \overline{f(\mathbf{u}_i)}\mathbf{u}_i$. ■

In Theorem 50.10 and in the definition of adjoint, we need only replace the word real by the word complex.

50.16 Theorem (Analogue of 50.13)

The following are equivalent for any linear operator $T:\mathbb{H}^n \to \mathbb{H}^n$:

(i) T is self-adjoint.
(ii) ${}_{\alpha^n}[T]_{\alpha^n}$ is Hermitian.
(iii) If β is any orthonormal basis for $\mathbb{H}^n$, then ${}_{\beta}[T]_{\beta}$ is Hermitian.

The proof is identical to that of 50.13 except that the inner product involved is $\cdot_2$ [Definition 45.19] and that Problem 45.5 replaces Problem 45.4 in the justification of the last line. ■

50.17 Definition

A linear operator T on a complex inner product space V is *normal* iff $T \circ T^* = T^* \circ T$ (i.e., iff T commutes with its adjoint under composition).

50.18 Theorem (Spectral Theorem for $\mathbb{H}^n$)

If T is a normal linear operator on $\mathbb{H}^n$, then $T = \Sigma_i \lambda_i I_{E_i}$, where $\lambda_1, \lambda_2, \ldots, \lambda_p$ are the distinct eigenvalues of T and $E_1, E_2, \ldots, E_p$ are the corresponding eigenspaces.

Proof outline: Use Theorems 50.8 and 50.16 to show that the Spectral Theorem for $\mathbb{H}^n$ is just a translation of the fact that a complex normal matrix is unitarily diagonalizable over $\mathbb{C}$. ■

Problem Set 50

50.1. Go back and do any of the exercises from Section 50 that you may have skipped.

*50.2. The function $g:\mathbb{E}^2 \to \mathbb{R}$ defined by $g(x_1, x_2) = 3x_1 - 2x_2$ is a linear functional on $\mathbb{E}^2$. By Theorem 50.9, there is a unique $\mathbf{w} \in \mathbb{E}^2$ such that $g = f_{\mathbf{w}}$. Find $\mathbf{w}$.
Hint: The proof of 50.9 provides a recipe for constructing $\mathbf{w}$.

51

The Jordan Canonical Form of an $n \times n$ Matrix

Let $T: V \to V$ be a linear operator on a finite-dimensional vector space V. We have seen that having a T-invariant direct-sum decomposition of V, $V = W_1 \oplus W_2 \oplus \cdots \oplus W_p$, reduces the study of T to the study of its restrictions $T \restriction W_i$. Something is gained by this only if the restrictions are of a particularly simple kind. The main result from Section 50 is that if T is a self-adjoint operator on $\mathbb{E}^n$ (or a normal operator on $\mathbb{H}^n$), then there is a T-inv. d.s.d. of dom T for which the restrictions are as simple as possible, that is, scalar multiples of the identity.

If T is not self-adjoint, what can be done along these lines? The final section of the text addresses this question.

We begin by restating four equivalent versions of diagonalizability in order to stress, one last time, that a single idea can be expressed in many different languages.

The following are equivalent for a linear operator T on a finite-dimensional vector space V:

(a) Operator-theoretic Language: There exists a T-invariant direct-sum decomposition of V, $V = W_1 \oplus W_2 \oplus \cdots \oplus W_p$, for which the restrictions $T \restriction W_i$ are scalar multiples of the identity.

(b) Descriptive Behavior of T: There exists a basis β for V such that, when β coordinates are used to label the points of V, T has the form $T(x_1, x_2, \ldots, x_n) = (c_1x_1, c_2x_2, \ldots, c_nx_n)$.

(c) Language of Matrices: There exists a basis β for V such that ${}_\beta[T]_\beta$ is a diagonal matrix.
(d) "Eigen-language": There exists a basis for V consisting of eigenvectors of T.

By analogy, we will be presenting four equivalent versions of the Jordan form theorem.

The fundamental object of investigation is the operator; but our preference is to deal with its matrix representations, because these are more tangible. The motivation for what is done in this section is to find, in the absence of a diagonal matrix representing T, some other matrix representing T that is as "simple" as possible. (Vaguely speaking, this means that it should have as many zero entries as possible.)

Recall that there are two possible obstructions to the diagonalizability over $\mathbb{R}$ of a real $n \times n$ matrix A:

Obstruction 1: A may not have n-many real eigenvalues, counting multiplicities.

Obstruction 2: Even when A has enough real eigenvalues, it may not have n-many linearly independent eigenvectors (i.e., the dimension of some eigenspace may be strictly less than the algebraic multiplicity of the corresponding eigenvalue).

We have seen that one way to overcome the first obstruction is to introduce complex numbers. There are other ways to overcome this while remaining within the real number system, but these methods are beyond the scope of this text. For this, you should consult the material on the Primary Decomposition Theorem and the Rational Canonical Form in a more advanced book, such as [HK] or [FIS]. The resolution of these problems depends on a considerably more detailed study of polynomials.

For the remainder of this section, therefore, we make the following assumption.

51.1 Assumption

V is an n-dimensional vector space over the scalar field $\mathbb{F}$, T is a linear operator on V, and the characteristic polynomial of T factors completely over $\mathbb{F}$ into a product of linear factors, say,

$$\chi(T) = (\lambda - c_1)^{m_1}(\lambda - c_2)^{m_2} \cdots (\lambda - c_p)^{m_p} \qquad (\dagger)$$

where $c_1, c_2, \ldots, c_p$ are the distinct roots of $\chi(T)$ in $\mathbb{F}$ and where $m_1 + m_2 + \cdots + m_p = n$.

In particular, the results of this section apply to *all* linear operators on complex vector spaces, as well as to those operators on an n-dimensional real vector space that have n-many real eigenvalues, counting multiplicities.

To provide a better focus on the main results and their applications, we will defer the proof of one result (Theorem 51.24) to the end of the section.

It is very easy to state the main result in language which describes the behavior of T.

51.2 Theorem (Jordan Canonical Form Theorem, Version (b))

If T satisfies 51.1, then there is a basis β for V such that, when points in V are labeled with their β coordinates, the description of T has the form

$$T(x_1, x_2, \ldots, x_n) = (a_1x_1 + b_2x_2, a_2x_2 + b_3x_3, \ldots, a_{n-1}x_{n-1} + b_nx_n, a_nx_n)$$

where, for $i = 2, 3, \ldots, n$, each b_i is 0 or 1 and where, if $b_i = 1$, then $a_i = a_{i-1}$.

The simplicity of this description resides in the fact that the ith coordinate of the output depends at most on the ith and $(i+1)^{\text{st}}$ coordinates of the input.

If T and β are as in 51.2, then the a_i must be eigenvalues of T: to see this, observe that

$$_{\beta}[T]_{\beta} = \begin{bmatrix} a_1 & b_2 & 0 & \cdot & \cdot & 0 \\ 0 & a_2 & b_3 & \cdot & \cdot & 0 \\ 0 & 0 & a_3 & \cdot & \cdot & 0 \\ 0 & \cdot & \cdot & a_{n-2} & b_{n-1} & 0 \\ 0 & \cdot & \cdot & 0 & a_{n-1} & b_n \\ 0 & \cdot & \cdot & \cdot & 0 & a_n \end{bmatrix}$$

and that its characteristic polynomial is therefore $\Pi_i(\lambda - a_i)$. Note that T is diagonalizable iff each b_i is 0.

To state equivalent versions of the Jordan Canonical Form Theorem in the language of operators, or matrices, or in "eigen-language," we need to introduce several new concepts.

51.3 Definition

If A is an $m \times m$ matrix and B is an $n \times n$ matrix, the *direct sum* of A and B is the $(m+n) \times (m+n)$ matrix

$$\left[\begin{array}{c|c} A & 0_{m \times n} \\ \hline 0_{n \times m} & B \end{array}\right]$$

It is denoted by $A \oplus B$ or by $\text{diag}(A, B)$.

This definition can be extended to finitely many matrices in an obvious way.

51.4 Definition

(i) If $A_1, A_2, \ldots, A_p$ are square matrices of order $k_1, k_2, \ldots, k_p$, respectively, then their *direct-sum* is the matrix

$$\left[\begin{array}{c:c:c:c:c:c}
A_1 & & & & & \\ \hdashline
 & A_2 & & & & \\ \hdashline
 & & & & & \\ \hdashline
 & & A_i & & & \\ \hdashline
 & & & & & \\ \hdashline
 & & & & & A_p
\end{array}\right]$$

in which all the blocks off the main diagonal are zero. Its size is $(\Sigma_i k_i) \times (\Sigma_i k_i)$, and it is denoted by $A_1 \oplus A_2 \oplus \cdots \oplus A_p$, by $\oplus_i A_i$, or by $\operatorname{diag}(A_1, A_2, \ldots, A_p)$.

(ii) A square matrix that can be written as a direct-sum of smaller square matrices is called a *block-diagonal* matrix.

51.5 Example

$$\left[\begin{array}{cc|ccc|c}
2 & 0 & 0 & 0 & 0 & 0 \\
3 & 2 & 0 & 0 & 0 & 0 \\ \hline
0 & 0 & 1 & 2 & 6 & 0 \\
0 & 0 & -1 & -1 & 0 & 0 \\
0 & 0 & 4 & 0 & 4 & 0 \\ \hline
0 & 0 & 0 & 0 & 0 & 7
\end{array}\right]$$

It is customary to display a block-diagonal matrix partitioned in a way that emphasizes its block-diagonal character. The preceding matrix is $\operatorname{diag}(A_1, A_2, A_3)$, where

$$A_1 = \begin{bmatrix} 2 & 0 \\ 3 & 2 \end{bmatrix}, \quad A_2 = \begin{bmatrix} 1 & 2 & 6 \\ -1 & -1 & 0 \\ 4 & 0 & 4 \end{bmatrix}, \quad A_3 = [7].$$

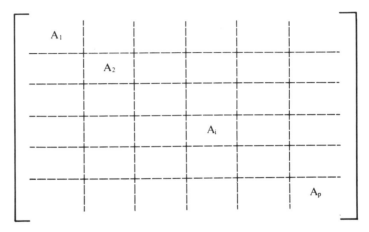

We will need the following result, which relates direct sums of matrices with direct sums of subspaces. Recall that if $V = W_1 \oplus W_2 \oplus \cdots \oplus W_p$, and if, for $i = 1, 2, \ldots, p$, β_i is a basis for W_i, then the union, $\beta = \beta_1 \cup \beta_2 \cup \cdots \cup \beta_p$, is a basis for V.

51.6 Exercise

With V, W_i, β, and β_i as above, show that this d.s.d. is T-invariant iff ${}_\beta[T]_\beta = \text{diag}(A_1, A_2, \ldots, A_p)$, where $A_i = {}_{\beta_i}[T \upharpoonright W_i]_{\beta_i}$.
Hint: The case $p = 2$ contains all the relevant ideas.

We also need the following result about determinants of block-diagonal matrices; the special case of this result for $p = 2$ was presented in Problem 28.1.

51.7 Exercise

If $A = \text{diag}(A_1, A_2, \ldots, A_p)$, then $|A| = |A_1|\,|A_2| \cdots |A_p|$.

Combining the two previous exercises yields an important fact about characteristic polynomials.

51.8 Exercise

If $V = \oplus_i W_i$ is a T-inv. d.s.d., then $\chi(T)$ is the product of the characteristic polynomials of its restrictions, $T \upharpoonright W_i$; that is,

$$\chi(T) = \chi(T \upharpoonright W_1)\chi(T \upharpoonright W_2) \cdots \chi(T \upharpoonright W_p).$$

The next step in the development is to define the concept of a Jordan matrix. These are matrices that can be expressed as direct sums of elementary Jordan blocks, so we must begin with this latter concept.

51.9 Definition

Where k is a positive integer and c is any scalar, the *elementary Jordan block*, $J[k; c]$, is the following $k \times k$ matrix:

$$\begin{bmatrix} c & 1 & 0 & \cdot & \cdot & \cdot & 0 \\ 0 & c & 1 & 0 & \cdot & \cdot & 0 \\ \vdots & \vdots & \vdots & \vdots & \vdots & \vdots & \vdots \\ 0 & \cdot & \cdot & \cdot & \cdot & \cdot & 0 \\ 0 & \cdot & \cdot & \cdot & \cdot & c & 1 \\ 0 & \cdot & \cdot & \cdot & \cdot & \cdot & c \end{bmatrix}$$

The main diagonal entries of $J[k; c]$ are all equal to c; the entries on the first superdiagonal are all equal to 1; all other entries are 0.

51.10 Examples

$$J[2;0] = \begin{bmatrix} 0 & 1 \\ 0 & 0 \end{bmatrix}, \qquad J[3;3] = \begin{bmatrix} 3 & 1 & 0 \\ 0 & 3 & 1 \\ 0 & 0 & 3 \end{bmatrix}$$

51.11 Definition

A matrix is said to be *in Jordan form* (or, synonymously, *is a Jordan matrix*) if it is a direct sum of elementary Jordan blocks.

51.12 Examples

It takes a great deal of space to display large matrices in Jordan form. Turn to item 51.30 and look at the solutions given there for examples of Jordan matrices.

Here are two observations concerning the characteristic polynomials of Jordan matrices.

51.13 Observations

(i) Since the determinant of an upper triangular matrix is the product of its main diagonal entries, $\chi(J[k;c]) = (\lambda - c)^k$. This observation, combined with 51.7, implies the following:

(ii) If $A = J[k_1;c_1] \oplus J[k_2;c_2] \oplus \cdots \oplus J[k_p;c_p]$, then $\chi(A) = (\lambda - c_1)^{k_1}(\lambda - c_2)^{k_2} \cdots (\lambda - c_p)^{k_p}$.

At this point, it becomes possible to state another version of the main result.

51.14 Theorem (Jordan Canonical Form Theorem, Version (c))

If T satisfies 51.1, then there exists a basis β for V such that ${}_\beta[T]_\beta$ is in Jordan form.

51.15 Exercise

Verify the equivalence of the two versions of the Jordan Form Theorem presented thus far.

To state the equivalent versions of this theorem in "eigen-language" and in the language of operators, we require a few more definitions. Recall that the problem we face is the lack of enough linearly independent eigenvectors; equivalently, the eigenspaces are so small that their direct sum does not add up to the whole of V. Consequently, we consider a collection of potentially larger subspaces.

If c is an eigenvalue of T, the eigenspace, E_c, is the kernel of the linear operator $T - cI$. That is, $E_c = \{\mathbf{v} \in V\colon (T - cI)\mathbf{v} = \mathbf{0}\}$. We consider the potentially larger subspace consisting of those vectors in V that are sent to $\mathbf{0}$ by *some power of* $(T - cI)$.

Where k is a positive integer and c is an eigenvalue of T, let $E_c^k = \{\mathbf{v} \in V\colon (T - cI)^k\mathbf{v} = \mathbf{0}\}$. It is clear that E_c^k is a subspace of V [it is the nullspace of the operator $(T - cI)^k$], and that $E_c = E_c^1 \subset E_c^2 \subset \cdots \subset E_c^k \subset \cdots$. Because V is finite-dimensional, this increasing chain of subspaces cannot continue to increase forever; thus there is a least k for which $E_c^k = E_c^{k+1}$; we now claim that in fact, once equality holds in this chain, it continues to hold from that point on.

51.16 Lemma

With the preceding notation, if k is the least positive integer for which $E_c^k = E_c^{k+1}$, then for any positive integer j, $E_c^k = E_c^{k+j}$.

Proof by induction on j: If $j = 1$, the result is true by assumption. Assuming as an induction hypothesis that $E_c^k = E_c^{k+j-1}$, we see that if $\mathbf{v} \in E_c^{k+j}$, then

$$(T - cI)^{k+j}(\mathbf{v}) = \mathbf{0}$$

which implies $(T - cI)^{k+1}((T - cI)^{j-1}(\mathbf{v})) = \mathbf{0}$
which implies $(T - cI)^{k}((T - cI)^{j-1}(\mathbf{v})) = \mathbf{0}$ [because $E_c^k = E_c^{k+1}$]
which implies $\mathbf{v} \in E_c^{k+j-1}$
which implies $\mathbf{v} \in E_c^k$ [by the induction hypothesis] ■

This argument justifies the following definition.

51.17 Definition

Let T be a linear operator on a finite-dimensional vector space V and let c be an eigenvalue of T. Let k be the least positive integer for which $E_c^k = E_c^{k+1}$.

(i) The *generalized eigenspace of c* is the set $K_c = \{\mathbf{v} \in V\colon (T - cI)^k(\mathbf{v}) = \mathbf{0}\}$.
(ii) The nonzero vectors in K_c are called the *generalized eigenvectors associated* with the eigenvalue c.

A few simple consequences of the definition should be mentioned immediately.

51.18 Observations

(i) The generalized eigenspace of c is at the end of a strictly increasing chain of subspaces that begins with the eigenspace of c.

$$E_c = E_c^1 \subsetneqq E_c^2 \subsetneqq E_c^3 \subsetneqq \cdots \subsetneqq E_c^k = K_c$$

Of course, it can happen that $k = 1$ and that $E_c = K_c$.

(ii) If $\mathbf{0} \neq \mathbf{v} \in E_c^q - E_c^{q-1}$, then q is the least integer for which $(T - cI)^q(\mathbf{v}) = \mathbf{0}$, and $\mathbf{0} \neq (T - cI)^{q-1}(\mathbf{v}) \in E_c$.
In words, this asserts that if $\mathbf{v}$ is a generalized eigenvector that enters K_c at the qth stage in the chain of subspaces leading up to K_c, then $(T - cI)^{q-1}(\mathbf{v})$ is an eigenvector of T. (To accommodate the case $q = 1$, we make the conventions that $E_c^0 = \{\mathbf{0}\}$ and that for any function f, f^0 is the identity.)

(iii) For every $\mathbf{v} \in K_c$, $(T - cI)^k(\mathbf{v}) = \mathbf{0}$. Moreover, k is the least integer for which this statement holds.

(iv) K_c is a T-invariant subspace, for if $\mathbf{v} \in K_c$ then

$$\begin{aligned}(T - cI)^k(T(\mathbf{v})) &= T\big((T - cI)^k(\mathbf{v})\big) \\ &= T(\mathbf{0}) && [\text{since } \mathbf{v} \in K_c] \\ &= \mathbf{0}.\end{aligned}$$

In the next theorem, we show that the generalized eigenspaces corresponding to the distinct eigenvalues of T are an independent collection of subspaces. To save one level of subscripting, we will simplify the notation by writing E_j and K_j in place of E_{c_j} and K_{c_j} from now on.

51.19 Lemma

If T is any linear operator, if c is an eigenvalue of T, if $\mathbf{v} \in E_c$, if q is a positive integer, and if s is a scalar, then $(T - sI)^q(\mathbf{v}) = (c - s)^q\mathbf{v}$.

The proof is a straightforward argument by induction on q; it is left as an exercise.

51.20 Theorem

Any collection $\{K_1, K_2, \ldots, K_p\}$ of generalized eigenspaces corresponding to distinct eigenvalues $c_1, c_2, \ldots, c_p$ of an operator T is an independent collection of subspaces.

Proof: By 43.22, it suffices to view the collection as an ordered set and prove for $i \geq 2$ that

$$K_i \cap \sum_{j \lneq i} K_j.$$

If this conclusion were false, there is a least integer i for which it fails; thus $(K_1, K_2, \ldots, K_{i-1})$ is independent, but $(K_1, K_2, \ldots, K_{i-1}, K_i)$ is dependent. So suppose that $\mathbf{0} = \mathbf{w}_1 + \mathbf{w}_2 + \cdots + \mathbf{w}_i$, where $\mathbf{w}_k \in K_k$ and where not all of the $\mathbf{w}_k$ are $\mathbf{0}$. Note that there must be at least two values of k for which $\mathbf{w}_k \neq \mathbf{0}$. Let q_i be the least integer such that $(T - c_i I)^{q_i}(\mathbf{w}_i) = \mathbf{0}$. We have

$$\mathbf{w}_1 + \mathbf{w}_2 + \cdots + \mathbf{w}_{i-1} = -\mathbf{w}_i \qquad (\dagger)$$

By applying $(T - c_i I)^{q_i}$ to both sides of ($\dagger$), we obtain

$$\begin{aligned}(c_1 - c_i)^{q_i}\mathbf{w}_1 &+ (c_2 - c_i)^{q_i}\mathbf{w}_2 + \cdots + (c_{i-1} - c_i)^{q_i}\mathbf{w}_{i-1} \\ &= (T - c_i I)^{q_i}(\mathbf{w}_1 + \mathbf{w}_2 + \cdots + \mathbf{w}_{i-1}) \\ &\qquad\qquad [\text{using the lemma and linearity}] \\ &= (T - c_i I)^{q_i}(-\mathbf{w}_i) \\ &= \mathbf{0}.\end{aligned}$$

Because the c's are distinct and $\mathbf{w}_k \neq \mathbf{0}$ for at least one $k \leq i - 1$, this contradicts the independence of $(K_1, K_2, \ldots, K_{i-1})$. ■

It is desirable at this point, since we now have the vocabulary at hand, to state two more versions of the Jordan Canonical Form Theorem.

51.21 Theorem (Jordan Canonical Form Theorem, Version (a))

If T satisfies 51.1, then the generalized eigenspaces corresponding to the distinct eigenvalues of T form a T-inv. d.s.d. of V; that is,

$$V = K_{c_1} \oplus K_{c_2} \oplus \cdots \oplus K_{c_p}$$

51.22 Theorem (Jordan Canonical Form Theorem, Version (d))

If T satisfies 51.1, then there exists a basis for V consisting of generalized eigenvectors of T.

It is not clear that these two versions are equivalent to the two presented earlier. The equivalence of version (a) requires further analysis of the restrictions $T \upharpoonright K_{c_i}$; that of (d) requires that the basis of generalized eigenvectors be ordered in a certain way so that the matrix that represents T with respect to this ordered basis will be in Jordan form. The purpose of stating these versions now is to illuminate the path ahead: we can anticipate, in view of 51.6, that a Jordan matrix representing T will be obtained as a direct sum of matrices representing the restrictions, $T \upharpoonright K_{c_i}$.

We continue with an explanation of the sense in which the restrictions of T to its generalized eigenspaces are "simple." Let c be an eigenvalue of T and let k be such that $K_c = E_c^k$. It is always true that $T = cI + (T - cI)$; thus $T \upharpoonright K_c = cI \upharpoonright K_c + (T - cI) \upharpoonright K_c$. The first summand is a scalar multiple of the identity; the second summand has the following property, which is a consequence of 51.18(iii): for every $\mathbf{v} \in K_c$, $(T - cI)^k(\mathbf{v}) = \mathbf{0}$, and k is the least integer for which this happens.

We will now present a general theorem concerning operators that have such a property; this theorem will then be applied to operators of the form $(T - cI) \upharpoonright K_c$.

51.23 Definition

An operator $S: W \to W$ is *nilpotent* if there exists a positive integer k such that $S^k(\mathbf{w}) = \mathbf{0}$ for all $\mathbf{w} \in W$. A nilpotent operator is *nilpotent of index* q if q is the least integer for which S^q is identically zero.

51.24 Theorem

Let $S: W \to W$ be a nilpotent operator of index q on a finite-dimensional vector space W of dimension m; then there exist generalized eigenvectors $\mathbf{y}_1, \mathbf{y}_2, \ldots, \mathbf{y}_t$ and integers $q = q_1 \geq q_2 \geq \cdots \geq q_t \geq 1$ such that

$$\gamma = \left(S^{q_1-1}(\mathbf{y}_1), \ldots, S(\mathbf{y}_1), \mathbf{y}_1, S^{q_2-1}(\mathbf{y}_2), \ldots, S(\mathbf{y}_2), \mathbf{y}_2, \ldots, S^{q_t-1}(\mathbf{y}_t), \ldots, S(\mathbf{y}_t), \mathbf{y}_t\right)$$

is a basis for W.

This is the result whose proof will be left to the end of the section.

The importance of this theorem is that, with respect to such a basis γ, the matrix ${}_{\gamma}[S]_{\gamma}$ has a very simple form. For $k = 1, 2, \ldots, t$, let $\gamma_k = \left(S^{q_k-1}(\mathbf{y}_k), \ldots, S(\mathbf{y}_k), \mathbf{y}_k\right)$ and let $W_k = \mathscr{L}(\gamma_k)$.

51.25 Exercise

(i) Show that the W_k are S-invariant subspaces of W.
(ii) Show that ${}_{\gamma_k}[S \upharpoonright W_k]_{\gamma_k} = \mathrm{J}[q_k;0]$.

The fact that γ is a basis guarantees that $m = \sum_{k=1}^{k=t} q_k$, which implies, using 43.20, that $W = W_1 \oplus W_2 \oplus \cdots \oplus W_t$ is an S-invariant direct-sum decomposition. Consequently

$$_\gamma[S]_\gamma = \mathrm{J}[q_1;0] \oplus \mathrm{J}[q_2;0] \oplus \cdots \oplus \mathrm{J}[q_t;0].$$

For this reason, an ordered basis of the type described in 51.24 is often called a *Jordan basis* or a *J-basis*.

A better picture can be obtained if we let $\gamma = (\mathbf{w}_1, \mathbf{w}_2, \ldots, \mathbf{w}_m)$ be a relabeling of the vectors in γ; suppose that with this relabeling, the vectors $\mathbf{y}_1, \mathbf{y}_2, \ldots, \mathbf{y}_t$ correspond to $\mathbf{w}_{k_1}, \mathbf{w}_{k_2}, \ldots, \mathbf{w}_{k_t}$. Then $_\gamma[S]_\gamma$ is the matrix which has 1's on the first superdiagonal except in columns $k_1, k_2, \ldots, k_t$; *all* other entries are 0.

51.26 Example

Consider the operator $S:\mathbb{R}^4 \to \mathbb{R}^4$ defined by $S(x_1, x_2, x_3, x_4) = (0, 3x_1, 0, -2x_2 + 5x_3)$. Observe that $S^2(x_1, x_2, x_3, x_4) = S(0, 3x_1, 0, -2x_2 + 5x_3) = (0, 0, 0, -6x_1)$ and that S^3 is identically zero. Thus S is nilpotent of index 3. The standard matrix for S is

$$\begin{bmatrix} 0 & 0 & 0 & 0 \\ 3 & 0 & 0 & 0 \\ 0 & 0 & 0 & 0 \\ 0 & -2 & 5 & 0 \end{bmatrix}$$

Let γ be the ordered basis $((0,0,0,1), (0,1,0,0), (1,0,0,0), (0,0,1,0))$. (You will have to await the proof of the theorem to understand how γ was obtained; for the moment, we are just illustrating the theorem.) You should verify that

$$_\gamma[S]_\gamma = \begin{bmatrix} 0 & 1 & 0 & 0 \\ 0 & 0 & 1 & 0 \\ 0 & 0 & 0 & 0 \\ 0 & 0 & 0 & 0 \end{bmatrix} = \mathrm{J}[3;0] \oplus \mathrm{J}[1;0]$$

This fits the general pattern in the theorem when we let $\mathbf{y}_1 = (1,0,0,0)$ and $\mathbf{y}_2 = (0,0,1,0)$.

An additional aspect of Theorem 51.24 should be mentioned.

51.27 Theorem 51.24 (Continued)

The number of blocks, t, and the sizes of the blocks, $q_1, q_2, \ldots, q_t$, do *not* depend on γ; they are completely determined by the operator S. Specifically, 0 is the only eigenvalue of S,

$$E_0 = E_0^1 \subsetneqq E_0^2 \subsetneqq \cdots \subsetneqq E_0^{q-1} \subsetneqq E_0^q = W$$

and the q_i and t are related, in a way that we will later explain, to the dimensions of the E_0^i; in particular, $t = \dim E_0$.

These additional aspects of Theorem 51.24 will also be proved at the end of the section. Their effect is the following: any two matrices in Jordan form that represent S are direct sums of the *same* elementary Jordan blocks; they differ at most in the order in which these elementary blocks appear down the main diagonal. In other words, there is an "essentially unique" matrix, J, in Jordan form representing a nilpotent operator S; the dimension of dom S determines the size of J, and the number and sizes of the elementary Jordan blocks determines J to within permutation of the blocks.

We turn next to the implications that this theorem has for the restrictions of an operator T to its generalized eigenspaces. The simplicity of $T \upharpoonright K_c$ is that it is the sum of a scalar multiple of the identity, $cI \upharpoonright K_c$, plus a nilpotent operator, $(T - cI) \upharpoonright K_c$.

It is true in general, for operators T_1 and T_2 on a vector space V and for an arbitrary basis β for V, that ${}_\beta[T_1 + T_2]_\beta = {}_\beta[T_1]_\beta + {}_\beta[T_2]_\beta$. Thus, if K_c is k-dimensional and if γ is a basis for K_c with respect to which the nilpotent part of $T \upharpoonright K_c$ is represented by $\mathrm{J}[q_1;0] \oplus \mathrm{J}[q_2;0] \oplus \cdots \oplus \mathrm{J}[q_t;0]$, then

$$\begin{aligned} {}_\gamma[T \upharpoonright K_c]_\gamma &= {}_\gamma[cI]_\gamma + {}_\gamma[T - cI]_\gamma \\ &= cI_k + (\mathrm{J}[q_1;0] \oplus \mathrm{J}[q_2;0] \oplus \cdots \oplus \mathrm{J}[q_t;0]) \\ &= \mathrm{J}[q_1;c] \oplus \mathrm{J}[q_2;c] \oplus \cdots \oplus \mathrm{J}[q_t;c] \end{aligned}$$

which is a direct sum of elementary Jordan blocks.

There is one more ingredient needed for the proof of the Jordan Canonical Form Theorem.

51.28 Theorem

If T satisfies 51.1, then for $i = 1, 2, \ldots, p$, the dimension of the generalized eigenspace, K_i, is the algebraic multiplicity, m_i, of the eigenvalue c_i.

Proof: (Adapted from [H]) Suppose that k_i is the least integer for which $E_{c_i}^{k_i} = E_{c_i}^{k_i+1}$; thus $E_{c_i}^{k_i} = K_i = \ker(T - c_iI)^{k_i}$. Let $H_i = \operatorname{ran}(T - c_iT)^{k_i}$. We claim that $V = H_i \oplus K_i$.

Proof of claim: If $\mathbf{v} \in K_i$, then $(T - c_iI)^{k_i}(\mathbf{v}) = \mathbf{0}$. If $\mathbf{v} \in H_i$, then $\mathbf{v} = (T - c_iI)^{k_i}(\mathbf{u})$ for some $\mathbf{u}$. So if $\mathbf{v} \in H_i \cap K_i$, $(T - c_iI)^{2k_i}(\mathbf{u}) = \mathbf{0}$. Because $E_i^{k_i} = \cdots = E_i^{2k_i} = \cdots$, this implies $\mathbf{u} \in E_i^{k_i}$, so $\mathbf{v} = (T - c_iI)^{k_i}(\mathbf{u}) = \mathbf{0}$. This proves that H_i and K_i are independent subspaces. The rank plus nullity theorem applied to $(T - c_iI)^{k_i}$ guarantees that $\dim H_i + \dim K_i = \dim V$. So by 43.20, $V = H_i \oplus K_i$.

This implies, by 51.8, that $\chi(T) = \chi(T \upharpoonright H_i)\chi(T \upharpoonright K_i)$. Our ultimate conclusion now requires two more facts.

(i) The only eigenvalue of $T \upharpoonright K_i$ is c_i. To see this, note that if c is an eigenvalue of T and if $\mathbf{0} \neq \mathbf{v} \in K_i$, then

$$\begin{aligned} \mathbf{0} &= (T - c_i I)^{k_i}(\mathbf{v}) && [\mathbf{v} \in K_i] \\ &= (c - c_i)^{k_i}\mathbf{v}. && [\text{Lemma } 51.19] \end{aligned}$$

(ii) c_i is not an eigenvalue of $T \upharpoonright H_i$. To see this, we will obtain a contradiction from the assumption that $\mathbf{0} \neq \mathbf{v} \in H_i$ and $T(\mathbf{v}) = c_i\mathbf{v}$. If this were true, then by definition of H_i, $\mathbf{v} = (T - c_i I)^{k_i}(\mathbf{u})$ for some $\mathbf{u}$, and we would have $\mathbf{0} = (T - c_i I)(\mathbf{v}) = (T - c_i I)^{k_i+1}(\mathbf{u})$. Since $E_i^{k_i} = E_i^{k_i+1}$, $\mathbf{0} = (T - c_i I)^{k_i}(\mathbf{u}) = \mathbf{v}$.

By assumption, $\chi(T) = (\lambda - c_1)^{m_1}(\lambda - c_2)^{m_2} \cdots (\lambda - c_p)^{m_p}$. But, also, $\chi(T) = \chi(T \upharpoonright H_i)\chi(T \upharpoonright K_i)$. By (i), $\chi(T \upharpoonright K_i) = (\lambda - c_i)^{d_i}$, where $d_i = \dim K_i$. By (ii), the polynomial $\chi(T \upharpoonright H_i)$ does *not* have c_i as a root. Hence $d_i = \dim K_i = m_i$. ■

Modulo 51.24, we can now prove (all four versions of) the Jordan canonical form theorem.

51.29 Proof of the Jordan Canonical Form Theorem

By 51.28, $\dim K_i = m_i$; by 51.20, the K_i are an independent collection of subspaces; so by 43.20, $\dim(\Sigma_i K_i) = \Sigma_i \dim K_i = \Sigma_i m_i = n = \dim V$. Consequently, $V = K_1 \oplus K_2 \oplus \cdots \oplus K_p$ and this is a T-invariant direct-sum decomposition of V. This proves version (a). To prove versions (b), (c), and (d), apply the procedure derived from 51.24 to each generalized eigenspace separately to produce bases, γ_i, for K_i such that $\mathrm{A}_i = {}_{\gamma_i}[T \upharpoonright K_i]_{\gamma_i}$ is a direct sum of elementary Jordan blocks. By 43.20, $\gamma = \gamma_1 \cup \gamma_2 \cup \cdots \cup \gamma_p$ is a basis for V; it consists of generalized eigenvectors of T and it follows from 51.6 and 51.19 that

$${}_\gamma[T]_\gamma = \mathrm{diag}(\mathrm{A}_1, \mathrm{A}_2, \ldots, \mathrm{A}_p)$$

which is a matrix in Jordan form. ■

Because this matrix is essentially unique (i.e., to within permutation of the elementary Jordan blocks), it is customary to speak of *the* Jordan form of a matrix A and to denote "it" (i.e., any one of them) by J(A). It is possible, although we will not do this, to introduce conventions concerning the ordering of the elementary Jordan blocks so that J(A) denotes a specific one among the matrices in Jordan form similar to A.

Let us summarize the presentation thus far. If the matrix A represents an operator T that satisfies 51.1, then $\mathrm{J(A)} = \mathrm{diag}(\mathrm{A}_1, \mathrm{A}_2, \ldots, \mathrm{A}_p)$ where for $i = 1, 2, \ldots, p$, m_i = algebraic multiplicity of $c_i = \dim K_i$ and the $m_i \times m_i$ matrix A_i is a direct sum of t_i-many elementary Jordan blocks, where $t_i = \dim E_i$. The information about J(A) that is still missing concerns the sizes of the elementary Jordan blocks within each of the A_i. To provide this information in general is tantamount to proving Theorem 51.24. We will proceed to this proof after giving some examples that illustrate the points just made.

51.30 Examples

Give a complete list (always to within permutation of the blocks) of the possible Jordan forms of the following matrices. The information given about the matrices is what you would have if you attempted to diagonalize them and found they are defective.

(i) Suppose that A is a 7×7 matrix, that $\chi(\mathrm{A}) = (\lambda - 4)^7$, and that $\dim E_4 = 3$.
Solution: J(A) is a direct sum of three elementary Jordan blocks; the sum of the sizes of these blocks is 7; the possible sizes of the blocks are therefore 5, 1, 1, or 4, 2, 1, or 3, 3, 1. Hence J(A) is equal to one of the following: J[5;4] ⊕ J[1;4] ⊕ J[1;4], or J[4;4] ⊕ J[2;4] ⊕ J[1;4], or J[3;4] ⊕ J[3;4] ⊕ J[1;4].

$$\begin{bmatrix} 4 & 1 & 0 & 0 & 0 & 0 & 0 \\ 0 & 4 & 1 & 0 & 0 & 0 & 0 \\ 0 & 0 & 4 & 1 & 0 & 0 & 0 \\ 0 & 0 & 0 & 4 & 1 & 0 & 0 \\ 0 & 0 & 0 & 0 & 4 & 0 & 0 \\ 0 & 0 & 0 & 0 & 0 & 4 & 0 \\ 0 & 0 & 0 & 0 & 0 & 0 & 4 \end{bmatrix} \quad \text{or} \quad \begin{bmatrix} 4 & 1 & 0 & 0 & 0 & 0 & 0 \\ 0 & 4 & 1 & 0 & 0 & 0 & 0 \\ 0 & 0 & 4 & 1 & 0 & 0 & 0 \\ 0 & 0 & 0 & 4 & 0 & 0 & 0 \\ 0 & 0 & 0 & 0 & 4 & 1 & 0 \\ 0 & 0 & 0 & 0 & 0 & 4 & 0 \\ 0 & 0 & 0 & 0 & 0 & 0 & 4 \end{bmatrix}$$

$$\text{or} \quad \begin{bmatrix} 4 & 1 & 0 & 0 & 0 & 0 & 0 \\ 0 & 4 & 1 & 0 & 0 & 0 & 0 \\ 0 & 0 & 4 & 0 & 0 & 0 & 0 \\ 0 & 0 & 0 & 4 & 1 & 0 & 0 \\ 0 & 0 & 0 & 0 & 4 & 1 & 0 \\ 0 & 0 & 0 & 0 & 0 & 4 & 0 \\ 0 & 0 & 0 & 0 & 0 & 0 & 4 \end{bmatrix}$$

(ii) Suppose that B is a 6×6 matrix, that $\chi(\mathrm{B}) = (\lambda - 2)^4(\lambda - 5)^2$, that $\dim E_2 = 2$, and that $\dim E_5 = 1$.

Solution: J(B) = J[3;2] ⊕ J[1;2] ⊕ J[2;5] or J(B) = J[2;2] ⊕ J[2;2] ⊕ J[2;5].

$$\begin{bmatrix} 2 & 1 & 0 & 0 & 0 & 0 \\ 0 & 2 & 1 & 0 & 0 & 0 \\ 0 & 0 & 2 & 0 & 0 & 0 \\ 0 & 0 & 0 & 2 & 0 & 0 \\ 0 & 0 & 0 & 0 & 5 & 1 \\ 0 & 0 & 0 & 0 & 0 & 5 \end{bmatrix} \text{ or } \begin{bmatrix} 2 & 1 & 0 & 0 & 0 & 0 \\ 0 & 2 & 0 & 0 & 0 & 0 \\ 0 & 0 & 2 & 1 & 0 & 0 \\ 0 & 0 & 0 & 2 & 0 & 0 \\ 0 & 0 & 0 & 0 & 5 & 1 \\ 0 & 0 & 0 & 0 & 0 & 5 \end{bmatrix}$$

(iii) Suppose that C is a 5 × 5 matrix, that $\chi(C) = (\lambda + 9)^3(\lambda - 5)^2$, that dim $E_{-9} = 1$, and that dim $E_5 = 2$.
Solution: J(C) = J[3; − 9] ⊕ J[1;5] ⊕ J[1;5]

$$= \begin{bmatrix} -9 & 1 & 0 & 0 & 0 \\ 0 & -9 & 1 & 0 & 0 \\ 0 & 0 & -9 & 0 & 0 \\ 0 & 0 & 0 & 5 & 0 \\ 0 & 0 & 0 & 0 & 5 \end{bmatrix}$$

Before proving Theorem 51.24, we restate it for convenience.

51.31 Theorem 51.24 (Repeated)

Let $S: W \to W$ be a nilpotent operator of index q on a finite-dimensional vector space W of dimension m; then there exist generalized eigenvectors $\mathbf{y}_1, \mathbf{y}_2, \ldots, \mathbf{y}_t$ and integers $q = q_1 \geq q_2 \geq \cdots \geq q_t \geq 1$ such that

$$\gamma = \big(S^{q_1-1}(\mathbf{y}_1), \ldots, S(\mathbf{y}_1), \mathbf{y}_1, S^{q_2-1}(\mathbf{y}_2), \ldots, S(\mathbf{y}_2), \mathbf{y}_2, \ldots, S^{q_t-1}(\mathbf{y}_t), \ldots, S(\mathbf{y}_t), \mathbf{y}_t\big)$$

is a basis for W.

Proof: (Adapted from [I]) The convention that $S^0 = I$, the identity operator, is in effect throughout. Let us mention immediately that although the q_i are listed in nonincreasing order, this feature is irrelevant for the proof of 51.24; it is relevant only for the additional aspects that relate to uniqueness, mentioned in 51.27, and which we will discuss subsequently. The next lemma is the heart of the proof of 51.24.

51.32 Lemma

Let S be any operator on a vector space W, let $\mathbf{y}_1, \mathbf{y}_2, \ldots, \mathbf{y}_t \in W$, and let $q_1 \geq q_2 \geq \cdots \geq q_t \geq 1$ be positive integers such that $\{S^{q_1-1}(\mathbf{y}_1), S^{q_2-1}(\mathbf{y}_2), \ldots, S^{q_t-1}(\mathbf{y}_t)\}$ is a linearly independent subset of

$\ker S$; then the set $\{S^{q_1-1}(\mathbf{y}_1), \ldots, S(\mathbf{y}_1), \mathbf{y}_1, S^{q_2-1}(\mathbf{y}_2), \ldots, S(\mathbf{y}_2), \mathbf{y}_2, \ldots, S^{q_t-1}(\mathbf{y}_t), \ldots, S(\mathbf{y}_t), \mathbf{y}_t\}$ is linearly independent.

Proof: Display the vectors in an array of rows and columns, as follows:

$$\begin{array}{ccccc} S^{q_1-1}(\mathbf{y}_1) & S^{q_1-2}(\mathbf{y}_1) & \cdots & S(\mathbf{y}_1) & \mathbf{y}_1 \\ S^{q_2-1}(\mathbf{y}_2) & S^{q_2-2}(\mathbf{y}_2) & \cdots & S(\mathbf{y}_2) & \mathbf{y}_2 \\ & & \vdots & & \\ S^{q_t-1}(\mathbf{y}_t) & S^{q_t-2}(\mathbf{y}_t) & \cdots & S(\mathbf{y}_t) & \mathbf{y}_t \end{array}$$

We have drawn this array as if it were rectangular, but in fact, the length of the ith row is q_i, so the lengths of the rows are not necessarily equal but only nonincreasing. One purpose of this display is to simplify the notation: let $\mathbf{w}_{ij}$ denote the vector in row i, column j of this display [i.e., $\mathbf{w}_{ij} = S^{q_i-j}(\mathbf{y}_i)$]. Observe that $S(\mathbf{w}_{i1}) = 0$ for $1 \le i \le t$, and that $S(\mathbf{w}_{ij}) = \mathbf{w}_{i,j-1}$ for $1 \le i \le t$ and $2 \le j \le q_i$. For $k = 1, 2, \ldots, q_1$, let Y_k consist of the vectors whose column index is less than or equal to k. Observe that Y_1 is the subset of $\ker S$ whose linear independence is assumed, and that Y_{q_1} is the set we wish to prove linearly independent. This is accomplished by showing, for arbitrary k, that if Y_k is linearly independent then so is Y_{k+1}. So assume that

$$\mathbf{0} = \sum_{i=1}^{i=t} \sum_{j=1}^{j=k+1} a_{ij}\mathbf{w}_{ij}$$

is a linear combination of the vectors from Y_{k+1}. Applying S to both sides of this equation and using the linearity of S, yields

$$\mathbf{0} = S(\mathbf{0}) = \sum_{i=1}^{i=t} \sum_{j=1}^{j=k+1} a_{ij}S(\mathbf{w}_{ij}) = \sum_{i=1}^{i=t} \sum_{j=2}^{j=k+1} a_{ij}\mathbf{w}_{i,j-1}$$

which is a linear combination of vectors from Y_k. The linear independence of Y_k now guarantees that $a_{ij} = 0$ for $1 \le i \le t$ and $2 \le j \le k+1$. From our original sum, we are now left with

$$\mathbf{0} = \sum_{i=1}^{i=t} a_{i1}\mathbf{w}_{i1}$$

and the linear independence of Y_1 finally implies that $a_{i1} = 0$ for $1 \le i \le t$. ■

51.33 Proof of Theorem 51.24

The proof is by induction on q, the index of nilpotency of S. For $q = 1$, the result is trivial; so assume $q > 1$. For any operator S, $\operatorname{ran} S$ is an S-invariant subspace, so we may consider the operator $S \restriction \operatorname{ran} S$. Since $S \restriction \operatorname{ran} S$

is nilpotent of index $q-1$, the induction hypothesis applies to produce vectors $\mathbf{x}_1, \mathbf{x}_2, \ldots, \mathbf{x}_r$ and positive integers $q-1 = p_1 \geq p_2 \geq \cdots \geq p_r \geq 1$ such that $S^{p_i}(\mathbf{x}_i) = \mathbf{0}$ and

$$\{S^{p_1-1}(\mathbf{x}_1), \ldots, S(\mathbf{x}_1), \mathbf{x}_1, S^{p_2-1}(\mathbf{x}_2), \ldots, S(\mathbf{x}_2), \mathbf{x}_2, \ldots, S^{p_r-1}(\mathbf{x}_r), \ldots, S(\mathbf{x}_r), \mathbf{x}_r\}$$

is a basis for ran S.

Let us exhibit these vectors in a "rectangular" array, as in the proof of the lemma.

$$\begin{array}{ccccc} S^{p_1-1}(\mathbf{x}_1) & S^{p_1-2}(\mathbf{x}_1) & \cdots & S(\mathbf{x}_1) & \mathbf{x}_1 \\ S^{p_2-1}(\mathbf{x}_2) & S^{p_2-2}(\mathbf{x}_2) & \cdots & S(\mathbf{x}_2) & \mathbf{x}_2 \\ \vdots & \vdots & \vdots & \vdots & \vdots \\ S^{p_r-1}(\mathbf{x}_r) & S^{p_r-2}(\mathbf{y}_r) & \cdots & S(\mathbf{x}_r) & \mathbf{x}_r \end{array}$$

Let $\mathbf{u}_{ij}$ denote the vector in row i, column j of this display; let X_k consist of those vectors whose column index is less than or equal to k. Thus, X_{q-1} $(= X_{p_1})$ is the entire array and is a basis for ran S.

We will now extend X_{q-1} to a basis for the whole of W. Each row in the array is extended by one vector as follows: for $1 \leq i \leq r$, $\mathbf{x}_i \in$ ran S, so we can choose a $\mathbf{y}_i \in W$ such that $S(\mathbf{y}_i) = \mathbf{x}_i$. The first column, X_1, of the array is a linearly independent subset of ker S; extend it by as many vectors as necessary (these will be labeled $\mathbf{y}_{r+1}, \mathbf{y}_{r+2}, \ldots, \mathbf{y}_{r+s}$) so that $X_1 \cup \{\mathbf{y}_{r+1}, \mathbf{y}_{r+2}, \ldots, \mathbf{y}_{r+s}\}$ is a basis for ker S.

We now claim that the set $\beta = X_{q-1} \cup \{\mathbf{y}_1, \mathbf{y}_2, \ldots, \mathbf{y}_r\} \cup \{\mathbf{y}_{r+1}, \mathbf{y}_{r+2}, \ldots, \mathbf{y}_{r+s}\}$ of vectors in this extended array is a basis for W. The linear independence of β follows from the lemma because, by construction, β satisfies the hypotheses of the lemma. The proof will be complete when we show that β has the right number of elements, that is, m $(= \dim V)$. To do this, combine the following six facts:

(i) $|\beta| = |X_{q-1}| + r + s = \dim(\operatorname{ran} S) + r + s$.
(ii) $\dim(\ker S) = |X_1| + s$.
(iii) $\dim(\operatorname{ran} S) + \dim(\ker S) = m$. [rank plus nullity theorem]
(iv) $\dim(\operatorname{ran} S + \ker S) = \dim(\operatorname{ran} S) + \dim(\ker S) - \dim(\operatorname{ran} S \cap \ker S)$. [Problem 8.12(iii)]
(v) X_1 is a basis for ran $S \cap$ ker S, and so $\dim(\operatorname{ran} S \cap \ker S) = |X_1| = r$.
(vi) $X_{q-1} \cup \{\mathbf{y}_{r+1}, \mathbf{y}_{r+2}, \ldots, \mathbf{y}_{r+s}\}$ is a basis for the sum, ran S + ker S, and so $\dim(\operatorname{ran} S + \ker S) = |X_{q-1}| + s$.

Of these, only (v) requires an argument: if $\mathbf{u} \in$ ran S, then

$$\mathbf{u} = \sum_{i=1}^{i=r} \sum_{j=1}^{j=p_i-1} b_{ij}\mathbf{u}_{ij}.$$

If $\mathbf{u} \in \ker S$ as well, then

$$\mathbf{0} = S(\mathbf{u}) = \sum_{i=1}^{i=r} \sum_{j=1}^{j=p_i-1} b_{ij} S(\mathbf{u}_{ij}) = \sum_{i=1}^{i=r} \sum_{j=2}^{j=p_i-1} b_{ij} \mathbf{u}_{i,j-1}.$$

Since X_{q-1} is linearly independent, we conclude that $b_{ij} = 0$ for $1 \le i \le r$ and $2 \le j \le p_i - 1$, so $\mathbf{u} \in \mathscr{L}(X_1)$. ■

51.34 Theorem 51.24 (Continued)

If S is a nilpotent operator of index q on an m-dimensional space W, and if β is any Jordan basis for W, then 0 is the only eigenvalue of W and

$$E_0 = E_0^1 \subsetneqq E_0^2 \subsetneqq \cdots \subsetneqq E_0^{q-1} \subsetneqq E_0^q = K_0 = W \qquad (\dagger)$$

Moreover, if β is displayed in an array as in Lemma 51.32, then for $k = 1, 2, \ldots, q$, the set Y_k is a basis for E_0^k.

Proof: If $c \neq 0$ and $\mathbf{v} \neq \mathbf{0}$ and $S\mathbf{v} = c\mathbf{v}$, then for every k, $S^k(\mathbf{v}) = c^k \mathbf{v} \neq \mathbf{0}$, contradicting the nilpotence of S; so 0 is the only eigenvalue of S. Thus ($\dagger$) holds by definition of q. Y_k is a linearly independent subset of E_0^k. The argument to show that it spans E_0^k is similar to that of (v) above: if

$$\mathbf{w} = \sum_{i=1}^{i=t} \sum_{j=1}^{j=q_i} a_{ij} \mathbf{w}_{ij} \in W$$

and if $S^k(\mathbf{w}) = \mathbf{0}$, then

$$\mathbf{0} = \sum_{i=1}^{i=t} \sum_{j=k+1}^{j=q_i} a_{ij} \mathbf{w}_{i,j-k}$$

So $a_{ij} = 0$ for $1 \le i \le t$ and $k + 1 \le j \le q_i$. Thus $\mathbf{w} \in \mathscr{L}(Y_k)$. ■

51.35 Corollary (Uniqueness of the Jordan Form for Nilpotent Operators)

If $S: W \to W$ is nilpotent of index q and W is finite-dimensional, and if

$$\gamma = \left(S^{q_1-1}(\mathbf{y}_1), \ldots, S(\mathbf{y}_1), \mathbf{y}_1, S^{q_2-1}(\mathbf{y}_2), \ldots, S(\mathbf{y}_2), \mathbf{y}_2, \ldots, S^{q_t-1}(\mathbf{y}_t), \ldots, S(\mathbf{y}_t), \mathbf{y}_t\right)$$

is a J-basis for W, then the integers q_i and t are determined by S itself and are therefore the same for any J-basis.

Proof: Display γ in an array as in Lemma 51.32. By the previous theorem, the number of columns in the array is q, the index of nilpotency of S. By the previous theorem, Y_1 is a basis for $E_0 = \ker S$; thus the number of vectors in column 1 of the array is the nullity of S. For $k \geq 2$, the number of vectors in column k is the nullity of S^k minus the nullity of S^{k-1}. Because the rows are listed so that their lengths $q_1 \geq q_2 \geq \cdots \geq q_t$ are nonincreasing, this information determines the lengths of the rows (which are the sizes of the elementary Jordan blocks). Since ${}_\gamma[S]_\gamma = \mathrm{J}[q_1;0] \oplus \mathrm{J}[q_2;0] \oplus \cdots \oplus \mathrm{J}[q_t;0]$, the number of elementary Jordan blocks and their sizes are determined by S. ■

This result not only establishes the essential uniqueness (to within permutation of the blocks) of J(A), but also provides a method for computing J(A) that does not require computing a specific basis of generalized eigenvectors. Recall that, if A is similar to a diagonal matrix D, then we can find D without having to find a matrix P such that $\mathrm{P}^{-1}\mathrm{AP} = \mathrm{D}$; the existence of D and its possible values are determined by the eigenvalues of A and by the relationship between their geometric and algebraic multiplicities. Similarly, J(A) is determined by the eigenvalues $c_1, c_2, \ldots, c_p$ of A and the dimensions of the $E_{c_i}^k$ for $k = 1, 2, \ldots, k_i$.

We conclude the text with a few applications of the Jordan form theorem.

The best known application explains why the word "canonical" is associated with it. The Jordan form is a canonical form for similarity; that is, every matrix is similar to a unique (always to within permutation of the blocks) matrix in Jordan form. This is important because it provides an algorithm for determining whether two matrices A and B are similar.

51.36 Theorem

A and B are similar iff J(A) = J(B).

Proof: ($\leftarrow$) Any matrix C is similar to its Jordan form, J(C), since J(C) represents T_{C} with respect to a Jordan basis for V. Since similarity is transitive, the proof is complete.

($\rightarrow$) Throughout the argument, we use the facts proved in 51.31 to 51.35 about the Jordan form of a nilpotent operator S of index q; it is the direct sum of elementary Jordan blocks, the number and sizes of which are determined by the dimensions of dom S and of $\ker S^k$ for $k = 1, 2, \ldots, q$. If A and B are similar, then $\chi(\mathrm{A}) = \chi(\mathrm{B})$; so A and B have the same eigenvalues, $c_1, c_2, \ldots, c_p$, with the same algebraic multiplicities, $m_1, m_2, \ldots, m_p$. Since a matrix is similar to its Jordan form, we also have $\chi(J(A)) = \chi(\mathrm{A}) = \chi(\mathrm{B}) = \chi(J(B))$. Thus J(A) and J(B) have the same distinct diagonal entries, $c_1, c_2, \ldots, c_p$; each of these is repeated the same number of times (c_i is repeated m_i-many times) in J(A) and in J(B). It remains to prove for arbitrary c that the c-blocks in J(A) and in J(B) (the blocks that have c down the main diagonal) are direct sums of the same number of elementary Jordan blocks of the same sizes. If $B = P^{-1}AP$, then $(B - cI) = P^{-1}(A - cI)P$, so $B - cI$ and $A - cI$ are similar; they consequently have the same nullity. This shows that the c-blocks in J(A) and in J(B) are direct sums of the same number of elementary Jordan blocks. If $(B - cI) = P^{-1}(A - cI)P$, then for every positive integer k, $(B - cI)^k = P^{-1}(A - cI)^k P$; this shows that the sizes of the elementary Jordan blocks are the same. ■

Computing powers of matrices is another important application of the Jordan form. In 33.7 and 33.8, it was explained how to obtain a general formula for the kth power, A^k, of a diagonalizable matrix A. If $D = P^{-1}AP$, then $A^k = PD^kP^{-1}$; if D is a diagonal matrix, then it is trivial to compute D^k and obtain a formula for A^k. Note that to exploit this we need to know both the eigenvalues of A (= the diagonal entries of D) *and* a basis of eigenvectors (= the columns of P).

If A is not diagonalizable, we can use the Jordan form of A to arrive at the same goal, with a bit more effort. The idea is the same: if $J(A) = P^{-1}AP$, then $A^k = P(J(A))^kP^{-1}$. The problem is to compute $(J(A))^k$.

To do this, let D be the diagonal matrix whose diagonal entries are those of J(A); let N be the strictly upper triangular matrix obtained from J(A) by changing all the diagonal entries to 0. Then $J(A) = D + N$. Now N is nilpotent, so that, if we could apply the binomial theorem,

$$\begin{aligned}(J(A))^k &= (D + N)^k \\ &= D^k + kD^{k-1}N + \frac{k(k-1)}{2}D^{k-2}N^2 + \cdots + kDN^{k-1} + N^k,\end{aligned}$$

we could get a relatively simple expression for $(J(A))^k$ because the terms in this sum that involve N^q for q greater than or equal to the index of nilpotency of N are all $0_{n \times n}$.

To apply the binomial theorem, D and N must commute [recall Problem 3.10(ii)]. Fortunately, in this circumstance, they do.

51.37 Lemma

Where D and N are as above, $DN = ND$.

Proof: Suppose that $J(A) = \text{diag}(A_1, A_2, \ldots, A_p)$, where A_i is the block associated with the eigenvalue c_i; then $D = \text{diag}(D_1, D_2, \ldots, D_p)$ and $N = \text{diag}(N_1, N_2, \ldots, N_p)$, where D_i and N_i are obtained from A_i as D and

N were obtained from A. Using the Block Multiplication Theorem, we have

$$\mathrm{DN} = \sum_{i=1}^{i=p} \mathrm{D}_i\mathrm{N}_i \quad \text{and} \quad \mathrm{ND} = \sum_{i=1}^{i=p} \mathrm{N}_i\mathrm{D}_i.$$

Because D_i is a scalar matrix (i.e., all its diagonal entries are the same), $\mathrm{D}_i\mathrm{N}_i = \mathrm{N}_i\mathrm{D}_i$. Since this is true for $i = 1, 2, \ldots, p$, the proof is complete. ■

51.38 Example

To illustrate these ideas, let us compute B^8 where

$$\mathrm{B} = \begin{bmatrix} 0 & 8 & 1 \\ 1 & 0 & 0 \\ 0 & 4 & -1 \end{bmatrix}$$

is the nondiagonalizable matrix in Example 35.6. We will also exhibit a general formula for B^k as a function of k. Refer to 35.6 for the following facts about B.

(i) $\chi(\mathrm{B}) = (\lambda - 3)(\lambda + 2)^2$.
(ii) $E_3 = \{(x_1, x_2, x_3) \in \mathbb{R}^3\colon x_1 = 3x_3, x_2 = x_3\}$.
(iii) $E_{-2}\{(x_1, x_2, x_3) \in \mathbb{R}^3\colon x_1 = \frac{1}{2}x_3, x_2 = -\frac{1}{4}x_3\}$.

Since E_{-2} is only one-dimensional, B is not diagonalizable. J(B) is already determined by the information at hand.

$$\mathrm{J(B)} = \begin{bmatrix} 3 & 0 & 0 \\ 0 & -2 & 1 \\ 0 & 0 & -2 \end{bmatrix} = \begin{bmatrix} 3 & 0 & 0 \\ 0 & -2 & 0 \\ 0 & 0 & -2 \end{bmatrix} + \begin{bmatrix} 0 & 0 & 0 \\ 0 & 0 & 1 \\ 0 & 0 & 0 \end{bmatrix} = \mathrm{D} + \mathrm{N}.$$

Since $\mathrm{N}^q = 0_{n\times n}$ for $q \geq 2$, we have, for arbitrary k, that

$$\begin{aligned}(\mathrm{J(B)})^k &= \mathrm{D}^k + k\mathrm{D}^{k-1}\mathrm{N} \\ &= \begin{bmatrix} (3)^k & 0 & 0 \\ 0 & (-2)^k & 0 \\ 0 & 0 & (-2)^k \end{bmatrix} + \begin{bmatrix} 0 & 0 & 0 \\ 0 & 0 & k(-2)^{k-1} \\ 0 & 0 & 0 \end{bmatrix} \\ &= \begin{bmatrix} (3)^k & 0 & 0 \\ 0 & (-2)^k & k(-2)^{k-1} \\ 0 & 0 & (-2)^k \end{bmatrix}.\end{aligned}$$

To obtain a formula for B^8, we also need a basis $\gamma = (\mathbf{u}_1, \mathbf{u}_2, \mathbf{u}_3)$ of generalized eigenvectors. For $\mathbf{u}_1$, we can take any vector in E_3; for example, (3, 1, 1). The vectors $\mathbf{u}_2$ and $\mathbf{u}_3$ must satisfy

$$\mathrm{B}\mathbf{u}_2 = -2\mathbf{u}_2 \quad \text{and} \quad \mathrm{B}\mathbf{u}_3 = -2\mathbf{u}_3 + \mathbf{u}_2. \qquad (\dagger)$$

One way to find such vectors is to rewrite (†) as a set of six equations in six unknowns: $(x_1, x_2, x_3) = \mathbf{u}_2$, $(x_4, x_5, x_6) = \mathbf{u}_3$ (see Problem 51.4 to understand why we proceed in this fashion). We leave it as an exercise to verify that this system has a two-dimensional solution space and that a particular solution is $\mathbf{u}_2 = (2, -1, 4)$ and $\mathbf{u}_3 = (-3, 1, 0)$. Thus

$$\mathrm{B}^8 = \begin{bmatrix} 3 & 2 & -3 \\ 1 & -1 & 1 \\ 1 & 4 & 0 \end{bmatrix} \begin{bmatrix} (3)^8 & 0 & 0 \\ 0 & (-2)^8 & 8(-2)^7 \\ 0 & 0 & (-2)^8 \end{bmatrix} \begin{bmatrix} 3 & 2 & -3 \\ 1 & -1 & 1 \\ 1 & 4 & 0 \end{bmatrix}^{-1}$$

$$= \begin{bmatrix} 3692 & 8260 & 347 \\ 804 & 3692 & 457 \\ 1828 & 1388 & -311 \end{bmatrix}.$$

Of course, for $k = 8$, it is less work to find B^8 by straightforward multiplication. The point is that we have obtained a formula for B^k as a function of k.

$$\mathrm{B}^k = \begin{bmatrix} 3 & 2 & -3 \\ 1 & -1 & 1 \\ 1 & 4 & 0 \end{bmatrix} \begin{bmatrix} (3)^k & 0 & 0 \\ 0 & (-2)^k & k(-2)^{k-1} \\ 0 & 0 & (-2)^k \end{bmatrix} \begin{bmatrix} 3 & 2 & -3 \\ 1 & -1 & 1 \\ 1 & 4 & 0 \end{bmatrix}^{-1}$$

Problem Set 51

51.1. Go back and do any of the exercises from Section 51 that you may have skipped.

*51.2. Assume that T satisfies 51.1, so that J(T) exists, and that $\chi(T) = \cdots(\lambda - c)^m \cdots$.

(i) Suppose $m = 5$, dim $E_c = 2$, and dim $E_c^2 = 3$. Explain why this information determines the Jordan form of $T \upharpoonright K_c$. What is J($T \upharpoonright K_c$)?

(ii) Suppose $m = 7$, dim $E_c = 3$, and dim $E_c^2 = 5$. Explain why this information does not fully determine the Jordan form of $T \upharpoonright K_c$. What are the possible values of J($T \upharpoonright K_c$)? What additional information is required to determine J($T \upharpoonright K_c$)?

51.3. If A is an $n \times n$ matrix, the number of 1's on the first superdiagonal in J(A) is the number of "missing" eigenvectors, that is, $n - \Sigma_{i=1}^{i=p} g_i$, where g_i is the geometric multiplicity of E_{c_i}.

*51.4. It is natural to wonder why, in Example 51.38, we did not *first* choose a specific vector $\mathbf{u}_2 \in E_{-2}$, say $\mathbf{u}_2 = (2, -1, 4)$, and *then* solve $\mathrm{B}\mathbf{u}_3 = -2\mathbf{u}_3 + (2, -1, 4)$ to find a generalized eigenvector $\mathbf{u}_3$. The fact is that this procedure, although it happens to work for the matrix B in 51.38, does not always work.

The general problem is to find $\mathbf{u}_i$ and $\mathbf{u}_{i+1}$ such that $\mathrm{A}\mathbf{u}_i = c\mathbf{u}_i$ and $\mathrm{A}\mathbf{u}_{i+1} = c\mathbf{u}_{i+1} + \mathbf{u}_i$. It could happen that if a specific $\mathbf{u}_i$ is chosen *first*, the resulting equation $\mathrm{A}\mathbf{u}_{i+1} = c\mathbf{u}_{i+1} + \mathbf{u}_i$ has no solution for $\mathbf{u}_{i+1}$. Here is a matrix for which this phenomenon is encountered.

$$\mathrm{A} = \begin{bmatrix} 5 & 0 & 1 \\ 1 & 4 & 1 \\ -1 & 0 & 3 \end{bmatrix}$$

(i) Calculate A^7 by finding the Jordan form of A and a basis for $\mathbb{R}^3$ of generalized eigenvectors of A.
(ii) Illustrate the phenomenon described.

51.5. The remarks in Problem 51.4 pertain to longer chains of generalized eigenvectors. If you seek a chain $\mathbf{u}_1, \mathbf{u}_2, \mathbf{u}_3$ of length 3 from the eigenspace of c satisfying $\mathrm{A}\mathbf{u}_1 = c\mathbf{u}_1$, $\mathrm{A}\mathbf{u}_2 = c\mathbf{u}_2 + \mathbf{u}_1$, $\mathrm{A}\mathbf{u}_3 = c\mathbf{u}_3 + \mathbf{u}_2$, you may not, in general, *begin* by finding specific vectors $\mathbf{u}_1$ and $\mathbf{u}_2$ for which $\mathrm{A}\mathbf{u}_1 = c\mathbf{u}_1$ and $\mathrm{A}\mathbf{u}_2 = c\mathbf{u}_2 + \mathbf{u}_1$, and *then* expect to find a $\mathbf{u}_3$ satisfying $\mathrm{A}\mathbf{u}_3 = c\mathbf{u}_3 + \mathbf{u}_2$. Let A be the standard matrix for the operator S from Example 51.26.
(i) Verify, by computing $\chi(\mathrm{A})$ and the dimensions of $E_0^1, E_0^2, \ldots$ and so on, that $\mathrm{J(A)} = \mathrm{J}[3;0] \oplus \mathrm{J}[1;0]$.
(ii) Find a basis for $\mathbb{R}^4$ consisting of generalized eigenvectors of A that is different from the one given in 51.26.
(iii) Let $\mathbf{u}_1 = (0, 0, 0, 1)$ and $\mathbf{u}_2 = (0, 1, 1, 0)$; show that $\mathrm{A}\mathbf{u}_1 = 0\mathbf{u}_1 = \mathbf{0}$ and that $\mathrm{A}\mathbf{u}_2 = 0\mathbf{u}_2 + \mathbf{u}_1 = \mathbf{u}_1$, but that there does not exist a $\mathbf{u}_3$ satisfying $\mathrm{A}\mathbf{u}_3 = 0\mathbf{u}_3 + \mathbf{u}_2 = \mathbf{u}_2$.

*51.6. Define $T{:}\mathbb{R}^2 \to \mathbb{R}^2$ by $T(x_1, x_2) = (x_1 + 2x_2, 3x_1 + 4x_2)$.
(i) Find the standard matrix for T.
(ii) Find the matrix that represents T with respect to the basis $\beta = ((-1, 1), (1, 0))$.
(iii) Observe the relationship between the answers to parts (i) and (ii); they are transposes of one another. Prove that for all $\mathrm{A} \in \mathbb{R}_{2\times 2}$, A and A^T are similar. (The general case of this result is in the next problem.)

51.7. Let P be the $k \times k$ matrix whose entries along the diagonal from upper right to lower left are equal to 1 and whose remaining entries are 0.

(i) Prove that $P^{-1}J[k;c]P = J[k;c]^T$.

(ii) Use part (i) together with the Jordan canonical form theorem to prove that, if $A \in \mathbb{R}_{n \times n}$, then A and A^T are similar.

51.8. Let $A = \begin{bmatrix} -3 & -12 & 2 \\ 2 & 7 & 0 \\ 0 & 0 & 1 \end{bmatrix}$.

[This is the matrix from Problem 37.2(iii).]

(i) Find the Jordan form of A.

(ii) Find a basis for $\mathbb{R}^3$ consisting of generalized eigenvectors of A.

(iii) Find a formula for A^k as a function of k.

(iv) Find A^9.

51.9. Repeat the previous problem for the matrix

$$B = \begin{bmatrix} 6 & -5 & -3 \\ 3 & -2 & -2 \\ 2 & -2 & 0 \end{bmatrix}.$$

[This is the matrix from Problem 37.2(vi).]

Appendix 1

One Hundred Supplementary Problems for Sections 1 to 30

1. True or False? Justify all your answers with a proof or with a counterexample.
   (i) If A is a diagonal, skew-symmetric matrix, then A is a zero matrix.
   (ii) There is a linear transformation $T:\mathbb{R}^3 \to \mathbb{R}^5$ such that $\text{rank}(T) = \text{nullity}(T)$.
   (iii) If $\{\mathbf{u}_1, \mathbf{u}_2, \mathbf{u}_3\}$ is a linearly dependent set of nonzero vectors, then there exist scalars s_2 and s_3, not both zero, such that $\mathbf{u}_1 = s_2\mathbf{u}_2 + s_3\mathbf{u}_3$.
   (iv) A nonzero 2×3 matrix has at least one right inverse.
   (v) If A, B, and C are $m \times n$ matrices and $A + C = B + C$, then $A = B$.
   (vi) If $T:\mathbb{R}^5 \to \mathbb{R}^7$ is a linear transformation and the dimension of the kernel of T is 3, then the dimension of the range of T is 4.
   (vii) If A is an invertible $n \times n$ matrix, then the matrix equation $A\mathbf{x} = \mathbf{0}$ has at least one nontrivial solution.
   (viii) If A is invertible, then $\det(\text{adj}\, A) \neq 0$.
   (ix) If A is an $m \times n$ matrix and $m \gneq n$, then GJ(A) must have a row of zeros.

(x) If a linear transformation $T:\mathbb{R}^7 \to \mathbb{R}^5$ is onto, then T is also one-to-one.

(xi) If A is an $n \times n$ matrix and the linear system $A\mathbf{x} = \mathbf{0}$ has nontrivial solutions, then A is invertible.

(xii) If the $n \times n$ matrix A is both symmetric and skew-symmetric, then $A = 0_{n\times n}$.

(xiii) If A is a 2×5 matrix, then the matrix equation

$$A\mathbf{x} = \begin{bmatrix} 2 \\ 5 \end{bmatrix}$$

has exactly 5 distinct solutions.

(xiv) If A is a 3×4 matrix, then the matrix equation

$$A\mathbf{x} = \begin{bmatrix} 1 \\ 1 \\ 1 \end{bmatrix}$$

will, if consistent, have infinitely many solutions.

(xv) If A is an $n \times n$ matrix, then $\text{rank}(A^2) = \text{rank}(A)$.

(xvi) There is a one-to-one linear transformation from $\mathbb{R}^2$ to $\mathbb{R}^2$ whose range does not contain the vector $(3, -5)$.

(xvii) If $A \in \mathbb{R}_{n\times n}$, then $|A| = |-A|$.

(xviii) If $A \in \mathbb{R}_{n\times n}$, then $|A| = 1$ iff $A = I_n$.

(xix) If A is a 3×3 matrix and if the matrix equation

$$A\mathbf{x} = \begin{bmatrix} 1 \\ 1 \\ 1 \end{bmatrix}$$

has a unique solution, then A is invertible.

(xx) If $\{\mathbf{u}, \mathbf{v}_1, \mathbf{v}_2\} \subset \mathbb{R}^2$, if $\mathbf{u} = s_1\mathbf{v}_1 + \mathbf{s}_2\mathbf{v}_2$ and $\mathbf{u} = t_1\mathbf{v}_1 + \mathbf{t}_2\mathbf{v}_2$, and if $s_2 \neq t_2$, then $\{\mathbf{v}_1, \mathbf{v}_2\}$ is not a basis for $\mathbb{R}^2$.

(xxi) Every nonzero 2×2 matrix is a product of elementary matrices.

(xxii) If $S = \{\mathbf{v}_1, \mathbf{v}_2\}$ is a set of nonzero vectors in $\mathbb{R}^3$, then S is linearly independent.

(xxiii) If A is invertible and $A^{-1} = A^T$, then $\det A = 0$.

(xxiv) If A is a skew-symmetric 5×5 matrix, then $\det A = 0$.

(xxv) $\mathbb{R}^2$ has an infinite number of subspaces.

(xxvi) If A is invertible, then $\text{rank}(A) = \text{rank}(A^{-1})$.

(xxvii) If A is a 3×2 matrix and if the matrix equation

$$A\mathbf{x} = \begin{bmatrix} 1 \\ 1 \\ 1 \end{bmatrix}$$

has a unique solution, then

$$Ax = \begin{bmatrix} 2 \\ 2 \\ 2 \end{bmatrix}$$

has a unique solution.

(xxviii) If $T:\mathbb{R}^n \to \mathbb{R}^m$ is a linear transformation and the dimension of the range of T is m, then T is one-to-one.

(xxix) If the $n \times n$ matrix A can be written as a product of elementary matrices, then $|A| \neq 0$.

(xxx) If $\{\mathbf{u}_1, \mathbf{u}_2\}$ is a linearly independent subset of $\mathbb{R}^n$, then any vector of the form $s_1\mathbf{u}_1 + s_2\mathbf{u}_2$ lies on the line in $\mathbb{R}^n$ joining $\mathbf{u}_1$ and $\mathbf{u}_2$.

(xxxi) If $A + A^2$ is invertible, then A is invertible.

(xxxii) If A is a 3×2 matrix and if the matrix equation

$$Ax = \begin{bmatrix} 1 \\ 1 \\ 1 \end{bmatrix}$$

has a unique solution, then

$$Ax = \begin{bmatrix} 2 \\ -2 \\ 2 \end{bmatrix}$$

has a unique solution.

2. Let $S:\mathbb{R}^3 \to \mathbb{R}^4$ be the linear transformation defined by

$$S(x, y, z) = (x, x + y, x + y + z, 2x - z)$$

and let $T:\mathbb{R}^4 \to \mathbb{R}^3$ be the linear transformation defined by

$$T(r, s, t, u) = (r + s, s + 2t, s + t - u).$$

(i) Find the formula for the composition $T \circ S$: $\mathbb{R}^3 \to \mathbb{R}^3$.

(ii) Find the formula for the composition $S \circ T$: $\mathbb{R}^4 \to \mathbb{R}^4$.

3. You are given that the 2×2 matrix A satisfies both

$$A\begin{bmatrix} 3 \\ -1 \end{bmatrix} = \begin{bmatrix} 4 \\ 4 \end{bmatrix} \quad \text{and} \quad A\begin{bmatrix} -1 \\ 3 \end{bmatrix} = \begin{bmatrix} 4 \\ 4 \end{bmatrix}$$

Is A invertible or not? You *must* give a reason for your answer.

4. If it is possible to do so, express the matrix

$$A = \begin{bmatrix} -2 & 8 \\ 1 & -4 \end{bmatrix}$$

as a product of elementary matrices. If this is not possible, explain why it is not.

5. $A = \begin{bmatrix} 1 & -5 & 0 \\ 0 & 1 & 0 \\ 0 & 0 & 1 \end{bmatrix}$

(i) Find A^{-1}.
(ii) Find A^{-2}.
(iii) Find $(4A)^{-1}$.

6. (i) Solve the system of linear equations

$$\begin{aligned} x_1 + x_2 - 3x_3 + x_4 &= 1 \\ x_1 + x_2 + x_3 - x_4 &= 2 \\ x_1 + x_2 - x_3 &= 0 \\ 2x_1 - x_2 + x_3 + x_4 &= 3 \end{aligned}$$

(ii) Is the coefficient matrix

$$A = \begin{bmatrix} 1 & 1 & -3 & 1 \\ 1 & 1 & 1 & -1 \\ 1 & 1 & -1 & 0 \\ 2 & -2 & 1 & 1 \end{bmatrix}$$

of the system from part (i) invertible or not?

7. (i) Find a system of equations in three unknowns whose solution set is the line in $\mathbb{R}^3$ that goes through the points $(0, 0, 0)$ and $(1, 1, 1)$.
(ii) Justify your answer to (i) by actually solving the linear system you gave as an answer to (i) to check that its solution set has the stated properties.

8. Does the set $S = \{(1, 0, -1, 0),\ (1, 1, 2, 0),\ (2, -1, 0, 0,),\ (3, 1, -1, 0)\}$ span $\mathbb{R}^4$ or not? You must justify your answer.

9. For what value(s) of t, if any, will the solution space of the linear system

$$\begin{aligned} x_1 + x_2 + x_3 &= 0 \\ 2x_1 + 8x_2 + (t + 2)x_3 &= 0 \\ 4x_1 - 2x_2 + 3x_3 &= 0 \end{aligned}$$

be (i) empty?
(ii) a single point?
(iii) one dimensional?
(iv) two dimensional?

For those values of t for which the system is consistent, describe the actual solution spaces in detail.

10. Let A be an $m \times n$ matrix. Let P be an invertible $m \times m$ matrix. Prove that the row-space of the product PA is the same as the row-space of A.
11. Is the set $S = \{(1, 2, -1), (3, 0, 4), (1, -4, 6)\}$ linearly dependent or independent? If it is linearly dependent, find a way to write the zero vector $\mathbf{0} = (0, 0, 0)$ as a nontrivial linear combination of the vectors in S.
12. You are given that the Gauss–Jordan algorithm applied to the matrix

$$A = \begin{bmatrix} 1 & 2 & 1 & 3 & -5 \\ 1 & 3 & 2 & 5 & -2 \\ 3 & 7 & 5 & 14 & -4 \end{bmatrix}$$

yields the matrix

$$\begin{bmatrix} 1 & 0 & 0 & 2 & -3 \\ 0 & 1 & 0 & -1 & -5 \\ 0 & 0 & 1 & 3 & 8 \end{bmatrix}$$

Find a basis for the solution space of the linear system $A\mathbf{x} = 0$.

13. Let $A = \begin{bmatrix} 1 & 0 & -1 & 0 \\ 2 & 1 & 1 & 1 \\ -1 & 3 & 2 & 4 \end{bmatrix}$.
    (i) Find a basis for the row-space of A.
    (ii) Find a basis for the column-space of A.
14. (i) Prove directly from the definition that the set

$$W = \{(x_1, x_2, x_3) \in \mathbb{R}^3 \colon x_2 = 10x_3\}$$

    is a linear subspace of $\mathbb{R}^3$.
    (ii) What is the dimension of W?
    (iii) Find a basis for W.
15. For each of the following sets of vectors, say whether it is linearly dependent or independent and justify your answer.
    (i) $\{(6, -9, 19)\}$
    (ii) $\{(4, 8, 2), (6, -9, 18), (0, 0, 0)\}$
    (iii) $\{(1, -1, 0, 3, 0), (0, -1, 1, 3, 0), (0, 3, 0, -1, 1)\}$
    (iv) $\{(3, 9), (-1, -3)\}$
    (v) $\{(3, 9), (6, 12), (9, 9)\}$
    (vi) $\{(6, -9, 18), (1, 1, 1)\}$
16. $H = \{(x_1, x_2, x_3) \in \mathbb{R}^3 \colon x_2 = 3x_1, x_3 = -3x_1 - 10\}$.
    (i) Find the linear subspace W of $\mathbb{R}^3$ and a vector $\mathbf{b} \in \mathbb{R}^3$ such that H is the translate of W by $\mathbf{b}$.
    (ii) Find a basis for W.
17.

$$A = \begin{bmatrix} 1 & 3 & 0 & 4 & 2 & 0 \\ 0 & 0 & 1 & 3 & 0 & 0 \\ 0 & 0 & 0 & 0 & 0 & 1 \\ 0 & 0 & 0 & 0 & 0 & 0 \end{bmatrix}$$

Let S_0 be the solution space of the homogeneous system whose coefficient matrix is A, and let S_1 be the solution space of the linear system whose augmented matrix is A. Describe S_0 and S_1 in detail.

18. A student solves the linear system

$$\begin{aligned} x_1 + x_2 + x_3 &= 3 \\ 8x_2 + 4x_3 &= 16 \\ 3x_1 - 3x_2 \qquad &= -3 \end{aligned}$$

and finds that the solution space is

$$\{(x_1, x_2, x_3) \in \mathbb{R}^3 \colon x_1 = -\tfrac{1}{2}x_3 + 1,\ x_2 = -\tfrac{1}{2}x_3 + 2\}$$

The same student then solves the linear system

$$\begin{aligned} x_1 + x_2 + x_3 &= 10 \\ 8x_2 + 4x_3 &= 16 \\ 3x_1 - 3x_2 \qquad &= 18 \end{aligned}$$

and finds that the solution space is $\{(7, 1, 2)\}$. Explain why the student should conclude, *without* doing any additional calculations, that an error must have been made somewhere.

19. Let W be a linear subspace of $\mathbb{R}^{15}$. Consider the following three subsets of W:

$$\begin{aligned} A &= \{\mathbf{u}_1, \mathbf{u}_2, \ldots, \mathbf{u}_7\} \\ B &= \{\mathbf{v}_1, \mathbf{v}_2, \ldots, \mathbf{v}_{10}\} \\ C &= \{\mathbf{v}_1, \mathbf{v}_2, \ldots, \mathbf{v}_{10}, \mathbf{v}_{11}, \mathbf{v}_{12}\} \end{aligned}$$

It is understood here that each of A, B, and C is a set of distinct vectors. Note that B is a subset of C but that there are no other known relationships among the vectors in A, B, and C. Here is a list of possible responses to the questions below:

(a) ... is linearly independent.
(b) ... is not a basis for W.
(c) ... is linearly dependent.
(d) ... spans W.
(e) ... does not span W.
(f) The hypothesis provides no information about the set mentioned in the conclusion.

Fill in the blanks with a list of *all* those possible responses that can be justifiably concluded.

(i) If A spans W, then B ____________________.
(ii) If B spans W, then C ____________________.
(iii) If C is a basis for W, then A ____________________.
(iv) If C is linearly independent, then B ____________________.
(v) If B is linearly independent, then C ____________________.

20. Give an example of a *nonzero* 2×2 matrix
    (i) that is not invertible.
    (ii) that is not skew-symmetric.
    (iii) whose transpose is not symmetric.
    (iv) that is not upper triangular.
    (v) that is symmetric but is not invertible.
    (vi) that is a diagonal matrix but is not invertible.
21. Prove that a 2×2 matrix that is obtained as the product of a 2×1 matrix and a 1×2 matrix cannot be invertible.
22. Identify the geometric nature of the following sets.
    (i) $\{(x_1, x_2, x_3, x_4) \in \mathbb{R}^4\colon x_2 = 3x_1 - x_4, x_3 = 3x_1 + x_4\}$
    (ii) $\{(x_1, x_2, x_3, x_4, x_5) \in \mathbb{R}^5\colon x_1 = 0,\ x_2 = 0,\ x_3 = 59,\ x_4 = x_5\}$
    (iii) $\{(x_1, x_2, x_3, x_4) \in \mathbb{R}^4\colon x_4 = 3x_1 + x_2 + 10\}$
23. Write down an expression of the form {_______:__________} in which the blanks are filled in such a way that the resulting set is
    (i) a plane in $\mathbb{R}^3$ containing the point $(1, 1, 59)$.
    (ii) a line in $\mathbb{R}^3$ that does not contain the origin.
    (iii) a plane in $\mathbb{R}^4$ containing the point $(0, 2, 0, 3)$ and that is perpendicular to the x_4-axis.
    (iv) a line in $\mathbb{R}^3$ that contains the point $(1, 1, 2)$.
    (v) a plane in $\mathbb{R}^4$ that contains the point $(1, 1, 2, 0)$ and is not perpendicular to the x_3-axis.
24. Give an example of a linear operator T on $\mathbb{R}^2$ such that rank (T) = nullity (T).
25. Let $\mathbf{u} = (2, 1, -4)$ and $\mathbf{v} = (-1, -2, -4)$. Find a vector $\mathbf{w}$ such that the set $\{\mathbf{u}, \mathbf{v}, \mathbf{w}\}$ is a basis for $\mathbb{R}^3$.
26. Given that the linear transformation $T\colon V \to W$ is represented by a 6×9 matrix A and also that the solution space of the homogeneous system $\mathrm{A}\mathbf{x} = \mathbf{0}$ is four-dimensional, what is the dimension of the range of T?
27.

$$\mathrm{A} = \begin{bmatrix} 1 & 2 & 1 \\ 1 & 1 & 3 \end{bmatrix} \quad \text{and} \quad \mathrm{B} = \begin{bmatrix} 1 & 1 \\ 1 & 2 \\ 1 & 3 \end{bmatrix}$$

    (i) What is the rank of $T_{\mathrm{B}} \circ T_{\mathrm{A}}$?
    (ii) What is the nullity of $T_{\mathrm{A}} \circ T_{\mathrm{B}}$?
28. A is an $m \times n$ matrix with the property that for every $\mathbf{b} \in \mathbb{R}^m$ there exists an $\mathbf{x} \in \mathbb{R}^n$ such that $\mathrm{A}\mathbf{x} = \mathbf{b}$. Prove that $m \leq n$.
29. $S = \{(1, -2, 2), (-4, -1, 1), (3, 5, -5)\}$. Which of the following belong to $\mathscr{L}(S)$?
    (i) $(0, 0, 0)$
    (ii) $(2, 0, 0)$
    (iii) $(3, -3, 1)$
    (iv) $(359, 359, -359)$

30. Let

$$A = \begin{bmatrix} 1 & 2 & 3 \\ 4 & 8 & 0 \\ -2 & -4 & 1 \end{bmatrix}$$

Find a basis for
(i) the column space of A.
(ii) the row space of A.
(iii) the nullspace of A.

31. For what value(s) of t does the following system have
(i) no solutions?
(ii) exactly one solution?
(iii) infinitely many solutions?
Give complete descriptions of the solutions sets in parts (ii) and (iii).

$$\begin{aligned} 2x_1 + 4x_2 + 2x_3 &= 0 \\ tx_1 + 6x_2 + x_3 &= 0 \\ 5x_1 + (t+7)x_2 + 7x_3 &= 0 \end{aligned}$$

32. Given that the linear transformation T: $\mathbb{R}^7 \to \mathbb{R}^5$ is represented by a 5×7 matrix A and also that the subspace spanned by the rows of A is three-dimensional, what is the dimension of the range of T?

33. Give an example of a linear transformation T:$\mathbb{R}^9 \to \mathbb{R}^6$ such that the dimension of $\{\mathbf{v}: T(\mathbf{v}) = \mathbf{0}\}$ is 2, or else explain why such an example is not possible.

34.

$$\begin{bmatrix} 1 & 0 & -1 & 3 & -1 \\ 1 & 0 & 0 & 2 & -1 \\ 2 & 0 & -1 & 5 & -1 \\ 0 & 0 & -1 & 1 & 0 \end{bmatrix} = A$$

Find all the following:
(i) Range of T_A
(ii) Kernel of T_A
(iii) A basis for the range of T_A
(iv) A basis for the kernel of T_A
(v) Rank of T_A
(vi) Nullity of T_A
Note: You may answer parts (i) to (vi) in any order you choose because there are several ways to obtain the desired information.

35.

$$A = \begin{bmatrix} 1 & 2 & -3 \\ 2 & -1 & 4 \\ 4 & 3 & -2 \end{bmatrix}$$

Find a basis for

(i) the column space of A.

(ii) the row space of A.

(iii) the nullspace of A.

36. Which of the following vectors belong to $\mathscr{L}(S)$, where $S = \{(3,9,1), (-2,-6,-4), (1,3,2)\}$?

(i) $(0,0,0)$

(ii) $(0,0,2)$

(iii) $(-4,-2,-6)$

(iv) $(55,165,105)$

37. Let $\mathbf{u} = (0,1,2)$ and $\mathbf{v} = (6,-1,-4)$. Find a vector $\mathbf{w}$ such that the set $\{\mathbf{u},\mathbf{v},\mathbf{w}\}$ is a basis for $\mathbb{R}^3$.

38. A is an $m \times n$ matrix with the property that the only solution to $\mathrm{A}\mathbf{x} = \mathbf{0}$ is the trivial solution $\mathbf{x} = \mathbf{0}$. Prove that $n \leq m$.

39. Compute the following determinant directly from the definition of the determinant as the sum of the signed elementary products.

$$\begin{vmatrix} -2 & 0 & 0 & 0 & 0 \\ 0 & 3 & 0 & 0 & 2 \\ 0 & 0 & -4 & 0 & 0 \\ 0 & 5 & 0 & 0 & -1 \\ 0 & 0 & 0 & 1 & 0 \end{vmatrix}$$

40. (i)

$$\mathrm{A} = \begin{vmatrix} 1 & -1 & 1 & -1 \\ 2 & 1 & 0 & 1 \\ 1 & 2 & 1 & -2 \\ 3 & -1 & -2 & 1 \end{vmatrix}$$

Compute det A using any method you choose.

(ii) Assuming your answer to part (i) is correct, then what are $|\mathrm{A}^{-1}|$, $|\mathrm{A}^T|$, $|\tfrac{1}{2}\mathrm{A}|$, and $|\mathrm{A}^2|$?

41.

$$\mathrm{A} = \begin{bmatrix} 1 & 3 & -2 & 0 & 2 & 0 \\ 2 & 6 & -5 & -2 & 4 & -3 \\ 0 & 0 & 5 & 10 & 0 & 15 \\ 2 & 6 & 0 & 8 & 4 & 18 \end{bmatrix}$$

You are *given* that

$$\mathrm{GJ(A)} = \begin{bmatrix} 1 & 3 & 0 & 4 & 2 & 0 \\ 0 & 0 & 1 & 2 & 0 & 0 \\ 0 & 0 & 0 & 0 & 0 & 1 \\ 0 & 0 & 0 & 0 & 0 & 0 \end{bmatrix}$$

(i) Find the solution space S of the 4×6 homogeneous system $\mathrm{A}\mathbf{x} = \mathbf{0}$.

(ii) Find a basis for S.

(iii) find a basis for the row space of A.

(iv) Is the set {(1, 0, 0, 0), (0, 1, 0, 0), (0, 0, 1, 0)} a basis for the column space of A? Answer yes or no and give a reason.

(v) Is the set {(1, 2, 0, 2), (−2, −5, 5, 0), (0, −2, 10, 8), (0, −3, 15, 18)} a basis for the column space of A? Answer yes or no and give a reason.

42. Define $S:\mathbb{R}^3 \to \mathbb{R}^2$ by

$$S(x_1, x_2, x_3) = (4x_1 - x_2 + 3x_3, 2x_1 + 6x_3)$$

Define $T:\mathbb{R}^2 \to \mathbb{R}^4$ by

$$T(x_1, x_2) = (x_1 + x_2, x_1 - x_2, 3x_2, 4x_1 - 2x_2)$$

Find the general formula for the composition $(T \circ S)$: $\mathbb{R}^3 \to \mathbb{R}^4$.

43. Write down the 4 × 4 matrix A whose entries are given by the formula

$$a_{ij} = \frac{1}{i + j - 1}$$

44. For what values of t is the solution space of the following system

(i) empty?

(ii) a single point in $\mathbb{R}^3$?

(iii) a line in $\mathbb{R}^3$?

(iv) a plane in $\mathbb{R}^3$?

$$\begin{aligned} 2x_1 \qquad + x_3 &= t \\ x_1 + x_2 + 3x_3 &= 2 \\ x_1 - x_2 - 2x_3 &= 2 \end{aligned}$$

For those value(s) of t, if any, for which the system is consistent, describe the solution space.

45. Let

$$A = \begin{bmatrix} 1 & 2 \\ 3 & t \end{bmatrix} \quad \text{and} \quad B = \begin{bmatrix} -t & 2 \\ 3 & -1 \end{bmatrix}$$

For what value(s) of t, if any, does $B = A^{-1}$?

46. Find a nonzero 2 × 2 matrix A other than I_2 that satisfies $A^2 = A$.

47. Let $B = \begin{bmatrix} 3 & 2 \\ -4 & -1 \end{bmatrix}$.

(i) Find B^{-1}.

(ii) Write B as a product of elementary matrices.

48. You are given that the Gauss–Jordan algorithm applied to the matrix

$$\begin{bmatrix} 1 & 2 & 1 & 3 & -5 \\ 1 & 3 & 2 & 5 & -2 \\ 3 & 7 & 5 & 14 & -4 \end{bmatrix}$$

yields the matrix

$$\begin{bmatrix} 1 & 0 & 0 & 2 & -3 \\ 0 & 1 & 0 & -1 & -5 \\ 0 & 0 & 1 & 3 & 8 \end{bmatrix}$$

Describe the solution spaces of each of the following systems of equations:

(i)

$$\begin{aligned} x_1 + 2x_2 + x_3 + 3x_4 - 5x_5 &= 0 \\ x_1 + 3x_2 + 2x_3 + 5x_4 - 2x_5 &= 0 \\ 3x_1 + 7x_2 + 5x_3 + 14x_4 - 4x_5 &= 0 \end{aligned}$$

(ii)

$$\begin{aligned} x_1 + 2x_2 + x_3 + 3x_4 &= -5 \\ x_1 + 3x_2 + 2x_3 + 5x_4 &= -2 \\ 3x_1 + 7x_2 + 5x_3 + 14x_4 &= -4 \end{aligned}$$

(iii)

$$\begin{aligned} x_1 + 2x_2 + x_3 &= 3 \\ x_1 + 3x_2 + 2x_3 &= 5 \\ 3x_1 + 7x_2 + 5x_3 &= 14 \end{aligned}$$

49. For what value(s) of k can there be a linear transformation $T:\mathbb{R}^5 \to \mathbb{R}^3$ with $\text{rank}(T) = k$?

50. Prove that if the $n \times n$ matrix A is skew-symmetric and n is odd then A is not invertible.

51. Write down the 2×3 matrix A whose $(i, j)^{\text{th}}$ entry, a_{ij}, is given by the formula

$$a_{ij} = \frac{i + 3j}{i - 4j}$$

52. If it is possible to do so, write the matrix

$$\mathrm{A} = \begin{bmatrix} 3 & 5 \\ 1 & -2 \end{bmatrix}$$

as a product of elementary matrices. If this is not possible, explain why it is not.

53. Let

$$A = \begin{bmatrix} 1 & 0 & 0 \\ 0 & 1 & 0 \\ -3 & 0 & 1 \end{bmatrix}$$

(i) Find A^{-1}.
(ii) Find A^{-2}.
(iii) Find $(3A)^{-1}$.

54. Find a matrix D such that DA = C, where

$$A = \begin{bmatrix} 1 & 2 & 0 & 2 \\ 2 & 3 & -1 & 4 \\ 6 & -1 & 4 & 1 \\ 0 & 2 & 1 & 7 \end{bmatrix} \quad \text{and} \quad C = \begin{bmatrix} 1 & 2 & 0 & 2 \\ 0 & -1 & -1 & 0 \\ 6 & -1 & 4 & 1 \\ 0 & 4 & 2 & 14 \end{bmatrix}$$

Note that C can be obtained from A by performing a sequence of two elementary row-operations on A.

55. Solve the following linear systems. Note that they each have the same coefficient matrix.

(i)

$$\begin{aligned} x_1 \phantom{{}+x_2} + 2x_3 &= 1 \\ -2x_1 + x_2 - 3x_3 &= 1 \\ -x_1 + 2x_2 - x_3 &= 2 \end{aligned}$$

(ii)

$$\begin{aligned} x_1 \phantom{{}+x_2} + 2x_3 &= 2 \\ -2x_1 + x_2 - 3x_3 &= -1 \\ -x_1 + 2x_2 - x_3 &= 1 \end{aligned}$$

(iii)

$$\begin{aligned} x_1 \phantom{{}+x_2} + 2x_3 &= 0 \\ -2x_1 + x_2 - 3x_3 &= 0 \\ -x_1 + 2x_2 - x_3 &= 0 \end{aligned}$$

(iv) Let A be the coefficient matrix of the preceding systems. Does the point (1, 1, 1) belong to the range of the linear transformation T_A?

56. (i) Find a system of equations in three unknowns whose solution space is the line in $\mathbb{R}^3$ through the points (0, 0, 0) and (1, 2, −3).

(ii) Justify your answer to part (i) by solving the linear system that you claim is an answer and checking that the stated conditions on the solution space are satisfied.

57. Given that $\operatorname{adj} A = \begin{bmatrix} 2 & -2 & 0 \\ 0 & 2 & -1 \\ 0 & 0 & 1 \end{bmatrix}$, find A.

58. For what value(s) of x, if any, is the matrix

$$\begin{bmatrix} x & 1 & 0 \\ 1 & x & 1 \\ 0 & 1 & x \end{bmatrix}$$

invertible?

59.

$$A = \begin{bmatrix} 1 & t & -1 & 2 \\ 2 & -1 & 3 & 5 \\ 1 & 10 & -6 & 1 \end{bmatrix}$$

For which value(s) of t does the rank of A equal
(i) three?
(ii) two?
(iii) one?
(iv) zero?

60. Evaluate the following determinant directly from the definition of determinants.

$$\begin{vmatrix} 0 & 0 & 1 & 0 & 0 \\ 3 & 0 & 0 & 5 & 0 \\ 0 & 6 & 0 & 0 & 0 \\ -2 & 0 & 0 & 4 & 0 \\ 0 & 0 & 0 & 0 & 2 \end{vmatrix}$$

61. Note that the two factors on the right are elementary matrices and use mental arithmetic to evaluate the product

$$\begin{bmatrix} 8 & -2 & 1 \\ 4 & 3 & 2 \\ -3 & -1 & 9 \\ -2 & -4 & -5 \end{bmatrix} \begin{bmatrix} 1 & 0 & 0 \\ 2 & 1 & 0 \\ 0 & 0 & 1 \end{bmatrix} \begin{bmatrix} 1 & 0 & 0 \\ 0 & 1 & 0 \\ 0 & 0 & 3 \end{bmatrix}$$

62. (i) Let

$$A = \begin{bmatrix} 3 & 1 & -1 \\ 1 & 2 & 1 \\ 5 & 0 & -3 \end{bmatrix}$$

What conditions, if any, must b_1, b_2, and b_3 satisfy if the linear system

$$A\begin{bmatrix} x_1 \\ x_2 \\ x_3 \end{bmatrix} = \begin{bmatrix} b_1 \\ b_2 \\ b_3 \end{bmatrix}$$

is to be consistent?

(ii) Let $T_A: \mathbb{R}^3 \to \mathbb{R}^3$ be the linear transformation determined by the matrix A from part (i). Does $(-1, 3, -5)$ belong to the range of T_A?

63. For what value(s), if any, of λ does the matrix $\begin{bmatrix} \lambda - 4 & -2 \\ -3 & \lambda - 3 \end{bmatrix}$ fail to be invertible?

64. Let $A = \begin{bmatrix} 2 & 6 & 6 \\ 2 & 7 & 6 \\ 2 & 7 & 7 \end{bmatrix}$. You are given that

$$A^{-1} = \begin{bmatrix} \frac{7}{2} & 0 & -3 \\ -1 & 1 & 0 \\ 0 & -1 & 1 \end{bmatrix}$$

(i) Find adj A.
(ii) Find $\det(A^3)$.
(iii) Find $\det(3A)$.
(iv) Find the cofactor of a_{23}.
(v) Find det(adj A).

65. Evaluate

$$\begin{vmatrix} 0 & 1 & 1 & 1 \\ 1 & 0 & 1 & 1 \\ 1 & 1 & 0 & 1 \\ 1 & 1 & 1 & 0 \end{vmatrix}$$

66. You are given that

$$\text{adj } A = \begin{bmatrix} -4 & 5 & -2 \\ -4 & 2 & 0 \\ 4 & -3 & 2 \end{bmatrix}$$

Find A.

67. Show that the two sets of vectors $\{(1, -1, 2), (3, 0, 1)\}$ and $\{(-1, -2, 3), (3, 3, -4)\}$ span the same subspace of $\mathbb{R}^3$.

68. Show that if the rank of the matrix

$$\begin{bmatrix} a_1 & a_2 & 1 \\ b_1 & b_2 & 1 \\ c_1 & c_2 & 1 \end{bmatrix}$$

is less than three, the points $(a_1, a_2), (b_1, b_2), (c_1, c_2) \in \mathbb{R}^2$ are collinear (i.e., lie on one line).

69. Which of the vectors

$$(4,-5,9,-7),\ (3,1,-4,4),\ (-1,1,0,1)$$

belong to the subspace of $\mathbb{R}^4$ spanned by the set

$$S = \{(1,1,-2,1),\ (3,0,4,-1),\ (-1,2,5,2)\}$$

70. Prove that

$$\begin{vmatrix} 0 & a & b & c \\ -a & 0 & 1 & -1 \\ -b & -1 & 0 & 1 \\ -c & 1 & -1 & 0 \end{vmatrix} = (a+b+c)^2$$

71. Let

$$\mathrm{A} = \begin{bmatrix} 1 & -1 \\ 2 & 2 \\ 1 & 0 \end{bmatrix}, \quad \mathrm{B} = \begin{bmatrix} 3 & 1 \\ -1 & 4 \end{bmatrix}$$

Find two *different* matrices C and D such that CA = B and DA = B.

72. Let P be an invertible $n \times n$ matrix, and let A be any $n \times n$ matrix.
(i) Prove that $\det(\mathrm{P}^{-1}\mathrm{AP}) = \det(\mathrm{A})$.
(ii) If

$$\mathrm{A} = \begin{bmatrix} 2 & 1 & -\frac{8}{3} \\ 0 & 3 & 1 \\ 2 & 0 & 5 \end{bmatrix}, \quad \mathrm{P} = \begin{bmatrix} 2 & -1 & 1 \\ 3 & 6 & 1 \\ 2 & 5 & 3 \end{bmatrix}$$

find $\det(\mathrm{P}^{-1}\mathrm{A}^2\mathrm{P})$.

73. For what value(s) of t does the linear system

$$\begin{aligned} (t+1)x_1 - \quad 2x_2 + \quad 3x_3 &= 1 \\ x_1 + (t-2)x_2 + \quad 3x_3 &= 5 \\ x_1 - \quad 2x_2 + (t+3)x_3 &= 3 \end{aligned}$$

have
(i) no solution?
(ii) exactly one solution?
(iii) infinitely many solutions?
For those value(s) of t for which the system is consistent, give a complete description of the solution set.

74. Suppose that $\{\mathbf{u}_1, \mathbf{u}_2, \dots \mathbf{u}_k\}$ and $\{\mathbf{v}_1, \mathbf{v}_2, \dots \mathbf{v}_{k+1}\}$ are each linearly independent subsets of $\mathbb{R}^n$. Prove that for some $i \le k+1$, the set $\{\mathbf{u}_1, \dots \mathbf{u}_k, \mathbf{v}_i\}$ is linearly independent.
75. Find, if possible, two distinct right inverses of the matrix

$$\begin{bmatrix} 1 & 0 \\ 3 & 1 \\ 1 & -2 \end{bmatrix}$$

76. Give a concrete example of each of the following (along with an argument to show that your example really is one), or explain why such an example cannot exist.
    (i) A symmetric matrix that is not invertible.
    (ii) A matrix whose main diagonal entries are all zero and which is row-reducible to the identity matrix.
    (iii) A skew-symmetric matrix A such that the homogeneous system $A\mathbf{x} = \mathbf{0}$ has exactly two solutions.
    (iv) A nonzero 2×2 matrix that has neither a right nor a left inverse.
    (v) A linear system of two equations in two unknowns which cannot be solved by Cramer's Rule.
    (vi) A symmetric matrix B such that B^2 is not symmetric.
    (vii) An upper triangular 3×3 matrix A such that the homogeneous linear system $A\mathbf{x} = \mathbf{0}$ has infinitely many solutions.
77. The fact that $|AB| = |BA|$ for $A, B \in \mathbb{R}_{n \times n}$ was proved in Section 26. Is this still true if, for example, A is 3×2 and B is 2×3? (*Caution*: AB and BA are then square matrices of different size. You may not argue that $|AB| = |A||B| = |B||A| = |BA|$, because neither $|A|$ nor $|B|$ is defined.
78. Prove for $A, B \in \mathbb{R}_{n \times n}$ that if $AB = -BA$ then at least one of A and B is singular.
79. Let $S = \{(2, 1, -1), (5, 2, 4), (3, -2, 0), (4, 4, 1), (3, -2, -4)\}$. Find a subset of S that is a basis for $\mathbb{R}^3$.
80. Without doing any calculations, explain why

$$GJ\left(\begin{bmatrix} \frac{39}{2} & 0 & 3 \\ -\frac{315}{2} & 0 & -24 \\ -99 & 0 & -15 \end{bmatrix}\right) = \begin{bmatrix} 1 & 0 & 0 \\ 0 & 0 & 1 \\ 0 & 0 & 0 \end{bmatrix}$$

81. (i) Find a nonzero 2×2 matrix A other than I_2 that satisfies $A^2 = A$.
    (ii) Find all such matrices.
82. Give an example of a linear operator T on $\mathbb{R}^2$ such that rank(T) = nullity(T).
83. Express the matrix

$$A = \begin{bmatrix} 3 & 2 \\ -4 & -1 \end{bmatrix}$$

as a product of elementary matrices.

84. Let

$$A = \begin{bmatrix} 1 & 2 & 1 & 3 & -5 \\ 1 & 3 & 2 & 5 & -2 \\ 3 & 7 & 5 & 14 & -4 \end{bmatrix}$$

Given that

$$GJ(A) = \begin{bmatrix} 1 & 0 & 0 & 2 & -3 \\ 0 & 1 & 0 & -1 & -5 \\ 0 & 0 & 1 & 3 & 8 \end{bmatrix}$$

write down by inspection the solution spaces of the following systems of equations:

(i) $$\begin{aligned} x_1 + 2x_2 + x_3 + 3x_4 - 5x_5 &= 0 \\ x_1 + 3x_2 + 2x_3 + 5x_4 - 2x_5 &= 0 \\ 3x_1 + 7x_2 + 5x_3 + 14x_4 - 4x_5 &= 0 \end{aligned}$$

(ii) $$\begin{aligned} x_1 + 2x_2 + x_3 + 3x_4 &= -5 \\ x_1 + 3x_2 + 2x_3 + 5x_4 &= -2 \\ 3x_1 + 7x_2 + 5x_3 + 14x_4 &= -4 \end{aligned}$$

(iii) $$\begin{aligned} x_1 + 2x_2 + x_3 &= 3 \\ x_1 + 3x_2 + 2x_3 &= 5 \\ 3x_1 + 7x_2 + 5x_3 &= 14 \end{aligned}$$

(iv) $$\begin{aligned} x_1 + 2x_2 + x_3 &= -5 \\ x_1 + 3x_2 + 2x_3 &= -2 \\ 3x_1 + 7x_2 + 5x_3 &= -4 \end{aligned}$$

(v) Give a theoretical justification for your answer to (iv).

85. (i) For what value(s) of t is the matrix

$$A = \begin{bmatrix} 2 & 2 & 3 \\ 3 & 2 & 4 \\ 1 & 0 & t \end{bmatrix}$$

invertible?

(ii) Find an expression for A^{-1} as a function of t that is valid for the values of t you found in part (i).

86. Suppose that $\{\mathbf{u}_1, \mathbf{u}_2, \ldots, \mathbf{u}_k\}$ and $\{\mathbf{v}_1, \ldots, \mathbf{v}_k, \mathbf{v}_{k+1}\}$ are each linearly independent subsets of a vector space V. Prove that for some $i \leq k + 1$, the set $\{\mathbf{u}_1, \ldots, \mathbf{u}_k, \mathbf{v}_1\}$ is linearly independent.

87. A is a 2×2 matrix satisfying $A = \operatorname{adj}(A)$. What is the general form of A?

88. One of the equations in a linear system of two equations in two unknowns is $x_1 - 2x_2 = 4$. By inspection, make up a second equation so that the resulting system will have

(i) no solution.

(ii) an infinite number of solutions.

(iii) a unique solution.

89. Assume that the reduced row-echelon form of the augmented matrix of a certain system of three linear equations in five

unknowns gives rise to three equations for the leading variables x_1, x_3, and x_5, respectively. Assume that $(2, 0, -1, 0, 4)$, $(4, 1, 6, 1, 4)$, and $(7, 1, -1, 0, 4)$ are known to be solutions of the system.

(i) Find the general solution.

(ii) Is the point $(1, 1, 0, 3, 4)$ a solution?

90. (i) Find a reduced description of the line in $\mathbb{R}^2$ determined by the two points $(1, 3)$ and $(7, 4)$.

(ii) Find a reduced description of the line in $\mathbb{R}^3$ determined by the two points $(1, 3, 2)$ and $(7, 4, -1)$.

91. If A is an invertible $n \times n$ matrix satisfying $A^2 = A$, then $A = I_n$.

92. Let

$$A = \begin{bmatrix} 1 & 2 & -3 \\ 2 & -1 & 4 \\ 4 & 3 & -2 \end{bmatrix}$$

Find a basis for

(i) the column space of A.

(ii) the row space of A.

(iii) the nullspace of A.

93. If A is an $n \times n$ matrix and $A^2 - A + I_n = 0_{n \times n}$, then A is invertible and $A^{-1} = I_n - A$.

94. Let $a_{ij}^{(r)}$ denote the $(i, j)^{\text{th}}$ entry in the rth power A^r of the matrix A. Let

$$A = \begin{bmatrix} 3 & 6 & 1 \\ -1 & 0 & 4 \\ -1 & 1 & 2 \end{bmatrix}$$

Find $a_{13}^{(2)}$ and $a_{23}^{(3)}$.

95. Given that the 2×2 matrices A and B are invertible, that

$$A^{-1} = \begin{bmatrix} 2 & 1 \\ 3 & 5 \end{bmatrix}$$

and that

$$B^{-1} = \begin{bmatrix} 2 & 4 \\ -1 & -3 \end{bmatrix}$$

(i) What is B?

(ii) What is $(AB)^{-1}$?

(iii) What is $A(BA)^{-1}(B^{-1})^T(A^{-1}B)^T$?

96. A is a 17×19 matrix. There are at least six linearly independent solutions of $A\mathbf{x} = \mathbf{0}$; A can be partitioned into

$$\left[\begin{array}{c|c} M & N \\ \hline P & Q \end{array}\right]$$

where P is a 13×13 matrix with $\det(P) = 13$. What is the rank of A?

97. Consider the two linear subspaces $W_1 = \mathscr{L}\{(1,2,3), (2,0,-4)\}$ and $W_2 = \{(x_1, x_2, x_3) \in \mathbb{R}^3 : 3x_1 - 2x_2 + x_3 = 0\}$. Find a reduced description and a basis for W_1, W_2, $W_1 \cap W_2$, and $W_1 + W_2$.

98. Prove that

$$\begin{bmatrix} a & b & c \\ a^2 & b^2 & c^2 \\ a^3 & b^3 & c^3 \end{bmatrix}$$

is invertible iff $a \neq 0$, $b \neq 0$, $c \neq 0$, $a - b \neq 0$, $a - c \neq 0$, and $b - c \neq 0$.

99. If $\{\mathbf{u}_1, \mathbf{u}_2, \mathbf{u}_3\}$ is a linearly dependent subset of $\mathbb{R}^n$ and A is any $m \times n$ matrix, then $\{A\mathbf{u}_1, A\mathbf{u}_2, A\mathbf{u}_3\}$ is linearly dependent.

100. Let S be a set of vectors in $\mathbb{R}^{100}$.
   (i) If $\dim \mathscr{L}(S) = 6$ and $\mathbf{u} \notin \mathscr{L}(S)$, what is $\dim \mathscr{L}(S \cup \{\mathbf{u}\})$?
   (ii) If $\dim \mathscr{L}(S) = 18$ and $\mathbf{v} \in \mathscr{L}(S - \{\mathbf{v}\})$, what is $\dim \mathscr{L}(S - \{\mathbf{v}\})$?
   (iii) If $\dim \mathscr{L}(S) = 39$, what is $\dim \mathscr{L}(S - \{\mathbf{0}\})$?

Appendix 2

Fifty Supplementary Problems for Sections 31 to 45

1. True or False? Justify all your answers with a proof or with a counterexample.
   (i) If 0 is an eigenvalue of A, then the determinant of A is 0.
   (ii) If λ is an eigenvalue of A, then λ^2 is an eigenvalue of A^2.
   (iii) If A and B are similar $n \times n$ matrices and $|A| \neq 0$, then $|B| \neq 0$.
   (iv) If two $n \times n$ matrices are each diagonalizable and have the same eigenvalues with the same multiplicities, then they are similar.
   (v) If A, B $\in \mathbb{R}_{n\times n}$ and A is row-reducible to B, then A and B are similar.
   (vi) For all A, B $\in \mathbb{R}_{n\times n}$, $\chi(AB) = \chi(A)\chi(B)$.
   (vii) If 0 is an eigenvalue of A, then A is diagonalizable.
2. $T:\mathbb{R}^2 \to \mathbb{R}^2$ is a linear operator which is represented with respect to the basis $\beta = ((1, -1), (1, 1))$ by the matrix

$$\begin{bmatrix} 3 & \frac{3}{2} \\ -4 & -1 \end{bmatrix}$$

   Find the matrix that represents T with respect to $\gamma = ((-1, 1), (0, -1))$.

3. $T:\mathbb{R}^3 \to \mathbb{R}^2$ is defined by $T(x_1, x_2, x_3) = (3x_1 + 2x_2, x_2 - 2x_3)$. Find ${}_\gamma[T]_\beta$, where $\beta = ((0,0,1), (0,1,1), (1,1,1))$ and $\gamma = ((1,0), (1,1))$.
4. Given that $T:\mathbb{R}^2 \to \mathbb{R}^2$ is a linear operator satisfying $T(3, -1) = (2,2)$ and $T(2,2) = (1,1)$,
   (i) Find $T(1,0)$.
   (ii) Find the general formula for $T(x_1, x_2)$.
   (iii) Is T an invertible linear operator?
   Note: you may answer parts (i), (ii), and (iii) in any order you choose.
5. Let $\beta = \{(1/2, 1/2, -1/2, 1/2), (0, 1/\sqrt{2}, 1/\sqrt{2}, 0), (1/2, -1/2, 1/2, 1/2), (1/\sqrt{2}, 0, 0, -1/\sqrt{2})\}$.
   (i) Show that β is an orthonormal basis for $\mathbb{R}^4$.
   (ii) Find the β-coordinates of the vector $(4, 3, 2, 1)$.
6. Find all linear transformations from $\mathbb{R}^2$ into $\mathbb{R}^2$ which send the line $x_2 = 0$ into the line $x_1 = 0$.
7. Find a linear transformation $T:\mathbb{R}^3 \to \mathbb{R}^3$ such that $\ker(T) = \{(x_1, x_2, x_3) \in \mathbb{R}^3\colon 2x_1 - x_2 + x_3 = 0\}$.
8. Find a linear transformation $T:\mathbb{R}^3 \to \mathbb{R}^3$ such that $\operatorname{ran}(T) = \{(x_1, x_2, x_3) \in \mathbb{R}^3\colon 2x_1 - x_2 + x_3 = 0\}$.
9. Find the kernel, range, nullity, and rank of each of the following linear transformations.
   (i) $T(x_1, x_2, x_3) = (x_1 - x_2 + x_3, -2x_1 + 2x_2 - 2x_3)$
   (ii) $T(x_1, x_2, x_3) = (x_1 - x_2 + 2x_3, 3x_1 + x_2 + 4x_3, 5x_1 - x_2 + 8x_3)$
   (iii) $T(x_1, x_2, x_3, x_4) = (x_1 - x_2 + 2x_3 + 3x_4, x_2 + 4x_3 + 3x_4, x_1 + 6x_2 + 6x_4)$
   (iv) $T(x_1, x_2) = (x_1 - 2x_2, 2x_1 + x_2)$
   (v) $T(x_1, x_2, x_3) = (2x_1 + x_2 + x_3, x_2 - 3x_3)$
10. (i) Where T is the fourth of the transformations in the preceding problem, find the matrix that represents T with respect to the basis $\beta = ((1, -2), (3, 2))$ for the domain and codomain.
    (ii) Where T is the fifth of the transformations in the preceding problem, find the matrix that represents T with respect to the bases $\beta = ((1,0,1), (1,1,0), (1,1,1))$ for the domain and $\gamma = ((1, -1), (2, 3))$ for the codomain.
11. $T:\mathbb{R}^2 \to \mathbb{R}^3$ is a linear transformation satisfying $T(5, -1) = (1,1,1)$ and $T(-1,5) = (-1, -1, -1)$. Find the general expression for $T(x_1, x_2)$, and find the kernel, range, nullity, and rank of T.
12. Given that the linear operator T is represented with respect to the basis $\beta = ((-1,1), (0, -1))$ by the matrix

$$\begin{bmatrix} 1 & 2 \\ -1 & 1 \end{bmatrix}$$

find the matrix that represents T with respect to the basis $\gamma = ((1, -1), (1,1))$.

13. Let $T:\mathbb{R}^3 \to \mathbb{R}^2$ be a linear transformation satisfying $T(1,1,0) = (-1,1)$, $T(0,-1,2) = (3,-4)$, and $T(1,1,1) = (0,0)$.
   (i) Find the general expression for $T(x_1, x_2, x_3)$.
   (ii) Find a basis for the kernel of T.
   (iii) What is the rank of T?
14. Let $T:\mathbb{R}^2 \to \mathbb{R}^2$ be the linear operator $T(x_1, x_2) = (-x_2, x_1)$, let β be any basis for $\mathbb{R}^2$, and let

$$\begin{bmatrix} a & b \\ c & d \end{bmatrix}$$

be the matrix that represents T with respect to the basis β. Prove that $b \neq 0$ and $c \neq 0$.
15. $T:\mathbb{R}^2 \to \mathbb{R}^2$ is a linear operator that satisfies $T(1,-3) = (3,-9)$ and $T(2,4) = (8,16)$.
   (i) Find the general expression for $T(x_1, x_2)$.
   (ii) What is the rank of T?
16. $T:\mathbb{R}^3 \to \mathbb{R}^2$ is defined by $T(x_1, x_2, x_3) = (3x_1 + 2x_2, x_2 - 2x_3)$. $\beta = ((1,0), (1,2,1), (0,1,2))$ and $\gamma = ((1,2), (-2,4))$. Find ${}_{\gamma}[T]_{\beta}$.
17. Fix $\mathbf{0} \neq \mathbf{u} \in \mathbb{E}^n$ and $t \in \mathbb{R}$. Show that $\{\mathbf{v} \in \mathbb{E}^n\colon \mathbf{v} \cdot \mathbf{u} = t\} = H = H(\mathbf{u}; t)$ is a hyperplane in $\mathbb{E}^n$.
18. Let $\mathbf{u} = (4,2,0)$, $\mathbf{v} = (-1,-2,-1)$, and $\mathbf{w} = (0,3,0)$.
   (i) Find $(\mathbf{u} \times (\mathbf{v} \times \mathbf{w})) \cdot \mathbf{w}$.
   (ii) Find $\mathbf{w} \times (\mathbf{u} \times (\mathbf{v} \times \mathbf{w}))$.
   (iii) Find $\mathbf{w} \cdot (\mathbf{u} \times (\mathbf{v} \times \mathbf{w}))$.
   (iv) Find $\|\mathbf{u} \times (\mathbf{v} \times \mathbf{w})\|$.
19. $T:\mathbb{R}^3 \to \mathbb{R}^4$ is a linear transformation satisfying $T(1,0,0) = (2,3,0,-1)$, $T(0,1,0) = (1,2,3,4)$ and $T(0,0,1) = (-4,-3,-2,0)$.
   (i) Find the general expression for $T(x_1, x_2, x_3)$.
   (ii) Find $T(1,1,1)$.
   (iii) Find the rank and nullity of T.
20. Find a basis for $\mathscr{P}_2$ that contains the vectors $x + 2$ and $6x^2 - x - 4$.
21. Verify that the function $\{a_0 + a_1x + a_2x^2, b_0 + b_1x + b_2x^2\} = a_0b_0 + 2a_1b_1 + 3a_2b_2$ is an inner product on $\mathscr{P}_2$ and find a basis for $\mathscr{P}_2$ that is orthonormal with respect to this inner product.
22. Let

$$A = \begin{bmatrix} 1 & 2 & 0 \\ 6 & 4 & 0 \\ -3 & 5 & -1 \end{bmatrix}$$

If it is possible to do so, find an invertible matrix P and a diagonal matrix D such that $P^{-1}AP = D$. If it is not possible, explain why.
23. Repeat the previous problem for

$$A = \begin{bmatrix} 2 & 0 & 3 \\ 5 & 2 & -2 \\ 0 & 0 & -1 \end{bmatrix}$$

24. Let $\mathbf{u} = (3, -2, 1)$ and $\mathbf{v} = (2, -1, -2)$.
   (i) Find vectors $\mathbf{w}_1$ and $\mathbf{w}_2$ such that $\mathbf{u} = \mathbf{w}_1 + \mathbf{w}_2$, where $\mathbf{w}_1$ is parallel to $\mathbf{v}$ and $\mathbf{w}_2$ is perpendicular to $\mathbf{v}$.
   (ii) Find vectors $\mathbf{w}_1$ and $\mathbf{w}_2$ such that $\mathbf{v} = \mathbf{w}_1 + \mathbf{w}_2$, where $\mathbf{w}_1$ is parallel to $\mathbf{u}$ and $\mathbf{w}_2$ is perpendicular to $\mathbf{u}$.
25. If $\mathbf{x} = (a - 1, 1, b, 0)$, $\mathbf{y} = (2, a - 2, -2, 2b)$, $\|\mathbf{x}\| = 3$, $\|\mathbf{y}\| = 5$, and $\mathbf{x} \cdot \mathbf{y} = 9$, what are a and b?
26. Let $\mathbf{u} = (1, 2, 3)$, $\mathbf{v} = (2, -3, 4)$ and $\mathbf{w} = (3, 1, -1)$.
   (i) Find $\mathbf{u} \times (\mathbf{v} \times \mathbf{w})$.
   (ii) Find $(\mathbf{u} \times \mathbf{v}) \times \mathbf{w}$.
   (iii) Find $\mathbf{u} \cdot (\mathbf{v} \times \mathbf{w})$.
   (iv) Find $\|\mathbf{u} \times \mathbf{w}\|$.
27. Let $\mathbb{R}$ be the scalar field and let $V = \{(x_1, x_2)\colon x_1, x_2 \in \mathbb{R}\}$ be the vectors. Define $\oplus$ for pairs of vectors $\mathbf{x}$, $\mathbf{y}$ by

$$(x_1, x_2) \oplus (y_1, y_2) = (5x_2 + 5y_2, -x_1 - y_1)$$

Define scalar multiplication by

$$t \odot (x_1, x_2) = (5tx_2, -tx_1)$$

Decide whether this structure is a vector space. If it is not, which axioms fail?
28. Let $\mathbb{R}$ be the scalar field and let $V = \{(x, x^2)\colon x \in \mathbb{R}\}$ be the vectors.
   (i) Note that V is a subset of $\mathbb{R}^2$. Is V a subspace of $\mathbb{R}^2$?
   (ii) Define $\oplus$ for pairs of vectors by

$$(x, x^2) \oplus (y, y^2) = \left(x + y, (x + y)^2\right)$$

   Define scalar multiplication by

$$t \odot (x, x^2) = (tx, t^2x^2)$$

   Decide whether this structure is a vector space. If it is not, which axioms fail?
29. Repeat the previous problem, changing the definition of scalar multiplication to the following:

$$t \odot (x_1, x_2) = \begin{cases} (0, 0), & \text{if } t = 0; \\ (tx_1, x_2/t), & \text{if } t \neq 0. \end{cases}$$

30. Note that the linear operator $T\colon\mathbb{R}^3 \to \mathbb{R}^3$ whose general formula is $T(x_1, x_2, x_3) = (x_1 + 2x_2 - x_3, x_1 + x_3, 4x_1 - 4x_2 + x_3)$ satisfies $T(-1, 1, 2) = (-1, 1, 2)$ and $T(-1, 1, 4) = (-3, 3, 12)$. Use this information to find values for a, b, c, d, e, f, and g such

that

$$\begin{bmatrix} -1 & a & -1 \\ 1 & b & 1 \\ 4 & c & 2 \end{bmatrix}^{-1} \begin{bmatrix} 1 & 2 & -1 \\ 1 & 0 & 1 \\ 4 & d & e \end{bmatrix} \begin{bmatrix} -1 & a & -1 \\ 1 & b & 1 \\ 4 & c & 2 \end{bmatrix} = \begin{bmatrix} f & 0 & 0 \\ 0 & 2 & 0 \\ 0 & 0 & g \end{bmatrix}$$

31. Let $\beta = \{(3, -6), (7, 6)\}$ and $\gamma = \{(2, 1), (0, 3)\}$. Find the γ-coordinates of the vector whose β-coordinates are $(2, 2)$.
32. $T: \mathbb{R}^2 \to \mathbb{R}^4$ is defined by $T(x_1, x_2) = (0, 3x_1 + 2x_1, -x_1, 4x_2)$.
    (i) Is T one-to-one?
    (ii) Is T onto?
33. Find a polynomial $p(x) = a_0 + a_1x + a_2x^2$ such that the set $\{1 - x^2, 1 + x^2, p(x)\}$ is a basis for $\mathscr{P}_2$.
34. Determine whether the matrix

$$\begin{bmatrix} 12 & 10 \\ -15 & -13 \end{bmatrix}$$

is diagonalizable or not. If it is, find the eigenvalues and a basis of eigenvectors.

35. Does the point $(29, 3, 8)$ belong to the plane determined by the three points $(3, 2, 1)$, $(2, 1, -1)$, and $(-1, 3, 2)$?
36. Let V be a vector space and let W_1 and W_2 be the subspaces spanned by $\{\mathbf{u}_1, \mathbf{u}_2\}$ and $\{\mathbf{v}_1, \mathbf{v}_2\}$, respectively.
    (i) Prove that if $\mathbf{u}_1, \mathbf{u}_2 \in W_2$ and $\mathbf{v}_1, \mathbf{v}_2 \in W_1$, then $W_1 = W_2$.
    (ii) Use part (i) to show that the two sets $S_1 = \{(1, -1, 2), (3, 0, 1)\}$ and $S_2 = \{(-1, 2, 3), (3, 3, -4)\}$ span the same subspace of $\mathbb{R}^3$.
37. $T: \mathbb{R}^5 \to \mathbb{R}^4$ is a linear transformation whose kernel is $\{(x_1, x_2, x_3, x_4, x_5) \in \mathbb{R}^5\colon\ x_1 = x_3 + x_5, x_2 = x_3, x_4 = -4x_5)$. What is the rank of T?
38. A is an $m \times n$ matrix. For every $\mathbf{y} \in \mathbb{R}^m$, there is an $\mathbf{x} \in \mathbb{R}^n$ such that $A\mathbf{x} = \mathbf{y}$. What is the rank of A?
39. Find an orthonormal basis for $\mathbb{R}^4$ containing the vectors $(1/2, 1/2, 1/2, 1/2)$ and $(1/6, 1/6, 1/2, -5/6)$.
40. Consider the ordered basis $\beta = ((4, 1), (2, 1))$ for $\mathbb{R}^2$. Let $\mathbf{u} = (1, 1)$, $\mathbf{v} = (0, 0)$, $\mathbf{w} = (4, 1)$, and $\mathbf{x} = (x_1, x_2)$. Find $(\mathbf{u})_\beta$, $(\mathbf{v})_\beta$, $(\mathbf{w})_\beta$, and $(\mathbf{x})_\beta$.
41. Suppose you compute the characteristic polynomial of the matrix

$$A = \begin{bmatrix} 8 & 9 & 9 \\ 3 & 2 & 3 \\ -9 & -9 & -10 \end{bmatrix}$$

and find that $\chi(A) = (\lambda + 1)(\lambda - 2)^2$. You should realize that this answer for $\chi(A)$ *must* be wrong. Why?

42. Suppose you are given that the operator T on $\mathbb{R}^2$ is represented with respect to the basis $\beta = ((-1, 1), (0, -1))$ by the matrix

$$\begin{bmatrix} 1 & 2 \\ -1 & 1 \end{bmatrix}$$

You are asked to find the matrix that represents T with respect to $\gamma = ((1, -1), (1, 1))$. You find that

$${}_{\gamma}[T]_{\gamma} = \begin{bmatrix} \frac{1}{2} & \frac{5}{2} \\ -\frac{1}{2} & \frac{3}{2} \end{bmatrix}$$

You should realize that this answer *must* be wrong. Why?

43. Each of the following questions has an "obvious" *yes* or *no* answer. In this context, "obvious" means that no calculations requiring pencil and paper are necessary.

(i) Does the vector $(3, 5, -9)$ belong to the column space of the matrix

$$\begin{bmatrix} 6 & 2 & -5 \\ 1 & 3 & -1 \\ 7 & 5 & -3 \\ -4 & 0 & -1 \end{bmatrix}$$

(ii) Is the set $\{x - 2x^2, x^2 - 4x, -7x + 8x^2)$ a linearly independent subset of $\mathscr{P}_2$?

(iii) Is there a homogeneous system of three equations in four unknowns whose solution set is

$$\{(x_1, x_2, x_3, x_4) \in \mathbb{R}^4 : x_1 = 3x_4, x_2 = 2x_4 + 4, x_3 = -2x_4\}?$$

(iv) Does the set

$$\left\{\begin{bmatrix} 2 & -1 \\ 4 & 0 \end{bmatrix}\begin{bmatrix} 0 & -3 \\ 1 & 5 \end{bmatrix}\begin{bmatrix} 4 & 1 \\ 7 & -5 \end{bmatrix}\right\}$$

span $\mathbb{R}_{2\times 2}$?

(v) Does $(3, -16)$ belong to the subspace of $\mathbb{R}^2$ spanned by the set $\{(5, -1), (13, -8)\}$?

(vi) Prove that $S = \{\mathbf{v}_1, \mathbf{v}_2, \ldots, \mathbf{v}_8\}$ is a basis for an eight-dimensional subspace of V and W is a subspace of V that contains the vectors $\mathbf{v}_1$, $\mathbf{v}_2$, and $\mathbf{v}_3$, then the dimension of W is greater than or equal to 3.

44. Find an orthonormal basis for $\mathscr{L}\{(1, 2, 2, 0), (1, 6, -2, 2), (1, 1, 0, 1)\}$.

45. Find a linear operator on $\mathbb{R}^3$ whose kernel is the line of intersection of the two planes $x_1 + x_2 = 0$ and $x_2 - x_3 = 0$.
46. (i) Let $\langle \cdot , \cdot \rangle$ be an inner product on the vector space V, let $\| \cdot \|$ be its associated norm, and let $\mathbf{u}, \mathbf{v} \in V$. Show that $\mathbf{u} + \mathbf{v}$ and $\mathbf{u} - \mathbf{v}$ are orthogonal with respect to this inner product iff $\|\mathbf{u}\| = \|\mathbf{v}\|$.
    (ii) What is the geometric interpretation of this result when $\langle \cdot , \cdot \rangle$ is the Euclidean inner product on $\mathbb{R}^2$?
47. Let $\mathbf{u}$ and $\mathbf{v}$ be nonzero vectors in $\mathbb{E}^2$ with $\|\mathbf{u}\| = s$ and $\|\mathbf{v}\| = t$. Show that the vector

$$\frac{1}{s+t}(t\mathbf{u} + s\mathbf{v})$$

bisects the angle between $\mathbf{u}$ and $\mathbf{v}$.
48. (i) Show that for any $t \in \mathbb{R}$, the set $\{1, x + t, (x + t)^2\}$ is a basis for $\mathscr{P}_2$.
    (ii) Find, as functions of t, the coordinates of the vector $3 - 4x + 5x^2$ with respect to this basis.
49. Let

$$B = \begin{bmatrix} 1 & -1 \\ -4 & 4 \end{bmatrix}$$

Define a linear operator on $\mathbb{R}_{2\times 2}$ by $T(A) = BA$. Find the dimension of and a basis for the kernel of T.
50. Let $\mathscr{P}_2$ denote the vector space of real polynomials of degree ≤ 2.
    (i) Find the transition matrix from the basis $\beta = (1, x, x^2)$ to the basis $\gamma = (1 - x, 1 + x + x^2, 3 - \frac{1}{2}x^2)$.
    (ii) Define $T\colon P_2 \to \mathbb{R}^2$ by

$$T(ax^2 + bx + c) = (a + b + c, a - 2c)$$

Find ${}_{\alpha_2}[T]_\beta$ and ${}_{\alpha_2}[T]_\gamma$.
    (iii) Find the rank and nullity of T.

Footnotes

[F]1.1 If you read the Preface, you will know that the exercises are to be done when encountered. You should get into the habit of doing this from the beginning. The answer to 1.3 is

$$\begin{aligned} x_1 &= 3z_1 - 18z_2 + 41z_3 \\ x_2 &= -35z_1 - 11z_2 + 19z_3 \end{aligned}$$

If you did this exercise by substitution, writing down all the x's, y's, and z's, you have the wrong attitude; repeat the exercise using rectangular arrays and following the pattern described after 1.2.

[F]1.2 The correct name for the operation defined in 1.8 should be "matrix composition." The reason for this will be explained in Section 18. Cayley himself used both words, "multiplication" and "composition," interchangeably in his original memoir. It is an unfortunate historical accident that "multiplication" came to predominate.

[F]2.1 As children, we humans instinctively learn to add before learning to multiply. This habit leads most contemporary authors first to define addition for matrices and to define multiplication afterwards, usually in an unmotivated fashion accompanied by an apologetic remark about how unnatural or unintuitive the definition is. The fact is, as we saw in Section 1, that matrices were born in 1858 with an instinct to multiply. It is addition for matrices that is the afterthought. In 1884, J. J. Sylvester wrote

> This revolution was effected by a forcible injection into the subject of the concept of addition, i.e., by choosing to regard matrices as

> susceptible of being added to one another; a notion, as it seems to me, quite foreign to the idea of substitution, the *nidus* (nest) in which that of multiple quantity (matrix) was laid, hatched, and reared.

Matrix addition was so unintuitive to nineteenth-century mathematicians that even Sylvester was unashamed to claim independent discovery of the concept. He writes,

> Much as I owe in the way of fruitful suggestion to Cayley's immortal memoir, the idea of subjecting matrices to the additive process and of their consequent amenability to the laws of functional operation was not taken from it, but occurred to me independently before I had seen the memoir or was acquainted with its contents.

[F]3.1 It may seem bizarre, at this early point in the text, to be emphasizing this particular list of properties. Why, for instance, is property (viii) even worth mentioning? The reason will be fully appreciated when we reach Example 41.7.

[F]3.2 With this terminology, the words *noninvertible* and *singular* have the same meaning. As well, the words *invertible* and *nonsingular* are synonyms. The reason for having two names for each of these concepts is the desire to avoid the use of English sentences containing double negations. Invertibility, and hence its negation, noninvertibility, are such pervasive concepts in this subject that it becomes very convenient to have a synonym for noninvertible that avoids the negative prefix "non."

[F]3.3 It is natural to ask whether the notion of invertibility can be extended to the case of matrices that are not square by declaring that an $m \times n$ matrix A is invertible if *both* $\mathrm{BA} = \mathrm{I}_n$ for some $n \times m$ matrix B *and* $\mathrm{AC} = \mathrm{I}_m$ for some $n \times m$ matric C. The reason we do not bother doing this is that we can prove that, if A is invertible in this ostensibly wider sense, then in fact $m = n$ (so A must be square) and B = C. But this is one of the deeper results of the subject and it will have to wait until Section 19. In particular, see Problem 19.4.

[F]10.1 We have used the letter r as a subscript denoting the number of nonzero rows of a matrix, A, in row-echelon form in anticipation of 15.7, which asserts that r is indeed the rank of A, in the sense of Definition 9.7.

[F]11.1 A much stronger result than 11.16 is in fact true.

Theorem: Every matrix is row-equivalent to a *unique* matrix in reduced row-echelon form.

The proof of uniqueness is quite technical and not very instructive. We will not prove this result since we will *never* make use of it. The extra strength of uniqueness is this: if $\mathrm{A} \underset{r}{\sim} \mathrm{B}$ and if B is in reduced row-echelon form, then B = GJ(A). Many authors use this result without proof. Some merely mention it in passing, while others base their entire development on its assumption. Because this result is extremely strong, "proofs" of the fundamental theorems of linear algebra that are based on it are deceptively oversimplified. We will incorporate the following very restricted special case of this result as part of Theorem 14.1: if A is a square matrix, say A is $n \times n$, and if $\mathrm{A} \underset{r}{\sim} \mathrm{I}_n$, then $\mathrm{I}_n = \mathrm{GJ(A)}$.

[F]16.1 The power set $\mathscr{P}(X)$ of a set X is the set of all subsets of X. That is, $\mathscr{P}(X) = \{S\colon\ S \subset X\}$. If $f\colon X \to Y$ is an arbitrary function, then there are two functions related to f, which we may denote by $f^{\to}$ and $f^{\leftarrow}$. $f^{\to}\colon \mathscr{P}(X) \to \mathscr{P}(Y)$ is defined for $S \subset X$ by $f^{\to}(S) = \{f(x)\colon\ x \in S\}$. $f^{\leftarrow}\colon \mathscr{P}(Y) \to \mathscr{P}(X)$ is defined for $S \subset Y$ by $f^{\leftarrow}(S) = \{x \in X: f(x) \in S\}$. It is a barbaric abuse of notation, regrettably sanctified by tradition, to use the symbol f^{-1} to denote $f^{\leftarrow}$

[F]19.1 We wish to emphasize, for the benefit of those familiar with these concepts, that none of the proofs we give in Section 19 depends on properties of the row-reduction process, the Gauss–Jordan algorithm, or on facts about matrices in reduced row-echelon form.

[F]30.1 The proof that we gave in Section 26 of the product rule for determinants (Theorem 26.1) made use of Problem 19.5 and, thereby, of the fundamental result from Section 19 that, for $A \in \mathbb{R}_{n\times n}$, if either $AB = I_n$ or $BA = I_n$, then A is invertible and $B = A^{-1}$. It is worth pointing out that we now have available an alternate proof of both these facts. This proof only uses properties of det proved prior to Section 25, plus the results about pseudodeterminants proved up to this point in Section 30.

Theorem 26.1 (revisited): For all $A, B \in \mathbb{R}_{n\times n}$, $\det(AB) = \det(A)\det(B)$.

Proof: For arbitrary fixed $B \in \mathbb{R}_{n\times n}$, consider the function $\Delta_B: \mathbb{R}_{n\times n} \to \mathbb{R}$ defined by $\Delta_B(A) = 1/\det(B)\det(AB)$. It is a routine exercise to check that Δ_B is a pseudodeterminant. So by Theorem 30.12, $1/\det(B)\det(AB) = \Delta_B(A) = \det(A)$. ■

Corollary: If B is either a right or a left inverse for A, then A is invertible. (This is the essential content of Theorem 19.5.)

Proof: Assume $AB = I_n$. Then,

$$\begin{aligned} 1 &= \det(I_n) \\ &= \det(AB) \\ &= \det(A)\det(B) \qquad \text{(by the preceding theorem)} \end{aligned}$$

Therefore, $\det(A) \neq 0$; so A is invertible by Theorem 25.1.

References

[A] Albert, A. A. (ed.), *Studies in Mathematics*, Vol. II, Mathematical Association of America, Washington, D.C., 1963.

[BM] Birkhoff, Garrett, and Saunders MacLane, *A Survey of Modern Algebra*, revised edition, Macmillan, Inc., New York, 1953.

[Ca] Cayley, A., "A Memoir on the Theory of Matrices," *Philosophical Transactions*, vol. CXVIII, 1858, 17–37.

[Cu] Curtis, Charles W., *Linear Algebra*, 2nd edition, Allyn and Bacon Inc., Boston, 1968.

[FS] Faddeev, D. K., and I. S. Sominskii, *Problems in Higher Algebra*, W. H. Freeman and Company, San Francisco, 1965.

[F] Finkbeiner, Daniel T., *Introduction to Matrices and Linear Transformations*, 2nd edition, W. H. Freeman and Co., San Francisco, 1966.

[FIS] Friedberg, Stephen H., Arnold J. Insel, and Lawrence E. Spence, *Linear Algebra*, Prentice-Hall, Inc., Englewood Cliffs, N.J., 1979.

[Gra] Graybill, Franklin A., *Introduction to Matrices with Applications in Statistics*, Wadsworth Publishing Co., Inc., Belmont, Calif., 1969.

[Gro] Grossman, Stanley I., *Applications for Elementary Linear Algebra*, Wadsworth Publishing Co., Inc., Belmont, Calif., 1980.

[Ha] Halmos, Paul R., *Finite-dimensional Vector Spaces*, Springer-Verlag, Inc., New York, New York, 1974.

[He] Helzer, Garry, *Applied Linear Algebra with APL*, Little, Brown and Company, Boston, 1982.

[HK] Hoffman, Kenneth, and Ray Kunze, *Linear Algebra*, Prentice-Hall, Inc., Englewood Cliffs, N.J., 1961.

[I] Insel, A. J., "Nilpotent Transformations and the Decomposition of a Vector Space," *American Mathematical Monthly*, February 1974, 160–162.

[M] Muir, Thomas, *A Treatise on the Theory of Determinants*, Dover Publications, Inc., New York, 1960.

[ND] Noble, Ben, and James W. Daniel, *Applied Linear Algebra*, Prentice-Hall, Inc., Englewood Cliffs, N.J., 1977.

[N] Nomizu, Katsumi, *Fundamentals of Linear Algebra*, 2nd edition, Chelsea Publishing Co., Inc., Bronx, N.Y., 1979.

[Pr] Proskuryakov, I. V., *Problems in Linear Algebra*, Mir Publishers, Moscow, 1978.

[Pu] Pullman, N. J., *Matrix Theory and Its Applications*, Marcel Dekker, Inc., New York, 1976.

[RA] Rorres, Chris, and Howard Anton, *Applications of Linear Algebra*, 2nd edition, John Wiley & Sons, Inc., New York, 1979.

[Sa] Satake, Ichiro, *Linear Algebra*, Marcel Dekker, Inc., New York, 1975.

[Se] Searle, Shayle R., *Matrix Algebra Useful for Statistics*, John Wiley & Sons, Inc., New York, 1982.

[St] Strang, Gilbert, *Linear Algebra and Its Applications*, 2nd edition, Academic Press, Inc., 1980.

[Sy] Sylvester, J. J., "Lectures on the Principles of Universal Algebra," *American Journal of Mathematics*, Vol. 6, 1884, 270–286.

[W] Whitelaw, Thomas A., *An Introduction to Linear Algebra*, Blackie & Son Limited, Glasgow, 1983.

Answers and Hints to Selected Problems

Problem Set 1

1.2 mn

1.3 120×1, 60×2, 40×3, 30×4, 24×5, 20×6, 15×8, 12×10, 10×12, 8×15, 6×20, 5×24, 4×30, 3×40, 2×60, 1×120

1.4 17×1 or 1×17

1.5 $$A = \begin{bmatrix} 0 & -1 & -2 & -3 & -4 \\ 1 & 0 & -1 & -2 & -3 \\ 2 & 1 & 0 & -1 & -2 \end{bmatrix}$$

1.6 $$B = \begin{bmatrix} 4 & -6 & 8 & -10 \\ -6 & 8 & -10 & 12 \\ 8 & -10 & 12 & -14 \\ -10 & 12 & -14 & 16 \\ 12 & -14 & 16 & -18 \end{bmatrix}$$

1.7 $$C = \begin{bmatrix} 1 & 0 & 1 \\ 9 & 4 & 1 \\ 25 & 16 & 9 \end{bmatrix}$$

1.8 BC is 5×4; AE is 5×5; (AC)D is 5×5; DA is 4×3; A(CD) is 5×5; (DA)E is 4×5; D(AE) is 4×5; (BC)E and AB are not defined.

1.9 $x_1 = 15z_1 - 13z_2 + 19z_3$
$x_2 = 5z_1 - 7z_2 + 6z_3$
$x_3 = -5z_1 - z_2 - 7z_3$

1.10 $x_1 = 7z_1 + 7z_2 - 5z_3 - 4z_4$
$x_2 = 4z_1 + 9z_2 + z_3 - 4z_4$

1.11 Suppose A and B are $n \times n$ upper triangular. We must show, for fixed i and j, that if $i \gneq j$, then the $(i, j)^{\text{th}}$ entry in AB is 0. The $(i, j)^{\text{th}}$ entry in AB is $a_{i1}b_{1j} + \cdots + a_{ij}b_{ij} + a_{i,j+1}b_{j+1,j} + \cdots + a_{in}b_{nj}$, which is equal, using the upper triangularity of A and B and the assumption that $i \gneq j$, to $0b_{1j} + \cdots + 0b_{ij} + a_{i,j+1}0 + \cdots + a_{in}0 = 0$.

1.12 $a = 3u + v + 11w$
$b = 11u + 5v - 3w$
$c = -6u - 7v + 11w$

1.13 (i) $\begin{bmatrix} 58 & 24 \\ 30 & 25 \\ -26 & 26 \\ 12 & 4 \end{bmatrix}$ (ii) 117 (iii) $\begin{bmatrix} 68 \\ 110 \\ -26 \end{bmatrix}$

(iv) $[-26 \quad -3]$

1.14 $w = -3,\ x = 3,\ z = -7,\ y = 0$

1.15 B commutes with A iff B has the form $\begin{bmatrix} a & b \\ 0 & a \end{bmatrix}$.

Problem Set 2

2.1 (vii) The Hadamard product of P and Q is $\begin{bmatrix} 8 & 63 & -7 \\ 12 & 0 & 0 \\ -12 & 3 & 2 \end{bmatrix}$.

(viii) The matrix product PQ = $\begin{bmatrix} 29 & -\frac{45}{2} & \frac{29}{2} \\ 34 & \frac{59}{2} & \frac{53}{2} \\ -14 & -40 & -12 \end{bmatrix}$.

2.2 $\begin{bmatrix} -\frac{4}{3} & \frac{23}{3} \\ \frac{1}{9} & -\frac{8}{9} \\ 0 & \frac{8}{3} \end{bmatrix}$

Problem Set 3

3.2 (i) $\begin{bmatrix} 17 & \frac{139}{4} & \frac{45}{2} \\ 30 & \frac{159}{4} & \frac{41}{2} \\ -22 & -11 & 0 \end{bmatrix}$ (ii) $\begin{bmatrix} 22 & 42 & -1 \\ \frac{38}{3} & 31 & \frac{1}{3} \\ 30 & 75 & \frac{15}{4} \end{bmatrix}$

(iii) $\begin{bmatrix} 172 & 71 & 57 \\ 26 & \frac{295}{2} & \frac{33}{2} \\ \frac{103}{2} & \frac{35}{4} & \frac{255}{4} \end{bmatrix}$

(iv) $\begin{bmatrix} 179 & -\frac{59}{2} & \frac{165}{2} \\ 70 & 150 & 58 \\ \frac{45}{2} & -\frac{329}{4} & -\frac{217}{4} \end{bmatrix}$

(v) $\begin{bmatrix} -127 & -\frac{7585}{4} & -\frac{1399}{4} \\ 1618 & -\frac{3819}{4} & \frac{3827}{4} \\ -1598 & -385 & -1119 \end{bmatrix}$

(vi) $\begin{bmatrix} \frac{7941}{2} & \frac{6975}{4} & \frac{10025}{4} \\ -202 & 1369 & 507 \\ 3022 & -25 & 1325 \end{bmatrix}$

Note that $P(Q + R) \neq QP + RP = (Q + R)P$.
Note that $(P + Q)^2 \neq P^2 + 2PQ + Q^2$.
Note that $(PQ)^2 \neq P^2Q^2$.

3.4 C^{-1}

3.5 $A^TB^TC^{-1}$

3.6 We need to find w, x, y, and z such that

$$\begin{bmatrix} a & b \\ c & d \end{bmatrix}\begin{bmatrix} w & x \\ y & z \end{bmatrix} = \begin{bmatrix} w & x \\ y & z \end{bmatrix}\begin{bmatrix} a & b \\ c & d \end{bmatrix} = \begin{bmatrix} 1 & 0 \\ 0 & 1 \end{bmatrix}$$

This is equivalent to the following 8 equations.

(1) $aw + by = 1$	(5) $wa + xc = 1$
(2) $ax + bz = 0$	(6) $wb + xd = 0$
(3) $cw + dy = 0$	(7) $ya + zc = 0$
(4) $cx + dz = 1$	(8) $yb + zd = 1$

When these equations are solved for w, x, y, and z as functions of a, b, c, and d, the conclusion is that $\begin{bmatrix} a & b \\ c & d \end{bmatrix}$ is invertible iff $ad - bc \neq 0$, in which case the inverse is given by

$$\begin{bmatrix} w & x \\ y & z \end{bmatrix} = \begin{bmatrix} a & b \\ c & d \end{bmatrix}^{-1} = \frac{1}{ad - bc}\begin{bmatrix} d & -b \\ -c & a \end{bmatrix}.$$

This result will be used routinely throughout the remainder of the text.

$$\begin{bmatrix} -4 & 2 \\ 3 & -1 \end{bmatrix}^{-1} = \begin{bmatrix} 1/2 & 1 \\ 3/2 & 2 \end{bmatrix}$$

3.9 $AB^{-1} = I_n AB^{-1}$ [property of I_n]
$= (B^{-1}B)AB^{-1}$ [property of B^{-1}]
$= B^{-1}(BA)B^{-1}$ [associative law]
$= B^{-1}(AB)B^{-1}$ [by assumption]
$= B^{-1}A(BB^{-1})$ [associative law]
$= B^{-1}AI_n$ [property of B^{-1}]
$= B^{-1}A$ [property of I_n]

3.14 (i) T; (ii) F; (iii) T; (iv) F; (v) T; (vi) F; (vii) T; (viii) F; (ix) T; (x) F; (xi) F; (xii) T.

3.17 *Hint:* Use Corollary 1.13.

3.19 By Problem 3.18, trace(AB − BA) = trace(AB) − trace(BA) = 0, whereas trace(I_n) = n.

Problem Set 4

4.2 (i) a line in $\mathbb{R}^3$; (ii) a plane in $\mathbb{R}^4$; (iii) a point in $\mathbb{R}^2$; (iv) a hyperplane in $\mathbb{R}^4$; (v) a three-dimensional affine subspace of $\mathbb{R}^5$; (vi) a three-dimensional affine subspace of $\mathbb{R}^5$; (vii) a line in $\mathbb{R}^3$.

Problem Set 5

5.2
(i) is a linear subspace.
(ii) is an affine subspace; it is the translate of the linear subspace in (i) by the vector $(0, 0, 17)$.
(iii) is not an affine subspace.
(iv) is an affine subspace; it is the translate of the linear subspace in (v) by the vector $(1, 0, 0)$.
(v) is a linear subspace.
(vi) is not an affine subspace.
(vii) is an affine subspace; it is the translate of the linear subspace $\{(x_1, x_2, x_3) \in \mathbb{R}^3\colon\ x_1 = -x_2 - x_3\}$ by the vector $(1, 0, 0)$.
(viii) is not an affine subspace.
(ix) is a linear subspace.
(x) is not an affine subspace.
(xi) is not an affine subspace.
(xii) is a linear subspace.
(xiii) is an affine subspace; it is the translate of the linear subspace $\{(x_1, x_2, x_3, x_4) \in \mathbb{R}^4\colon\ x_2 = x_1,\ x_3 = x_1\}$ by the vector $(0, 2, 3, 0)$.

Problem Set 6

6.3 (i)
$$\begin{aligned}
2x_1 + \phantom{2x_2 +{}} 4x_3 - 3x_4 + 6x_5 &= 0\\
x_1 - 2x_2 + 9x_3 + x_4 + 4x_5 &= 0\\
x_2 - 4x_3 + x_4 \phantom{{}+ 4x_5} &= 0\\
4x_1 + x_2 + 7x_3 + 7x_4 - 3x_5 &= 0
\end{aligned}$$

$$\begin{aligned}\text{(ii) } 2x_1 + 4x_3 - 3x_4 &= 6\\ x_1 - 2x_2 + 9x_3 + x_4 &= 4\\ x_2 - 4x_3 + x_4 &= 0\\ 4x_1 + x_2 + 7x_3 + 7x_4 &= -3\end{aligned}$$

6.5 $x_1 = -4$, $x_2 = 4$; (ii) $x_1 = 2$, $x_2 = -2$

6.6 (i) $-4\begin{bmatrix} 2 \\ -4 \end{bmatrix} + 4\begin{bmatrix} 3 \\ -2 \end{bmatrix} = \begin{bmatrix} 4 \\ 8 \end{bmatrix}$

(ii) $2\begin{bmatrix} 7 \\ 9 \end{bmatrix} - 2\begin{bmatrix} 6 \\ 8 \end{bmatrix} = \begin{bmatrix} 2 \\ 2 \end{bmatrix}$

6.7 $\begin{bmatrix} 6 \\ 7 \end{bmatrix} = 4\begin{bmatrix} 3 \\ 3 \end{bmatrix} - \begin{bmatrix} 6 \\ 5 \end{bmatrix}$

Problem Set 7

7.2 (i) independent.
(ii) dependent; it contains the zero vector.
(iii) dependent; the second vector is -3 times the first.
(iv) independent; it is a reduced set of vectors.
(v) dependent; any set of five 4-tuples is dependent.
(vi) independent.

7.3 Suppose that $t_1\mathbf{A}\mathbf{u}_1 + t_2\mathbf{A}\mathbf{u}_2 + \cdots + t_n\mathbf{A}\mathbf{u}_n = \mathbf{0}$; then by Theorem 3.1, $\mathbf{A}(t_1\mathbf{u}_1 + t_2\mathbf{u}_2 + \cdots + t_n\mathbf{u}_n) = \mathbf{0}$. By Theorem 6.11, $t_1\mathbf{u}_1 + t_2\mathbf{u}_2 + \cdots + t_n\mathbf{u}_n = \mathbf{A}^{-1}\mathbf{0} = \mathbf{0}$. The linear independence of $\{\mathbf{u}_1, \mathbf{u}_2, \ldots, \mathbf{u}_n\}$ now implies that $t_1 = t_2 = \cdots = t_n = 0$.

Problem Set 8

8.2 When S is affine but not linear, the associated linear subspace W was given in the answer to Problem 5.2.
(i) $\dim S = 1$; $\{(1,1,0)\}$
(ii) $\dim W = 1$; $\{(1,1,0)\}$
(iv) $\dim W = 2$; $\{(0,1,0),(0,0,1)\}$
(v) $\dim S = 2$; $\{(0,1,0),(0,0,1)\}$
(vii) $\dim W = 2$; $\{(-1,1,0),(-1,0,1)\}$
(ix) $\dim S = 3$; $\{(1,1,0,0),(-1,0,1,0),(1,0,0,1)\}$
(xii) $\dim S = 2$; $\{(1,0,3,0,0),(0,1,4,0,2)\}$
(xiii) $\dim W = 2$; $\{(1,1,1,0),(0,0,0,1)\}$

8.3 (i) *Hint:* the problem is to find x_1 and x_2 such that

$x_1\begin{bmatrix} 1 \\ 2 \\ 1 \end{bmatrix} + x_2\begin{bmatrix} 2 \\ 3 \\ 4 \end{bmatrix} = \begin{bmatrix} 4 \\ 7 \\ 6 \end{bmatrix}$. This is equivalent, by Observation 6.14, to solving the linear system

$$\begin{aligned} x_1 + 2x_2 &= 4\\ 2x_1 + 3x_2 &= 7\\ x_1 + 4x_2 &= 6\end{aligned}$$

(ii) Show that the system

$$x_1 + 2x_2 = 2$$
$$2x_1 + 3x_2 = 9$$
$$x_1 + 4x_2 = 5$$

is inconsistent.

(iii) We may take $\mathbf{w} = (2, 9, 5)$. The set $\{(1, 2, 1), (2, 3, 4)\}$ is linearly independent by inspection; by 7.15, if we add to this set any vector that is not in its span, the larger set is still linearly independent. From part (ii), $(2, 9, 5) \notin \mathscr{L}\{(1, 2, 1), (2, 3, 4)\}$, so the set $\{(1, 2, 1), (2, 3, 4), (2, 9, 5)\}$ is a linearly independent set of size 3 in $\mathbb{R}^3$. Since $\dim \mathbb{R}^3 = 3$, this set is a basis for $\mathbb{R}^3$ by exercise 8.5(i).

8.4 (i) Yes. The set $\{(-3, 9), (2, -5)\}$ is linearly independent, by inspection. It therefore spans $\mathbb{R}^2$ by Exercise 8.5(i); so *every* vector in $\mathbb{R}^2$, $(4, -18)$ in particular, belongs to $\mathscr{L}\{(-3, 9), (2, -5)\} = \mathbb{R}^2$.

(ii) No. The only four-dimensional subspace of $\mathbb{R}^4$ is the whole of $\mathbb{R}^4$.

(iii) Yes. Since $\{\mathbf{v}_1, \mathbf{v}_2, \mathbf{v}_5\}$ is a subset of the linearly independent set β, it is linearly independent by 7.8(i). Since W contains this linearly independent subset of size 3, $\dim W$ is at least 3.

(iv) No. The set $\{(1, 2, 2), (2, 4, 4)\}$ is linearly dependent, by inspection. Any set containing this will also be linearly dependent by 7.8(ii).

8.5 $ad - bc \neq 0$

8.6 (i) d; (ii) c, e; (iii) f; (iv) e; (v) a, d; (vi) c, e.

8.9 (i) → (ii) follows from 7.13. (ii) → (iii) is trivial. For (iii) → (i), there exists, by 8.8, a subset $(\mathbf{u}_{i_1}, \mathbf{u}_{i_2}, \ldots, \mathbf{u}_{i_p})$ that is a basis for $\mathscr{L}(\mathbf{u}_1, \mathbf{u}_2, \ldots, \mathbf{u}_m)$. Because $\mathscr{L}(\mathbf{u}_1, \mathbf{u}_2, \ldots, \mathbf{u}_m, \mathbf{v})$ is also p-dimensional, the ordered set $(\mathbf{u}_{i_1}, \mathbf{u}_{i_2}, \ldots, \mathbf{u}_{i_p}, \mathbf{v})$ is linearly dependent, so, by 7.16, some vector in it is a linear combination of its predecessors. Since $(\mathbf{u}_{i_1}, \mathbf{u}_{i_2}, \ldots, \mathbf{u}_{i_p})$ is linearly independent, it must be that $\mathbf{v} \in \mathscr{L}(\mathbf{u}_{i_1}, \mathbf{u}_{i_2}, \ldots, \mathbf{u}_{i_p})$.

Problem Set 9

9.2 (i) 4; (ii) 12; (iii) 9.

9.3 (→) If $A\mathbf{x} = \mathbf{b}$ is consistent, then $\mathbf{b} \in CS(A)$ by 9.4. Hence $CS([A \mid \mathbf{b}]) = CS(A)$. So $\text{rank}([A \mid \mathbf{b}]) = \dim CS([A \mid \mathbf{b}]) = \dim CS(A) = \text{rank}(A)$.

(←) This follows by Problem 8.9 and Observation 9.4.

9.4 $r(A) = \dim RS(A) = \dim CS(A^T) = r(A^T)$

9.5 (i) 2; (ii) 1; (iii) 2; (iv) 2; (v) 2; (vi) 4; (vii) 2; (viii) 1.

9.7 $r(AC) \leq r(C)$ by Theorem 9.12. Since $C = A^{-1}(AC)$, we can apply Theorem 9.12 once more to obtain $r(C) \leq r(AC)$. The argument for BA is similar.

10.2 Part 1, assuming the matrices are augmented matrices of linear systems.

(i) $\{(1, 2, 3)\}$, a point in $\mathbb{R}^3$
(ii) Not in row-echelon form
(iii) Inconsistent
(iv) $\{(18, 17)\}$, a point in $\mathbb{R}^2$
(v) $\{(x_1, x_2, x_3, x_4) \in \mathbb{R}^4\colon\ x_1 = -2x_3 + 4,\ x_2 = -3x_3 + 6,\ x_4 = -1\}$, a line in $\mathbb{R}^4$
(vi) $\{(5, -5)\}$, a point in $\mathbb{R}^2$
(vii) Not in row-echelon form
(viii) Inconsistent
(ix) $\{(x_1, x_2, x_3, x_4) \in \mathbb{R}^4\colon\ x_1 = 44x_4 + 1,\ x_2 = 17,\ x_3 = -4x_4 + 2\}$, a line in $\mathbb{R}^4$
(x) $\{(x_1, x_2, x_3, x_4, x_5) \in \mathbb{R}^5\colon\ x_1 = -4x_2 + x_3 - 2x_5 + 2,\ x_4 = 2x_5 + 1\}$, a three-dimensional affine subspace of $\mathbb{R}^5$
(xi) $\{(x_1, x_2, x_3, x_4, x_5) \in \mathbb{R}^5\colon\ x_1 = -4x_2 - 2x_5 + 2,\ x_4 = 2x_5 + 1\}$, a three-dimensional affine subspace of $\mathbb{R}^5$
(xii) $\{(x_1, x_2, x_3, x_4, x_5) \in \mathbb{R}^5\colon\ x_1 = -4x_2 + x_3 - 2x_5,\ x_4 = 2x_5\}$, a three-dimensional linear subspace of $\mathbb{R}^5$
(xiii) $\{(x_1, x_2, x_3, x_4, x_5) \in \mathbb{R}^5\colon\ x_2 = -4x_3 - 2x_5 + 2,\ x_4 = 2x_5 + 1\}$, a three-dimensional affine subspace of $\mathbb{R}^5$
(xiv) $\{(x_1, x_2, x_3, x_4) \in \mathbb{R}^4\colon\ x_1 = 3/2,\ x_2 = -7x_3,\ x_4 = 0\}$, a line in $\mathbb{R}^4$
(xv) Not in row-echelon form
(xvi) $\{(34.33, -13.14, 4.4)\}$, a point in $\mathbb{R}^3$
(xvii) Inconsistent
(xviii) Not in row-echelon form

Part 2, assuming the matrices are coefficient matrices of homogeneous linear systems.

(i) $\{(x_1, x_2, x_3) \in \mathbb{R}^3\colon\ x_1 = -x_4,\ x_2 = -2x_4,\ x_3 = -3x_4\}$, a line in $\mathbb{R}^4$
(ii) Not in row-echelon form
(iii) $\{(0, 0, 0)\}$, a point in $\mathbb{R}^3$
(iv) $\{(x_1, x_2, x_3) \in \mathbb{R}^3\colon\ x_1 = -18x_3,\ x_2 = -17x_3\}$, a line in $\mathbb{R}^3$
(v) $\{(x_1, x_2, x_3, x_4, x_5) \in \mathbb{R}^5\colon\ x_1 = -2x_3 - 4x_5,\ x_2 = -3x_3 - 6x_5,\ x_4 = x_5\}$, a plane in $\mathbb{R}^5$
(vi) $\{(x_1, x_2, x_3) \in \mathbb{R}^3\colon\ x_1 = -5x_3,\ x_2 = 5x_3\}$, a line in $\mathbb{R}^3$
(vii) Not in row-echelon form
(viii) $\{(x_1, x_2, x_3, x_4) \in \mathbb{R}^4\colon\ x_1 = -8x_3,\ x_2 = -2x_3,\ x_4 = 0\}$, a line in $\mathbb{R}^4$
(ix) $\{(x_1, x_2, x_3, x_4, x_5) \in \mathbb{R}^5\colon\ x_1 = 44x_4 - x_5,\ x_2 = -17x_5,\ x_3 = -4x_4 - 2x_5\}$, a plane in $\mathbb{R}^5$
(x) $\{(x_1, x_2, x_3, x_4, x_5, x_6) \in \mathbb{R}^6\colon\ x_1 = -4x_2 + x_3 - 2x_5$

$-2x_6,\ x_4 = 2x_5 - x_6\}$, a four-dimensional linear subspace of $\mathbb{R}^6$

(xi) $\{(x_1, x_2, x_3, x_4, x_5, x_6) \in \mathbb{R}^6\colon\ x_1 = -4x_2 - 2x_5 - 2x_6,\ x_4 = 2x_5 - x_6\}$, a four-dimensional linear subspace of $\mathbb{R}^6$

(xii) $\{(x_1, x_2, x_3, x_4, x_5, x_6) \in \mathbb{R}^6\colon\ x_1 = -4x_2 + x_3 - 2x_5,\ x_4 = 2x_5\}$, a four-dimensional linear subspace of $\mathbb{R}^6$

(xiii) $\{(x_1, x_2, x_3, x_4, x_5, x_6) \in \mathbb{R}^6\colon\ x_2 = -4x_3 - 2x_5 - 2x_6,\ x_4 = 2x_5 - x_6\}$, a four-dimensional linear subspace of $\mathbb{R}^6$

(xiv) $\{(x_1, x_2, x_3, x_4, x_5) \in \mathbb{R}^5\colon x_1 = -\frac{3}{2}x_5,\ \ x_2 = -7x_3,\ x_4 = 0\}$, a plane in $\mathbb{R}^5$

(xv) Not in row-echelon form

(xvi) $\{(x_1, x_2, x_3, x_4) \in \mathbb{R}^4\colon\ x_1 = -34.33x_4,\ x_2 = 13.14x_4,\ x_3 = -4.4x_4\}$, a line in $\mathbb{R}^4$

(xvii) $\{(x_1, x_2, x_3, x_4) \in \mathbb{R}^4\colon x_1 = -2x_3,\ x_2 = 3x_3,\ x_4 = 0\}$, a line in $\mathbb{R}^4$

(xviii) Not in row-echelon form

10.4 A has n-many rows, so if the number of leading ones in A is strictly less than n, A has an all-zero row. If the number of leading ones is exactly n, then because no column of a matrix in reduced row-echelon form can contain two leading ones, each column contains a leading one, so $A = I_n$.

10.5 $A = \begin{bmatrix} 1 & -1 \\ 0 & 0 \end{bmatrix}$

Problem Set 11

11.2 (i) $\mathcal{O}_1(A) = \begin{bmatrix} 4 & 0 & -\frac{1}{2} \\ -5 & -1 & \frac{27}{4} \\ -3 & \frac{1}{3} & 6 \end{bmatrix}$ $\mathcal{O}_2(A) = \begin{bmatrix} 4 & 0 & -\frac{1}{2} \\ 1 & -1 & 6 \\ 5 & \frac{1}{3} & 7 \end{bmatrix}$

$\mathcal{O}_3(A) = \begin{bmatrix} 1 & 0 & -\frac{1}{8} \\ 1 & -1 & 6 \\ -3 & \frac{1}{3} & 6 \end{bmatrix}$ $\mathcal{O}_4(A) = \begin{bmatrix} 4 & 0 & -\frac{1}{2} \\ -3 & \frac{1}{3} & 6 \\ 1 & -1 & 6 \end{bmatrix}$

$\mathcal{O}_1(B) = \begin{bmatrix} 2 & 1 & 0 & 1 \\ 0 & -\frac{1}{2} & 0 & \frac{3}{2} \end{bmatrix}$ $\mathcal{O}_3(B) = \begin{bmatrix} \frac{1}{2} & \frac{1}{4} & 0 & \frac{1}{4} \\ 3 & 1 & 0 & 3 \end{bmatrix}$

$\mathcal{O}_1(C)$, $\mathcal{O}_2(C)$, $\mathcal{O}_3(C)$, and $\mathcal{O}_4(C)$ are, respectively,

$$\begin{bmatrix} -2 & 8 & 8 \\ 12 & -7 & -11 \\ 4 & 8 & -1 \\ 0 & 4 & 4 \end{bmatrix} \quad \begin{bmatrix} -2 & 8 & 8 \\ 6 & 5 & 1 \\ 0 & 24 & 15 \\ 0 & 4 & 4 \end{bmatrix}$$

$$\begin{bmatrix} -\frac{1}{2} & 2 & 2 \\ 6 & 5 & 1 \\ 4 & 8 & -1 \\ 0 & 4 & 4 \end{bmatrix} \quad \begin{bmatrix} -2 & 8 & 8 \\ 4 & 8 & -1 \\ 6 & 5 & 1 \\ 0 & 4 & 4 \end{bmatrix}$$

(ii) $\begin{bmatrix} 4 & 0 & -\frac{1}{2} \\ -5 & 1 & \frac{27}{4} \\ 5 & \frac{1}{3} & 7 \end{bmatrix}$ and $\begin{bmatrix} 4 & 0 & -\frac{1}{2} \\ -5 & 1 & \frac{27}{4} \\ 5 & \frac{1}{3} & 7 \end{bmatrix}$

(iii) $\begin{bmatrix} 1 & 0 & -\frac{1}{8} \\ -5 & -1 & \frac{27}{4} \\ -3 & \frac{1}{3} & 6 \end{bmatrix}$ and $\begin{bmatrix} 1 & 0 & -\frac{1}{8} \\ -\frac{1}{2} & -1 & \frac{99}{16} \\ -3 & \frac{1}{3} & 6 \end{bmatrix}$

(iv) $\begin{bmatrix} 16 & 0 & -2 \\ 1 & -1 & 6 \\ -3 & \frac{1}{3} & 6 \end{bmatrix}$ $\begin{bmatrix} -2 & 8 & 8 \\ 6 & 5 & 1 \\ 8 & -8 & -17 \\ 0 & 4 & 4 \end{bmatrix}$

$\begin{bmatrix} 2 & 1 & 0 & 1 \\ 6 & \frac{5}{2} & 0 & \frac{9}{2} \end{bmatrix}$

$\mathcal{O}_2'(\mathcal{O}_2(\mathrm{C})) = \mathcal{O}_2(\mathcal{O}_2'(\mathrm{C})) = \mathrm{C}$.

11.3 (i) and (ii) have the same answer: $\begin{bmatrix} 1 & 0 & 0 & 2 \\ 0 & 1 & 0 & -3 \end{bmatrix}$.

11.5 (i) $\begin{bmatrix} -4 & 1 & 0 & -3 \\ 5 & 0 & 1 & 4 \end{bmatrix}$ (ii) $\begin{bmatrix} 1 & -\frac{1}{4} & 0 & \frac{3}{4} \\ 0 & \frac{5}{4} & 1 & \frac{1}{4} \end{bmatrix}$

(iii) $\begin{bmatrix} 1 & 0 & \frac{1}{5} & \frac{4}{5} \\ 0 & 1 & \frac{4}{5} & \frac{1}{5} \end{bmatrix}$

Problem Set 12

12.2 $\{(\frac{2}{5}, -\frac{6}{5}, -\frac{3}{5})\}$
12.3 Inconsistent
12.4 Inconsistent
12.5 $\{(x_1, x_2, x_3) \in \mathbb{R}^3: x_1 = \frac{2}{3}x_3 + \frac{5}{3},\ x_2 = \frac{1}{3}x_3 - \frac{5}{12}\}$
12.6 $\{(-\frac{6}{5}, 2, -\frac{2}{5})\}$
12.7 $\{(0, 0, 0)\}$
12.8 $\{(x_1, x_2, x_3) \in \mathbb{R}^3: x_1 = -2x_2 - 3x_3 + 4\}$
12.9 $\{(x_1, x_2, x_3, x_4, x_5) \in \mathbb{R}^5: x_1 = x_4 + 1,\ x_2 = -2x_4 - 1,\ x_3 = 0,\ x_5 = 2\}$
12.10 $\{(\frac{20}{3}, -\frac{4}{3})\}$
12.11 Inconsistent
12.12 $\{(x_1, x_2, x_3, x_4) \in \mathbb{R}^4: x_1 = -5x_3 - 13,\ x_2 = 2x_3 + 3,\ x_4 = 5\}$
12.13 $\{(x_1, x_2, x_3) \in \mathbb{R}^3: x_1 = 3x_3,\ x_2 = 6x_3 + 2\}$
12.14 Inconsistent
12.15 $\{(x_1, x_2, x_3, x_4) \in \mathbb{R}^4: x_1 = -\frac{1}{8}x_3 - \frac{27}{16}x_4 + 2,\ x_2 = -\frac{1}{12}x_3 + \frac{3}{8}x_4\}$
12.16 Inconsistent
12.17 $\{(x_1, x_2, x_3) \in \mathbb{R}^3: x_1 = -6x_2 - 12x_3 + \frac{24}{5}\}$
12.18 $\{(\frac{13}{5}, -\frac{3}{5}, \frac{9}{5})\}$
12.19 $\{(x_1, x_2, x_3) \in \mathbb{R}^3: x_1 = 5x_3 + 5,\ x_2 = -3x_3 - 2\}$

12.20 $\{(x_1, x_2, x_3) \in \mathbb{R}^3\colon x_1 = 0,\ x_2 = 2x_3 + 1\}$
12.21 $\{(x_1, x_2, x_3) \in \mathbb{R}^3\colon x_1 = -\tfrac{1}{2}x_3 + 1,\ x_2 = -5\}$
12.22 $\{(x_1, x_2, x_3) \in \mathbb{R}^3\colon x_1 = -2x_2 - \tfrac{1}{2}x_3 + \sqrt{2}\}$
12.23 $\{(x_1, x_2, x_3, x_4, x_5) \in \mathbb{R}^5\colon x_1 = -x_4 + 3,\ x_2 = x_4 + 2x_5 + 1,\ x_3 = 3x_4 - 4x_5\}$
12.24 (i) *Hint:* If your system has more than two equations, you have done more work than necessary. The system must also, of course, be homogeneous since its solution space is to contain the origin.
12.25 (i) The system is inconsistent if $t = -4$.
(ii) For values of t different from 0 and -4, the solution is the single point

$$\left\{\left(\frac{t+9}{t+4}, \frac{4}{t+4}, \frac{1}{t+4}\right)\right\}$$

(iii) When $t = 0$, the solution is the line $\{(x_1, x_2, x_3) \in \mathbb{R}^3\colon x_1 = -3x_3 + 3,\ x_2 = 1\}$.

Problem Set 13

13.4 Perform the operations $3C_2$ and $C_3 - 2C_1$ on the given 4×3 matrix to obtain

$$\begin{bmatrix} 2 & 3 & 3 \\ -3 & 27 & 16 \\ -2 & -12 & 10 \\ 8 & 24 & -17 \end{bmatrix}$$

Problem Set 14

14.2 (i) E_1, E_2, and E_3 are, respectively,

$$\begin{bmatrix} 1 & 0 & 0 & 0 \\ 0 & 1 & 0 & 0 \\ 0 & 0 & 1 & 0 \\ 0 & 0 & 0 & 2 \end{bmatrix}, \begin{bmatrix} 1 & 0 & 0 & 0 \\ -2 & 1 & 0 & 0 \\ 0 & 0 & 1 & 0 \\ 0 & 0 & 0 & 1 \end{bmatrix}, \begin{bmatrix} 1 & 0 & 0 & 0 \\ 2 & 1 & 0 & 0 \\ 0 & 0 & 1 & 0 \\ 0 & 0 & 0 & 1 \end{bmatrix}$$

(ii) To obtain $(E_1)^{-1}$, change the (4, 4)-entry in E_1 from 2 to $\tfrac{1}{2}$; $(E_2)^{-1} = E_3$; $(E_3)^{-1} = E_2$.
(iii) $E_2E_1A = C$; hence

$$D = E_2E_1 = \begin{bmatrix} 1 & 0 & 0 & 0 \\ -2 & 1 & 0 & 0 \\ 0 & 0 & 1 & 0 \\ 0 & 0 & 0 & 2 \end{bmatrix}$$

14.3 (i) $\begin{bmatrix} \frac{5}{8} & -\frac{7}{8} & \frac{1}{8} \\ \frac{3}{8} & -\frac{1}{8} & -\frac{1}{8} \\ -\frac{15}{8} & \frac{13}{8} & \frac{5}{8} \end{bmatrix}$ (ii) $\begin{bmatrix} 1 & -1 & 1 \\ -6 & 8 & -5 \\ -4 & 5 & -3 \end{bmatrix}$

(iii) Not invertible (iv) $\begin{bmatrix} \frac{1}{4} & -\frac{3}{4} & -\frac{1}{4} \\ \frac{1}{4} & \frac{3}{4} & \frac{1}{4} \\ -\frac{1}{16} & \frac{7}{16} & \frac{1}{16} \end{bmatrix}$

(v) Not invertible (vi) $\begin{bmatrix} \frac{3}{2} & -1 \\ -\frac{5}{2} & 2 \end{bmatrix}$

(vii) $\begin{bmatrix} \frac{1}{4} & \frac{3}{16} & 0 & \frac{3}{8} \\ 0 & \frac{1}{2} & 0 & 0 \\ -\frac{1}{8} & -\frac{7}{32} & \frac{1}{2} & \frac{1}{16} \\ \frac{1}{4} & -\frac{1}{16} & 0 & -\frac{1}{8} \end{bmatrix}$

(viii) Not invertible

14.4 (12.6) $\begin{bmatrix} -\frac{19}{5} & \frac{21}{5} & \frac{11}{5} \\ 2 & -2 & -1 \\ \frac{7}{5} & -\frac{8}{5} & -\frac{3}{5} \end{bmatrix}$ (12.7) $\begin{bmatrix} \frac{1}{5} & \frac{11}{35} & \frac{1}{35} \\ 0 & \frac{1}{7} & \frac{2}{7} \\ \frac{1}{5} & -\frac{4}{35} & \frac{6}{35} \end{bmatrix}$

(12.8) Not invertible (12.10) $\begin{bmatrix} \frac{8}{5} & -\frac{2}{15} \\ -\frac{1}{5} & \frac{4}{15} \end{bmatrix}$

(12.16) Not invertible (12.21) Not invertible

14.5 $\begin{bmatrix} 4 & 0 \\ 0 & 1 \end{bmatrix}\begin{bmatrix} 1 & 0 \\ 5 & 1 \end{bmatrix}\begin{bmatrix} 1 & 0 \\ 0 & \frac{1}{2} \end{bmatrix}\begin{bmatrix} 1 & \frac{1}{2} \\ 0 & 1 \end{bmatrix}$

14.6 (i) $\begin{bmatrix} -5 & -4 & 2 \\ -1 & -1 & 1 \\ 3 & 2 & -1 \end{bmatrix}$

(ii) $z_1 = -9y_1 - 7y_2 + 4y_3$
$z_2 = 9y_1 + 7y_2 - 4y_3$

(iii) No

14.7 (i) $\begin{bmatrix} -\frac{5}{4} \\ \frac{5}{4} \\ \frac{9}{16} \end{bmatrix}, \begin{bmatrix} -\frac{17}{4} \\ \frac{23}{4} \\ \frac{41}{16} \end{bmatrix}, \begin{bmatrix} -\frac{11}{4} \\ \frac{29}{4} \\ \frac{35}{16} \end{bmatrix}$

(ii) $x_1 = \frac{1}{4}b_1 - \frac{3}{4}b_2 - \frac{1}{4}b_3$
$x_2 = \frac{1}{4}b_1 + \frac{3}{4}b_2 + \frac{1}{4}b_3$
$x_3 = -\frac{1}{16}b_1 + \frac{7}{16}b_2 + \frac{1}{16}b_3$

14.8 (i) $t \neq 6$ (ii) $\begin{bmatrix} \dfrac{3}{6-t} & \dfrac{t}{36-6t} \\ \dfrac{1}{t-6} & \dfrac{1}{3t-18} \end{bmatrix}$

14.9 (i) $t \neq 0,\ t \neq -\frac{5}{3}$

(ii) $$\begin{bmatrix} \dfrac{-15-15t}{10+6t} & \dfrac{2}{5+3t} & \dfrac{13+9t}{10+6t} \\ \dfrac{5}{5t+3t^2} & \dfrac{2}{5t+3t^2} & \dfrac{-1}{5t+3t^2} \\ \dfrac{5+6t}{5+3t} & \dfrac{-2}{5+3t} & \dfrac{-4-3t}{5+3t} \end{bmatrix}$$

14.10 (i) $t \neq 0,\ t \neq 3$ (ii) $$\begin{bmatrix} \dfrac{t}{t-3} & \dfrac{-1}{t-3} & \dfrac{-2-2t}{5t-15} \\ 0 & 0 & -1/5 \\ \dfrac{-3}{t^2-3t} & \dfrac{1}{t^2-3t} & \dfrac{8}{5t^2-15t} \end{bmatrix}$$

Problem Set 15

15.2 r(A) = 3. $\{(1,0,-5,5,0),(0,1,-3,3,0),(0,0,0,0,1)\}$
15.3 r(B) = 4. $\{(1,1,2,3),(2,-1,1,2),(-1,2,4,0),(3,1,2,3)\}$
15.4 Yes
15.5 No
15.6 Yes: $(-2,1,-3) = 3(-3,-2,1) - \frac{5}{3}(1,1,1) + \frac{13}{3}(2,2,-1)$ is the only way.
15.7 $\{(b_1, b_2, b_3, b_4) \in \mathbb{R}^4\colon\ b_3 = b_2 - b_1,\ b_4 = 3b_2 - 8b_1\}$. Its dimension is 2.
15.8 No
15.9 $\{(b_1, b_2, b_3) \in \mathbb{R}^3\colon\ b_3 = b_1 + b_2\}$

Problem Set 16

16.2 (i) No. For example, $f(2) = f(4) = 4$.
(ii) Yes
(iii) Let $g_1(1) = 10,\ g_1(2) = 8,\ g_1(3) = 6,\ g_1(4) = 4$; and let $g_2(1) = 10,\ g_2(2) = 8,\ g_2(3) = 0,\ g_2(4) = 4$. Both g_1 and g_2 are right inverses for f; there are also others.

16.3 (i) Yes
(ii) No. For example, $0 \notin \text{ran}(g)$.
(iii) Let $f_1(0) = 1,\ f_1(2) = 1,\ f_1(4) = 2,\ f_1(6) = 3,\ f_1(8) = 4,\ f_1(10) = 3$; and let $f_2(0) = 2,\ f_2(2) = 1,\ f_2(4) = 2,\ f_2(6) = 3,\ f_2(8) = 4,\ f_2(10) = 4$. Both f_1 and f_2 are left inverses for g; there are also others.

16.4 (iii) Let $X = \{a\}$, $Y = \{b, c\}$, and $Z = \{d\}$. Let $f(a) = b$, $g(b) = d$, and $g(c) = d$.

16.5 (i) $\{0, 2, 4, 6, \dots\}$
(ii) $\{5, 6, 7, 8, \dots\}$
(iii) $\{5, 7, 9, 11, \dots\}$
(iv) $\{10, 12, 14, 16, \dots\}$

16.6 (i) $[0, \infty)$
(ii) $[-1, 1]$
(iii) $[-1, 1]$
(iv) $[0, 1]$

Problem Set 17

17.3 $(T \circ S)(\mathbf{u} + \mathbf{v}) = T(S(\mathbf{u} + \mathbf{v}))$ [definition of composition]
$= T(S(\mathbf{u}) + S(\mathbf{v}))$ [linearity of S]
$= T(S(\mathbf{u})) + T(S(\mathbf{v}))$ [linearity of T]
$= (T \circ S)(\mathbf{u}) + (T \circ S)(\mathbf{v})$
[definition of composition]
For similar reasons, $(T \circ S)(t\mathbf{u}) = T(S(t\mathbf{u})) = T(tS(\mathbf{u})) = tT(S(\mathbf{u})) = t(T \circ S)(\mathbf{u})$.

17.4 dim(CS(A)) = 3; dim(RS(A)) = 3; dim(NS(A)) = 5 − 3 = 2.

17.5 By inspection, dim(NS(A)) = 2; so dim(CS(A)) = dim(RS(A)) = 5 − 2 = 3.

17.6 ($\rightarrow$) Suppose $\mathbf{v} \in \ker T$; so $T(\mathbf{v}) = \mathbf{0}$. By linearity, $T(\mathbf{0}) = \mathbf{0}$ as well; so $\mathbf{v} = \mathbf{0}$ because T is one-to-one.
($\leftarrow$) Suppose $\mathbf{u} \neq \mathbf{v}$; so $\mathbf{v} - \mathbf{u} \neq \mathbf{0}$. Then $\mathbf{v} - \mathbf{u} \notin \ker T = \{\mathbf{0}\}$, so $T(\mathbf{v} - \mathbf{u}) \neq \mathbf{0}$. By linearity, $T(\mathbf{v} - \mathbf{u}) = T(\mathbf{v}) - T(\mathbf{u})$; hence $T(\mathbf{u}) \neq T(\mathbf{v})$.

17.7 No. Using 17.2(i), the observation that $f(\mathbf{0}) \neq \mathbf{0}$ is enough to justify this conclusion.

17.8 No. For example, $g((1,0) + (0,1)) = g(1,1) = 1$, whereas $g(1,0) + g(0,1) = 0 + 0 = 0$.

Problem Set 18

18.2 (ii) $(T_2 \circ T_1)(x, y, z) = (4x - 4y, 6x + 16y + 6z)$

Problem Set 19

19.2 If A is invertible, then (i) holds by 6.11; if A is not invertible, then (ii) holds by the remarks following 19.6; thus either (i) or (ii) holds. Since $A\mathbf{x} = \mathbf{0}$ always has the trivial solution, $\mathbf{x} = \mathbf{0}$, (i) and (ii) cannot both hold.

19.3 (i) T; (ii) T; (iii) F; (iv) F; (v) F; (vi) T; (vii) F; (viii) F.

19.4 Parts (i) and (ii) are immediate consequences of 19.1 and 19.3, respectively, since r(A) = dim CS(A) = dim RS(A).

19.5 If AB is invertible, then it has a right inverse, C. Since $I_n = (AB)C = A(BC)$, A has a right inverse, BC, so A is invertible by 19.6. A similar argument with left inverses shows that B is invertible.

Problem Set 20

20.2 Since row equivalence is transitive, this follows from 14.1.

20.3 Because T_A is linear, $T_A(4,4) = T_A(2(2,2)) = 2T_A(2,2) = 2(1,1) = (2,2)$. Since $T_A(3,-1) = (2,2)$ also, T_A is not one-to-one. Therefore, it is not invertible by 19.6.

20.4 Since the range of T_A includes the set $\{(4,-2),(3,6)\}$, which is a basis for $\mathbb{R}^2$, ran $T_A = \mathbb{R}^2$. So the answers to both (i) and (ii) are Yes.

20.5 Since the range of T_A includes the set $\{(1,0),(0,1)\}$, which is a basis for $\mathbb{R}^2$, T_A is onto, so A is invertible. Using 19.7 and 17.8, we have $A^{-1} = \begin{bmatrix} 1 & 2 \\ -2 & -3 \end{bmatrix}$.

20.6 We need a 2×2 matrix whose column space is $\{(x_1, x_2) \in \mathbb{R}^2: x_2 = 3x_1 - 4\}$. $A = \begin{bmatrix} 0 & 0 \\ -4 & 0 \end{bmatrix}$ is one possible answer.

Problem Set 21

21.2 Negative

21.3 Negative

21.4 Two square units

Problem Set 22

22.2 (i) Odd; (ii) even; (iii) even.

Problem Set 23

23.2 The permutation $(5,4,2,1,3,6)$ contains $4+3+1=8$ inversions; so $|A| = +a_{15}a_{24}a_{32}a_{41}a_{53}a_{66}$.

23.3 Note that

$$A = \begin{bmatrix} 0 & 0 & \cdot & \cdot & \cdot & a_{1k} \\ 0 & 0 & \cdot & \cdot & a_{2\,k-1} & 0 \\ \cdot & \cdot & \cdot & \cdot & \cdot & \cdot \\ \cdot & \cdot & \cdot & \cdot & \cdot & \cdot \\ \cdot & \cdot & \cdot & \cdot & \cdot & \cdot \\ a_{k1} & 0 & \cdot & \cdot & \cdot & 0 \end{bmatrix}$$

Since the number of inversions in the permutation $(k, k-1, k-2, \ldots, 2, 1)$ is

$$k-1+k-2+\cdots+1 = \sum_{n=1}^{n=k-1} n = \frac{(k-1)k}{2},$$

$$|A| = (-1)^{(k-1)k/2} a_{1k}a_{2\,k-1} \cdots a_{k1}.$$

In particular, when $k=6$, $|A| = -a_{16}a_{25}a_{34}a_{43}a_{52}a_{61}$. When $k=99$, $|A| = -a_{1\,99}a_{2\,98} \cdots a_{99\,1}$.

23.4 Observe that every elementary product from tA is of the form

$$(ta_{1j_1})(ta_{2j_2}) \cdots (ta_{nj_n}) = t^n a_{1j_1}a_{2j_2} \cdots a_{nj_n}$$

Problem Set 24

24.2 If the pth and q^{th} rows of A are the same, then A = EA, where E is the elementary matrix corresponding to the operation $R_p \leftrightarrow R_q$. Now

$$\begin{aligned} |A| &= |EA| \\ &= |E||A| && [24.4] \\ &= (-1)|A| && [24.1(i)] \end{aligned}$$

So $|A| = 0$.

24.3 (i) 8; (ii) -4; (iii) 54; (iv) 0; (v) 200; (vi) 0.

Problem Set 25

25.1 The matrix is invertible for all t except $t = 0$ and $t = 17$.

25.2 The matrix is singular iff $t \in \{-2, -\frac{1}{5}, 0, 2\}$.

25.4 $A = \begin{bmatrix} b_{11} \\ b_{21} \end{bmatrix} \begin{bmatrix} c_{11} & c_{12} \end{bmatrix} = \begin{bmatrix} b_{11}c_{11} & b_{11}c_{12} \\ b_{21}c_{11} & b_{21}c_{12} \end{bmatrix}$

So $|A| = b_{11}c_{11}b_{21}c_{12} - b_{11}c_{12}b_{21}c_{11} = 0$.

25.5 The correct answer is no. From the fact that $|A| = 0$, you may conclude that the columns of A are linearly dependent. This does not imply that the fourth column is a linear combination of the first three; it may be, as in the present case, that the third column is a combination of the first two; it may also be that the second column is a multiple of the first.

Problem Set 26

26.2 (i) $|AB| = |-BA| = (-1)^n|BA|$; so if n is odd, $|AB| = 0$. Thus either A or B is singular by 25.1 and 26.1.

(ii) $A = \begin{bmatrix} 1 & 0 \\ 0 & -1 \end{bmatrix}$ and $B = \begin{bmatrix} 0 & 1 \\ 1 & 0 \end{bmatrix}$ is one of many possible answers.

Problem Set 27

27.1 $\frac{5}{4}$; 270; $-\frac{1}{10}$; 2700; 72,900; $\frac{2}{25}$.

27.2 6; 750; 36; 2304; 2304; -22; -22.

Problem Set 28

28.1 The proof for $k = 1$ is trivial. For arbitrary k, we will let A_{ij}^c denote the cofactor of a_{ij} viewed as an element of M and will let a_{ij}^c denote the cofactor of a_{ij} viewed as an element of the submatrix A. In this language, the induction hypothesis is the fact that $A_{ij}^c = a_{ij}^c|D|$ for $1 \le i \le k$ and $1 \le j \le k$. So $|M| = a_{1k}A_{1k}^c + \cdots + a_{kk}A_{kk}^c = a_{1k}a_{1k}^c|D| + \cdots + a_{kk}a_{kk}^c|D| = |A||D|$.

28.2 96

Problem Set 29

29.3 $\{(\frac{11}{8}, -\frac{3}{8}, -\frac{37}{16}, \frac{67}{16})\}$

29.4 It follows from the proof of 29.3 that $A \operatorname{adj} A = |A|I_n$. If A is invertible, then $\operatorname{adj} A = |A|A^{-1}$, so $|\operatorname{adj} A| = ||A|A^{-1}| = |A|^n|A^{-1}| = |A|^{n-1}$. If A is not invertible, then $A \operatorname{adj} A = 0I_n = 0_{n\times n}$, so if $A \neq 0_{n\times n}$, then adj A cannot be invertible; hence $|\operatorname{adj} A| = 0$ as well. Finally, if $A = 0_{n\times n}$, the result is trivial.

29.5 $\operatorname{adj} A = \begin{bmatrix} -1 & -1 & 3 \\ 4 & 4 & -1 \\ -9 & 2 & 5 \end{bmatrix}$; $|A| = 11$; $A^{-1} = \frac{1}{11}\operatorname{adj} A$;

$|2A^2 \operatorname{adj} A| = 2^3|A^2||\operatorname{adj} A| = (8)(11)^2(11)^2 = 117{,}128$

29.6 Use 29.3 plus the fact that the entries in adj A are integers because they are sums of products of integers.

29.7 $A^{-1} = (1/|A|)\operatorname{adj} A$; hence $(A^{-1})^T = (1/|A|) \operatorname{cof} A$. So the $(i, j)^{\text{th}}$ entry of $A \circ (A^{-1})^T$ is $(1/|A|)a_{ij}a^c_{ij}$.

Problem Set 31

31.2 (i) (5, 6, 2); (ii) $(3x_1 + 2x_2, x_1 + 3x_2, -2x_1 + 8x_2)$; (iii) 2.

31.3 (i) $(\frac{3}{2}x_1 + 2x_2, -\frac{1}{2}x_1 - x_2)$; (ii) (2, −1).

31.4 (i) to (iv) can be answered by inspection. (i) (0, 10, 4);
(ii) $\{\mathbf{0}\}$; (iii) 2; (iv) $(2x_1 - x_2, -3x_1 + 4x_2, 6x_1 - 2x_2)$; (v) and (vi) are answered using the procedure from 15.21.
(v) $\{(b_1, b_2, b_3) \in \mathbb{R}^3\colon b_3 = \frac{2}{5}b_2 + \frac{18}{5}b_1\}$; (vi) No.

31.5 (i) $\begin{bmatrix} 1 & 0 & 1 & -1 \\ 0 & -1 & -2 & 4 \\ 2 & -1 & 0 & 2 \end{bmatrix}$

(ii) $\{(1, 0, 2), (0, -1, -1)\}$ is a basis for ran T; $\{(-1, -2, 1, 0), (1, 4, 0, 1)\}$ is a basis for ker T.
(iii) $r(T) = 2$; $n(T) = 2$.

31.6 (i) $\begin{bmatrix} -9 & -5 & 2 \\ -4 & -2 & 1 \\ 10 & 5 & -2 \end{bmatrix}$ (ii) $\begin{bmatrix} 1 & 0 & 1 \\ -2 & 2 & -1 \\ 0 & 5 & 2 \end{bmatrix}$

(iii) (−2, −1, 3)

31.7 (i) $\begin{bmatrix} 2 & 1 & 1 \\ -3 & 2 & -2 \\ 5 & 3 & -3 \end{bmatrix}$ (ii) (1, −2, 1) (iii) (0, −3, 5)

31.8 (i) $(2x_1 + x_2 + 3x_3, -3x_1 + 2x_2 - x_3, -6x_1 - 6x_3, -x_1 - 2x_2 - 3x_3)$
(ii) $\{(2, -3, -6, -1), (1, 2, 0, -2)\}$ is a basis for ran T; $\{(-1, -1, 1)\}$ is a basis for ker T.
(iii) $r(T) = 2$; $n(T) = 1$.

31.9 (i) All three vectors belong to ran T since ran $T = \mathbb{R}^3$.
(ii) Only (39, −24, 9, 3)

Problem Set 32

32.2 (i) $\begin{bmatrix} 2 & 3 \\ 1 & -1 \end{bmatrix}$ (ii) $\begin{bmatrix} -30 & 25 \\ -37 & 31 \end{bmatrix}$
(iii) (5, 0); (iv) (15, 20); (v) (−5, −6).

32.3 $\begin{bmatrix} -1 & \frac{3}{4} \\ -8 & 3 \end{bmatrix}$

32.4 No, since the matrix is not invertible.

32.5 If $B = P^{-1}AP$, then $|B| = |P^{-1}AP| = |P^{-1}||A||P| = (1/|P|)|A||P| = |A|$.

32.6 Using P3.18(iv), we have $\text{trace}(B) = \text{trace}(P^{-1}AP) = \text{trace}(APP^{-1}) = \text{trace}(A)$.

Problem Set 33

33.2 Recall (P3.23) that, if D is an invertible diagonal matrix, then so is D^{-1}. If $D = P^{-1}AP$, then $D^{-1} = P^{-1}A^{-1}P$.

Problem Set 34

34.2 You need to know which eigenvector belongs to which eigenspace.

34.3 $P = \begin{bmatrix} 0 & 1 \\ 1 & 0 \end{bmatrix}$

Problem Set 35

35.2 A is diagonalizable; $D = \text{diag}(3, 2, -1)$; $P = \begin{bmatrix} 2 & -6 & 0 \\ 2 & -3 & 0 \\ 1 & 1 & 1 \end{bmatrix}$.

35.3 B is not diagonalizable. 2 is a double root of $\chi(B)$, but $\dim E_2 = 1$.

35.4 $\chi_A(\lambda) = |\lambda I_n - A|$, so $\chi_A(0) = |0I_n - A| = |-A| = (-1)^n|A|$.

35.5 Just before 35.3, we saw that

$$\begin{aligned}\chi(A) = |\lambda I_n - A| &= \textstyle\sum \text{ signed elementary products from } \lambda I_n - A \\ &= (\lambda - a_{11})(\lambda - a_{22}) \cdots (\lambda - a_{nn}) \\ &\quad + \text{other signed elementary products.}\end{aligned}$$

Note that any elementary product from $\lambda I_n - A$ other than the one, $(\lambda - a_{11})(\lambda - a_{22}) \cdots (\lambda - a_{nn})$, formed by the entries on the main diagonal can involve at most $(n-2)$-many factors of the form $(\lambda - a_{ii})$. So λ occurs at most to the power $n-2$ in any of these other elementary products. Thus the coefficient of λ^{n-1} in $\chi(A)$ is the coefficient of λ^{n-1} in $(\lambda - a_{11})(\lambda - a_{22}) \cdots (\lambda - a_{nn})$, which is $-\Sigma_i a_{ii}$.

Problem Set 36

36.2 (i) $-1-5i$; (ii) $7+i$; (iii) $-18-i$; (iv) $\dfrac{-6+17i}{25}$;

(v) 13; (vi) $-6+17i$; (vii) $\dfrac{-18+i}{325}$; (viii) $5-12i$;

(ix) $-9-46i$; (x) $\dfrac{3+2i}{13}$; (xi) $-13i$; (xii) $6-2i$;

(xiii) $32-24i$; (xiv) $\dfrac{12-5i}{13}$

36.5 $z_1 = \dfrac{4+19i}{24}$

36.6 $z_1 = \dfrac{25+3i}{5}$ and $z_2 = \dfrac{9+13i}{5}$

Problem Set 37

37.2 (i) Diagonalizable over $\mathbb{R}$ with $D = \text{diag}(4, -2)$ and

$$P = \begin{bmatrix} -1 & -\frac{5}{2} \\ 1 & 1 \end{bmatrix}$$

(ii) Diagonalizable over $\mathbb{C}$ with $D = \text{diag}(2+3i, 2-3i)$ and

$$P = \begin{bmatrix} 4-3i/5 & 4+3i/5 \\ 1 & 1 \end{bmatrix}$$

(iii) Not diagonalizable

(iv) Diagonalizable over $\mathbb{R}$ with $D = \text{diag}(3, 3, -2)$ and

$$P = \begin{bmatrix} -1 & -1 & 1 \\ 1 & 0 & 0 \\ 0 & 1 & 0 \end{bmatrix}$$

(v) Diagonalizable over $\mathbb{R}$ with $D = \text{diag}(\frac{3}{2}, 1, -\frac{1}{2})$ and

$$P = \begin{bmatrix} -1 & 0 & -\frac{3}{5} \\ 2 & 1 & \frac{14}{5} \\ 1 & 1 & 1 \end{bmatrix}$$

(vi) Not diagonalizable.

37.3 $\begin{bmatrix} -56{,}489 & -37{,}830 \\ 94{,}575 & 63{,}306 \end{bmatrix}$

Problem Set 38

38.2 Suppose $S = \{\mathbf{u}_1, \mathbf{u}_2, \ldots, \mathbf{u}_p\}$ is an orthogonal set of nonzero vectors. Assume $\mathbf{0} = \Sigma_i a_i \mathbf{u}_i$; we must show for arbitrary j that $a_j = 0$. We have

$$\begin{aligned} 0 &= \mathbf{0} \cdot \mathbf{u}_j && [38.6] \\ &= \left(\sum_i a_i \mathbf{u}_i\right) \cdot \mathbf{u}_j && [\text{by assumption}] \\ &= a_j \|\mathbf{u}_j\|^2 && [\text{orthogonality of } S] \end{aligned}$$

Since $\mathbf{u}_j \neq \mathbf{0}$, $\|\mathbf{u}_j\| \neq 0$ by 38.5(i), so we conclude $a_j = 0$.

38.4 (i) $(1,-6,1)$; (ii) $(-1,6,-1)$; (iii) $(1,1,5)$; (iv) $(4,-1,4)$; (v) $(4,-1,4)$ (vi) $(0,-7,0)$; (vii) $(0,-7,0)$; (viii) 7; (ix) 7; (x) $\sqrt{38}$. Note that there are three pairs of answers that are the same: (iv) and (v), (vi) and (vii), and (viii) and (ix). Decide whether this happened by accident or whether some general results are true, and, in the latter case, prove the results.

38.5 $(\sqrt{2}/2, 0, \sqrt{2}/2)$ and $(-\sqrt{2}/2, 0, -\sqrt{2}/2)$

38.6 (i) Yes; (ii) no; (iii) yes.

38.8 (i) -7 and 3; (ii) none.

38.9 (i) $5\sqrt{2}$; (ii) 0; (iii) -0.1.

Problem Set 39

39.2 (i) $(\frac{2}{7}, -\frac{3}{7}, \frac{6}{7})$; (ii) $(1/\sqrt{3}, 1/\sqrt{3}, -1/\sqrt{3})$;
(iii) $\mathbf{w}_1 = (-\frac{2}{7}, \frac{3}{7}, -\frac{6}{7})$ and $\mathbf{w}_2 = (\frac{9}{7}, \frac{4}{7}, -\frac{1}{7})$;
(iv) $\mathbf{w}_1 = (-\frac{7}{3}, -\frac{7}{3}, \frac{7}{3})$ and $\mathbf{w}_2 = (\frac{13}{3}, -\frac{2}{3}, \frac{11}{3})$.

39.3 (i) $\|\mathbf{u}\| = \sqrt{(\frac{4}{5})^2 + (\frac{1}{5})^2 + (-\frac{2}{5})^2 + (-\frac{2}{5})^2} = 1$; similarly, $\|\mathbf{v}\| = 1$. Also, $\mathbf{u} \cdot \mathbf{v} = 0$. (ii) $(6,-7,-6,2)$;
(iii) $(-2,2,-4,1)$.

39.4 For Exercise 39.32: $\mathbf{v}_1 = \mathbf{u}_1/\|\mathbf{u}_1\| = 1/\sqrt{3}(1,1,0,1)$. $\mathbf{u}_2 \cdot \mathbf{v}_1 = 3/\sqrt{3}$, so $(\mathbf{u}_2 \cdot \mathbf{v}_1)\mathbf{v}_1 = (1,1,0,1)$; $\mathbf{w}_2 = \mathbf{u}_2 - (\mathbf{u}_2 \cdot \mathbf{v}_1)\mathbf{v}_1 = (1,4,3,-5)$ and $\mathbf{v}_2 = \mathbf{w}_2/\|\mathbf{w}_2\| = 1/\sqrt{51}(1,4,3,-5)$. Since we happen to know from the exercise that CS(A) is two-dimensional, we can quit; if we did not know this, we would find it out when we compute the next step in the GS algorithm and obtain $\mathbf{v}_3 = \mathbf{0}$. So $(1/\sqrt{3}(1,1,0,1), 1/\sqrt{51}(1,4,3,-5)) = (\mathbf{v}_1, \mathbf{v}_2)$ is an orthonormal basis for CS(A). $\text{Proj}_{CS(A)}\mathbf{b} = (\mathbf{b} \cdot \mathbf{v}_1)\mathbf{v}_1 + (\mathbf{b} \cdot \mathbf{v}_2)\mathbf{v}_2 = 0\mathbf{v}_1 + (3/\sqrt{51})(1/\sqrt{51}(1,4,3,-5)) = \frac{3}{51}(1,4,3,-5)$. Now, to solve $A\mathbf{x} = \text{proj}_{CS(A)}\mathbf{b}$, we apply the Gauss–Jordan algorithm to the matrix

$$\begin{bmatrix} 1 & 2 & 1 & \frac{3}{51} \\ 1 & 5 & 0 & \frac{12}{51} \\ 0 & 3 & -1 & \frac{9}{51} \\ 1 & -4 & 3 & -\frac{15}{51} \end{bmatrix}$$

and conclude that the set of least squares solutions is

$$\left\{(x_1, x_2, x_3) \in \mathbb{E}^3 : x_1 = -\tfrac{5}{3}x_3 - \tfrac{1}{17},\ x_2 = \tfrac{1}{3}x_3 + \tfrac{1}{17}\right\}$$

39.5 (i) $\{(\frac{2}{3}, 0, \frac{1}{3}, \frac{2}{3}), (-\frac{2}{3}, 0, \frac{2}{3}, \frac{1}{3}), (\frac{1}{3}, 0, \frac{2}{3}, -\frac{2}{3})\}$
(ii) $\mathbf{w}_1 = (1,0,1,1)$ and $\mathbf{w}_2 = (0,1,0,0)$

39.6 (i) Yes; (ii) you should realize, *without doing any calculations* that, because S is orthogonal, the Gram–Schmidt algorithm applied to S will simply normalize the vectors. So $GS(S) = \{1/\sqrt{11}(-1,3,-1), 1/\sqrt{22}(3,2,3), 1/\sqrt{2}(1,0,-1)\}$.

39.7 $\mathbf{w}_1 = (\frac{1}{5}, \frac{32}{5}, -\frac{52}{5}, -\frac{89}{5})$ and $\mathbf{w}_2 = (\frac{4}{5}, -\frac{32}{5}, \frac{42}{5}, -\frac{36}{5})$

Problem Set 40

40.2 $|A| = |A^T| = |A^{-1}| = 1/|A|$, so $|A| = \pm 1$.

40.3 $\begin{bmatrix} 1 & 1 \\ -2 & 2 \end{bmatrix}$ is a counterexample. The statement would be true with the word orthonormal replacing the word orthogonal.

40.5 The matrix is equal either to $\begin{bmatrix} \cos\theta & -\sin\theta \\ \sin\theta & \cos\theta \end{bmatrix}$ or to $\begin{bmatrix} \sin\theta & \cos\theta \\ \cos\theta & -\sin\theta \end{bmatrix}$ for some value of θ. The geometric interpretation of this result is important. Recall that $\begin{bmatrix} \cos\theta & -\sin\theta \\ \sin\theta & \cos\theta \end{bmatrix}$ represents the transformation that rotates every vector through the angle θ. Note too that

$$\begin{bmatrix} \sin\theta & \cos\theta \\ \cos\theta & -\sin\theta \end{bmatrix} = \begin{bmatrix} 0 & 1 \\ 1 & 0 \end{bmatrix}\begin{bmatrix} \cos\theta & -\sin\theta \\ \sin\theta & \cos\theta \end{bmatrix}$$ and that $\begin{bmatrix} 0 & 1 \\ 1 & 0 \end{bmatrix}$ represents reflection about the line $y = x$. Thus an orthogonal transformation of $\mathbb{R}^2$ must either be a rotation (these are the ones whose determinant is $+1$) or a rotation followed by a reflection about the line $y = x$ (these are the ones whose determinant is -1).

Problem Set 41

41.2 Only axiom (8) fails. Note that axiom (3) is satisfied with $\mathbf{0} = (0,1)$ and that axiom (4) is satisfied with $\ominus(x, y) = (-x/y^2, 1/y)$. This example constitutes a typical "independence proof" in mathematics. It establishes the independence of axiom (8) from the first seven axioms [i.e., axiom (8) cannot be proved as a consequence of axioms (1) through (7)]. Compare this, for instance, with Theorem 41.13, which asserts that $\ominus\mathbf{v} = (-1)\mathbf{v}$ *can* be proved from axioms (1) through (8).

41.3 Only axiom (6) fails. Note that axiom (3) is satisfied with $\mathbf{0} = (1,0)$ and that axiom (4) is satisfied with $\ominus(x, y) = (1/x, -y/x^2)$.

41.5 (i) To prove $\mathbf{f} = \Sigma_i \mathbf{f}(\mathbf{v}_i)\mathbf{f}_i$, we need to show, for arbitrary $\mathbf{v} \in V$, that $\mathbf{f}(\mathbf{v}) = \Sigma_i f_i(\mathbf{v}_i)\mathbf{f}(\mathbf{v})$. Suppose $\mathbf{v} = t_1\mathbf{v}_1 + t_2\mathbf{v}_2 + \cdots + t_n\mathbf{v}_n \in V$. Observe that, by definition, $t_i = ((\mathbf{v})_\beta)_i$. We have

$$\begin{aligned} \mathbf{f}(\mathbf{v}) &= \Sigma_i t_i \mathbf{f}(\mathbf{v}_i) && \text{[linearity of } \mathbf{f}\text{]} \\ &= \Sigma_i ((\mathbf{v})_\beta)_i \mathbf{f}(\mathbf{v}_i) && \text{[as just observed]} \\ &= \Sigma_i \mathbf{f}(\mathbf{v}_i)\mathbf{f}_i(\mathbf{v}) && \text{[definition of } \mathbf{f}_i \text{ and commutativity of multiplication in } \mathbb{F}\text{]} \end{aligned}$$

(ii) By part (i), the $\mathbf{f}_i$ span V^*. To show that they are linearly independent, suppose that $\mathbf{0} = s_1\mathbf{f}_1 + s_2\mathbf{f}_2 + \cdots + s_n\mathbf{f}_n$; when we recall that $(\mathbf{v}_i)_\beta = \mathbf{e}_i^n$, we see, for arbitrary i, that $0 = \mathbf{0}(\mathbf{v}_i) = \Sigma_i s_i \mathbf{f}_i(\mathbf{v}_i) = s_i$.

41.6 If $\mathbf{v} = (v_1, v_2) \in V$, then

$$[\mathbf{v}]_\beta = {}_\beta[I]_{\alpha^2}\,[\mathbf{v}]_{\alpha^2} = \begin{bmatrix} 2 & -1 \\ 1 & 4 \end{bmatrix}^{-1} \begin{bmatrix} v_1 \\ v_2 \end{bmatrix} = \begin{bmatrix} \frac{4}{9} & \frac{1}{9} \\ -\frac{1}{9} & \frac{2}{9} \end{bmatrix} \begin{bmatrix} v_1 \\ v_2 \end{bmatrix}$$

Thus $\mathbf{f}_1(\mathbf{v}) = 1^{\text{st}}$ component of $(\mathbf{v})_\beta = \frac{4}{9}v_1 + \frac{1}{9}v_2 = \mathbf{v} \cdot (\frac{4}{9}, \frac{1}{9})$ and $\mathbf{f}_2(\mathbf{v}) = 2^{\text{nd}}$ component of $(\mathbf{v})_\beta = -\frac{1}{9}v_1 + \frac{2}{9}v_2 = \mathbf{v} \cdot (-\frac{1}{9}, \frac{2}{9})$.

Problem Set 42

42.2 (i), (ii), and (iii) are subspaces.

42.3 (i) and (iii) are subspaces.

Problem Set 43

43.2 (i), (ii), and (iv)

43.3 (i) $\left\{\begin{bmatrix} a & b \\ c & d \end{bmatrix} \in \mathbb{R}_{2\times 2}: d = a + c\right\}$

(ii) The first three matrices belong to this subspace.

43.4 None

43.5 (i), (ii), and (iv) are independent; (iii) is dependent.

43.6 The dimensions of the subspaces described in P42.2(i), (ii), and (iii) are 2, 3, and 2, respectively. Each of the subspaces described in P42.3(i) and (iii) is five-dimensional.

43.7 $f(x) = \frac{10}{3}\mathbf{p}_1 + 4\mathbf{p}_2 - \frac{2}{3}\mathbf{p}_3$
$g(x) = \frac{5}{3}\mathbf{p}_1 + 2\mathbf{p}_2 - \frac{1}{3}\mathbf{p}_3$
$h(x) = \frac{4}{3}\mathbf{p}_1 + 2\mathbf{p}_2 - \frac{5}{3}\mathbf{p}_3$.

43.9 $\mathscr{L}(S_1)$ is all of $\mathscr{P}_2$.

$$\mathscr{L}(S_2) = \left\{\begin{bmatrix} a & b \\ c & d \end{bmatrix} \in \mathbb{R}_{2\times 2}: a = -\tfrac{3}{8}b + \tfrac{19}{8}c + 2d\right\}$$

43.10 Yes

Problem Set 44

44.2 (i) $\begin{bmatrix} 1 & -1 & 0 \\ 0 & 0 & 2 \end{bmatrix}$

(ii) $\ker T = \{a_0 + a_1x + a_2x^2 \in \mathscr{P}_2: a_0 = a_1, a_2 = 0\}$; $\operatorname{ran} T = \mathscr{P}_1$

(iii) $\mathrm{r}(T) = 2$; $\mathrm{n}(T) = 1$

44.3 (i) F; (ii) T; (iii) T; (iv) F; (v) T; (vi) T.

44.4 $\begin{bmatrix} 1 & -2 \\ 2 & 1 \\ 1 & 1 \end{bmatrix}$ (ii) $\begin{bmatrix} 2 & -4 \\ 3 & -3 \\ 1 & 2 \end{bmatrix}$ (iii) (2, 3, 1)

44.5 (i) $\left\{\begin{bmatrix} -2 & 0 \\ 1 & 1 \end{bmatrix}\right\}$ is a basis for ker T;

$\left\{\begin{bmatrix} 1 & 2 \\ 3 & 4 \end{bmatrix}, \begin{bmatrix} 2 & 3 \\ 0 & 5 \end{bmatrix}, \begin{bmatrix} 3 & 2 \\ 1 & -2 \end{bmatrix}\right\}$ is a basis for ran T.

(ii) Yes

44.6 (i) $\begin{bmatrix} -5 & -3 & 2 \\ -3 & 0 & 3 \\ -2 & -1 & 1 \end{bmatrix}$

(ii) $\ker T = \{a_0 + a_1x + a_2x^2 \in \mathscr{P}_2\colon\ a_0 = a_2,\ \ a_1 = -a_2);$
$\operatorname{ran} T = \{a_0 + a_1x + a_2x^2 \in \mathscr{P}_2\colon a_2 = \frac{1}{9}a_1 + \frac{1}{3}a_0\}$

(iii) No

Problem Set 45

45.2 It is trivial to verify that $\langle \mathbf{u}, \mathbf{v}\rangle$ is symmetric and linear in the first variable; the fact that the t_i are positive is needed to prove that $\langle\,\cdot\,,\cdot\rangle$ is positive definite.

45.4
$$\begin{aligned}
\langle \mathrm{A}, \mathrm{B}\rangle &= \operatorname{trace}(\mathrm{AB}^T) \\
&= \operatorname{trace}((\mathrm{AB}^T)^T) && \text{[P3.18(iii)]} \\
&= \operatorname{trace}(\mathrm{BA}^T) && \text{[3.15]} \\
&= \langle \mathrm{B}, \mathrm{A}\rangle \\
\langle \mathrm{A} + \mathrm{B}, \mathrm{C}\rangle &= \operatorname{trace}((\mathrm{A} + \mathrm{B})\mathrm{C}^T) \\
&= \operatorname{trace}(\mathrm{AC}^T + \mathrm{BC}^T) && \text{[3.1]} \\
&= \operatorname{trace}(\mathrm{AC}^T) + \operatorname{trace}(\mathrm{BC}^T) && \text{[P3.18(ii)]} \\
&= \langle \mathrm{A}, \mathrm{C}\rangle + \langle \mathrm{B}, \mathrm{C}\rangle \\
\langle t\mathrm{A}, \mathrm{B}\rangle &= \operatorname{trace}((t\mathrm{A})\mathrm{B}^T) \\
&= \operatorname{trace}(t(\mathrm{AB}^T)) && \text{[3.1]} \\
&= t \operatorname{trace}(\mathrm{AB}^T) && \text{[P3.18(i)]} \\
&= t\langle \mathrm{A}, \mathrm{B}\rangle \\
\langle \mathrm{A}, \mathrm{A}\rangle &= \operatorname{trace}(\mathrm{AA}^T) \\
&= a_{11}^2 + a_{12}^2 + \cdots + a_{1n}^2 + a_{21}^2 + a_{22}^2 + \cdots + a_{2n}^2 \\
&\quad + \cdots + a_{n1}^2 + a_{n2}^2 + \cdots + a_{n\times n}^2
\end{aligned}$$

Thus $\langle \mathrm{A}, \mathrm{A}\rangle \geq 0$; and $\langle \mathrm{A}, \mathrm{A}\rangle = 0$ iff $\mathrm{A} = 0_{n\times n}$.

45.5 (i) $(\mathrm{A}\mathbf{u}) \cdot_1 \mathbf{v} = [\mathrm{A}\mathbf{u}]^H[\mathbf{v}] = [\mathbf{u}]^H\mathrm{A}^H[\mathbf{v}] = \mathbf{u} \cdot_1 (\mathrm{A}^H\mathbf{v})$. The arguments for (ii) and for the rest of (i) are similar.

45.6 We will prove it for $\cdot_1$; the arguments for $\cdot_2$ are similar. ($\leftarrow$) Assuming $\mathrm{A} = \mathrm{A}^H$, we have $(\mathrm{A}\mathbf{u}) \cdot_1 \mathbf{v} = (\mathrm{A}\mathbf{u})^H\mathbf{v} = \mathbf{u}^H\mathrm{A}^H\mathbf{v} = \mathbf{u}^H\mathrm{A}\mathbf{v} = \mathbf{u} \cdot_1 \mathrm{A}\mathbf{v}$. ($\rightarrow$) By assumption, we have for arbitrary i and j that $(\mathrm{A}\mathbf{e}_i^n) \cdot_1 \mathbf{e}_j^n = \mathbf{e}_i^n \cdot_1 (\mathrm{A}\mathbf{e}_j^n)$. Since $(\mathrm{A}\mathbf{e}_i^n) \cdot_1 \mathbf{e}_j^n = (\mathrm{A}\mathbf{e}_i^n)^H\mathbf{e}_j^n = (\mathbf{e}_i^n)^H\mathrm{A}^H\mathbf{e}_j^n = \bar{a}_{ji}$ and $\mathbf{e}_i^n \cdot_1 (\mathrm{A}\mathbf{e}_j^n) = (\mathbf{e}_i^n)^H\mathrm{A}\mathbf{e}_j^n = (\mathbf{e}_i^n)\mathrm{A}\mathbf{e}_j^n = a_{ij}$, this implies that $\mathrm{A} = \mathrm{A}^H$.

45.8 (i) is trivial.

(ii) Let $\langle\,\cdot\,,\cdot\rangle$ be an inner product on a finite-dimensional vector space V. Verify that the proof of the Projection Theorem given in Section 39 goes through when $\mathbb{E}^n$ and the dot product are replaced by V and $\langle\,\cdot\,,\cdot\rangle$.

(iii) The proof that $W \subset W^{\perp\perp}$ does not require finite-dimensionality: assume $\mathbf{u} \in W$; then for any $\mathbf{w} \in W^\perp$, $\langle \mathbf{u}, \mathbf{w}\rangle =$

0; and hence $\mathbf{u} \in W^{\perp\perp} = (W^{\perp})^{\perp} = \{\mathbf{v} \in V\colon \langle \mathbf{v}, \mathbf{w} \rangle = 0$ for all $\mathbf{w} \in W^{\perp}\}$.

To prove the reverse inclusion, we will make use of part (ii): $V = W \oplus W^{\perp}$. This implies that if $\dim W = k$, then $\dim W^{\perp} = n - k$ and that if $\beta = (\mathbf{v}_1, \mathbf{v}_2, \ldots, \mathbf{v}_k)$ and $\gamma = (\mathbf{v}_{k+1}, \mathbf{v}_{k+2}, \ldots, \mathbf{v}_n)$ are orthonormal bases for W and $W^{\perp}$, respectively, then $\beta \cup \gamma$ is an orthonormal basis for V. Choose such a pair of orthonormal bases β and γ. For an arbitrary $\mathbf{u} \in V$, $\mathbf{u} = \langle \mathbf{u}, \mathbf{v}_1 \rangle \mathbf{v}_1 + \cdots + \langle \mathbf{u}, \mathbf{v}_n \rangle \mathbf{v}_n$ because $\beta \cup \gamma$ is an orthonormal basis. Now if $\mathbf{u} \in W^{\perp\perp} = (W^{\perp})^{\perp}$, then $\langle \mathbf{u}, \mathbf{w} \rangle = 0$ for all $\mathbf{w} \in W^{\perp}$, so the expression for $\mathbf{u}$ reduces to $\mathbf{u} = \langle \mathbf{u}, \mathbf{v}_1 \rangle \mathbf{v}_1 + \cdots + \langle \mathbf{u}, \mathbf{v}_k \rangle \mathbf{v}_k$, which shows that $\mathbf{u} \in W$.

45.9 (i) See the answer to P45.8(iii).

(ii) $W^{\perp} \subset W^{\perp\perp\perp}$ by (i) applied to $W^{\perp}$. If $\mathbf{u} \in W^{\perp\perp\perp} = \{\mathbf{v}\colon \langle \mathbf{v}, \mathbf{w} \rangle = 0$ for all $\mathbf{w} \in W^{\perp\perp}\}$, then since $W \subset W^{\perp\perp}$, we get $\mathbf{u} \in W^{\perp}$.

45.13 Use the fact that the Triangle Inequality holds for each of the four pairs of vectors $\mathbf{u}$ and $\mathbf{v}$, $\mathbf{u}$ and $-\mathbf{v}$, $\mathbf{u}$ and $i\mathbf{v}$, and finally $\mathbf{u}$ and $-i\mathbf{v}$. Assume conjugate linearity in the first variable.

TI for $\mathbf{u}, \mathbf{v}$ yields $\mathrm{Re}\langle \mathbf{u}, \mathbf{v} \rangle \le \|\mathbf{u}\|\|\mathbf{v}\|$.
TI for $\mathbf{u}, -\mathbf{v}$ yields $-(\mathrm{Re}\langle \mathbf{u}, \mathbf{v} \rangle) \le \|\mathbf{u}\|\|\mathbf{v}\|$.
TI for $\mathbf{u}, i\mathbf{v}$ yields $\mathrm{Im}\langle \mathbf{u}, \mathbf{v} \rangle \le \|\mathbf{u}\|\|\mathbf{v}\|$.
TI for $\mathbf{u}, -i\mathbf{v}$ yields $-(\mathrm{Im}\langle \mathbf{u}, \mathbf{y} \rangle) \le \|\mathbf{u}\|\|\mathbf{v}\|$.

Combine these to get $|\mathrm{Re}\langle \mathbf{u}, \mathbf{v} \rangle| \le \|\mathbf{u}\|\|\mathbf{v}\|$ and $|\mathrm{Im}\langle \mathbf{u}, \mathbf{v} \rangle| \le \|\mathbf{u}\|\|\mathbf{v}\|$. Conclude $|\langle \mathbf{u}, \mathbf{v} \rangle| = [(\mathrm{Re}\langle \mathbf{u}, \mathbf{v} \rangle)^2 + (\mathrm{Im}\langle \mathbf{u}, \mathbf{v} \rangle)^2]^{1/2} \le \|\mathbf{u}\|\|\mathbf{v}\|$.

45.14 (i) Suppose that $\mathbf{0} = \mathbf{w}_1 + \mathbf{w}_2 + \cdots + \mathbf{w}_p$, where $\mathbf{w}_i \in W_i$. Then, for each fixed i, we have $0 = \langle \mathbf{0}, \mathbf{w}_i \rangle = \Sigma_j \langle \mathbf{w}_i, \mathbf{w}_j \rangle = \|\mathbf{w}_i\|^2$ because the subspaces are orthogonal; so $\mathbf{w}_i = \mathbf{0}$ because $\langle \cdot, \cdot \rangle$ is positive definite.

(ii) $(\rightarrow)$ β is a basis for W by 43.21. It consists of unit vectors by hypothesis. Suppose $\mathbf{u}, \mathbf{v} \in \beta$: if $\mathbf{u}$ and $\mathbf{v}$ belong to the same β_i, then $\langle \mathbf{u}, \mathbf{v} \rangle = 0$ because β_i is orthonormal; if $\mathbf{u} \in \beta_i$ and $\mathbf{v} \in \beta_j$, where $i \ne j$, then $\langle \mathbf{u}, \mathbf{v} \rangle = 0$ because the collection of subspaces is orthogonal. $(\leftarrow)$ If β fails to be an orthonormal basis for W, then, by 43.21, this can only happen if $\langle \mathbf{u}, \mathbf{v} \rangle \ne 0$ for some $\mathbf{u} \in \beta_i$ and $\mathbf{v} \in \beta_j$ with $i \ne j$, contradicting the orthogonality of the subspaces.

45.15 This follows easily from the bilinearity of $\langle \cdot, \cdot \rangle$.

45.16 (i) is easy. For (ii), you can take $\mathbf{u} = (1, 0)$ and $\mathbf{v} = (0, 1)$.

Problem Set 46

46.2 Given that $\mathrm{A}\mathbf{u} = \lambda \mathbf{u}$ iff $\mathrm{A}^H \mathbf{u} = \bar{\lambda} \mathbf{u}$, if, in particular, $\mathrm{A} = \mathrm{A}^H$ and λ is an eigenvalue of A and $0 \ne \mathbf{u} \in E_{\lambda}$, then $\lambda \mathbf{u} = \bar{\lambda} \mathbf{u}$; since $\mathbf{u} \ne \mathbf{0}$, we conclude $\lambda = \bar{\lambda}$.

46.3 $(\rightarrow)$ If A is Hermitian, then A is normal because $AA^H = A^2 = A^HA$; it has real eigenvalues by Theorem 46.11. $(\leftarrow)$ A is unitarily diagonalizable over $\mathbb{C}$ by 46.29. If $P^{-1}AP = P^HAP = D$, then $D = D^H$ because the eigenvalues of A are real. So $A = PDP^{-1} = PDP^H = PD^HP^H = (PDP^H)^H = A^H$.

46.4 False: A is normal, and hence diagonalizable over $\mathbb{C}$; but if any of its eigenvalues are not real, A will not be symmetric by 46.11.

Problem Set 47

47.2 We need $[\mathbf{u}]_\alpha^T\,{}_\alpha[I]_\beta = [\mathbf{u}]_\beta^T$ for all $\mathbf{u} \in \mathbb{R}^m$. This happens iff ${}_\alpha[I]_\beta$ is orthogonal, as you are asked to show in the next problem.

47.3 $(\leftarrow)\ [\mathbf{u}]_\alpha^T\,{}_\alpha[I]_\beta = [\mathbf{u}]_\alpha^T\,{}_\beta[I]_\alpha^{-1} = [\mathbf{u}]_\alpha^T\,{}_\beta[I]_\alpha^T = ({}_\beta[I]_\alpha\,[\mathbf{u}]_\alpha)^T = [\mathbf{u}]_\beta^T$. $(\rightarrow)$ Let $\beta = (\mathbf{u}_1, \mathbf{u}_2, \ldots, \mathbf{u}_n)$ so that $[\mathbf{u}_i]_\alpha = i^{\text{th}}$ column of ${}_\alpha[I]_\beta$ and $(\mathbf{u}_i)_\beta = \mathbf{e}_i^n$. We prove that ${}_\alpha[I]_\beta$ is orthogonal by showing that its columns form an orthonormal set.

$$\begin{aligned}(\mathbf{u}_i)_\alpha \cdot (\mathbf{u}_j)_\alpha &= [\mathbf{u}_i]_\alpha^T\,[\mathbf{u}_j]_\alpha \\ &= [\mathbf{u}_i]_\alpha^T\,{}_\alpha[I]_\beta[\mathbf{u}_j]_\beta = [\mathbf{u}_i]_\beta^T[\mathbf{u}_j]_\beta = [\mathbf{e}_i^n]^T[\mathbf{e}_j^n] \\ &= \mathbf{e}_i^n \cdot \mathbf{e}_j^n = \delta_{ij}\end{aligned}$$

Problem Set 48

48.2

	Standard matrix:	Associated quadratic form:
(i)	$\begin{bmatrix} 3 & 2 \\ 2 & 5 \end{bmatrix}$	$3x_1^2 + 4x_1x_2 + 5x_2^2$
(ii)	$\begin{bmatrix} 2 & 0 \\ 0 & -5 \end{bmatrix}$	$2x_1^2 - 5x_2^2$
(iii)	$\begin{bmatrix} 1 & -\frac{3}{2} & -\frac{5}{2} \\ -\frac{3}{2} & -3 & \frac{3}{2} \\ -\frac{5}{2} & \frac{3}{2} & 7 \end{bmatrix}$	$x_1^2 - 3x_2^2 + 7x_3^2 - 3x_1x_2 + 3x_2x_3 - 5x_1x_3$
(iv)	$\begin{bmatrix} 0 & 0 & -\frac{1}{2} & 0 \\ 0 & -3 & \frac{3}{2} & -4 \\ -\frac{1}{2} & \frac{3}{2} & 0 & 0 \\ 0 & -4 & 0 & 2 \end{bmatrix}$	$2x_4^2 - 3x_2^2 - x_1x_3 + 3x_2x_3 - 8x_2x_4$

48.3 $((1 + \sqrt{2}, 1), (1 - \sqrt{2}, 1))$

Problem Set 49

49.2 $a > 0$ and $ad - bc > 0$

49.3 (i) $\mathbf{x}^T(B^TB)\mathbf{x} = (B\mathbf{x})^TB\mathbf{x} = (B\mathbf{x}) \cdot (B\mathbf{x}) = \|B\mathbf{x}\|^2 \geq 0$ for all $\mathbf{0} \neq \mathbf{x} \in \mathbb{R}^n$, so B^TB is positive semidefinite. The argument for BB^T is similar.

(ii) $\|B\mathbf{x}\|^2 > 0$ for all $\mathbf{0} \neq \mathbf{x} \in \mathbb{R}^n$, iff $B\mathbf{x} \neq \mathbf{0}$ for all $\mathbf{0} \neq \mathbf{x} \in \mathbb{R}^n$, iff B has a left inverse, iff $r(B) = n$ (recall Theorem 19.3).

49.4 (ii) → (i):

$$
\begin{aligned}
\overline{\mathbf{x}^H\mathrm{A}\mathbf{x}} &= \overline{\mathbf{x}^H\mathrm{A}^H\mathbf{x}} && [\text{A is Hermitian}]\\
&= \overline{\overline{\mathbf{x}^T\mathrm{A}^T\mathbf{x}}} && [36.15]\\
&= \mathbf{x}^T\mathrm{A}^T\overline{\mathbf{x}} && [36.2(\text{ii}) \text{ and } (\text{iii})]\\
&= (\mathbf{x}^T\mathrm{A}^T\overline{\mathbf{x}})^T && [\text{because } \mathbf{x}^T\mathrm{A}^T\overline{\mathbf{x}} \text{ is } 1 \times 1]\\
&= \mathbf{x}^H\mathrm{A}\mathbf{x} && [3.15 \text{ and } 36.15]
\end{aligned}
$$

(i) → (ii): By P36.3, there is a Hermitian matrix B and a skew-Hermitian matrix C such that A = B + C. We have $\mathbf{x}^H\mathrm{A}\mathbf{x} = \mathbf{x}^H(\mathrm{B} + \mathrm{C})\mathbf{x} = \mathbf{x}^H\mathrm{B}\mathbf{x} + \mathbf{x}^H\mathrm{C}\mathbf{x}$. Using (ii) → (i), which was just proved, $\mathbf{x}^H\mathrm{B}\mathbf{x}$ is real; thus if $\mathbf{x}^H\mathrm{A}\mathbf{x}$ is real, then so is $\mathbf{x}^H\mathrm{C}\mathbf{x}$. Therefore,

$$
\begin{aligned}
\mathbf{x}^H\mathrm{C}\mathbf{x} &= \overline{\mathbf{x}^H\mathrm{C}\mathbf{x}}\\
&= \overline{(\mathbf{x}^H\mathrm{C}\mathbf{x})^T} && [\mathbf{x}^H\mathrm{C}\mathbf{x} \text{ is } 1 \times 1]\\
&= (\mathbf{x}^H\mathrm{C}\mathbf{x})^H && [36.15]\\
&= \mathbf{x}^H\mathrm{C}^H\mathbf{x} && [36.16]\\
&= -\mathbf{x}^H\mathrm{C}\mathbf{x} && [\text{C is skew-Hermitian}]
\end{aligned}
$$

Thus $\mathbf{x}^H\mathrm{C}\mathbf{x} = 0$ for all $\mathbf{x} \in \mathbb{C}^n$, so $\mathrm{C} = 0_{n\times n}$ and A = B, which is Hermitian.

Problem Set 50

50.2 $\mathbf{w} = (3, -2)$

Problem Set 51

51.2 (i) If we display a Jordan basis for $T \upharpoonright K_c$ in a "rectangular" array, as in the proofs of 51.32 and 51.34, the array must contain five entries. The fact that dim $E_c = 2$ implies that the first column of this array contains two elements. The fact that dim $E_c^2 = 3$ implies that the first two columns of this array contain a total of three elements. Thus the second column contains a single element and the first two columns of the array look like

```
**
*
```

There is only one way to continue this array to one that respects the constraints described in the proofs of 51.32 and

51.34 and that contains five elements:

```
****
*.
```

Thus $J(T \upharpoonright K_c) = J[4; c] \oplus J[1; c]$.

(ii) The given information implies that a Jordan basis for $T \upharpoonright K_c$ displayed in an array as above must contain seven elements and must have

```
**
**
*
```

as the form of its first two columns. This does not fully determine the Jordan form of $T \upharpoonright K_c$ because there are two ways,

```
****        ***
**    and   ***
*           *
```

in which this might be extended to an array that respects the constraints described in the proofs of 51.32 and 51.34 and that contains seven elements. The additional information needed is dim E_c^3. The possibilities are dim $E_c^3 = 6$ or 7; these correspond, respectively, to the first and second of the two preceding patterns; in the first case, $J(T \upharpoonright K_c) = J[4; c] \oplus J[2; c] \oplus J[1; c]$; in the second case, $J(T \upharpoonright K_c) = J[3; c] \oplus J[3; c] \oplus J[1; c]$.

51.4 (i) $A^7 = \begin{bmatrix} 45{,}056 & 0 & 28{,}672 \\ 28{,}672 & 16{,}384 & 28{,}672 \\ -28{,}672 & 0 & -12{,}288 \end{bmatrix}$

(ii) $\chi(A) = (\lambda - 4)^3$ and $E_4 = \{(x_1, x_2, x_3) \in \mathbb{R}^3\colon x_1 = -x_3)$. So $J(A) = J[2; 4] \oplus J[1; 4]$. If we choose $\mathbf{u}_1 = (-1, 0, 1)$, there is no $\mathbf{u}_2$ satisfying $A\mathbf{u}_2 = 4\mathbf{u}_2 + \mathbf{u}_1$. For consistency, $\mathbf{u}_1$ must have the form $(-a, -a, a)$.

51.6 (i) $\begin{bmatrix} 1 & 2 \\ 3 & 4 \end{bmatrix}$ (ii) $\begin{bmatrix} 1 & 3 \\ 2 & 4 \end{bmatrix}$

(iii) The question is whether, given an arbitrary $A = \begin{bmatrix} a & b \\ c & d \end{bmatrix}$, there exists an invertible $P = \begin{bmatrix} p & q \\ r & s \end{bmatrix}$, such that $P^{-1}AP = A^T$, or, equivalently, $AP = PA^T$. Write out the equations that arise from these conditions and solve them for p, q, r, and s as functions of a, b, c, and d. You will find that there are, in fact, infinitely many solutions.

Index

Concepts which are first introduced for $\mathbb{R}^n$ and subsequently generalized to arbitrary vector spaces have references of the form $m(n)$ where m is the location of the initial introduction and n is the location of the generalized concept. There is an index of notation inside the front and back covers.

Index of Notation

In this index the symbols are listed in pseudoalphabetical order. See the inside front cover for a listing in the order in which they appear in the text.

Symbol	*Meaning*	*Numbered Item Containing Definition of Symbol*
A, B, C, D, E, I, J, M, N, P, Q, X, Y, Z	matrices (with rare exceptions)	1.4
A^H	the conjugate transpose of the matrix A	36.15
A^T	the transpose of A	3.13
$\bar{A}$	the complex conjugate of the matrix A	36.11
A^{-1}	the inverse of the matrix A (when A is invertible)	3.6
$\hat{A}_{ij}$	the matrix obtained by deleting the ith row and j^{th} column from A	28.1
$A \oplus B$ or diag(A, B)	the direct sum of the matrices A and B	51.3
$A[j; \mathbf{b}]$	the result of substituting **b** for the jth column of A	Problem 29.1
Arg z	the principal argument of the complex number z	36.3
a, **b**, **c**, . . . , **x**, **y**, **z**	vectors	5.1
adj A	the adjugate (or classical adjoint) of A	29.1
a_{ij}	the $(i, j)^{th}$ entry in the matrix A	1.6
a^c_{ij}	the cofactor of a_{ij}	28.1
a^m_{ij}	the minor of a_{ij}	28.1
$\alpha, \beta, \gamma, \delta$	bases for vector spaces	8.10
α^n	$(\mathbf{e}^n_1, \mathbf{e}^n_2, \ldots, \mathbf{e}^n_n)$, the standard basis for $\mathbb{R}^n$	7.22
β^*	the basis dual to the basis β	Problem 41.5
CS(A)	the column space of A	9.1
$C(\mathbb{R})$	the space of continuous functions from $\mathbb{R}$ to $\mathbb{R}$	42.5
codom f	the codomain of the function f	16.1
cof A or A^c	the matrix of cofactors of A	29.1
det A or $\|A\|$	the determinant of the matrix A	23.5
diag$(d_1, d_2, \ldots, d_n)$	the diagonal matrix with main diagonal entries $d_1, d_2, \ldots, d_n$	1.9
dim W	the dimension of W	8.2
dom f	the domain of the function f	16.1
Δ	a pseudodeterminant	30.1
E	an elementary matrix	13.1
E^k_c	the nullspace of $(T - cI)^k$	51.16
E_λ	the eigenspace of λ	34.4
$\mathbb{E}^n$	$\mathbb{R}^n$ with the Euclidean inner product	38.18
$\mathbf{e}^m_j$	the m-tuple whose j^{th} coordinate is 1 and whose remaining coordinates are 0	7.22
$f \upharpoonright S$	the restriction of the function f to the set S	16.1
f^{-1}	the inverse of the function f (when f is invertible)	16.13
$\mathbb{F}_{n \times n}$	the set of $n \times n$ matrices over the field $\mathbb{F}$	34.1
G(A)	the Gaussian form of A	12.11
GJ(A)	the Gauss–Jordan form of A	11.9 and 12.11
GS(S)	the output of the Gram–Schmidt algorithm applied to S	39.14
$g \circ f$	f composed with g	16.4
$\mathbb{H}^n$	$\mathbb{C}^n$ with the Euclidean inner product	45.26
Im z	the imaginary part of the complex number z	36.1